Medical Mycology

Current Trends and Future Prospects

Medical Mycology

Current Trends and Future Prospects

Editors

Mehdi Razzaghi-Abyaneh
Professor and head
Department of Mycology
Pasteur Institute of Iran
Tehran
Iran

Masoomeh Shams-Ghahfarokhi
Department of Mycology
Faculty of Medical Sciences
Tarbiat Modares University
Tehran
Iran

Mahendra Rai
Department of Biotechnology
S.G.B. Amravati University
Amravati – 444 602
Maharashtra
India

CRC Press
Taylor & Francis Group
Boca Raton London New York

CRC Press is an imprint of the
Taylor & Francis Group, an **informa** business
A SCIENCE PUBLISHERS BOOK

Cover illustrations have been provided by the editors of the book and are reproduced with their kind permission.

CRC Press
Taylor & Francis Group
6000 Broken Sound Parkway NW, Suite 300
Boca Raton, FL 33487-2742

First issued in paperback 2020

ISBN-13: 978-1-4987-1421-1 (hbk)
ISBN-13: 978-0-367-73801-3 (pbk)

Library of Congress Cataloging-in-Publication Data

Medical mycology (Razzaghi-Abyaneh)
Medical mycology : current trends and future prospects / editors, Mehdi Razzaghi-Abyaneh, Masoomeh Shams-Ghahfarokhi, Mahendra Rai.
p. ; cm.
Includes bibliographical references and index.
ISBN 978-1-4987-1421-1 (hardcover : alk. paper)
I. Razzaghi Abyaneh, M., editor. II. Shams-Ghahfarokhi, Masoomeh, editor. III. Rai, Mahendra, editor. IV. Title.
[DNLM: 1. Mycoses. 2. Fungi. 3. Microbiological Techniques. 4. Mycology--methods. WC 450]

QR245
616.9'69--dc23 2015018287

Visit the Taylor & Francis Web site at
http://www.taylorandfrancis.com

and the CRC Press Web site at
http://www.crcpress.com

Preface

Medical Mycology has been an important field since the dawn of civilization as fungi play a pivotal role in causing infections in human beings and animals. In addition to the more commonly encountered fungi, life-threatening fungal infections due to emerging fungi that had previously rarely been reported in clinical practice have dramatically increased in recent years. The relationship of fungi with human health was known before the work of Louis Pasteur and Robert Koch on bacteria in the middle of 19th century. In 1941 a Hungarian physician, David Gruby (1810–1894) described for the first time the etiologic agent of fungal infection of scalp (Favus) as *Trichophyton schoenleinii*, a discovery that was independent of the findings of Johann Lukas Schönlein (1793–1864). In the current years, the ever increasing opportunistic fungal pathogens which are difficult to detect and treat have warranted new challenges for the diagnosis and treatment of fungal infections, especially in immunocompromised patients. Such infections are increasing at an alarming rate. Moreover, another reason for the increasing incidence of fungal infections is the development of resistance to different antifungal agents.

The identification of medically important fungi has been an important area of research that warrants further extensive research. We need to use both traditional and novel methods of identification such as PCR and immunoassays. These methods provide new insights into differentiation of species and eventually the line of treatment can be determined. The proposed book is a unique combination of contributions from mycologists, microbiologists and clinical experts from around the world and provides in-depth comprehensive data on the biology and pathogenesis of diverse groups of medically important fungi and related mycoses including common dermatophytes, candidiasis, onychomycosis, coccidioidomycosis, paracoccidioidomycosis, mycotic keratitis, sporotrichosis, histoplasmosis, fungal infections in otorhinolaryngological diseases and kidney transplantation. It also elaborates on the application of modern techniques such as PCR and MALDI-TOF as rapid and new approaches in fungal diagnosis and differentiation.

This book can be used as a comprehensive textbook by students, researchers and teachers of mycology, microbiology and biotechnology, fungal taxonomists, clinical experts and pathologists.

Mehdi Razzaghi-Abyaneh, Ph.D.
Masoomeh Shams-Ghahfarokhi, Ph.D.
Mahendra Rai, Ph.D.

Contents

Section III: Classic Mycoses Caused by Dimorphic Fungi

Section IV: Fungal Pathogenesis in Biofilm and Allergy

Section V: Novel Diagnostic Methods, Susceptibility Testing and Miscellaneous Mycoses

SECTION I

Superficial Mycoses Caused by Molds and Yeasts

CHAPTER 1

Dermatophyte Infections in Humans: Current Trends and Future Prospects

Mateja Dolenc-Voljč

Introduction

Dermatophytes are a group of closely related filamentous fungi that have the capacity to invade the keratinized tissue of skin, hair and nails in humans and animals. They produce superficial infections termed "dermatophytoses" (Crissey et al. 1995). In clinical dermatology, the terms "tinea" and "ringworm" are used for these infections. Dermatophytes are the most common causative pathogens responsible for fungal infections worldwide (Havlickova et al. 2008). The prevalence of these infections has been observed to be on the rise in recent decades. This is in part due to aging of the population, the changes in immune response that occur with age, an increased number of immunocompromised patients, HIV infected persons and those who have diabetes or other chronic diseases. Changes in lifestyle have also contributed to the rising incidences of these fungal infections. Human migration, mass tourism and international sports acitivities have contributed to the dissemination of dermatophyte species throughout different geographical areas (Havlickova et al. 2008). Increased urbanisation and ready access to communal sports and bathing facilities are also among the reasons responsible for the high prevalence of anthropophilic dermatophytes (Havlickova et al. 2008; Borman et al. 2007). Crowded living conditions provide multiple opportunities for interhuman contact. Living in close proximity to animals

Department of Dermatovenereology, University Medical Centre Ljubljana, Zaloška 2, 1000 Ljubljana, Slovenia; Faculty of Medicine, University of Ljubljana, Vrazov trg2, 1000 Ljubljana, Slovenia.
E-mail: mateja.dolenc-voljc@mf.uni-lj.si

enables the spread of infection from animals to their owners in both rural and urban environments. Poor medical care in the undeveloped countries further increases the epidemic spread of these common infections.

Like other keratinophilic fungi, dermatophytes are capable of destroying keratin by means of some keratinolytic enzymes. They only grow in dead keratinized tissue, within the stratum corneum of the epidermis, within the keratinized hair shaft and in the nail plate and keratinized nail bed (Hay and Ashbee 2010). Dermatophyte infections are therefore localized superficially on the body. They do not usually cause infections of the mucous membranes or systemic infections which involve the internal organs. Such cases are considered to be an exceptional rarity (Marconi et al. 2010).

Although not life-threatening, their increasing prevalence and associated morbidity make them an important public health problem. Dermatophyte infections show a low tendency towards self-limitation. If not diagnosed and treated properly, infections may develop a chronic and progressive course and may involve large skin areas. From the superficial layers of the epidermis, they may proceed deeper into the dermis and can cause severe acute infections. They may penetrate along the hair shafts into deeper layers of the dermis, inducing deep follicular and perifollicular inflammation (Brasch 2010). In case of untreated infections, tissue damage due to inflammation may lead to permanent hair loss and scarring (Korting 2009). Additionally these infections also spread from the infected person to other people. Some dermatophytoses may take an endemic course. In addition, damage of the epidermal barrier function caused by dermatophytes enables other microorganisms to enter the skin (Nenoff et al. 2014a). Staphylococci, streptococci or gram-negative bacteria may act as a co-pathogen and can induce aggravation of the primary fungal infection, consequently causing some serious systemic complications. In some patients, dermatophytes can induce an allergic response with morphologically diverse allergic eruptions (dermatophytide reactions) (Brasch 2010). Dermatophytoses also cause physical discomfort in those affected and a fear of transmitting the infections to others. In some infections, the quality of life may be significantly reduced (Nenoff et al. 2014a; Whittam and Hay 1997).

Classification of Dermatophytes

Dermatophytes are mainly present in asexual states. On the basis of the morphological characteristics of the macroconidia, they are classified into three asexual genera: *Microsporum, Trichophyton* and *Epidermophyton*. In the genus *Microsporum* (M.), macroconidia are spindle-shaped, they have a thick wall and 1–12 septa. The genus *Trichophyton* (T.) has oblong and rectangular macroconidia with a thin wall and up to 12 septa. In *Epidermophyton* (E.), macroconidia are broader, rounded or oval, thin walled and with up to 5 septa. There are many representatives within each genus. About 40 different dermatophyte species have so far been recognised (Crissey et al. 1995). In addition, there are many keratinophilic dermatophytes, which are soil dwellers and are considered to be non-pathogenic (Hay and Ashbee 2010).

Some dermatophytes have also been found in their perfect (sexual) form. These representatives are termed *hyphomycetes* and are classified among the *Athrodermaceae*. Sexual states have only been observed in geophilic and some zoophilic dermatophytes, not in zoophilic dermatophytes that infect large animals, or in anthropophilic dermatophytes (Hay and Ashbee 2010).

For epidemiological and clinical reasons, it is useful to classify dermatophytes according to their natural habitats into three different groups: geophilic, zoophilic and anthropophilic dermatophytes. Representatives of all of these groups can cause infections in humans (Crissey et al. 1995).

Geophilic dermatophytes grow in the soil and are transmitted to humans through infected soil. They may also be present on vegetables (Korting 2009). Infections can occur in professional gardeners or when children play outdoors. They can also be transmitted from soil to humans indirectly via animals. Geophilic dermatophytoses are diagnosed worldwide and are usually observed in the spring and summer. The most common geophilic pathogen of worldwide distribution is *M. gypseum*. *Microsporum fulvum* is also geophilic and can cause infections in humans (Korting 2009; Nenoff et al. 2014) (Table 1).

Zoophilic dermatophytes primarily cause infections in certain mammals and have also been found on the feathers of birds (Crissey et al. 1995). After direct contact with sick animals, the infection may spread to human skin. Farmers and veterinarians, and also other people who come in close contact with infected domestic pets, risk being infected. Close association between humans and companion animals contributes to the spread of zoophilic dermatophytes. Indirect spread of infection via infected clothes, towels, brushes or other infected objects should also be considered. Infections may also be transmitted via interhuman skin-to-skin contacts. Outbreaks may occur at school or in families. Infections with zoophilic dermatophytes are observed in both rural and urban populations (Dolenc-Voljč 2005).

Table 1. Geophilic and anthropophilic dermatophytes and their geographical distribution.

Dermatophyte	**Ecology**	**Distribution**
Microsporum gypseum	geophilic	worldwide
Microsporum fulvum	geophilic	South America
Trichophyton rubrum	anthropophilic	worldwide
Trichophyton mentagrophytes var. *interdigitale*	anthropophilic	worldwide
Trichophyton violaceum	anthropophilic	Middle East, India, Western China, North Africa
Trichophyton schoenlenii	anthropophilic	worldwide, more common in Mediterranean
Trichophyton tonsurans	anthropophilic	worldwide, more common in North America
Trichophyton concentricum	anthropophilic	Indonesia, Latin America
Trichophyton soudanense	anthropophilic	Africa
Trichophyton megninii	anthropophilic	Mediterranean
Epidermophyton floccosum	anthropophilic	worldwide
Microsporum audouinii	anthropophilic	worldwide

Microsporum canis, *T. mentagrophytes* var. *mentagrophytes* and *T. verrucosum* are the most common zoophilic dermatophytes (Korting 2009). Cats, dogs, rodents and cattle are the most common sources of infections (Table 2). Recently, a new zoophilic pathogen, *Trichophyton* species of *Arthroderma benhamiae*, which corresponds to the zoophilic *T. mentagrophytes* isolates, has been recognized. The sources are small rodents, especially guinea pigs (Nenoff et al. 2014a). Zoophilic infections are more commonly diagnosed in children and adolescents, on the uncovered parts of the body, with a higher affinity to hairy regions of the skin. They are usually presented clinically as acute inflammatory lesions, pustule formation and deep infiltrates. Infection is usually observed in otherwise healthy and immunocompetent individuals.

Table 2. Zoophilic dermatophytes, their hosts and geographical distribution.

Dermatophyte	Hosts	Distribution
Microsporum canis	cats, dogs	worldwide, more common in Central and Southern Europe
Trichophyton mentagrophytes var. *mentagrophytes*	rodents: hamster, guinea pig	worldwide
Trichophyton mentagrophytes var. *erinacei*	hedgehogs	Europe, New Zeland
Trichophyton mentagrophytes var. *quinckeanum*	mice	worldwide
Trichophyton verrucosum	cattle	worldwide
Trichophyton gallinae	fowl	worldwide
Trichophyton equinum	horse	worldwide
Trichophyton simii	monkey	India
Microsporum nanum	pigs	worldwide

Anthropophilic dermatophytes have become highly specialized pathogens restricted to human keratinized tissues and they parasitize humans exclusively. These infections are more common on covered parts of the body in adult patients and have a chronic course. Anthropophilic dermatophytes are commonly diagnosed on the feet, toenails, in the groin and on the trunk. These infections are more common in developed countries. Transmission normally occurs through infected warm and humid floor areas in communal bathing and sports facilities and only rarely via direct personal contact. It can occur in hotels and mosques. A transmission is also common among family members and the source of infection is mainly the bath at home (Nenoff et al. 2014a). Household dust may also serve as a reservoir of anthropophilic dermatophytes, preserving their spores for years (Havlickova et al. 2008). The most common anthropophilic dermatophytes are *T. rubrum* and *T. mentagrophytes* var. *interdigitale* (Table 1).

Based on the site of infection in the hair shaft, dermatophytes are classified into two major groups. Infections of the outer layer of the hair shaft are designated *ectothrix*. Infections in which spores are produced within the hair shaft are of the *endothrix* type (Table 3).

Table 3. Most common ectothrix and endothrix dermatophytes species.

ectothrix		endothrix
small-spored	large-spored	
Microsporum audouinii	*T. verrucosum*	*T. violaceum*
Microsporum canis	*T. mentagrophytes* var. *mentagrophytes*	*T. tonsurans*
Microsporum gypseum	*T. mentagrophytes* var. *erinacei*	*T. soudanense*
other *Microsporum* species	*T. rubrum (rare)*	*T. rubrum (rare)*

On the basis of molecular biological analysis of dermatophyte DNA, various changes have been recommended to the nomenclature of dermatophyte species (Hay and Ashbee 2010; Nenoff et al. 2014a). These changes have not yet been internationally accepted in the clinical practice and the terminology used currently is not uniform.

Epidemiology

The distribution of dermatophytes varies with geographical region and with a wide range of environmental and socio-economic conditions, as well as cultural factors (Havlickova et al. 2008). Considerable inter- and intra-continental differences in epidemiological data have been observed. *T. rubrum* and *T. mentagrophytes* var. *interdigitale* are the most common anthropophilic dermatophytes reported in published surveys. Both are distributed worldwide. Some other dermatophytes are restricted to particular geographic regions or continents. In Northern Europe, as well as in other developed countries, *T. rubrum* is the predominant dermatophyte. In Central and Southern Europe, *M. canis* and *E. floccosum* have been reported more commonly than in Northern Europe. In the Middle East, *T. violaceum, T. tonsurans* and *M. canis* have been reported more commonly than in European countries. In Africa, *T. violaceum, T. soudanense* but also *M. canis, M. audouinii* and *E. floccosum* have been observed in high percentages (Nenoff et al. 2014a). In India, *T. rubrum* and *T. mentagrophytes* var. *interdigitale* along with *T. violaceum* and *M. audouinii* have been found in significant proportions (Havlickova et al. 2008). In North and Central America, both *T. rubrum* and *T. tonsurans* have become the common causative dermatophytes in recent decades (Borman et al. 2007). In South America, *T. rubrum* and *M. canis* have most commonly been reported (Borman et al. 2007). In China, Malaysa, Singapore, Japan and Australia, *T. rubrum* and *T. mentagrophytes* var. *interdigitale* have been reported as the predominant pathogens (Havlickova et al. 2008; Borman et al. 2007).

The epidemiological situation has been changing constantly with time. At the beginning of the 20th century, *T. rubrum* was restricted mainly to Southeast Asia, Indonesia, Northern Australia and West Africa (Thomas 2010). Dramatic changes were observed after the two world wars in Europe. Since then, *T. rubrum* has prevailed over other anthropophilic dermatophytes. Important worldwide changes in distribution have also been observed in the last three decades. In some European countries, the frequency of zoophilic dermatophytes has decreased but the incidence of *T. violaceum* and *T. tonsurans* in scalp infections in urban areas has increased. The proportion of *T. rubrum* and *T. mentagrophytes* var. *interdigitale* in foot infections has also increased

(Borman et al. 2007). In the USA, a dramatic increase in *T. tonsurans* infections has been reported (Havlickova et al. 2008).

Differences on the global scale probably also reflect different personal hygiene levels, as well as different availability of therapeutic measures (Borman et al. 2007). To some extent, differences are probably due to different diagnostic possibilities and different requirements to notify fungal diseases. Data for some countries or geographic areas are not well known or have not been reported. Notification of zoonotic dermatophytoses and epidemiological survey is regulated in some countries by law and has been performed continuously, while it is not so strict in some other countries. In addition, not all types of dermatophyte infections need to be reported. The epidemiological data collected and reported therefore, do not in themselves necessarily reflect the real epidemiologic situation. The true prevalence of these infections is probably much higher than reported.

Pathogenesis of Dermatophyte Infection

The complexity of the host-fungus relationship in dermatophyte infection has still not been explained in detail (Achterman and White 2012). Dermatophytes can induce both immunostimulating as well as immunosuppressive reactions in the host. Their pathogenic potential may be different (Brasch 2010). On the other hand, considerable individual variations in non-immune and immune host responses are possible. The clinical course of dermatophyte infection may therefore vary substantially in infections with the same dermatophyte.

Initially, the adhesion of vital spores to keratinocytes and the formation of fibrillar projections take place. This phase has already been observed *in vitro* within the first hours (Vermout et al. 2008). The adherence is mediated by mannan glycoproteins in the cell wall of the fungus (Kasperova et al. 2013). Damage to the protective barrier of the stratum corneum may facilitate the adherence of fungi. Maceration, occlusion, skin trauma and a warm and moist climate enable the entry of fungi into the epidermis (Brasch 2010; Korting 2009). After the first day, germination with the formation of hyphae follows. Hyphae grow in multiple directions and invade the lower layers of the stratum corneum (Vermout et al. 2008). Dermatophytes proceed through keratinocytes as well between them. They release many proteolytic enzymes (keratinolytic proteases), which degrade and utilise keratin and other proteins of the stratum corneum (Brasch 2010; Vermout et al. 2008). Degradation of keratin is considered to be a major virulene factor. Genomic analysis of dermatophytes showed that dermatophytes contain genes for various proteases, needed in the process of keratolysis (Achterman and White 2012). Dermatophytes secrete more than 20 proteases when grown *in vitro* (Achterman and White 2012). Cysteine dioxygenase and a sulphite efflux pump have recently been recognised as new virulence factors in the process of keratin degradation (Grumbt et al. 2013; Kasperova et al. 2013). Disulfide bridges in epidermal keratins have a protective role against proteolitytic enzymes. Dermatophytes are able to break these bridges by the enzyme cysteine dioxygenase (Nenoff et al. 2014a). Sulfite probably also accelerates keratin degradation. Cysteine dioxygenase and sulfite efflux pump enable dermatophytes to form sulfite from cysteine found in keratin (Nenoff et al. 2014).

Additional virulent factors play a role in the process of infection. Nonprotease genes encoding for opsin-related protein and enzymes of the glyoxylate cycle are upregulated during interaction with keratinocytes (Achterman and White 2012). Differentially regulated synthesis of secondary metabolites may play a significant role in adaptation of dermatophytes to environmental conditions (Nenoff et al. 2014). Fungal-specific genes that code for kinases and pseudokinases may be involved in phosphorylation (Nenoff et al. 2014a). Some toxins probably also play a pathogenic role (Brasch 2010).

Advances in sequencing genome of several dermatophytes enabled genetic studies of dermatophyte virulence factors. Genetic manipulation of dermatophytes by inducing deletion and a complementation of the mutation will be able to definitely assess the role of specific genes in the pathogenesis. Differences in growth between the mutant and the wild type of *Arthroderma benhamiae* have been studied in guinea pig infections and have confirmed an important role of the gene encoding malate synthase (Grumbt et al. 2011; Achterman and White 2012).

Animal virulence models cannot completely mimic infections *in vivo* caused by anthropophilic dermatophytes. Human epidermis tissues have already been used as a new virulence model to study the initial stages of dermatophyte infections *ex vivo* (Achterman and White 2012; Vermout et al. 2008). These new virulence models will provide more reliable information on dermatophytes virulence factors in the human skin.

Host response is both non-immunologic (unspecific) and immunologic (specific). Fatty acids from the sebaceous glands possess fungistatic properties. Younger children with dormant sebaceous glands are therefore more prone to scalp infections. Skin also contains various antimicrobial peptides. Unsaturated transferrin is considered to be a serum inhibitory factor, which has been presumed to play a protective role by binding iron, needed for fungal growth (Hay and Ashbee 2010). Epidermal turnover can to some extent inhibit penetration of dermatophytes into the deeper layers of the stratum corneum (Brasch 2010). Additionally, dermatophytes are termosensitive and grow optimally between 25 and 28°C and therefore prefer to spread superficially. Penetration from the stratum corneum into deeper layers usually occurs along the hair shaft. UV radiation can also promote deeper spread of dermatophytes (Brasch 2010).

Keratinocytes and Langerhans cells play a key role in the process of recognition of dermatophytes. Keratinocyte cells express Toll-like receptors that can recognize the pathogen. Via these receptors, signals activate the unspecific immune reaction, by releasing proinflammatory cytokines, such as IFN-γ, TNFα, IL-13, IL8 and IL-16 (Brasch 2010). Dermatophytes are chemotactic and activate the complement pathway (Hay and Ashbee 2010). Complement mechanisms attract neutrophilic granulocytes and monocytes, which are capable of damaging or killing dermatophyte conidia. Natural killer cells probably also play a protective role. An inflammatory reaction at the site of infection increases epidermal turnover and helps to eliminate fungal elements (Achterman and White 2012).

The immunologic response plays a crucial role in defense against dermatophytes. Various dermatophyte antigens have been identified, capable of inducing both acute (type I) and delayed (type IV) immune reactions (Korting 2009). However, a humoral reaction with specific antibodies has not been found to have a protective role (Korting 2009). Increased levels of antibodies may persist for years and probably do not protect predisposed individuals against reinfection (Hay and Ashbee 2010). A cellular immune response via T lymphocytes has an indispensable role in the final healing of the infection. Langerhans cells initiate this process by recognition of dermatophytes and presentation of their antigens to T-cells. A delayed type reaction to trichopytin, the fungal antigen, can be demonstrated by a positive skin trichophytin test (Brasch 2010; Vermout et al. 2008) and is considered to be a marker of a good cellular immunity. On the other hand, dermatophytes are capable of producing some immunosuppressive factors, such as mannan, which inhibits T lymphocytes, resulting in chronic infection with mild clinical signs (Achterman and White 2012; Vermout et al. 2008).

Typical annular erythematous lesions, observed clinically, develop within 1 to 3 weeks. Due to the host response in the affected lesions, fungi expand peripherally and centrifugally, forming characteristic annular erythematous and scaly lesions. In the center of the lesions, fungi are destroyed and eliminated, which consequently leads to regression of inflammation, erythema and scaling (Korting 2009). Dermatophytes are not part of the normal skin microflora. If isolated, they should be considered as pathogens.

Clinical Presentation

Dermatophytes may induce various types of skin lesions in humans, from discrete superficial scaling without any associated symptoms to deep inflammatory infiltrates with purulent discharge, accompained by enlarged regional lymph nodes and systemic symptoms with fever. The clinical picture may therefore mimic many infectious and non-infectious skin diseases, causing difficulties in diagnosis. The type of skin lesion depends on the causative pathogen, the localisation of infection and the host immune reaction (Hay and Ashbee 2010; Korting 2009). Previous topical or systemic treatments may alter the clinical course. Topical corticosteroids modify the clinical picture by reducing the signs of inflammation. Steroid-modified tinea is difficult to recognise and is called “tinea incognita” (Korting 2009). An atypical clinical course is often observed in immunosuppressed patients. On the other hand, some irritative external factors, such as UV radiation and cosmetic products, may worsen erythema.

Tinea Capitis

Various dermatophytes may cause scalp infections but some species have a higher affinity for hair invasion. *M. canis*, *T. mentagrophytes* var. *mentagrophytes*, *T. verrucosum*, *T. violaceum*, *T. schoenleinii* and *M. audouinii* are the most common causative pathogens. The epidemiologic situation of tinea capitis varies in different countries. In Europe, epidemiologic situation has changed in recent decades (Ginter-Hanselmayer et al. 2007). *M. canis* has become the most common isolated pathogen in recent decades. In central and southern European countries, it causes up to 90% of all scalp infections (Ginter-Hanselmayer et al. 2007; Dolenc-Voljč 2005). In some other countries, a rising incidence in anthropophilic dermatophytes in urban areas has been reported (Hay and Ashbee 2010). The observed changing patterns of tinea capitis are mainly due to population movements and immigration from Africa and Asia to Europe. In USA, *T. tonsurans* has been the most commonly isolated pathogen (Havlickova et al. 2008).

If not treated, tinea capitis may have a chronic course, leading to destruction of the follicles with irreversible scarring alopecia (Korting 2009). Children are especially prone to this infection (Nenoff et al. 2014b). Spontaneous regression may sometimes occur if the infection begins at puberty.

The infection starts in the stratum corneum; after three weeks, clinical signs of hair shaft invasion may be noticed. Initially, infection is localised superficially in the stratum corneum. In tinea capitis superficialis, one or more lesions are present with mild scaling; erythema may be mild or absent. Hairs may be broken a few mm above the skin surface. Such a type of scalp infection is usually observed with *M. canis, M. audouinii* and *T. tonsurans* (Korting 2009; Nenoff et al. 2014b) (Fig. 1). In tinea capitis profunda, follicular papules and pustules are associated, sometimes with signs of deep inflammation and purulent discharge. On rare occasions, tumorous infiltrate can appear, called kerion Celsi (Korting 2009). Deeper infections are more commonly caused by *T. verrucosum* and *T. mentagrophytes* var. *mentagrophytes*. Regional lymphadenopathy or even systemic signs of infection can be associated. Secondary bacterial infections are possible, resulting in mixed fungal and bacterial infection (Nenoff et al. 2014b). Favus (tinea capitis favosa) is a special entity of tinea capitis, which is by definition caused by *T. schoenleinii*. It has become rare nowadays but is still present in some endemic areas with poor hygiene and malnutrition (Ginter-Hanselmayer et al. 2007). A familial spread of infection appears, with the involvement of many family members

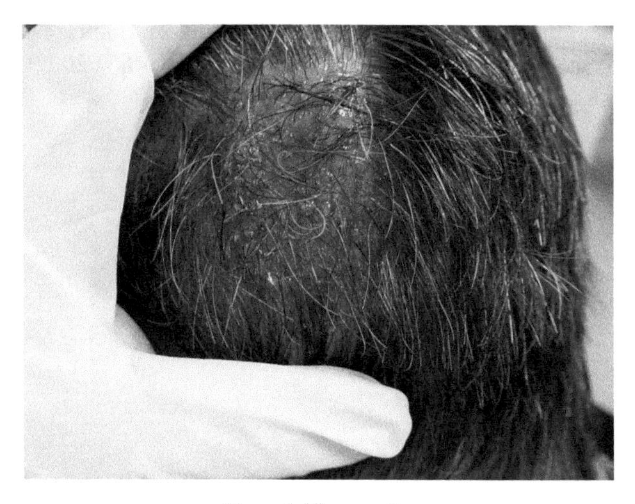

Figure 1. Tinea capitis.

of different ages. Infection is very contagious. If not treated, it has a chronic course. Typical sulfur-yellowish crusts (scutula) appear on perifollicular erythematous lesions. There is also erosion and formation of scarring alopecia under the scutulum (Korting 2009).

Tinea Faciei

Tinea faciei is by definition a dermatophyte infection of the glabrous skin on the face. It has similar epidemiological characteristics as tinea corporis. It may be caused by all known dermatophytes. Zoophilic dermatophytes *M. canis* and *T. mentagrophytes* var. *mentagrophytes* are more common in children while anthropophilic species predominate in adults (Nenoff et al. 2014b). Tinea faciei may also occur as a consequence of fungi inoculation from a pre-existing foot infection.

Clinically, skin lesions often have an untypical course, without sharp margins and with a lack of scaling. Exogenous influences from cosmetic products and UV radiation may mask or worsen the inflammation and induce atypical skin lesions. Tinea faciei may therefore mimic other facial skin diseases and is often misdiagnosed (Korting 2009). Because of its atypical clinical characteristics, it is presented as a separate clinical entity among dermatophyte infections.

Tinea Barbae

Tinea barbae is a typical zoophilic infection, localized on the hairy skin of the beard in men. The most common causative agents are *T. verrucosum* and *T. mentagrophytes* var. *mentagrophytes*. Farmers and veterinarians are most commonly infected. Infection is often transmitted through cattle or rodents (Korting 2009).

This tinea is one of the most severe dermatophyte infections and difficult to treat. Deep infiltrates with follicular papules and pustules, painful furunculoid nodules and inflammatory discharge and crusts are present (Fig. 2). Hairs are easily removed and fall out. Regional lymph nodes are often enlarged (Korting 2009). If not treated

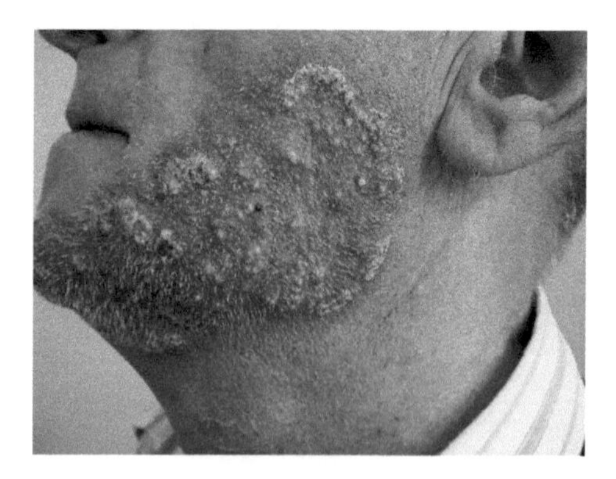

Figure 2. Tinea barbae.

properly, lesions heal with scarring and alopecia. A fungal origin of infection is often overlooked and diagnosis made with delay.

Tinea Corporis

Tinea corporis is one of the most common dermatophyte infections. Lesions are by definition localised on the trunk and extremities. It may be caused by all known dermatophytes and their prevalence reflects the epidemiologic situation in each country. In children and adolescents, acute infections caused by zoophilic dermatophytes are more common. In adults, anthropophilic fungi predominate, most commonly *T. rubrum* (Nenoff et al. 2014b). It may also be a complication of a neglected and untreated tinea pedis.

Initially, a small erythematous macule or papule arises. In about one to three weeks, typical erythematous annular and ciricinar lesions develop, with sharp borders and a central regression of erythema. At the borders, erythema is more marked and scaly (Korting 2009) (Fig. 3). In the hairy parts of the trunk and limbs, it may involve deeper layers of the skin, with a follicular pattern of inflammation.

Some special types of tinea corporis can be distinguished. Infection may spread in a wrestling team due to transmission of dermatophytes through close personal contact. Such infections are termed tinea gladiatorum. *T. tonsurans* has been reported in this type of infection (Korting 2009; Nenoff et al. 2014b). Tinea caused by *M. canis* has typical coin sized erythematous annular lesions with sharp borders. It is usually localized on the exposed parts of the body. It is also termed tinea microsporica or microsporia. Microsporia has been often observed in small endemic areas. It is present in both rural and urban areas. Cats are the most common source of the infection but dogs and rodents should be also considered. In our patients, it has more often been observed in small children. One quarter of all infected patients were below 5 years. It exhibits a typical seasonal variation, with a higher incidence in the period from July to October

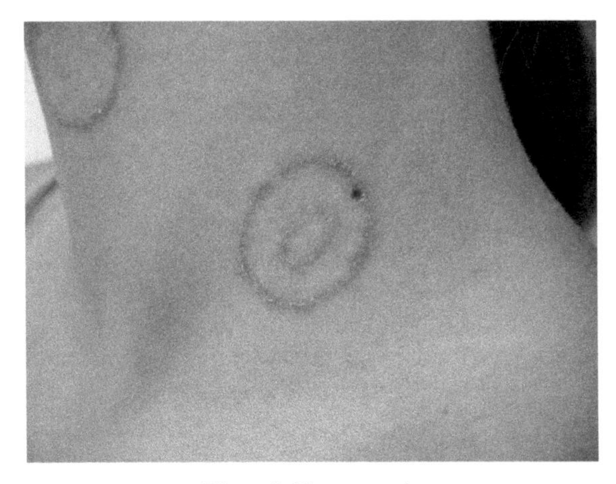

Figure 3. Tinea corporis.

(Dolenc-Voljč 2005). Close contact with infected animals is usually required for infection. Additionally infections may also be transmitted indirectly via interpersonal contacts. Similar clinical presentations are observed with *M. gypseum* infections but infections with this geophilic dermatophyte are rarer. Infections caused by *T. rubrum* induce a different pattern of erythematous lesions, which progress slowly to large erythematous macules or plaques involving large areas of the body. Tinea imbricata is a special entity caused by *T. concentricum* in endemic regions with a tropical climate. Initial annular erythematous and scaly erythema spread centrifugally, while new rings develop in the center of the lesion, forming characteristic concentric rings. It often has a chronic course and affects large areas of the body (Hay and Ashbee 2010).

Tinea Inguinalis

Synonyms: Tinea cruris (incorrect term), ringworm of the groin, jock itch

The causative dermatophytes are anthropophilic, most commonly *T. rubrum*, rarely *E. floccosum* and *T. mentagrophytes* var. *interdigitale*.

Tinea inguinalis is distributed worldwide but is more prevalent in warm and humid climates. Men are affected more commonly. It usually results from autoinfection from the foot. Obesity, diabetes, inadequate personal hygiene, synthetic clothing and hyperhidrosis may cause this infection (Hay and Ashbee 2010). This infection may take a chronic course and is very rare in children.

Sharply margined erythematous and itchy lesions or plaques are present in the groin and inner parts of the thighs. Scaling and vesicles may appear at the borders of the lesions. Distribution may be unilateral or bilateral. Infection often spreads to the scrotal and perianal area, perineum and gluteal region.

Tinea Manus

It is caused by anthropophilic dermatophytes, most commonly by *T. rubrum*, rarely by *T. mentagrophytes* var. *interdigitale* and *E. floccosum*. In many cases, infection of the hand is a consequense of a pre-existing foot infection and transmission of the dermatophyte from the foot (Korting 2009).

On the inner parts of the hand, the clinical presentation is similar to that in tinea pedis of the sole. In infections caused by *T. rubrum*, only mild superficial scaling may be seen, which is usually neglected or attributed to other causes. Only one hand is initially affected. In this case, it may be a part of a chronic palmoplantar dermatophyte infection, called "two feet, one hand syndrome" (Nenoff et al. 2014b). Both hands may later be infected with concomitant finger nail onychomycosis. On the dorsal site of the hand, the infection has a similar inflammatory pattern to that in tinea coporis, with annular erythematous lesions with central regression and marked erythematous borders. In this location, zoophilic dermatophytes should be also considered.

Tinea Pedis

Synonyms: Foot ringworm, Athlete's foot

This infection is almost always caused by anthropophilic dermatophytes. *T. rubrum* and *T. mentagrophytes* var. *interdigitale* are the most common aetiologic fungi. *E. floccosum* is a less common cause (Dolenc-Voljč 2005). *T. violaceum* can cause tinea pedis in countries in which this dermatophyte is common. Sporadically, zoophilic dermatophytes can also cause tinea pedis on the dorsal site of the foot. Since 1980, *T. rubrum* foot infections have been rising in incidence (Hay and Ashbee 2010).

Tinea pedis is the most commom dermatophyte infection in humans in developed countries and one of the most common diseases in humans generally. The estimated prevalence is around 10% for the general population (Hay and Ashbee 2010). Some epidemiological studies have shown that the prevalence in adults may reach 20% (Korting 2009; Burzykowski et al. 2003). In special patient populations (soldiers, athletes, miners), it may affect up to 70% of individuals (Korting 2009). More than 50% of sports-active individuals have clinical signs of foot disease, which are in most cases of fungal origin (Caputo et al. 2001). Infection is transmitted most commonly in sports facilities, swimming bath resorts, hospital wards and among family members. It is spread mainly indirectly via vital arthrospores. Transmission via direct personal contact is rare. The sources of infection are usually individuals who have a foot infection. In an appropriate warm and humid environment, arthrospores may be infective for months or even longer (Havlickova et al. 2008). This infection is therefore difficult to prevent.

Infection is more common in middle and old age and is more frequent in men. Certain patient populations are more prone to this infection. Occlusive footwear and skin maceration may facilitate infection between the toes. It is rare in those who habitually go barefoot (Hay and Ashbee 2010). Among other risk factors, diabetes, obesity, immunosuppression, peripheral vascular disease, trauma, osteoarticular pathology, participation in sports and hyperhidrosis should be considered (Burzykowski et al. 2003). It is more common in warm climates. Children are rarely infected and usually get the infection from their parents or at swimming facilities.

If untreated, tinea pedis can have a chronic course. From the skin, it spreads to the toenails, which are often infected concomittantly. Tinea pedis may also function as an entrance to secondary bacterial infection. Bacteria aggravate the clinical picture and may cause erysipelas or cellulitis.

Clinically, the infection may present itself as various types. Interdigital tinea pedis is the most common type and usually caused by *T. rubrum,* rarely by *T. mentagrophytes* var. *interdigitale* or other anthropophilic dermatophytes. It starts as mild superficial scaling in the toe webs between the third and the fifth toes (Korting 2009). Skin lesion can be very discreet and without any symptoms, so many infected individuals are unaware of this infection and are therefore not treated. Erosions, fissures and maceration may develop, which may cause itching or pain. Erythema is mild or absent (Fig. 4). From the interdigital spaces, infection extends to the undersurface of the toes and rarely to the dorsal site of the fingers and foot. In the dyshidrotic type, pruritic grouped vesicles are present on the sole, which may coalesce to bulla. After the vesicles rupture, erosions and scaling follows. Mild erythema is usually associated. *T. mentagrophytes* var. *interdigitale* is the most common causative pathogen (Korting

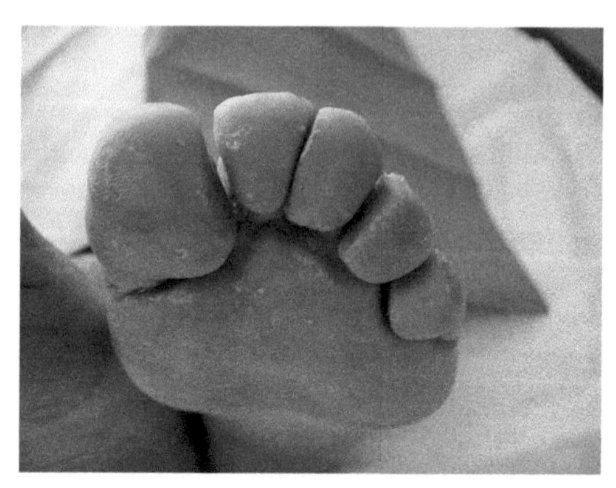

Figure 4. Tinea pedis.

2009). The hyperkeratotic type of tinea pedis is rare. Diffuse scaling is prominent on the sole, heel and sides of the foot, with slight erythema at the borders. The erythematosquamous type is observed on the dorsal site of the foot and has similar clinical characteristics as tinea corporis.

Tinea Unguium (onychomycosis)

The term "tinea unguium" is used to describe fungal nail infection caused by dermatophytes. Onychomycosis is a broader term that also includes nail infections caused by yeasts and non-dermatophyte fungi.

Toenail onychomycosis

The epidemiology of toenail onychomycosis is similar to that of tinea pedis. The most common causative dermatophytes are *T. rubrum* and *T. mentagrophytes* var. *interdigitale*, while *E. floccosum*, *T. tonsurans* and *T. violaceum* are rarely involved. Dermatophytes account for at least 90% of toenail onychomycosis.

Similar to tinea pedis, toenail onychomycosis is one of the most common fungal infections in humans. It is often associated with tinea pedis. Its prevalence in developed countries has been on the increase in recent decades (Thomas 2010). The first epidemiological studies, performed some decades ago, reported a prevalence rate between 2.2 to 8.4% (Roberts 1992; Hekkilä and Stubb 1995). A larger study performed later in European countries found a prevalence of 23% in the adult population (Burzykowski et al. 2003). In the USA, the prevalence of onychomycosis is thought to have increased sevenfold (Gräser et al. 2012). In East Asia, onychomycosis has been reported as being found in 22% of the population (Thomas 2010). Toenail onychomycosis is present in at least 20% of people aged more than 60 years and up to 50% in people older than 70 years (Thomas 2010).

Predisposing factors are similar to those for tinea pedis. Adults are infected much more often than children. In young people, sports-active individuals are especially predisposed. Nail trauma, wearing of occlusive footwear, osteoarticular pathology, peripheral occlusive arterial diseases, chronic venous insufficiency, lymphoedema in the lower extremities, peripheral neuropathy, immunosuppresion and HIV infection, diabetes and nail psoriasis are considered to be predisposing factors. Toenail onychomycosis affects one third of the patients with diabetes (Gupta et al. 1998). Genetic predisposition is also considered to be important. Clinical observations suggest autosomal dominant inheritance of susceptibility (Nenoff et al. 2014b). Some studies performed in the last decade have found a HLA-DR4 and HLA-DR6 genetic constellation to play a protective role in onychomycosis (Asz-Sigall et al. 2010; Nenoff et al. 2014b).

The importance of onychomycosis is usually neglected and it is still considered to be a cosmetic problem. Studies have shown that it can significantly lower the quality of life (Whittam and Hay 1997). It can cause pain, inhibit the mobility of infected persons and it enables the evolution of some severe complications. Erosions or ulceration of the skin, secondary infections and gangrene are more common in diabetic patients with onychomycosis (Gupta et al. 1998; Cathcart et al. 2009).

Different types of toenail onychomycosis can be identified clinically. The distolateral type is the most common. The first signs of infection begin at the distal and lateral portion of the nail plate with discoloration and detachment from the nail bed. With further evolution, nails become brittle at the free edge and may crack. Spontaneous healing does not occur. The signs of infection progress slowly in a proximal direction towards the lunula and nail matrix. Involvement of the entire nail length usually occurs in a few years. At the beginning, the first toenail is most commonly infected, with the infection later spreading to other nails (Fig. 5). Superficial white onychomycosis (leukonychia trichophytica) is a rare pattern of onychomycosis. Infection is localised at the dorsal parts of the nail plate, with white patches that may proceed to deeper layers of the nail plate. *T. mentagrophytes* var. *interdigitale* is usually the causative dermatophyte. Proximal subungual onychomycosis is even rarer; however, it is more frequent in HIV positive patients. Fungi enter the nail plate proximally from the cuticle

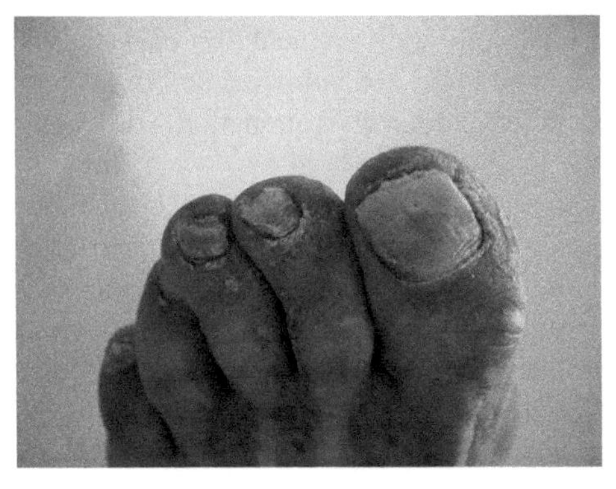

Figure 5. Toenail onychomycosis.

and proceed in a distal direction under the nail plate. White lesions are noticed at the proximal part of the nail plate. Growing of the nail is impaired in the advanced stage of infection due to matrix damage. Dystrophic onychomycosis is the most severe type of onychomycosis (Nenoff et al. 2014b). The whole nail plate is thickened and destroyed. It is the end stage of all other types of onychomycosis. The nail matrix may become irreversibly damaged and nail growth is consequently impaired. Treatment of this type of onychomycosis is very difficult. Mixed forms of onychomycosis can also be recognised.

Fingernail onychomycosis

The epidemiologic situation in fingernail onychomycosis is quite different to that in toenail onychomycosis. The percentage of dermatophytes is considerably lower in fingernail infection. In our patients, it was found to be only 17% (Dolenc-Voljč 2005). Fingernail onychomycosis is more commonly caused by yeasts of *Candida* species. Fingernail onychomycosis caused by dermatophytes is usually transmitted from a pre-existing infection of the foot. Clinically, it most commonly presents as a distolateral type of nail infection.

Onychomycosis is sometimes difficult to diagnose on the basis of the clinical picture. Nail trauma, psoriasis and dystrophic nails for other reasons can cause similar clinical presentation. Accurate laboratory examination and confirmation of the diagnosis is therefore recommended, since 50% of cases are misdiagnosed when relying on the clinical appearance only (Gräser et al. 2012).

Diagnostics

The patient's history and the clinical picture are the basis in the diagnostic approach. In many cases, dermatophyte infection can be suspected clinically. A typical clinical presentation with centrifugally spreading annular erythema with sharp erythematous scaly borders and asymmetrical distribution is highly suggestive of tinea. However, in previously treated or irritated skin lesions, in hairy skin areas and in nail infections, the diagnosis may be challenging. Because the source of the infection varies substantially in different types of infections and treatment approaches are different, it is highly advisable to perform a mycological examination before treatment and to identify the causative dermatophyte. Mycological examination also enables assessment of the treatment response.

Direct microscopy and culture have traditionally been the mainstays in laboratory diagnostics in dermatologic practice and are still considered to be the "gold standard". Preparation of skin, hair or nail specimens requires experienced personnel to perform the task correctly. Specimens are obtained by scraping the scales from the border of the skin lesion, from the nailbed and nails or by plucking hairs. A sterile scalpel, curette or tweezers can be used. The scales are placed on a glass slide with a drop of 10–20% KOH solution (Nenoff et al. 2014b). This test is also called "native examination". The KOH dissolves the keratin and enables better transparancy of hyphae and spores. If special dyes are added (Parkers blue or lactophenol blue), the hyphae are visible more easily. The sensitivity of microscopic examination can be enhanced by using fluorescent dyes and examination under fluorescent microscope (Korting 2009). Direct microscopy is an easy and readily available method. It is a useful aid to a clinician in choosing the appropriate treatment. Spores and hyphae can be clearly seen. The arrangement of spores along the hair shaft can be assessed. In ectothrix species, the spores are located in the outer layer of the hair shaft, while in endothrix species the spores are found within the hair shaft (Hay and Ashbee 2010).

However, microscopic examination can only detect the presence or absence of fungal elements, without information on their viability. Within the keratinized tissue, dermatophytes exist only as mycelium and arthroconidia. In all dermatophyte infections, hyphae and arthrospores produced by hyphae are morphologically indistinguishable (Crissey et al. 1995). In addition, false-negative results of microscopic examination should be considered. Topical antimycotic drugs should be stopped for at least 1 week before mycological examination to avoid false-negative results (Korting 2009).

Culture is an essential adjunct to direct microscopy because it enables identification of dermatophytes. The most common medium used for isolation of dermatophytes is Sabouraud glucose agar, amended with chloramphenicol and cycloheximide to inhibit bacterial and saprophytic fungal contamination. The results of laboratory fungal culture are usually not available for 3–6 weeks (Nenoff et al. 2014b). Identification of dermatophytes is based on gross colony characteristics and their microscopic morphology. Dermatophyte cultures show great variability in colour and growth rate. Macroconidia are species specific, while microconidia are similar in most species. The morphology of hyphae may also help in identification; they may be spiral, raquet or pectinate (Crissey et al. 1995; Hay and Ashbee 2010). As the culture matures, the colour of the culture, gross appearance and morphological characteristics may change. Exact differentiation between different dermatophyte species on the basis of morphological criteria may be difficult. Some physiological tests can be additionally used. Subculture on lactritmel agar or rice grains may stimulate macroconidial production. Urea agar, peptone agar, potato dextrose agar and a hair perforation test can be of additional help in differentiation between difficult dermatophytes species (Hay and Ashbee 2010).

An additional diagnostic tool is Wood's lamp (UVA light examination), which can detect some fluorescent dermatophyte species. *M. canis*, *M. audouinii*, *M. distortum* and *M. ferrugineum* induce green fluorescence. *T. schoenleinii* demonstrates yellow fluorescence (Korting 2009). Wood's lamp is also often used in veterinarian medicine.

In case of infections difficult to diagnose, a small skin biopsy is advised to detect hyphae by using periodic-acid-Schiff stain. It can confirm the diagnosis but does not allow identification of the species. Nail fragments can also be examined histopathologically (Nenoff et al. 2014b).

Dermatophyte test medium (DTM) may be used as a simplified diagnostic test for confirmation of dermatophytes in the specimen. The medium contains a pH indicator (phenol red), which converts from yellow to red under alkaline condition associated with growth of dermatophytes (Taplin et al. 1969; Rich et al. 2003). Positive DTM can confirm the presence of dermatophyte organism within 3–7 days but does not enable the identification of the species.

Serological tests have no relevant role in the diagnostics of dermatophyte infections. Serum antibodies to dermatophytes remain positive for years and cannot be used as a criterion of acute infection. The trichophytin test can only be used as a marker of the cellular immune response, without playing a significant role in confirmation of the diagnosis (Korting 2009).

New molecular techniques with polymerase chain reaction (PCR) represent an important progress in the diagnostics of dermatophyte infection. Analysis of fungi specific nucleic acid sequences enables accurate species identification of dermatophytes. A nucleic acid sequence of the DNA topoisomerase II gen, internal transcriber spacer regions (ITS) in ribosomal DNA, mitochondrial DNA, chitinsynthase, superoxide dismutase, actin and tubulin have been used in primers (Brasch 2012). Real time PCR assays is a highly specific and sensitive method for the amplification and quantification of dermatophyte DNA (Nenoff et al. 2014b). Results are already available within 1–2 days. It also allows detection of multiple pathogens. In-house PCR assays have been developed to identify dermatophytes directly from skin scales and nail materials (Gräser et al. 2012; Jensen and Arendrup 2012; Nenoff et al. 2014b).

Recently, mass spectrometry (MALDI-TOF MS) has been implemented as a novel and time saving method for identification of dermatophytes, grown in culture (Gräser et al. 2012; Nenoff et al. 2014b; L'Ollivier et al. 2013). It is mainly used as an additional diagnostic tool for differentiation between dermatophytes. This method enables identification of up to 64 dermatophyte strains and its specificity is estimated to be high (Nenoff et al. 2014). The most widespread dermatophyte species can be rapidly identified with this method.

Novel molecular diagnostic techniques represent an important supplementation to classical diagnostic possibilities. Their main advantages are improved sensitivity, specificity and fast identification. However, they are associated with higher costs and specific technical requirements and are not used routinely in clinical practice (Gräser et al. 2012). False-positive and false-negative results should also be considered. False-positive results are mainly due to contamination of specimens. False-negative findings result from inappropriate DNA extraction or inhomogeneous distribution of dermatophyte DNA in skin or nails samples (Gräser et al. 2012; Nenoff et al. 2014b). Sample preparation and carefully controlled laboratory conditions are essential for optimal results. Findings need to be interpreted cautiously and in correlation with the clinical picture. Conventional diagnostic methods and good clinical practice still have an indispensable role in the diagnosis of dermatophyte infections.

Treatment

Treatment of dermatophyte infection depends on the location, stage of infection, depth of the skin involved, age of the patient, the causative pathogen and contagiousness. Azoles, allylamines, morpholines and pyridons are available for treatment nowadays (Hay and Ashbee 2010; Korting 2009). They interfere with the ergosterol biosynthesis of the fungus.

Topical treatment is the basic approach in fungal skin infections. It is used for mild, localised and superficial infections of the glabrous skin. Topical antifungal agents are available in cream, ointment, lotion, gel, spray and powder form and some also in nail laquer and shampoo. Among older topical antifungal drugs, tolnaftate and undecylenic acid are still used in some countries. Whitfield's ointment, with an active compound of benzoic acid, has a long history of use in dermatology and possesses both antiseptic and antifungal properties. Sulphur in creams or ointments were also used in the past to treat fungal and parasitic skin infections. Castellani solution (margenta paint, fuchsine) is still used for inflammatory tinea pedis (Hay and Ashbee 2010). The main advantage of older antifungal compounds is the low cost. They have largely been replaced by modern topical antifungal drugs.

The azoles are the largest group among topical antifungal agents. They have a broad spectrum of antifungal activity. Clotrimazole, miconazole, ketoconazole, bifonazole, sertaconazole, econazole and tioconazole are well established and widely prescribed. They also have anti-inflammatory properties and express activity against gram-positive bacteria. The allylamines include naftifine and terbinafine, which have good activity particularly against dermatophytes. Terbinafine can be used only once daily in tinea pedis. A new film-forming solution of 1% terbinafine allows single-shot treatment of interdigital tinea pedis (Korting 2009). The morpholines also have a broader spectrum of antifungal activity. Amorolfine is used in a nail laquer to treat onychomycosis. Among the pyridones, ciclopiroxolamine has been used as an antifungal of broad spectrum activity and good ability to penetrate deeper into the epidermis. It is also registered in a nail laquer.

Once or twice daily topical application for several weeks is usally sufficient to treat acute and superficial infections of the glabrous skin. For interdigital tinea pedis, 1–2 weeks of treatment with terbinafine may be sufficient. In tinea corporis, 2–6 weeks of treatment is usually needed, while in infections due to *M. canis*, at least 6 weeks of treatment is recommended. In onychomycosis, topical antimycotics are applied for many months as the nail re-grow. In toenail onychomycosis, at least 9 months of topical treatment is needed.

Systemic treatment is needed in tinea capitis, tinea barbae, chronic dermatophyte infections, which are unresponsive to topical treatment, hyperkeratotic tinea pedis and manus and widespread tinea corporis. Toenail onychomycosis is treated systemically if more than half of the nail plate is infected, if more than 3 nails are affected and in the case of nail matrix involvement (Tietz and Nenoff 2012; Roberts et al. 2003).

Griseofulvin and ketoconazole are representatives of the older generation of systemic antifungal drugs, which are rarely used nowadays. Griseofulvin is still considered to be the drug of choice for tinea capitis caused by *M. canis* in children (Ginter-Hanselmayer and Seebacher 2011). Its use has been abandoned for other dermatophyte infections. It possesses fungistatic activity against dermatophytes and has a low incidence of serious side effects. Ketoconazole has broad-spectrum antifungal activity. It is recommended as a second-line treatment option and only for shorter treatment periods of less than 1 month. Its use has been restricted because of hepatotoxicity (Hay and Ashbee 2010; Korting 2009).

Among the new generation of systemic antifungal drugs, terbinafine can be recommended as a first-line treatment option for dermatophyte infections. It has well demonstrated fungicidal activity against dermatophytes *in vitro*. As a lipophilic drug, it accumulates and persists in nails in a high concentration for months. It is also rapidly taken up into the stratum corneum and is found in sebum (Hay and Ashbee 2010). Tinea corporis, tinea pedis and tinea manus are treated for 2–6 weeks with terbinafin, depending on the stage and depth of infection. In tinea capitis, at least 6 weeks of treatment is needed. Fingernail onychomycosis is treated for 6 weeks and toenail onychomycosis for 12 weeks. Some patients may need longer treatment. Terbinafine has a low potential of severe side effects and drug interactions (Korting 2009). Itraconazole is a broad-spectrum azole antifungal systemic drug suitable for use in a wide range of dermatophyte infections. It is a lipophilic drug, bound in keratinized tissues and in nails and found in sebum. It may persist in nails for several months after cessation of therapy. Pulsed treatment regimes are often used, with only one week of therapy per month in a 400 mg daily dose. One pulse is usually sufficient for tinea corporis and tinea pedis, while 2 pulses are needed for fingernail onychomycosis and 3 pulses for toenail onychomycosis. In oral solution, it has also been used off-label for tinea capitis in children, with good results (Ginter-Hanselmayer and Seebacher 2011). Itraconazole is a safe antifungal drug with exceptionally rare reports of hepatic side reaction (Hay and Ashbee 2010). Fluconazole has been used mainly for *Candida* infections but, as a broad spectrum azole antifungal, it also shows good activity in dermatophyte infections. In a pulsed treatment regime, it is used in a 150 mg dose once weekly, for 2–4 weeks for tinea corporis and for several months in onychomycosis. It has a low incidence of gastrointestinal side effects.

The new broad spectrum systemic azole antifungal drugs, posaconazole and voriconazole, are reserved for the treatment of systemic fungal infections. Clinical experiences are therefore lacking in dermatology. Both may be considered only for dermatophyte infections that cannot be treated by other therapies (Brasch 2012).

Systemic antifungal treatment is contraindicated in pregnancy and in lactation. Neither itraconazole nor fluconazole have been officially approved for use in children. In severe hepatic disease and renal insufficiency, systemic antifungal drugs have to be used very cautiously and may even be contraindicated. Interactions with other drugs are rare with terbinafine. With azoles, interactions with statins, benzodiazepines, coumarin anticoagulants, ciclosporin, digoxin, rifampicin, astemizole and terfenadin can occur (Hay and Ashbee 2010), especially in older patients. Other side effects are usually mild, including headache, nausea and allergic skin reactions. Rare cases of severe hepatic damage and severe allergic skin reactions have been reported with azoles and terbinafine (Hay and Ashbee 2010; Korting 2009).

Antifungal agents have high success rates in the treatment of dermatophyte infections. Lack of clinical response may be due to many host and drug-related factors (Sarifakioglu et al. 2007). Arthroconidia are considered more resistant to antifungals than hyphye (Martinez-Rossi et al. 2008; Sarifakioglu et al. 2007). The formation of biofilm can also increase antifungal resistance (Costa-Orlandi et al. 2014; Martinez-Rossi et al. 2008). Aquired resistance of dermatophytes to antifungal drugs is rare and of low clinical relevance (Brasch 2012). However, terbinafine-resistant *T. rubrum* strains have been reported (Ghannoum et al. 2004; Martinez-Rossi et al. 2008). Dermatophytes may also exhibit different susceptibility to various antifungal drugs (Sarifakioglu et al. 2007). Antifungal susceptibility of dermatophytes can be measured using a reference broth dilution method with excellent reproducibility of minimal inhibitory concentrations (MIC) (Sarifakioglu et al. 2007). Studies confirmed that terbinafine has the highest antifungal activity for dermatophytes and the lowest MIC values. Itraconazole was the second most active antifungal agent while fluconazole was the least active and with greatest variation in MICs (Sarifakioglu et al. 2007; Silva et al. 2014). Testing for antifungal susceptibility has an important role in detecting resistance strains of dermatophytes. However, correlation between *in vitro* activity of antifungal drugs and situation *in vivo* is difficult (Sarifakioglu et al. 2007; Silva et al. 2014). Antifungal susceptibility testing is therefore of limited value and has not been routinely used in clinical practise.

Preventive measures

The control of tinea capitis depends on the causative pathogen. In the anthropophilic type of infection, spread occurs in kindergartens, schools and among the family. All family members and school goers should be examined for signs of infection and appropriately treated. Selenium sulphide and ketoconazole shampoo can be recommended. Prevention of anthropophilic dermatophytes in shower facilities, swimming baths and in household baths is difficult. Washing of shower floors with an antiseptic may be useful (Hay and Ashbee 2010). Additional individual preventive measures are also needed. Personal hygiene, keeping the skin of the foot dry and use of personal towels are advisable.

In zoophilic infections, the source of infection in domestic pets or cattle should be sought and treated if possible. Cooperation between dermatologists and veterinarians and epidemiologists is needed (Ginter-Hanselmayer et al. 2007). Notification of these infections for the purpose of disease control is mandatory in some countries.

Conclusion and Future Prospects

Dermatophyte infections will certainly remain an important public health problem in both developed and undeveloped countries in the future. On the basis of current epidemiological trends, the prevalence of dermatophyte infections in both developed and undeveloped countries is expected to remain high. The observed dynamic in the epidemiologal pattern of dermatophyte infections will probably continue. A further rise in the incidence of anthropophilic infections in humans can be expected, especially in the older patient population. Tinea pedis and toenail onychomycosis are especially common problem in adults in developed countries. The anthropophilic dermatophytes *T. violaceum*, *T. tonsurans* and *M. audouinii* are expected to remain responsible for outbreaks of tinea capitis and tinea corporis in endemic regions as well as in some European countries with immigrant families. In addition, the target populations for anthropophilic infections seem to be expanding in the future. Onychomycosis due to *T. rubrum* has recently been recognised as a new trend in infancy (Nenoff et al. 2014b). Zoophilic dermatophytes remain frequent causative pathogens and should be considered especially in children and adolescents with tinea capitis, tinea faciei and tinea corporis. In the last decade, an increase in zoophilic dermatophyte of *Arthroderma benhamiae* has been recognised in children with tinea capitis, tinea corporis and tinea faciei (Nenoff et al. 2014a). Thorough surveillance of dermatophyte species and the characteristics of the patients infected have to be continued. Treatment on the basis of mycological examination should be encouraged to reduce the number of patients misdiagnosed and not treated properly. Notification of zoophilic and geophilic dermatophytoses should be improved in countries with deficient data.

The understanding of the pathogenesis of dermatophyte infection has improved in the last few years. Genomic analysis of dermatophytes enables identification of the candidate genes that encode synthesis of certain keratolytic enzymes responsible for the initiation of infection. It opened possibilities of research by genetic manipulation of dermatophytes to assess the role of specific genes. *In vivo* animal models and *ex vivo* human epidermis tissues represent more appropriate infection models for testing virulence factors of dermatophytes (Achterman and White 2012). Recent new insights in the pathophysiological mechanisms provide better knowledge of the complexity of dermatophyte virulence factors and are a good basis for the development of new antifungal treatments in the future.

Recently, the ability of *T. rubrum* and *T. mentagrophytes* to form biofilms has been demonstrated *in vitro* (Costa-Orlandi et al. 2014). Characteristics of biofilms of both dermatophytes have been described for the first time. This ability may explain treatment resistance in special types of chronic dermatophyte infections, especially in onychomycosis. Biofilms may also contribute in the research of new drugs and in the revision of antifungal treatment with currently available drugs (Costa-Orlandi et al. 2014).

Genetic predisposition in the host has long been presumed to play an important role as a risk factor for onychomycosis; however, until recently, genetic analyses were not able to confirm this clinical observation. Some studies performed in the last decade have found a HLA-DR4 and HLA-DR6 genetic constellation to play a protective role in onychomycosis (Asz-Sigall et al. 2010; Nenoff et al. 2014a). In very rare cases of patients with deep dermatophytosis of a life-threatening course, autosomal recessive CARD9 deficiency has been found (Lanternier et al. 2013). Genetic background enables better understanding of individual susceptibility to special types of dermatophyte infections.

In diagnostics, new molecular techiques for DNA analysis of dermatophytes using PCR has become an important diagnostic tool for quick and reliable identification of dermatophyte strains. They allow even identification of some dermatophyte subspecies. Mass spectrometry has also improved identification of dermatophytes grown in culture. Recent advances in the molecular diagnostics have improved speed, specificity and sensitivity of the diagnosis of dermatophytes. On the basis of molecular biological analysis of dermatophyte DNA, new taxonomic division of dermatophytes has been introduced. New classification of dermatophytes will probably be implemented also in clinical mycology in the future.

In relation to the field of treatment, clinical experiences with the new azole antifungal drugs, posaconazole and voriconazole, have to date been limited in dermatology (Brasch 2012). These new broad spectrum systemic antifungal agents still have good therapeutic potential for some difficult to treat dermatophyte infections. Among topical drugs, abafungin (arylguanidines) seems to be promising for the future. It has a different mode of action to other antifungal drugs (Korting 2009).

Resistance against antifungal drugs used nowadays has been found to be rare among dermatophytes (Silva et al. 2014). However, this may change in the future. Rational use of topical and systemic antifungal drugs is necessary to prevent the evolution of resistant strains of dermatophyte species. Antifungal resistance mechanisms in dermatophytosis have become a new area of research. Resistance to a particular drug can be achieved by various biochemical pathways. In *T. rubrum,* modification of the enzyme squalene epoxidase, a key enzyme in the ergosterol biosynthesis in fungi was found. Mutations in the squalene epoxidase gen can cause the mutant to become resistant to terbinafine, probably by inhibition of terbinafine binding to the target enzyme (Martinez-Rossi et al. 2008). Furthermore, inhibition of ergosterol synthesis induced by terbinafine can be overcome by overexpression of genes involved in the ergosterol synthesis or by gene amplification. Another mechanism demonstrated in *T. rubrum* is increased drug efflux achieved by overexpression of genes that encode for membrane protein transporters. Changes in the expression of cell stress genes which enable adaptation of fungi to toxic effects of drugs were found in *T. rubrum* and might be important in the survival of fungi under stress conditions (Martinez-Rossi et al. 2008). Knowledge of genes and the proteins they code is essential in searching for new targets of antifungal drugs with fewer side effects in the host.

Dermatophyte infections will certainly remain a challenging topic for further research in diagnostics and treatment. Good clinical practice and medical education are of utmost importance for ensuring recognition and treatment of dermatophyte infections according to modern professional standards.

References

Achterman, R.R. and T.C. White. 2012. Dermatophyte virulence factors: Identifying and analyzing genes that may contribute to chronic or acute skin infections. Int J Microbiol. Article ID 358305. doi: 10.1155/2012/358305.

Asz-Sigall, D., L. López-García and M.E. Vega-Memije. 2010. HLA-DR6 association confers increased to *T. rubrum* onychomycosis in Mexican Mestizos. Int J Dermatol. 49: 1406–1409.

Borman, A.M., C.K. Campbell, M. Fraser and E.M. Johnson. 2007. Analysis of the dermatophyte species isolated in the British Isles between 1980 and 2005 and review of worldwide dermatophyte trends over the last three decades. Medical Mycology 45: 131–141.

Brasch, J. 2010. Pathogenesis of tinea. J Dtsch Dermatol Ges. 8: 780–786.

Brasch, J. 2012. Neues zu Diagnostik und Therapie bei Mykosen der Haut. Hautarzt 63: 390–395.

Burzykowski, T., G. Molenberghs, D. Abeck, E. Haneke, R. Hay, A. Katsambas et al. 2003. High prevalence of foot diseases in Europe: results of the Achilles Project. Mycoses 46: 496–505.

Caputo, R., K. Boulle, J. Rosso and R. Nowicki. 2001. Prevalence of superficial fungal infections among sports-active individuals: results from the Achilles survey, a review of the literature. J Eur Acad Dermatol Venereol. 15: 312–316.

Cathcart, S., W. Cantrell and B.E. Elewski. 2009. Onychomycosis and diabetes. J Eur Acad Dermatol Venereol. 23: 1119–1122.

Costa-Orlandi, C.B., J.C. Sardi, C.T. Santos, A.M. Fusco-Almeida and M.J. Mendes-Giannini. 2014. *In vitro* characterization of Trichophyton rubrum and T. mentagrophytes biofilms. Biofouling 30(6): 719–727.

Crissey, J.T., H. Lang and L.C. Parish. 1995. Medical Mycology. Blackwell Science, Cambridge, pp. 36–82.

Dolenc-Voljč, M. 2005. Dermatophyte infections in the Ljubljana region, Slovenia, 1995–2002. Mycoses 48: 181–186.

Ghannoum, M.A., V. Chaturvedi, A. Espinel-Ingroff, M.A. Pfaller, M.G. Rinaldi, W. Lee-Yang et al. 2004. Intra- and interlaboratory study of a method for testing the antifungal susceptibilities of dermatophytes. J Clin Microbiol. 42: 2977–2979.

Ginter-Hanselmayer, G. and C. Seebacher. 2011. Treatment of tinea capitis—a critical appraisal. J Dtsch Dermatol Ges. 9: 109–114.
Ginter-Hanselmayer, G., W. Weger, M. Ilkit and J. Smolle. 2007. Epidemiology of tinea capitis in Europe: current state and changing patterns. Mycoses 50(Suppl. 2): 6–13.
Gräser, Y., V. Czaika and T. Ohst. 2012. Diagnostic PCR of dermatophytes—an overview. J Dtsch Dermatol Ges. 19: 721–726.
Grumbt, M., V. Defaweux, B. Mignon, M. Monod, A. Burmester et al. 2011. Targeted gene deletion and *in vivo* analysis of putative virulence gene function in the pathogenic dermatophyte Arthroderma benhamiae. Eukaryotic Cell 10: 842–853.
Grumbt, M., M. Monod, T. Yamada et al. 2013. Keratin degradation by dermatophytes relies on cysteine dioxygenase and a sulfite efflux pump. J Invest Dermatol. 133: 1550–1555.
Gupta, A.K., N. Konnikov and P. MacDonald. 1998. Prevalence and epidemiology of toenail onychomycosis in diabetic subjects: a multicentre survey. Br J Dermatol. 139: 665–671.
Havlickova, B., V.A. Czaika and M. Friedrich. 2008. Epidemiological trends in skin mycoses worldwide. Mycoses 51(Suppl. 4): 2–15.
Hay, R.J. and H.R. Ashbee. 2010. Mycology. pp. 36.18–36.51. *In*: Burns, T., S. Breathnach, N. Cox and C. Griffiths (eds.). Rook's Textbook of Dermatology, 8th ed. Blackwell, Oxford.
Hekkilä, H. and S. Stubb. 1995. The prevalence of onychomycosis in Finland. Br J Dermatol. 133: 699–703.
Jensen, R.H. and M.C. Arendrup. 2012. Molecular diagnosis of dermatophyte infections. Curr Opin Infect Dis. 25: 126–134.
Kasperova, A., J. Kunert and M. Raska. 2013. The possible role of dermatophyte cysteine dioxygenase in keratin degradation. Med Myc. 51: 449–454.
Korting, H.C. 2009. Fungal infections. pp. 205–222. *In*: Burgdorf, W.H.C., G. Plewig, H.H. Wolff and M. Landthaler (eds.). Braun Falco's Dermatology, 3rd ed. Springer, Heidelberg.
L'Ollivier, C., C. Cassagne, A.C. Normand, J.P. Bouchara, N. Contet-Audonneau, M. Hendrickx et al. 2013. A MALDI-TOF MS procedure for clinical dermatophyte species identification in the routine laboratory. Med Mycol. 51: 713–720.
Lanternier, F., S. Pathan, Q.B. Vincent, L. Liu, S. Cypowyj, C. Prando et al. 2013. Deep dermatophytosis and inherited CARD9 deficiency. N Engl J Med. 369(18): 1704–1714.
Marconi, V.C., R. Kradin, F.M. Marty, D.R. Hospenthal and C.N. Kotton. 2010. Disseminated dermatophytosis in a patient with hereditary hemochromatosis and hepatic cirrhosis: case report and review of the literature. Med Myc. 48: 518–517.
Martinez-Rossi, N.M., N.T.A. Peres and A. Rossi. 2008. Antifungal resistance mechanisms in dermatophytes. Mycopathologia 166: 369–383.
Nenoff, P., C. Krüger, G. Ginter-Hanselmayer and H.J. Tietz. 2014a. Mycology—an update. Part 1. Dermatomycoses: causative agents, epidemiology and pathogenesis. J Dtsch Dermatol Ges. 2014; 12 (3): 188–209.
Nenoff, P., C. Krüger, J. Schaller, G. Ginter-Hanselmayer, R. Schulte-Beerbühl and H.J. Tietz. 2014b. Mycology—an update. Part 2. Dermatomycoses: Clinical picture and diagnostics. J Dtsch Dermatol Ges. 2014; 12: 749–777.
Rich, P., L.B. Harkless and E.S. Atillasoy. 2003. Dermatophyte test medium culture for evaluating toenail infections in patients with diabetes. Diabetes Care 26: 1480–1484.
Roberts, D.T. 1992. Prevalence of dermatophyte onychomycosis in the United Kingdom: results of an omnibus survey. Br J Dermatol. 126(Suppl 39): 23–27.
Roberts, D.T., W.D. Taylor and J. Boyler. 2003. Gudelines for treatment of onychomycosis. Br J Dermatol. 148: 402–410.
Sarifakioglu, E., D. Seckin, M. Demirbilek and F. Can. 2007. *In vitro* antifungal susceptibility patterns of dermatophyte strains causing tinea unguium. Clin Exp Dermatol. 32: 675–679.
Silva, L.B., D.B.C. de Oliveira, B.V. da Silva, R.A. de Souza, P.R. da Silva, K. Ferreira-Palm et al. 2014. Identification and antifungal susceptibility of fungi isloated from dermatomycoses. J Eur Acad Dermatol Venereol. 28: 633–640.
Taplin, D., N. Zaias, G. Rebell and H. Blank. 1969. Isolation and recognition of dermatophytes on a new medium (DTM). Arch Dermatol. 99: 203–209.
Thomas, J., G.A. Jacobson and C.K. Narkowicz. 2010. Toenail onychomycosis: an important global disease burden. J Clin Phar Ther. 35: 497–519.
Tietz, H.J. and P. Nenoff. 2012. Onychomykose. Ein Kronjuwel der Dermatologie. Hautarzt 63: 842–847.

Vermout, S., J. Tabart, A. Baldo, A. Mathy, B. Losson and B. Mignon. 2008. Pathogenesis of Dermatophytosis. Mycopathologia 166: 267–275.

Whittam, L.R. and R.J. Hay. 1997. The impact of onychomycosis on qualty of life. Clin Exp Dermatol. 22: 87–89.

CHAPTER 2

Onychomycosis: Diagnosis and Therapy

*Shari R. Lipner** and *Richard K. Scher*

Introduction

Onychomycosis is the fungal infection of the nail unit by dermatophytes, yeasts and nondermatophyte molds. It is a common disorder and poses many challenges to both patients and physicians. It can be hard to manage due to difficulty in diagnosis, long treatment periods, potential side effects of systemic medications, drug-drug interactions and frequent recurrences. Onychomycosis is not just a cosmetic problem, but is an important medical condition that may be progressive and warrants therapy. The discovery of onychomycosis as a medical condition dates back to the 19th century. In 1853, Meissner, a German medical student, first described and reported this condition (Haas and Sperl 2001). Until modern times, *Trichophyton rubrum* was found only in Southeast Asia, Indonesia, Northern Australia and West Africa. While there were reports of chronic tinea corporis in this region, tinea pedis was not found until the arrival of European colonists and soldiers, who wore occlusive footwear, creating a favorable environment for *T. rubrum* (Rippon 1988).

Due to world wars, mass migration and recreational travel, *T. rubrum* was transported from its original endemic locations to new regions in Europe and America (Rippon 1988). Tinea pedis was first reported in the United States (US) soon after World War I and the first documented case of onychomycosis was reported in 1928 (Salgo 2003). There are a number of factors that led to the increased prevalence of tinea pedis and onychomycosis in the 20th century, namely World War II, the Korean

Weill Cornell Medical College, New York, New York/ 1305 York Avenue, 9th floor, NY, NY 10021.
* Corresponding author: shl9032@med.cornell.edu

and Vietnam wars, popularization of sports activities, use of closed shoes and increased travel (Elewski 1993a). After the Vietnam War, *T. rubrum* became the most commonly isolated dermatophyte worldwide, surpassing *T. mentagrophytes* (Elewski 1993a).

Nails serve an important function in everyday life. They augment fine touch and tactile sensitivity. They are helpful in picking up small items, like coins. They protect the fingertips and can be used as weapons. Finally, they are used for scratching. Therefore, when onychomycosis is present, these functions may be impaired. For instance, when there is onychomycosis of the toenails, the resulting dystrophy may result in improperly fitting shoes and impact walking and sports activities (Scher 1994). This is a significant problem in the elderly as infection may aggravate existing foot problems and lead to decreased mobility. It is known that fungal nail infections may contribute to the severity of diabetic foot problems. In a patient with diabetes mellitus, progressive disability, cellulitis, osteomyelitis and tissue necrosis may occur with the end result being limb amputation in some patients (Levy 1997). Onychomycosis also affects quality of life in a number of ways. For example, in some patients the infection may be painful. In addition, untreated or partially treated infections may limit social interactions due to embarrassment and fear of contagion. This may also be problematic in job situations that involve direct contact with the public such as for sales people, restaurant employees, and health care workers (Scher 1996).

There are also enormous financial implications in treating onychomycosis. In just the one-year period (1989 to 1990), the total reported cost was over $43 million (US dollars, 1997 values) for 1.3 million treatment visits by 6,62,000 patients. If a similar study were performed today, the expense would certainly be much greater (Rosenbach 1989).

The aim of this chapter is to review the epidemiology and etiology of onychomycosis, as well as the diagnosis and differential diagnosis of this important condition. Finally, we will discuss treatment options and prevention of recurrences. We will briefly mention treatments that were used in the past, and then review currently available treatments and therapies in development.

Epidemiology

Reported prevalence rates of onychomycosis in the US and worldwide are varied and there have been no scientifically thorough large-scale studies done so far. In a literature study spanning the years 1950–2012, the authors found that the mean prevalence for the population-based studies in Europe and North America was 4.3%. Of note, the authors also listed the prevalence rates cited in the literature during the year 2012. The mean prevalence cited in all studies was 11.4% (mean lower limit 5.0%, mean upper limit 17.7%) (Sigurgeirsson and Baran 2013). It is believed to be the most common nail disorder, accounting for up to 50% of all nail diseases (Scher and Daniel 2005).

Certain groups deserve special mention. For example, numerous studies have showed that onychomycosis is more prevalent with increasing age, particularly in those more than 50 years old (Heikkila and Stubb 1995; Mercantini et al. 1996; Elewski and Charif 1997; Velez et al. 1997). In fact, fungal infections of the nail are amongst the most common infections that affect older adults with approximately 40% of elderly patients being diagnosed with this condition (Elewski and Charif 1997; Smith et al. 2001). In contrast, onychomycosis is far less common in children under 18 years old with an estimated prevalence of 0.4% in North America (Gupta et al. 1997b). It is also commonly cited that this is an important problem in diabetics as they have a much higher prevalence of onychomycosis than the general population (22–51%) (Saunte et al. 2006; Wang and Margolis 2006; Eckhard et al. 2007; Chang et al. 2008). Additionally, onychomycosis is more common in immunosuppressed individuals (reported prevalence 23%–31%) (Cribier et al. 1998; Gupta et al. 2000). The prevalence of onychomycosis in patients with psoriasis is elevated as well (13–21.5%) (Gupta et al. 1997a; Larsen et al. 2003).

Etiology

Dermatophytes cause most cases of onychomycosis (1). They are ubiquitous and are found in soil, animals and humans. *T. rubrum*, which also causes tinea pedis, is responsible for the majority of toenail onychomycosis in the US. *Trichophyton mentagrophytes* is the second most common organism. Non-dermatophyte molds, such as *Fusarium*, *Acremonium* and yeasts including *Candida parapsilosis* are responsible for the remaining cases (Ghannoum et al. 2000).

Table 1. Onychomycosis in the United States: Causal Organisms (Ghannoum et al. 2000).

Type of Organism	Most common isolates	Less common isolates	Least common isolates
Dermatophytes	*T. rubrum*	*T. mentagrophytes*	*T. tonsurans* *Microsporum canis* *Epidermophyton floccosum*
Nondermatophytes	*Fusarium, Acremonium*	*Scopulariopsis*	*Scytalidium* spp., *Aspergillus* spp.
Yeasts	*C. parapsilosis*	*C. albicans,* *C. guilliermondii*	*C. tropicalis, C. lusitaniae*

Pathophysiology

Onychomycosis is a fungal infection of the nail apparatus most commonly caused by dermatophytes (sometimes yeasts and non-dermatophyte molds). However, the pathogenesis is not well understood. It is known that the nail unit lacks effective cell-mediated immunity, and is therefore, susceptible to invasion by fungal organisms. The infecting organism adheres to the nail apparatus and then invades into the sublayers. Dermatophytic fungi have been shown to have keratinolytic, proteolytic and lipolytic activities (Monod et al. 2002). While onychomycosis can involve the fingernails, it more commonly involves the toenails (Elewski 2000). Tinea pedis occurs in about 50% of patients with onychomycosis (Jennings et al. 2006). It is also known that in predisposed individuals, many cases of onychomycosis begin as tinea pedis (Hainer 2003). Tinea pedis and onychomycosis are spread by direct or indirect contact with infected skin, fomites, or surfaces (Borgers et al. 2005; Seebacher et al. 2008).

A common source of infection is the home environment, and transmission between family members is a common route (Ghannoum et al. 2013). Onychomycosis can also be transmitted through showers in gyms, locker rooms at public pools, and mats in sports facilities, such as gymnasiums or martial arts facilities (Pleacher and Dexter 2007). Table 2 summarizes the major risk factors for onychomycosis.

Table 2. Summary of Risk Factors for Onychomycosis (Elewski 2000).

Increased age
Genetics
Family history
Poor general health
Frequent nail trauma
Environmental contact with pathogens
Warm, humid climates
Communal bathing facilities
Occlusive shoes
Immunosuppression
Tinea pedis prevalence
Psoriasis
Diabetes

Clinical Diagnosis

To diagnose onychomycosis, the clinician should perform a physical examination closely inspecting all the fingernails and toenails. Documentation should include which nails are involved as well as the percentage of involvement for each nail. Photographs are a helpful aid to document changes in the nails over time. Table 3 summarizes the clinical signs of onychomycosis. Onychomycosis is characterized by hyperkeratosis of the nail bed, which may lead to distal detachment of the nail plate from the nail bed (onycholysis). Frequently, there are subungual debris and a white or yellow discoloration (uncommonly brown) of the nail plate (Fig. 1). A common associated finding is tinea pedis, characterized by scale in the web spaces and plantar feet. While not specific for onychomycosis, the nail may be dystrophic with nail

Table 3. Clinical Signs of Onychomycosis.

Nail bed hyperkeratosis
Onycholysis
Subungual debris
Nail plate dyschromia
Associated tinea pedis
Nail dystrophy
Nail crumbling
Nail loss
Tenderness of nail bed and surrounding skin

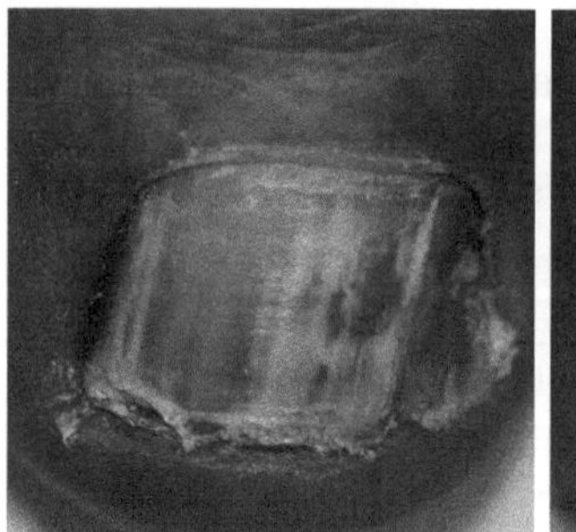

Figure 1. Clinical Appearance of Onychomycosis. Great toenails with hyperkeratosis, subungual debris, yellowing and ridging.

thickening, crumbling, ridging and there may be partial nail loss. There may also be associated tenderness of the nail bed or surrounding skin.

The main subtypes of onychomycosis recognized under the current classification are distal lateral subungual, proximal subungual, superficial, endonyx, mixed pattern, totally dystrophic and secondary onychomycosis (Hay and Baran 2011).

Distal lateral subungual onychomycosis (DLSO) is the most common presentation and commences when the organism infects the cornified layer of the hyponychium and distal or lateral nail bed. This is followed by proximal invasion of the nail bed and ventral invasion of the nail plate. *T. rubrum* and *T. mentagrophytes* are most commonly the causative organisms, but *C. parapsilosis* and *C. albicans* are also possibilities. The DLSO form presents with an isolated distal or lateral focus of onycholysis, yellow dyschromia of the nail plate, and hyperkeratosis of the nail bed. Mild inflammation in the hyponychium and nail bed results in further hyperkeratosis with subungual debris, dyschromia, and onycholysis. This may lead to nail plate dystrophy as the normal nail contour is lost (Scher and Daniel 2005).

Superficial onychomycosis (SO) is less common than DSLO, and in the former, the fungus invades the nail plate and subsequently invades the nail bed and hyponychium. The pattern of nail plate invasion may present with superficial patches or be organized as transverse striate leukonychia (Baran et al. 2007a). It is more common in the toenails than in the fingernails and is usually caused by *T. mentagrophytes*, whose enzymatic activity allows it to digest and directly invade the nail plate. *Microsporum persicolor* is rarely the causative organism, and *C. albicans* may affect infants. The nondermatophyte molds, such as *Aspergillus terreus*, *Fusarium oxysporum*, and *Acremonium* species may cause SO and *T. rubrum* has even reported to cause SO in children (Ploysangam and Lucky 1997).

Proximal subungual onychomycosis (PSO) is an uncommon pattern that affects fingernails and toenails equally, and is primarily caused by *T. rubrum*. The organism invades the proximal nail fold stratum corneum and then penetrates the newly growing nail plate. Clinically it presents with a white discoloration under the proximal nail plate over the lunula, while the distal nail plate appears normal and remains intact. However, in advanced disease, subungual hyperkeratosis, onychomadesis, destruction and shedding of the entire nail plate may occur (Scher and Daniel 2005). This pattern is far more common in HIV patients and usually involves the toenails but spares the fingernails. Since PSO may be the presenting sign in HIV patients, HIV testing should be performed in these patients (Elewski 1993b). While this pattern mainly occurs in immunocompromised patients, when it occurs in immunocompetent patients there is usually a history of preceding trauma (Elewski 1992).

For endonyx onychomycosis (EO), there is lamellar splitting of the nail, dyschromia of the nail plate (e.g., milky patches), and internal nail plate invasion. There is no nail bed invasion. It is thought that fungal hyphae penetrate the distal nail plate directly. On histopathology, there is no inflammation or fungi in the nail bed and no subungual hyperkeratosis, but there are numerous fungal hyphae in the nail plate (Hay and Baran, 2011). Both *T. soudanense* and *T. violaceum* have been reported to cause endonyx onychomycosis (Fletcher et al. 2001).

In a recently described classification system for onychomycosis, mixed pattern onychomycosis (MPO) was proposed to describe different patterns of nail plate infection in the same individual and in the same nail (Hay and Baran 2011). It is thought to be less common than DLSO and the most common patterns are PSO with SO or DLSO with SO (Gupta and Summerbell 1999).

Totally dystrophic onychomycosis (TDO) is the end stage of fungal nail plate invasion. It is usually the end result of DLSO, but PSO may have the same outcome. The nail plate typically crumbles and the nail bed is thickened, ridged and covered with debris (Baran 1981).

Secondary onychomycosis occurs when fungi invade the nail plate and surrounding tissues secondary to other non-fungal pathologies, such as psoriasis and traumatic nail dystrophy. Clinically, the nail usually looks more typical of the underlying condition (Hay and Baran 2011). For example, with underlying nail psoriasis and secondary onychomycosis, there is hyperkeratosis resulting in thickening of the nail plate and nail pitting.

Differential Diagnosis

While onychomycosis accounts for 50% of nail diseases (Scher and Daniel 2005) it is important to know that there are other conditions that affect the nail and have similar presentations. Table 4 is a summary of the differential diagnosis in adults. These conditions include trauma, bacterial infections such as *Pseudomonas aeruginosa* and *Proteus mirabilis*, and inflammatory skin diseases such as lichen planus and psoriasis. Malignant neoplasms, such as squamous cell carcinoma or rarely amelanotic melanoma are also considerations. In addition, exostosis, warts and benign neoplasms such as onychomatricoma can present in this fashion. Finally, yellowing, discoloration and ridging may occur with aging (Cockerell and Odom 1995; Daniel 1991).

One must also consider exogenous substances. Nail polish can cause yellow staining of the nail plate. In addition, self-tanning creams can cause brown staining of the nail. Chemicals in the household or workplace can cause irritant or allergic contact dermatitis that resembles onychomycosis (Rich et al. 2013).

Systemic medications can cause nail changes that have features similar to onychomycosis. For example, chemotherapy agents may cause onycholysis and tetracycline derivatives in combination with sun exposure may cause photo-onycholysis. Oral retinoids, such acitretin and isotretinoin often result in brittleness of the nails during therapy (Rich et al. 2013).

Onychomycosis is uncommon in children and the differential diagnosis is somewhat dissimilar to that in adults (Table 5). Psoriasis, exostosis, and warts are included in both tables, but children may also have congenital malalignment of the large toenail, paronychia secondary to finger sucking and nail biting, and parakeratosis pustulosa. If household members have tinea pedis and/or onychomycosis, the diagnosis of onychomycosis in the child is more likely (Tosti et al. 2003).

Table 4. Differential Diagnosis of Onychomycosis in Adults.

Trauma
Bacteria (*Pseudomonas aeruginosa* and *Proteus mirabilis*)
Psoriasis
Lichen planus
Squamous cell carcinoma
Amelanotic melanoma
Subungual exostosis
Onychomatricoma
Benign and malignant melanonychia
Verruca
Aging
Exogenous substances (nail polish, self tanner)
Contact irritants
Medications (chemotherapy, tetracyclines, retinoids)

Table 5. Differential Diagnosis of Onychomycosis in Children.

Psoriasis
Subungual exostosis
Verruca
Congenital malalignment of large toenail
Paronychia secondary to finger sucking or nail biting
Parakeratosis pustulosa
Benign and malignant neoplasms

Laboratory Diagnosis

While onychomycosis accounts for half of all nail diseases, it is imperative to obtain laboratory confirmation of the diagnosis before initiating therapy. In the past it was common practice to make the diagnosis of onychomycosis and the subsequent treatment was based on the physical exam alone, but that is no longer considered good practice due to the consequences of treating inappropriately, the long term complications of onychomycosis, and the overlooking of the diagnosis of other nail disorders and the associated morbidity (i.e., psoriasis, neoplasm). Interestingly, in an international study in which physicians completed questionnaires about onychomycosis, one of the more disturbing findings was that diagnostic tests were performed in only a minority of the cases (only 3.4% of GPs and 39.6% of dermatologists) (Effendy et al. 2005). For a definitive diagnosis, fungal hyphae in the nail unit (penetrance) must be validated, the fungi must be viable, and the species should be identified.

Direct Microscopy

The diagnosis of onychomycosis can be made simply in the medical office using a potassium hydroxide (KOH) stain (Fig. 2). The advantage of this technique is rapid determination of the presence or absence of fungi (several minutes in experienced hands), and therefore expedited initiation of treatment. The main disadvantages of KOH, are that the viability of the fungus and the identity of the causative organism cannot be assessed. Finally, since interpretation of the results is based on the expertise of the clinician, there may be false positives and negatives, which can adversely affect treatment.

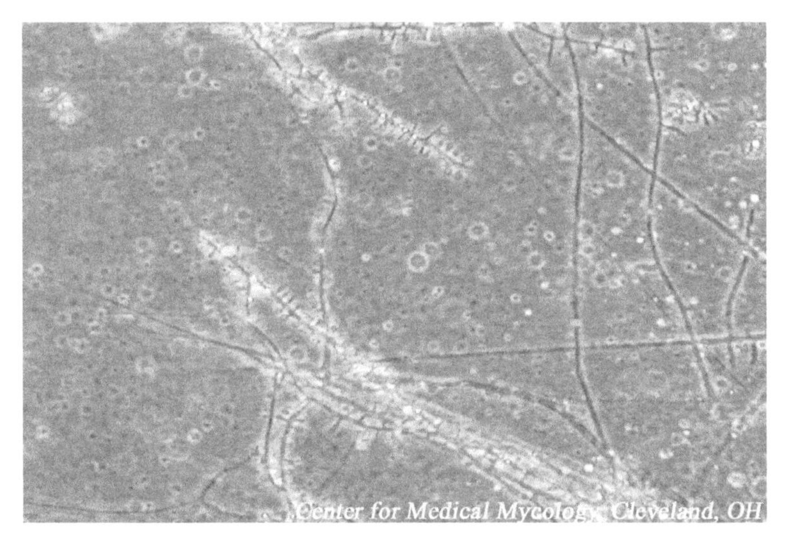

Figure 2. KOH, *T. rubrum*.

The nail and subungual skin should be cleansed thoroughly with soap and water, followed by alcohol, to decrease bacteria and other fungi that are contaminants and not pathogenic. For patients with the most common form (DSLO), the nail is cut back as far as possible, and the subungual debris and nail bed are scraped with a #1 curette. Highest yield scrapings are at the advancing infected edge closest to the cuticle, where the likelihood of fungi is the greatest. Scrapings should be done gently to avoid cutting the nail bed and causing bleeding. The technique is slightly different for patients with

PSO, as the fungus is in the proximal nail bed. The healthy nail plate is pared down with a no. 15 blade, and debris from the proximal nail bed scraped with a sharp curette. In patients with SO, since the fungi invade the nail plate, the nail can be scraped with a no. 15 blade or a sharp curette.

The scrapings can be separated into two samples, one for direct microscopy and the other for sending a culture. The sample intended for culture should be packed into a sterile container or urine cup, a sheet of white paper folded and sealed with tape, or a mailer such as a Dermapak™ (Scher and Daniel 2005). It is important to avoid placing the scrapings intended for culture into broth, saline, or other moist media as contaminants can proliferate in these media, resulting in difficulty in growing the pathogenic organisms.

For direct microscopy, the 10% to 15% KOH solution is added to scrapings on a glass slide, which dissolves the keratin and leaves the fungal elements untouched. Adding dimethyl sulfoxide to the KOH solution is helpful in degrading larger debris. Chlorazol black E is a counterstain that is highly specific for chitin in the fungal cell wall, and will therefore selectively accentuate hyphae and aids in visualizing rare fungal elements. In addition, the counterstain does not highlight contaminants such as cotton or elastic fibers, which can help reduce false-positives (Burke and Jones 1984). The Parker's blue-black ink counterstain is used as well, but its disadvantage is that it is not chitin specific. Another counterstain is the fluorescent dye calcofluor white, which also stains chitin, but it is used more often in reference laboratories since a fluorescent microscope is necessary (Weinberg et al. 2003). For visualization, the microscope should be set on medium power with the ×20 objective, and the light should be turned down. Since the KOH solution contains glycerol, further degradation of debris can proceed overnight and the slide can be reanalyzed the following day in the event that no hyphae are visualized (Haley 1990).

Culture

Fungal culture is currently the gold standard of diagnosis for onychomycosis, as it is the only technique at this time that can identify the causative organism and its viability. This technique is very important in the less common cases when yeasts or other nondermatophyte organisms are the suspected pathogens causing onychomycosis. Identification of the pathogen allows the clinician to choose the most effective treatment. For culturing nail specimens, two types of media are employed. One agar contains the antifungal agent, cycloheximide as an additive, for its ability to inhibit saprophytic molds and grow dermatophytes. Media with cycloheximide include dermatophyte test media (DTM), Mycosel (BBL), and Mycobiotic (DIFCO). Chloramphenicol and gentamicin are antibiotics added to Sabouraud's glucose agar or potato dextrose agar to reduce bacterial contaminants (Elewski 1995). In contrast, media without cycloheximide, such as Sabouraud's glucose agar and potato dextrose agar allow isolation of nondermatophyte nail pathogens that are sensitive to cycloheximide, such as *Scopulariopsis, Scytalidium* spp. and *Candida* spp. other than *C. albicans* (Elewski and Greer 1991). The culture is grown at 25–30° for up to a month. While mycological culture is a highly specific technique, it has a high false negative rate, which limits its sensitivity. It is important to note that yeasts and nondermatophytes grow faster than dermatophytes, so it may take significant time (2–6 weeks) for the pathogenic organism to grow and allow for identification (Fig. 3).

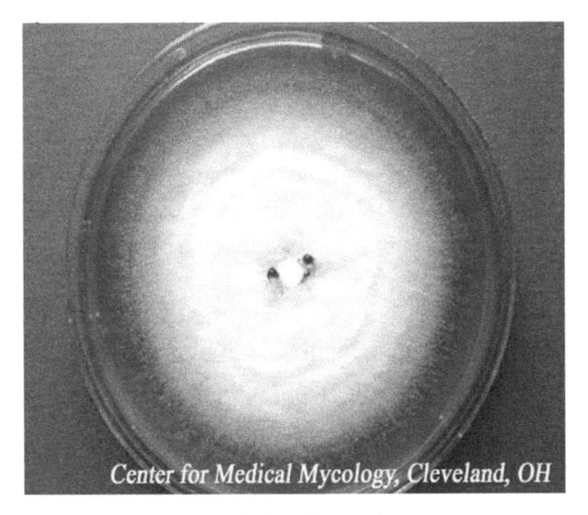

Figure 3. *T. rubrum* colony.

Histopathology

Another frequently used method for diagnosis of onychomycosis is histopathology. The clinician can send a nail plate clipping in a 10% buffered formalin container with a request for fungal stains, such as periodic acid-Schiff (PAS) or Grocott's methanamine silver (GMS). For this technique, the nails are fixed, dehydrated, embedded in paraffin and sectioned prior to staining. For PAS staining, periodic acid oxidizes hydroxyl groups of cell wall polysaccharides into aldehyde, which then reacts with the Schiff reagent. The background is green, and in contrast fungi are stained red. For GMS staining, chromic acid oxidizes cell wall polysaccharides, converting them to aldehydes, resulting in methenamine silver nitrate reduced to metallic silver. The cell walls are stained a dark- brown against a pale -green background. Stains also employed include Fontana-Masson, hematoxylin and eosin, immunofluorescence, and Mayer's mucicarmine (Smith 2008). It should be noted that a PAS revealing septate hyphae is diagnostic of onychomycosis, however, a result showing only yeast forms is inconclusive because the yeast may not be pathogenic. Histopathology has several advantages over KOH or culture. First, while it does not provide as rapid results as KOH, it has a much faster turnaround time than culture. PAS results are usually available in just a few days, while culture results can take several weeks. In addition, PAS may be more sensitive than KOH and culture. For example, in one study, PAS was found to be more sensitive than KOH preparation or culture alone (92% versus 80% or 59%, respectively) (Mahoney et al. 2003). Disadvantages of this technique are that the pathogen cannot be identified and neither can the viability of the organism.

Polymerase Chain Reaction

Polymerase chain reaction (PCR) is a DNA-based strategy that can be used to identify dermatophytes and nondermatophyte molds, that is not widely used at the present time, but is gaining popularity. Two genes are usually amplified for fungal identification using PCR. These are the internal transcribed spacer region of ribosomal DNA (ITS) or the chitin synthase 1 (CHS1) gene (Graser et al. 2012). As opposed to traditional PCR, real time PCR can be utilized to quantify the relative levels of the transcript and assess the probable viability of the sample (Arabatzis et al. 2007). It should be noted that PCR will amplify all fungal DNA, regardless of whether it is from the causative organism or a contaminant, so care must be taken to clean properly before taking the sample, and thus avoid false positives. Advantages of the PCR technique are that results show the presence or absence of fungi, the identification of the species, and give rapid test results. It has the disadvantage of being a costlier technique and cannot unquestionably assess the viability of the organism.

Confocal Microscopy

The reflectance confocal microscopy technique utilizes an 830-nm laser in reflectance mode to generate horizontal sections of different depths from the surface of the nail plate down to the nail bed. Data analysis and software are then employed to generate images of the nail plate and identify the species, based on morphological characteristics. This technique is not widely used at the present time. Advantages are that it is office based, rapid, and specific. Disadvantages of this technique are that it requires specialized training, is costlier than regular light microscopy, and based on the available studies it has varying sensitivity 53%–80% (Rothmund et al. 2013; Pharaon et al. 2014).

Dermoscopy

Dermoscopy is now frequently used for diagnosis of nail diseases (Richert et al. 2009). It can also be used as a tool to aid in the diagnosis of onychomycosis. Certain findings can help to distinguish DSLO from traumatic onycholysis, another frequently encountered nail pathology. In DSLO, the proximal edge of the onycholytic area is jagged with indentations caused by sharp structures ('spikes'). In addition, the detached nail plate has an irregular matte pigmentation that is distributed in striae (called 'Aurora pattern') (Piraccini et al. 2013).

Flow Cytometry

Flow cytometry is a technique that utilizes granulosity, cell size, DNA and protein markers along with the stains propidium iodide (PI) and fluorescein isothiocyanate (FITC), to identify pathogens in the nail plate. Nail plates are treated with Tween 40, centrifuged, and then filtered to collect the fungal sample. This technique has the advantage of confirming fungi in the nail plate and identification of the pathogen.

However, due to the high cost of the machine, it is not practical for clinical use and it is more often used for research purposes (Gupta and Simpson 2013).

Infrared Imaging

A new technique in development for the diagnosis of onychomycosis is based on the analysis of medium-range infrared images. The authors found that nails affected by onychomycosis emit less energy than nails free of disease. Advantages of this technique are that it is rapid, low cost and noninvasive. Disadvantages are that the species and viability can not be determined and results are affected by ointments, nail polish and onycholysis (Villasenor-Mora et al. 2013).

If standard techniques fail to show evidence of onychomycosis, but there is still clinical suspicion, nail biopsy should be performed (Rich 2001).

Table 6 summarizes the techniques currently used for the diagnosis of onychomycosis and those in development.

Table 6. Diagnosis of Onychomycosis (Gupta and Simpson 2013).

Technique	Nail plate penetrance	Fungal viability	Identification of pathogen	Commonly used
KOH	No	No	Yes*	Yes
Fungal culture**	No	Yes	Yes	Yes
Histopathology	Yes	No	Yes*	Yes
PCR	No	No	Yes	Sometimes
Confocal microscopy	Yes	No	Yes*	In development
Dermoscopy	No	No	No	As an aid
Flow cytometry	No	No	Yes	Research only
Infrared imaging	Yes	No	No	In development

* Can identify probable pathogen, but only mycological culture, PCR and flow cytometry can be used to definitively identify the causative organism.
** Mycological culture is considered the gold standard for diagnosis.

Treatment of Onychomycosis

The goal of treatment is to eliminate the fungus and to restore the nail to its normal state when it fully grows out. Patients should be counseled that the treatment process is long (i.e., ≥6 months for fingernails [growth rate, 2–3 mm per month]; 12–18 months for toenails [growth rate, 1–2 mm per month]) (Scher et al. 2013). Nails grow fastest during adolescence and slow down with advancing age (Abdullah and Abbas 2011). It should be noted that advanced cases of onychomycosis involving the nail matrix may result in permanent nail dystrophy even if the causative organism is eliminated. Several definitions are necessary to understand the results of onychomycosis studies. Complete cure is defined as a negative KOH preparation and negative fungal culture as well as a completely normal appearance of the nail. Mycological cure is defined as KOH microscopy and culture negative. Clinical cure is stated as 0% nail plate involvement, but at times is reported as <5% and <10% involvement.

Topical therapies were used exclusively until the mid-1900s, when oral agents were developed. In 1938, there is a report of a topical treatment involving the use of sandpaper, potassium permanganate soaks and Castellani carbofuchsin paint. Formaldehyde vapor was used to disinfect shoes and gloves (Stokes 1999).

In 1939, griseofulvin was isolated from the mold *Penicillium griseofulvum* Dierckx and by the late 1950s and early 1960s, it was commonly used as an oral agent to treat onychomycosis. While griseofulvin is FDA approved for the treatment of onychomycosis, because its affinity for keratin is low, long-term treatment is required (at least 4 months for fingernails and at least 6 months for toenails), thus it is uncommonly used (Insert January 2007). In 1981, ketoconazole became the first azole to be commonly used for the treatment of onychomycosis in the US. Ketoconazole has better absorption and a shorter course of treatment than griseofulvin. However, since there is an increased risk of severe hepatic side effects and there are newer oral antifungals, it is no longer used as first line treatment for onychomycosis (Gupta and Ryder 2003).

Current Treatment Strategies

As modern science progressed, newer antifungals were developed which had the ability to directly target the molecular structures of dermatophyte fungi. These antifungals are typically oral agents that provided greater efficacy with shorter durations of therapy than historical topical or oral treatments. These oral therapies have been in long-term use and have well-known efficacy and safety profiles.

Treatment of onychomycosis can be challenging for a number of reasons. First, hyperkeratosis and/or the fungal mass may limit delivery of topical and systemic drugs to the source of the infection. In addition, high rates of recurrence after treatment may occur due to residual hyphae or spores that were not previously eradicated (Scher and Baran 2003). Finally, the prolonged length of treatment period may be a barrier to patient compliance, and many patients may be unwilling to forego wearing nail cosmetics during the treatment course. To improve patient care, physicians should understand the current therapies for onychomycosis and those awaiting FDA approval, as the newer therapies show improved efficacy with shorter treatment courses.

Nail Avulsion

One treatment for onychomycosis is nail avulsion, which is the chemical or surgical separation of the nail plate from the nail bed. One potential side effect of nail plate avulsion is an ingrown nail due to lack of counter pressure on the nail bed. This can be avoided by use of a prosthetic nail to maintain the normal width of the nail bed. Nail avulsion, while rarely use today, can be used alone for the treatment of onychomycosis, but it is more effective in combination with antifungals (Gupta et al. 2013c).

Chemical nail avulsion is infrequently used in the US, but is often used in Europe. Using this procedure, the nail plate is softened by dissolving the bond between the nail plate and the nail bed (Pandhi and Verma 2012). Urea ointments are typically prescribed under occlusion for 1 to 2 weeks, followed by use of a nail elevator and clipper to remove the nail. The procedure is relatively painless and there is low risk of bleeding and infection. It should be noted that chemical avulsion is not very effective alone but can be used in combination with topical antifungals for increased efficacy (Rollman 1982).

Surgical nail avulsion can also be performed for the treatment of onychomycosis. Local anesthesia is used, as is a tourniquet to reduce bleeding. Side effects include possible reduction in the width of the nail bed, distal paronychia and infection (Lai et al. 2011). To improve efficacy, this treatment should be used in combination with topical antifungal agents. This technique is not commonly used for treatment of onychomycosis due to the high dropout rate and poor patient compliance (Grover et al. 2007).

Currently Available Systemic and Topical Therapies

Currently, there are four approved classes of antifungal drugs used for the treatment of onychomycosis: namely, the allylamines, azoles, morpholines, and hydroxypyridones (Welsh et al. 2010). The benzoxaboroles are a new class of antifungal drugs that is currently in development (Rock et al. 2007). These drugs and their respective mechanisms of action will be discussed in the sections that follow. Figure 4 shows the mechanism of action of the antifungal medications. Table 7 summarizes the systemic therapies and Table 8 summarizes the topical treatments available for onychomycosis.

Table 7. Summary of Systemic Therapy for Onychomycosis for Adults.

Therapy	**Fingernails**	**Toenails**
Terbinafine (insert, 2012a)	250 mg daily for 6 weeks	250 mg daily for 12 weeks
Itraconazole* (insert, 2012b)	2 pulses: 200 mg bid for 1 week/ month	200 mg daily for 12 weeks
Fluconazole** (Gupta et al. 2013a)	150 mg weekly for greater than 6 months	150 mg weekly for greater than 6 months

* The pulse regimen is FDA approved for fingernails but not toenails.
** Not FDA approved for treatment of onychomycosis.

Table 8. Summary of Current Topical Therapy for Onychomycosis.

Therapy	**Fingernails**	**Toenails**
Ciclopirox 8% nail lacquer* (Laboratories, 2004)	Daily for 24 weeks	Daily for 48 weeks
Amorolfine 5% nail lacquer** (2006, insert, 2011)	Once or twice weekly for 24 weeks	Once or twice weekly for 36–48 weeks
Efinaconazole	No indication	Daily for 48 weeks

* Patients must also file and trim nail weekly, removal of infected nail by health care professional monthly.
** Patients must also file and trim nail weekly. Available in Europe, not in US.

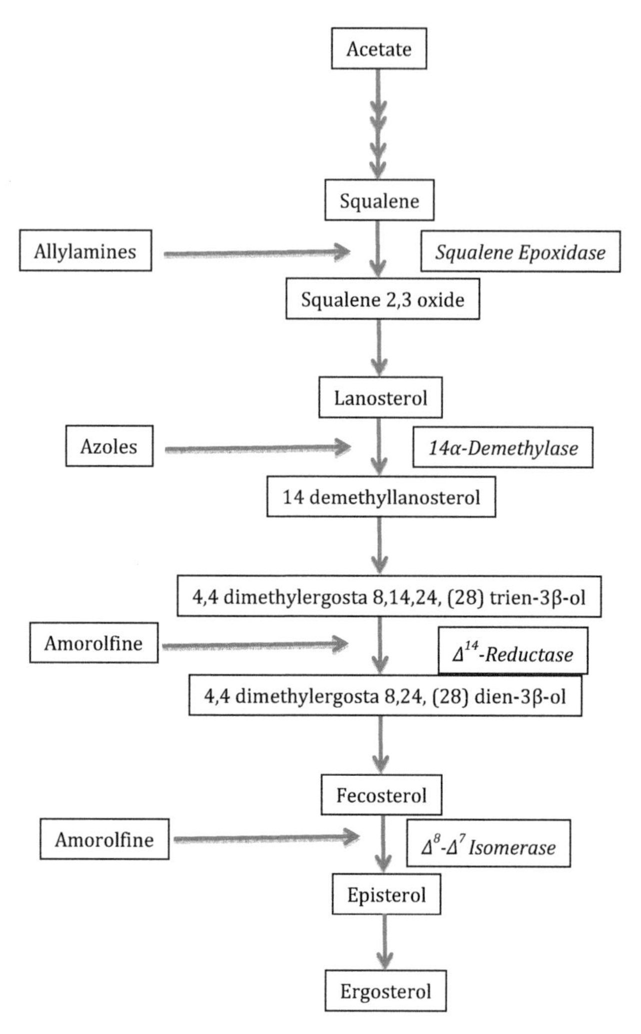

Figure 4. Mechanism of Action of Antifungal Agents (Adapted from Gupta et al., *Antifungal agents: an overview. Part II.* J Am Acad Dermatol. 1994. 30(6): 911–33.

Systemic Treatments

In general oral systemic antifungal are more effective in achieving onychomycosis cure than topical agents. However, there are more significant side effects with oral agents and there is a higher potential for drug-drug interactions than with topical agents. It is important to remember that onychomycosis is more prevalent among older patients, diabetics and the immunocompromised, who often have comorbidities and are taking other medications (Thomas et al. 2010).

Terbinafine is an allylamine commonly used in the treatment of onychomycosis, whose mechanism of action is inhibition of squalene epoxidase, a step in the synthesis of ergosterol (Fig. 4). It is fungicidal as squalene is concentrated intracellularly and fungistatic as ergosterol is depleted (Gupta et al. 1994). *In vitro*, terbinafine has broad-spectrum activity against dermatophytes and some nondermatophytes. It should be noted that it is fungicidal against *C. parapsilosis* but fungistatic against *C. albicans* (Clayton 1989), which must be taken into consideration when choosing an appropriate therapy for candida onychomycosis. Oral terbinafine hydrochloride (250 mg) taken once daily for 6 weeks for fingernails and 12 weeks for toenails is currently the gold standard for the treatment of onychomycosis (Gupta et al. 2013b), with FDA package insert complete cure rates 59% for fingernails and 38% for toenails (Insert 2012a).

In children, the daily dose is adjusted based on weight, namely 62.5 mg in children who weigh less than 20 kg, 125 mg for those who weigh between 20 and 40 kg, and 250 mg in those who weigh more than 40 kg (Gupta and Skinner 2004). Two small studies have been performed, one reporting a mycological cure rate of 82% (Jones 1995) and the other reporting a complete cure rate of 88% (Ginter-Hanselmayer et al. 2008).

Pulsed-dose therapy with terbinafine for the treatment of onychomycosis is not FDA approved, but may be desirable for reasons such as cost, compliance, adverse events and fungal resistance. A recent meta-analysis of the literature showed that treatment with continuous terbinafine is significantly superior to a pulsed terbinafine regimen for mycological cure. One caveat is that some pulse terbinafine regimens were shown to be as effective as continuous terbinafine regimens for complete cure. For example, one intermittent regimen, namely 2 pulse regimens of 250 mg/day of terbinafine for 4 weeks on/4 weeks off, had the greatest efficacy of all pulse regimens, in that mycological and complete cure rates were comparable to that of 250 mg/day terbinafine for 12 continuous weeks (Gupta et al. 2013b).

Side effects include nausea, vomiting, diarrhea, headache, skin rashes, and taste and smell disturbances. More serious side effects include liver failure, depressive symptoms, severe neutropenia, Stevens-Johnson syndrome, toxic epidermal necrolysis and systemic lupus erythematosus. Thus far, no embryonic or fetal toxicity was noted in rats or rabbits at high doses (Villars and Jones 1992). Its bioavailability is not significantly affected whether it is taken on an empty stomach or with a meal (Balfour and Faulds 1992). Terbinafine is FDA pregnancy category B, however because there have been no well-controlled studies in pregnant women, it should not be taken during pregnancy. It is also excreted into breast milk (Balfour and Faulds 1992). In case of liver dysfunction or renal disease there is decreased clearance of the drug (Shear et al. 1991).

In vivo studies have shown that terbinafine inhibits the CYP450 2D6 isozyme. Therefore, caution must be taken with tricyclic antidepressants, selective serotonin reuptake inhibitors, beta-blockers, antiarrhythmics class 1C (e.g., flecainide and propafenone) and monoamine oxidase inhibitors Type B. It is important to note that terbinafine is metabolized by the hepatic cytochrome P450 system, thus plasma levels of terbinafine may be affected by other drugs. For example, rifampin and phenobarbital induce these enzymes, so there will be increased clearance of terbinafine (Breckenridge 1992). In contrast, cimetidine inhibits this system, so there will be increased concentrations of terbinafine (Back and Tjia 1991). It should be noted that terbinafine therapy is safer than itraconazole for diabetics. As opposed to itraconazole, it does not cause hypoglycemia in diabetic patients who are taking hypoglycemic medications metabolized by the CYP3A4 pathway (Matricciani et al. 2011). Terbinafine should not be used in patients with a history of allergy to oral terbinafine. The FDA recommends checking liver function tests before initiating therapy with terbinafine.

The azoles inhibit lanosterol 14α-demethylase, a step in the ergosterol biosynthesis pathway (Fig. 4). Fungistatic activity is achieved by the depletion of ergosterol (Elewski 1993c). Two members of this class that are widely used in treating onychomycosis are the triazoles, oral itraconazole (Insert 2012b) and oral fluconazole (Scher et al. 1998).

Itraconazole is used to treat both fingernail and toenail onychomycosis. Dosing may be given daily (continuous) or given on an intermittent (pulsed) schedule since it accumulates in the skin and mucous membranes (Korting and Schollmann 2009). It is usually given for 2 to 3 months for fingernail infections and 3 to 4 months for toenail infections (Insert 2012b). The first dosing schedule, known as pulse dosing, is 400 mg/day for 1 week out of each month for 4 months. The second dosing schedule is 200 mg/day continuously for 3 months. Pulsed dosing is approved for fingernail but not toenail onychomycosis. The complete cure rate for the continuous regimen for toenails is 14% as per the package insert and 47% for the pulsed treatment for fingernails (Insert 2012b). It should be noted that in a study comparing the efficacy of pulsed dosing of itraconazole vs. terbinafine, somewhat higher cure rates for itraconazole than the package insert were reported, namely 25% complete cure in three cycles; 28% complete cure in four cycles (Evans and Sigurgeirsson 1999).

Itraconazole has also been used successfully in childhood onychomycosis. In a study of children treated with 200 mg of itraconazole daily for 12 weeks, 83% to 89% had a complete cure. This is in contrast to the complete cure rate of 63% in the children who were treated with itraconazole 1 week per month for 2 to 5 months (Heikkila and Stubb 2002).

One advantage of itraconazole is that it has broader spectrum activity than terbinafine, so it is a useful agent in suspected mixed infections or in Candida onychomycosis (Korting and Schollmann 2009). It is important to be aware of drug interactions with this medication. Since itraconazole is a potent inhibitor of CYP3A4, serious cardiovascular events may occur if cisapride, pimozine, quinidine, or levomethadyl are taken concurrently. When patients have comorbidities such as congestive heart failure or other ventricular dysfunction, itraconazole should be used with caution (Elewski et al. 2013a). Itraconazole should preferably be taken with a fatty meal for optimal bioavailability (Van Peer et al. 1989). Its bioavailability is not ideal in situations in which gastric acidity is impaired as in the case with antacids or histamine-2 receptor blockers (Cleary et al. 1992). These medications should be taken at least 2 hours after itraconazole is ingested. Side effects include nausea, abdominal pain, vomiting, diarrhea and rash (Graybill and Sharkey 1990). Elevation of liver function tests may occur in some patients and hepatic injury is rarely reported (Lavrijsen et al. 1992). When given long term, peripheral neuropathy has been described (Insert 2012b). Itraconazole must be used with caution in patients taking other medications as they may affect the P450 system. Treatment failure may occur with concomitant administration of itraconazole with rifampin, isoniazid, phenobarbital, carbamazepine, and phenytoin, as these drugs decrease levels of itraconazole. Cimetidine, on the other hand, inhibits the P450 system leading to high levels of itraconazole (Hay et al. 1988). Since itraconazole is a CYP3A4 inhibitor, drug interactions may occur with cisapride, oral midazolam, nisoldipine, terfenadine, felodipine, pimozide, quinidine, dofetilide, triazolam, levacetylmethadol (levomethadyl), lovastatin, simvastatin, eletriptan, ergot alkaloids (e.g., dihydroergotamine), ergometrine (ergonovine), ergotamine, methylergometrine (methylergonovine), and methadone. Warfarin must be used with caution in patients taking itraconazole (Insert 1993). Itraconazole should be avoided during pregnancy as it causes embryotoxicity and malformations in the skeletal system (FDA Pregnancy Class C) (Van Cauteren 1987).

Fluconazole is another member of the azole class, and while it is not FDA approved for the treatment of onychomycosis in the US, it is approved in Europe and used off label in the US, Canada and Australia. It has broad spectrum antifungal activity against dermatophytes, *Candida* species, and some nondermatophyte molds. While oral itraconazole absorption is dependent on food, gastric pH, antacids and cimetidine, oral fluconazole is independent of these factors (Brammer et al. 1990). In a double blind randomized study, with 362 patients, complete cure rates were 48% in patients who received 450 mg per week, 46% in those who received 300 mg per week, and 37% in those who received 150 mg per week for up to 9 months. Notably, the clinical relapse rate over 6 months of follow-up was very low, namely 4% (Scher et al. 1998). A recent evidence-based analysis of fluconazole dosing regimens showed that 150 mg once per week for at least 6 months has optimal efficacy against dermatophyte onychomycosis. Specifically, it was recommended that the drug is administered until the whole nail grows out, which is 6–9 months for fingernails and 12–18 months for toenails (Gupta et al. 2013a). The most common side effects are headache, nausea, vomiting and diarrhea. It can also cause elevation of liver function tests (Munoz et al. 1991). Fluconazole is not mutagenic and is not embryotoxic or teratogenic in animals except at extremely high doses (Stevens 1988). However, since there are case reports of fetal anomalies in humans, it is classified as FDA Pregnancy Class D (Insert 2014). Fluconazole inhibits the human cytochrome P450 system (potent inhibitor of CYP2C and moderate inhibitor of CYP3A4), so caution must be taken when the patient is taking other medications. Since elimination of fluconazole is dependent on renal function, the dose should be adjusted in patients with reduced creatinine clearance (Grant and Clissold 1990). An important disadvantage of fluconazole is that it has a shorter residual concetration in the nails, necessitating a longer treatment course than both terbinafine and itraconazole (Brown 2009). Fluconazole can be considered when a patient has failed both terbinafine and itraconazole or when these drugs are contraindicated.

Topical Treatment

Topical therapy offers several potential advantages over systemic therapy. First, systemic side effects and laboratory monitoring would be avoided. Furthermore, the potential for drug-drug interactions is significantly lower than for oral therapies. In addition, if adverse reactions occurred they would be localized and easier to treat. There are several reasons for the lack of effective topical antifungal agents for nails. First, since the nail plate is thick with a compact structure, it is difficult to design drugs with proper molecular weight, lipophilicity, and keratin-binding properties to adequately penetrate the nail plate and reach the nail bed (Elewski et al. 2013a). This problem is compounded by the hyperkeratosis that is common in nail infections. In addition, when there is onychomycosis, functional cells of the immune system are incapable of accessing the fungal mass (Murdan 2008). Another problem is that there are residual hyphae or spores present after the patient has completed treatment, and this may lead to relapse (Scher and Baran 2003). Adherence to topical treatment is also problematic as patients are often unwilling to apply topical medications for the long treatment periods often required. Finally, patients may not be compliant with therapy due to the restrictions on wearing nail polish, as they will not be able to camouflage the problem.

Some of the topical therapeutic agents for onychomycosis are nail lacquers. These lacquers have been shown to reduce transonychial water loss and may cause the germination of dormant and drug-resistant spores, which help to eliminate the pathogenic organism and thus prevent recurrences of onychomycosis (Flagothier et al. 2005). Topical monotherapy is indicated for superficial onychomycosis, and it is commonly used as therapy when there is distal subungual onychomycosis affecting less than 50% of the surface area of the nail without matrix involvement. It can also be attempted when few (i.e., three or four) nails are affected (Lecha et al. 2005). Topical therapy can also be used in children due to its efficacy on thin, rapidly growing nails (Singal and Khanna 2011).

Topical amorolfine is a member of the morpholine class, which is approved for use in Europe but not in the United States (Baran et al. 2007b). Amorolfine inhibits two enzymes, namely Δ14 reductase and Δ7–Δ8 isomerase (Fig. 4), thus depleting ergosterol, a component of the fungal cell membrane (Polak 1992). It has fungistatic and fungicidal activity against dermatophytes, nondermatophyte molds, and yeasts (Gupta et al. 2003). It has been shown that amorolfine is more highly concentrated in the upper layers of the nail plate than the lower layers, and that the total amount of amorolfine in the nail is highly dependent on the thickness of the nail (Polak 1993). Amorolfine 5% nail lacquer is applied once or twice weekly for 6 months for fingernails and 9 to 12 months for toenails (Insert 2011). Complete cure rates in randomized trials with 6 months of active treatment followed by 3 months of follow up were 38% to 54% in patients without matrix involvement (Gupta et al. 2003). In one randomized controlled study, the combination of amorolfine hydrochloride nail lacquer 5% and oral terbinafine (250 mg once daily) resulted in a higher clinical cure rather than oral terbinafine alone (59.2% vs. 45.0%; complete cure rate was not reported) (Baran et al. 2007b). Some advantages of amorolfine are that it remains in the nail plate longer than ciclopirox 8% nail lacquer, it is more cost-effective than ciclopirox (Lecha et al. 2005), and that the weekly dosing regimen may improve patient compliance as compared with the ciclopirox daily dosing regimen (Kruk and Schwalbe 2006). Important disadvantages of the drug are that the nail must be degreased with alcohol and filed down before each application of the lacquer and nail polishes should not be applied during the treatment (Insert 2011), which may limit patient compliance. Since there are no clinical trials in pregnant women and embryotoxicity was observed in laboratory animals at high doses, the medication should be avoided during pregnancy. It can be used during breastfeeding, since the systemic exposure to the mother is negligible. Adverse effects are localized and include burning, itching, redness, irritation, and pain (Gupta et al. 2003).

Ciclopirox is a member of the hydroxypyridone class and has both antifungal and antibacterial activity. It inhibits dermatophytes, *Candida* species, and some nondermatophyte molds, as well as a number of Gram-positive and Gram-negative aerobic and anaerobic bacteria (Bohn and Kraemer 2000). Its mechanism of action is not entirely understood, but it inhibits several metal-dependent enzymes (e.g., cytochromes) (Lee et al. 2005), is involved in oxidative damage, affects nutrient uptake and the synthesis of proteins and nucleic acids (Belenky et al. 2013). Ciclopirox also has anti-inflammatory and anti-allergic properties. The FDA approved Ciclopirox 8% nail lacquer in 1999, and the recommended application is 24 weeks for fingernails and 48 weeks for toenails. Complete cure rates of 5.5% to 8.5% are reported compared to placebo, which are 0% to 0.9% (Laboratories 2004). Some disadvantages of this therapy are the long treatment course and that it is labor intensive. The lacquer must be applied daily for 1 week, then removed with alcohol, and the nail must be trimmed and filed with an emery board. It is also recommended that the patient go to a healthcare professional monthly for removal of the infected nail. It is FDA Pregnancy category B, and while there were no fetal malformations in laboratory animals given high doses of ciclopirox, it is not recommended for use by pregnant women unless the potential benefit justifies the risk. It is not known whether this drug is excreted into breast milk, so caution must be taken with breastfeeding women. Adverse effects are localized and include periungual erythema, application site reactions and burning (Laboratories 2004).

Efinaconazole, a member of the azole class of drugs, was approved for use in Canada in October 2013 and FDA approved in the US in June 2014. Patients in two randomized studies were treated with either efinaconazole solution 10% or vehicle for 48 weeks followed by a 4-week washout period. Complete cure rates were 17.8% and 15.2% in the each of the treated groups, and 3.3% and 5.5% in each of the control groups. The side-effect profile was minimal with the most common being application-site dermatitis and vesiculation, with no statistically significant difference between the treated and control groups (Elewski et al. 2013c). There are some marked differences between ciclopirox and efinaconazole that may impact patient compliance. First, treatment with ciclopirox includes monthly nail debridement, which is not required with efinaconazole. Second, although ciclopirox nail lacquer must be removed weekly, efinaconazole is a solution, so no removal is necessary.

Oral and Topical Therapies on the Horizon

Based on the poor efficacy of many of the onychomycosis treatments that are currently available as well as time-consuming treatment courses, there is a clear need for alternative and novel therapies. Several oral triazole antifungals, namely albaconazole (Sigurgeirsson et al. 2013) posaconazole (Elewski et al. 2012), and ravuconazole (Gupta et al. 2005), have undergone phase I and phase II studies and have shown efficacy in the treatment of onychomycosis. There has been a greater emphasis on topical agents due to their more favorable side-effect profile and lower risk for drug-drug interactions.

Posaconazole is an extended-spectrum triazole, that is approved in US and Europe for prophylaxis of invasive fungal infection and for treatment of oropharyngeal candidiasis. *In vitro*, it has activity against dermatophyte and nondermatophyte molds. A phase IIB, proof-of-concept study was performed to evaluate the efficacy of oral posaconazole regimens compared with terbinafine and placebo in the treatment of toenail onychomycosis. Complete cure rates at 48 weeks were 54% for posaconazole 200 mg/24 weeks compared to 37% for terbinafine (Elewski et al. 2012).

Terbinafine nail solution (TNS) has undergone three phase III clinical trials (2 vehicle controlled and 1 active comparator). The first trial compared TNS and vehicle applied daily for 24 weeks, the second study repeated the same for 48 weeks, and the third study compared TNS to amorolfine nail lacquer 5% daily for 48 weeks. Overall, complete cure was not different between TNS and vehicle, nor did TNS result in any treatment benefit over and above amorolfine 5% nail lacquer. The most common side effects were headaches, nasopharyngitis, and influenza (Elewski et al. 2013b).

Tavaborole is member of the new benzoxaborole class, which inhibits protein synthesis by blocking aminoacyl transfer RNA synthetase (Rock et al. 2007). Topical tavaborole solution was engineered to have improved penetration through the nail plate, and *in vitro* studies have shown better penetration than both ciclopirox and amorolfine (Hui et al. 2007). Two identical phase III, randomized, double-blind, vehicle-controlled studies evaluating tavaborole solution 5% daily compared to vehicle for 48 weeks followed by a 4-week washout period showed complete cure rates of 6.5% and 27.5% for tavaborole and 0.5% and 14.6% for vehicle. The incidence of treatment-related side effects was comparable to the vehicle. The most common side effects were exfoliation, erythema, and dermatitis, all occurring at the application site. The New Drug Application (NDA) was accepted by the FDA for review in October 2013 (Elewski and Wiltz 2013).

Luliconazole is an imidazole antifungal with fungicidal activity against dermatophytes and nondermatophytes (Uchida et al. 2003). In fact, *in vitro*, luliconazole has higher potency than amorolfine, ciclopirox, and terbinafine against dermatophytes (Wiederhold et al. 2014). Luliconazole has been on the market in Japan since 2005 for the treatment of superficial mycoses including dermatophytosis, candidiasis and pityriasis versicolor (Watanabe et al. 2007). To date, approximately 12 million patients have been treated with luliconazole in Japan, where it is the preferred topical prescription antifungal treatment (Scher et al. 2014). Luliconazole 1% cream was recently approved by the FDA in the US for the treatment of superficial fungal infections caused by the *T. rubrum* and *E. floccosum*. A randomized, double blind, phase II/III study is underway to determine the efficacy and safety of luliconazole 10% solution in the treatment of adults with mild-to-moderate distal lateral subungual onychomycosis of the toenail. Study enrollment has been completed, with 334 patients in the study, with results expected in the latter part of 2014 (Trial 2013b).

Devices

Laser treatment is a developing area for the treatment of onychomycosis. The appeal stems from the ability to selectively deliver energy to the target tissue, thus avoiding systemic side effects and drug-drug interactions. In addition, since treatments are done in the office, there is better patient compliance (Gupta and Simpson 2012b). Since 2010, the FDA has approved specific Nd:YAG 1064-nm lasers for onychomycosis therapy. However, it should be noted that this approval is different than the authorization given for medications, as the lasers are approved for the temporary cosmetic improvement of onychomycosis (Gupta and Simpson 2012a). It was previously thought that the mechanism of action for the fungicidal effect of lasers was achieved with heat (Vural et al. 2008), but newer *in vitro* studies have shown that the amount of time and level of heat required to kill *T. rubrum* would not be tolerable to patients (Carney et al. 2013). Although the mechanism of action is poorly understood, some clinical trials have shown success using the Nd:YAG 1064-nm laser for treatment of onychomycosis; however, in a recent study of 8 patients treated with the Nd:YAG 1064-nm laser for 5 treatment sessions, no mycological clinical cure was achieved and there was only mild clinical improvement. Additionally, most patients reported pain and burning during the treatments and required many short breaks (Carney et al. 2013). Other types of lasers that are being studied, but not yet FDA approved, include CO_2, near-infrared diode, and femtosecond infrared lasers (Elewski et al. 2013a).

Photodynamic therapy (PDT) is another therapy that is being investigated for use in the treatment of onychomycosis. Using this method, visible light is used to excite a topically applied photosensitizing agent, such as the heme precursors, aminolevulinic acid (ALA), and methyl-aminolevulinic acid (MAL), which cause an accumulation of protoprophyrin IX (PpIX). PpIX in combination with red light results in the formation of reactive oxygen species, and produces selective tissue destruction (Heinemann et al. 2008). While traditionally used for the treatment of actinic keratosis, PDT was shown to inhibit *T. rubrum* when used in conjunction with red light (same as next sentence). There are no randomized controlled studies published to date, but in the largest study including 30 patients with *T. rubrum*-infected toenails, the clinical cure rate was 36.6% at 18 months (Sotiriou et al. 2010). It should be noted that at the present time, PDT is not FDA approved and is not practical in an office setting given the time commitment and discomfort experienced by patients. For example, in this study, nail plates were pretreated with urea for ten days. They were then avulsed in the office and 20% 5-ALA was applied to the nail bed under occlusion for three hours followed by irradiation with 570–670 nm red light. The patients were treated for three times every two weeks. In addition, all patients experienced pain, some requiring several minute breaks in the middle of treatment (Sotiriou et al. 2010).

Plasma therapy is a developing area for the treatment of onychomycosis. Plasma was shown to be fungicidal to *T. rubrum* in an *in vitro* model, and a clinical trial to evaluate the safety, tolerability, and efficacy of plasma in human subjects is currently ongoing (Trial 2013a).

Over the Counter (OTC) Treatments

Health care professionals will frequently encounter patients who self treat their onychomycosis by using OTC therapies either as monotherapy or as an adjunct to prescription medications. Popular remedies are foot soaks with hydrogen peroxide,

Listerine, beer soaks, household chlorine bleach and applications of salicylic acid, tea tree oil and Vicks VapoRub (The Proctor & Gamble Company, Cincinnati, OH). While there are no randomized controlled studies evaluating these treatments, there is theoretical scientific basis for some of these treatments, anecdotes of efficacy by patients and clinicians and a small number of published reports. For example, nanocapsules containing tea tree oil were shown to inhibit the growth of *T. rubrum* in two different *in vitro* models of dermatophyte nail infection (Flores et al. 2013). In addition, in a small pilot study involving 18 patients with onychomycosis treated with Vicks VapoRub ointment (containing thymol, menthol, camphor, and oil of Eucalypus) at least once daily, 27.8% had a complete cure at 48 weeks. Interestingly, they found, however, that the clinical effect was more likely when the cultured organism was *C parapsilosis* and *T. mentagrophytes* as opposed to *T. rubrum* (Derby et al. 2011). While evidence-based medicine can not be used to recommend these products, since they are not harmful or costly, they can be attempted by patients, provided they understand this caveat and that definitive treatments are available to them if they so choose (Elewski et al. 2013a).

Reinfection and Recurrence

Several studies have shown that recurrence of onychomycosis is common after a full course of treatment, year?? (Piraccini et al. 2010), however it is not known whether these are new infections or whether the original pathogen was not completely eradicated. Certain authors use one or two years as cutoffs for recurrence vs. reinfection. Some theories on the inability to obtain long-term cure for onychomycosis, include strain type switching, emergence of drug-resistant fungal strains, and relapse or reinfection through continual contact with fungal material shed from the patient at the time of active infection (Gupta et al. 2001; Scher and Baran 2003; Tosti et al. 2005).

Before undergoing treatment for onychomycosis, patients should be counseled that they may need to be treated several times over the long term (Elewski et al. 2013a). It should be noted that a meta-analysis was performed comparing long term recurrences (follow up greater than 2 years) of toenail onychomycosis after successful treatment with continuous terbinafine versus itraconazole (intermittent or continuous) and it was found that terbinafine is superior to itraconazole in maintaining mycological cure over a long term (Yin et al. 2012).

Measures can be taken to prevent recurrences. It is generally recommended that patients keep feet clean and dry, apply antifungal creams daily, spray shoes with antifungal spray or use antifungal powder, and trim nails short. Patients are instructed to wear footwear in public pools and locker rooms (Tosti et al. 2005). In addition, when ozone gas was studied to prevent recurrences, the authors found that it was able to kill *T. rubrum* and *T. mentagrophytes in vitro* (Gupta and Brintnell 2014) and that it was also effective in sanitizing footwear (Gupta and Brintnell 2013). In another study, the authors found that a commercial ultraviolet C sanitizing device was able to inhibit *T. rubrum* and *T. mentagrophytes* in shoes (Ghannoum et al. 2012).

Conclusion

In sum, we have reviewed current diagnosis and treatment of onychomycosis. Clinical exam, combined with direct microscopy, fungal culture, and histopathology are presently the techniques used for diagnosis. Terbinafine and itraconazole are the oral therapies for onychomycosis that are available, but fluconazole is sometimes used off-label. Ciclopirox, efinaconazole, and tavabarole are options for topical treatment of onychomycosis. Onychomycosis is an active area of research. PCR and confocal microscopy are examples of techniques that will be used for the diagnosis of onychomycosis in the near future. In addition, new topicals, orals and devices may be therapeutic options in the next few years.

References

Abdullah, L. and O. Abbas. 2011. Common nail changes and disorders in older people: Diagnosis and management. Can Fam Physician 57: 173–81.

Arabatzis, M., Bruijnesteijn Van Coppenraet, L.E., E.J. Kuijper, G.S. De Hoog, A.P. Lavrijsen, K. Templeton, E.M. Van Der Raaij-Helmer, A. Velegraki, Y. Graser and R.C. Summerbell. 2007. Diagnosis of common dermatophyte infections by a novel multiplex real-time polymerase chain reaction detection/identification scheme. Br J Dermatol. 157: 681–9.

Back, D.J. and J.F. Tjia. 1991. Comparative effects of the antimycotic drugs ketoconazole, fluconazole, itraconazole and terbinafine on the metabolism of cyclosporin by human liver microsomes. Br J Clin Pharmacol. 32: 624–6.

Balfour, J.A. and D. Faulds. 1992. Terbinafine. A review of its pharmacodynamic and pharmacokinetic properties, and therapeutic potential in superficial mycoses. Drugs 43: 259–84.

Baran, R. 1981. Onychia and paronychia of mycotic microbial and parasitic origin. pp. 39–45. *In*: Pierre, M. (ed.). The Nail. Churchill Livingstone, Edinburgh.

Baran, R., J. Faergemann and R.J. Hay. 2007a. Superficial white onychomycosis—a syndrome with different fungal causes and paths of infection. J Am Acad Dermatol. 57: 879–82.

Baran, R., B. Sigurgeirsson, D. De Berker, R. Kaufmann, M. Lecha, J. Faergemann, N. Kerrouche and F. Sidou. 2007b. A multicentre, randomized, controlled study of the efficacy, safety and cost-effectiveness of a combination therapy with amorolfine nail lacquer and oral terbinafine compared with oral terbinafine alone for the treatment of onychomycosis with matrix involvement. Br J Dermatol. 157: 149–57.

Belenky, P., D. Camacho and J.J. Collins. 2013. Fungicidal drugs induce a common oxidative-damage cellular death pathway. Cell Rep. 3: 350–8.

Bohn, M. and K.T. Kraemer. 2000. Dermatopharmacology of ciclopirox nail lacquer topical solution 8% in the treatment of onychomycosis. J Am Acad Dermatol. 43: S57–69.

Borgers, M., H. Degreef and G. Cauwenbergh. 2005. Fungal infections of the skin: infection process and antimycotic therapy. Curr Drug Targets 6: 849–62.

Brammer, K.W., P.R. Farrow and J.K. Faulkner. 1990. Pharmacokinetics and tissue penetration of fluconazole in humans. Rev Infect Dis. 12(Suppl 3): S318–26.

Breckenridge, A. 1992. Clinical significance of interactions with antifungal agents. Br J Dermatol. 126(Suppl 39): 19–22.

Brown, S.J. 2009. Efficacy of fluconazole for the treatment of onychomycosis. Ann Pharmacother. 43: 1684–91.

Burke, W.A. and B.E. Jones. 1984. A simple stain for rapid office diagnosis of fungus infections of the skin. Arch Dermatol. 120: 1519–20.

Carney, C., W. Cantrell, J. Warner and B. Elewski. 2013. Treatment of onychomycosis using a submillisecond 1064-nm neodymium:yttrium-aluminum-garnet laser. J Am Acad Dermatol. 69: 578–82.

Chang, S.J., S.C. Hsu, K.J. Tien, J.Y. Hsiao, S.R. Lin, H.C. Chen and M.C. Hsieh. 2008. Metabolic syndrome associated with toenail onychomycosis in Taiwanese with diabetes mellitus. Int J Dermatol. 47: 467–72.

Clayton, Y.M. 1989. *In vitro* activity of terbinafine. Clin Exp Dermatol. 14: 101–3.

Cleary, J.D., Taylor, J.W. and S.W. Chapman. 1992. Itraconazole in antifungal therapy. Ann Pharmacother. 26: 502–9.
Cockerell, C. and R. Odom. 1995. The differential diagnosis of nail disease. AIDS Patient Care 9(Suppl 1): S5–10.
Cribier, B., M.L. Mena, D. Rey, M. Partisani, V. Fabien, J.M. Lang and E. Grosshans. 1998. Nail changes in patients infected with human immunodeficiency virus. A prospective controlled study. Arch Dermatol. 134: 1216–20.
Daniel, C.R., 3rd. 1991. The diagnosis of nail fungal infection. Arch Dermatol. 127: 1566–7.
Derby, R., P. Rohal, C. Jackson, A. Beutler and C. Olsen. 2011. Novel treatment of onychomycosis using over-the-counter mentholated ointment: a clinical case series. J Am Board Fam Med. 24: 69–74.
Eckhard, M., A. Lengler, J. Liersch, R.G. Bretzel and P. Mayser. 2007. Fungal foot infections in patients with diabetes mellitus—results of two independent investigations. Mycoses 50(Suppl 2): 14–9.
Effendy, I., M. Lecha, M. Feuilhade De Chauvin, N. Di Chiacchio, R. Baran and O. European Onychomycosis. 2005. Epidemiology and clinical classification of onychomycosis. J Eur Acad Dermatol Venereol. 19(Suppl 1): 8–12.
Elewski, B. 1993a. Tinea pedis and tinea manuum. *In*: Demis, J.D. (ed.) Clinical dermatology. JB Lippincott Co., Philadelphia.
Elewski, B., R. Pollak, S. Ashton, P. Rich, J. Schlessinger and A. Tavakkol. 2012. A randomized, placebo- and active-controlled, parallel-group, multicentre, investigator-blinded study of four treatment regimens of posaconazole in adults with toenail onychomycosis. Br J Dermatol. 166: 389–98.
Elewski, B., D. Pariser, P. Rich and R.K. Scher. 2013a. Current and emerging options in the treatment of onychomycosis. Semin Cutan Med Surg. 32: S9–12.
Elewski, B.E. 1992. Cutaneous Fungal Infections. Igaku-Shoin Medical Publishers, New York.
Elewski, B.E. 1993b. Clinical pearl: proximal white subungual onychomycosis in AIDS. J Am Acad Dermatol. 29: 631–2.
Elewski, B.E. 1993c. Mechanisms of action of systemic antifungal agents. J Am Acad Dermatol. 28: S28–S34.
Elewski, B.E. 1995. Clinical pearl: diagnosis of onychomycosis. J Am Acad Dermatol. 32: 500–1.
Elewski, B.E. 2000. Onychomycosis. Treatment, quality of life, and economic issues. Am J Clin Dermatol. 1: 19–26.
Elewski, B.E. and D.L. Greer. 1991. Hendersonula toruloidea and Scytalidium hyalinum. Review and update. Arch Dermatol. 127: 1041–4.
Elewski, B.E. and M.A. Charif. 1997. Prevalence of onychomycosis in patients attending a dermatology clinic in northeastern Ohio for other conditions. Arch Dermatol. 133: 1172–3.
Elewski, B.E., M.A. Ghannoum, P. Mayser, A.K. Gupta, H.C. Korting, R.J. Shouey, D.R. Baker, P.A. Rich, M. Ling, S. Hugot, B. Damaj, J. Nyirady, K. Thangavelu, M. Notter, A. Parneix-Spake and B. Sigurgeirsson. 2013b. Efficacy, safety and tolerability of topical terbinafine nail solution in patients with mild-to-moderate toenail onychomycosis: results from three randomized studies using double-blind vehicle-controlled and open-label active-controlled designs. J Eur Acad Dermatol Venereol. 27: 287–94.
Elewski, B.E., P. Rich, R. Pollak, D.M. Pariser, S. Watanabe, H. Senda, C. Ieda, K. Smith, R. Pillai, T. Ramakrishna and J.T. Olin. 2013c. Efinaconazole 10% solution in the treatment of toenail onychomycosis: Two phase III multicenter, randomized, double-blind studies. J Am Acad Dermatol. 68: 600–8.
Elewski, B.E. R., P.; Wiltz, H. 2013. Tavaborole. Womens and Pediatric Dermatology Seminar October.
Evans, E.G. and B. Sigurgeirsson. 1999. Double blind, randomised study of continuous terbinafine compared with intermittent itraconazole in treatment of toenail onychomycosis. The LION Study Group. BMJ 318: 1031–5.
Flagothier, C., C. Pierard-Franchimont and G.E. Pierard. 2005. New insights into the effect of amorolfine nail lacquer. Mycoses 48: 91–4.
Fletcher, C.L., M.K. Moore and R.J. Hay. 2001. Endonyx onychomycosis due to Trichophyton soudanense in two Somalian siblings. Br J Dermatol. 145: 687–8.
Flores, F.C., J.A. De Lima, R.F. Ribeiro, S.H. Alves, C.M. Rolim, R.C. Beck and C.B. Da Silva. 2013. Antifungal activity of nanocapsule suspensions containing tea tree oil on the growth of Trichophyton rubrum. Mycopathologia 175: 281–6.
Ghannoum, M.A., R.A. Hajjeh, R. Scher, N. Konnikov, A.K. Gupta, R. Summerbell, S. Sullivan, R. Daniel, P. Krusinski, P. Fleckman, P. Rich, R. Odom, R. Aly, D. Pariser, M. Zaiac, G. Rebell, J. Lesher, B.

Gerlach, G.F. Ponce-De-Leon, A. Ghannoum, J. Warner, N. Isham and B. Elewski. 2000. A large-scale North American study of fungal isolates from nails: the frequency of onychomycosis, fungal distribution, and antifungal susceptibility patterns. J Am Acad Dermatol. 43: 641–8.
Ghannoum, M.A., N. Isham and L. Long. 2012. Optimization of an infected shoe model for the evaluation of an ultraviolet shoe sanitizer device. J Am Podiatr Med Assoc. 102: 309–13.
Ghannoum, M.A., P.K. Mukherjee, E.M. Warshaw, S. Evans, N.J. Korman and A. Tavakkol. 2013. Molecular analysis of dermatophytes suggests spread of infection among household members. Cutis 91: 237–45.
Ginter-Hanselmayer, G., W. Weger and J. Smolle. 2008. Onychomycosis: a new emerging infectious disease in childhood population and adolescents. Report on treatment experience with terbinafine and itraconazole in 36 patients. J Eur Acad Dermatol Venereol. 22: 470–5.
Grant, S.M. and S.P. Clissold. 1990. Fluconazole. A review of its pharmacodynamic and pharmacokinetic properties, and therapeutic potential in superficial and systemic mycoses. Drugs 39: 877–916.
Graser, Y., V. Czaika and T. Ohst. 2012. Diagnostic PCR of dermatophytes—an overview. J Dtsch Dermatol Ges. 10: 721–6.
Graybill, J.R. and P.K. Sharkey. 1990. Fungal infections and their management. Br J Clin Pract. Suppl. 71: 23–31.
Grover, C., S. Bansal, S. Nanda, B.S. Reddy and V. Kumar. 2007. Combination of surgical avulsion and topical therapy for single nail onychomycosis: a randomized controlled trial. Br J Dermatol. 157: 364–8.
Gupta, A.K. and R.C. Summerbell. 1999. Combined distal and lateral subungual and white superficial onychomycosis in the toenails. J Am Acad Dermatol. 41: 938–44.
Gupta, A.K. and J.E. Ryder. 2003. The use of oral antifungal agents to treat onychomycosis. Dermatol Clin. 21: 469–79, vi.
Gupta, A.K. and A.R. Skinner. 2004. Onychomycosis in children: a brief overview with treatment strategies. Pediatr Dermatol. 21: 74–9.
Gupta, A.K. and F.C. Simpson. 2012a. Medical devices for the treatment of onychomycosis. Dermatol Ther. 25: 574–81.
Gupta, A.K. and F.C. Simpson. 2012b. New therapeutic options for onychomycosis. Expert Opin Pharmacother. 13: 1131–42.
Gupta, A.K. and F.C. Simpson. 2013. Diagnosing onychomycosis. Clin Dermatol. 31: 540–3.
Gupta, A.K. and W.C. Brintnell. 2013. Sanitization of contaminated footwear from onychomycosis patients using ozone gas: a novel adjunct therapy for treating onychomycosis and tinea pedis? J Cutan Med Surg. 17: 243–9.
Gupta, A.K. and W. Brintnell. 2014. Ozone gas effectively kills laboratory strains of Trichophyton rubrum and Trichophyton mentagrophytes using an in vitro test system. J Dermatolog Treat. 25: 251–5.
Gupta, A.K., D.N. Sauder and N.H. Shear. 1994. Antifungal agents: an overview. Part II. J Am Acad Dermatol. 30: 911–33; quiz 934–6.
Gupta, A.K., C.W. Lynde, H.C. Jain, R.G. Sibbald, B.E. Elewski, C.R. Daniel, 3rd, G.N. Watteel and R.C. Summerbell. 1997a. A higher prevalence of onychomycosis in psoriatics compared with non-psoriatics: a multicentre study. Br J Dermatol. 136: 786–9.
Gupta, A.K., R.G. Sibbald, C.W. Lynde, P.R. Hull, R. Prussick, N.H. Shear, P. De Doncker, C.R. Daniel, 3rd and B.E. Elewski. 1997b. Onychomycosis in children: prevalence and treatment strategies. J Am Acad Dermatol. 36: 395–402.
Gupta, A.K., P. Taborda, V. Taborda, J. Gilmour, A. Rachlis, I. Salit, M.A. Gupta, P. Macdonald, E.A. Cooper and R.C. Summerbell. 2000. Epidemiology and prevalence of onychomycosis in HIV-positive individuals. Int J Dermatol. 39: 746–53.
Gupta, A.K., Y. Kohli and R.C. Summerbell. 2001. Variation in restriction fragment length polymorphisms among serial isolates from patients with Trichophyton rubrum infection. J Clin Microbiol. 39: 3260–6.
Gupta, A.K., J.E. Ryder and R. Baran. 2003. The use of topical therapies to treat onychomycosis. Dermatol Clin. 21: 481–9.
Gupta, A.K., C. Leonardi, R.R. Stoltz, P.F. Pierce, B. Conetta and G. Ravuconazole Onychomycosis. 2005. A phase I/II randomized, double-blind, placebo-controlled, dose-ranging study evaluating the efficacy, safety and pharmacokinetics of ravuconazole in the treatment of onychomycosis. J Eur Acad Dermatol Venereol. 19: 437–43.
Gupta, A.K., C. Drummond-Main and M. Paquet. 2013a. Evidence-based optimal fluconazole dosing regimen for onychomycosis treatment. J Dermatolog Treat. 24: 75–80.

Gupta, A.K., M. Paquet, F. Simpson and A. Tavakkol. 2013b. Terbinafine in the treatment of dermatophyte toenail onychomycosis: a meta-analysis of efficacy for continuous and intermittent regimens. J Eur Acad Dermatol Venereol. 27: 267–72.

Gupta, A.K., M. Paquet and F.C. Simpson. 2013c. Therapies for the treatment of onychomycosis. Clin Dermatol. 31: 544–54.

Haas, N. and H. Sperl. 2001. A medical student discovers onychomycosis. Hautarzt 52: 64–7.

Hainer, B.L. 2003. Dermatophyte infections. Am Fam Physician 67: 101–8.

Haley, L.D., R.C. 1990. Fungal infections. pp. 106–119. *In*: R.K. Scher and C.R. Daniel (ed.). Nails: Therapy, Diagnosis, Surgery. WB Saunders, Philadelphia.

Hay, R.J. and R. Baran. 2011. Onychomycosis: a proposed revision of the clinical classification. J Am Acad Dermatol. 65: 1219–27.

Hay, R.J., Y.M. Clayton, M.K. Moore and G. Midgely. 1988. An evaluation of itraconazole in the management of onychomycosis. Br J Dermatol. 119: 359–66.

Heikkila, H. and S. Stubb. 1995. The prevalence of onychomycosis in Finland. Br J Dermatol. 133: 699–703.

Heikkila, H. and S. Stubb. 2002. Onychomycosis in children: treatment results of forty-seven patients. Acta Derm Venereol. 82: 484–5.

Heinemann, I.U., M. Jahn and D. Jahn. 2008. The biochemistry of heme biosynthesis. Arch Biochem Biophys. 474: 238–51.

Hui, X., S.J. Baker, R.C. Wester, S. Barbadillo, A.K. Cashmore, V. Sanders, K.M. Hold, T. Akama, Y.K. Zhang, J.J. Plattner and H.I. Maibach. 2007. *In Vitro* penetration of a novel oxaborole antifungal (AN2690) into the human nail plate. J Pharm Sci. 96: 2622–31.

Insert, P. 1993. Itraconazole. Med Lett Drugs Ther. 35: 7–9.

Insert, P. 2011. Amorolfine 5% W/V Nail Lacquer [Online]. Medicines and Healthcare Products Regulatory Agency. Available: http://www.mhra.gov.uk/home/groups/par/documents/websiteresources/con129125.pdf2011.

Insert, P. 2012a. LAMISIL (terbinafine hydrochloride) Tablets, 250 mg Drugs@FDA: FDA Approved Drug Products [Online]. Available: http://www.accessdata.fda.gov/drugsatfda_docs/label/2012/020539s021lbl.pdf [Accessed 21 June 2014.

Insert, P. 2012b. SPORANOX! (itraconazole) Capsules. Drugs@FDA: FDA Approved Drug Products [Online]. Available: http://www.accessdata.fda.gov/drugsatfda_docs/label/2012/020083s048s049s050lbl.pdf [Accessed 21 June 2014.

Insert, P. 2014. DIFLUCAN® (fluconazole) tablets [Online]. Available: http://www.accessdata.fda.gov/drugsatfda_docs/label 2014/019949s058,019950s062,020090s042lbl.pdf [Accessed 21 June 2014.

Insert, P. January 2007. Gris-PEG [package insert]. Pedinol Pharmacal, Farmingdale, NY.

Jennings, M.B., R. Pollak, L.B. Harkless, F. Kianifard and A. Tavakkol. 2006. Treatment of toenail onychomycosis with oral terbinafine plus aggressive debridement: IRON-CLAD, a large, randomized, open-label, multicenter trial. J Am Podiatr Med Assoc. 96: 465–73.

Jones, T.C. 1995. Overview of the use of terbinafine (Lamisil) in children. Br J Dermatol. 132: 683–9.

Korting, H.C. and C. Schollmann. 2009. The significance of itraconazole for treatment of fungal infections of skin, nails and mucous membranes. J Dtsch Dermatol Ges 7: 11-9, 11–20.

Kruk, M.E. and N. Schwalbe. 2006. The relation between intermittent dosing and adherence: preliminary insights. Clin Ther. 28: 1989–95.

Laboratories, D. 2004. Penlac Nail Lacquer (Ciclopirox) Topical Solution, 8% [Online]. Available: http://www.accessdata.fda.gov/drugsatfda_docs/label/2004/21022s004lbl.pdf2004 [Accessed 21 June 2014.

Lai, W.Y., W.Y. Tang, S.K. Loo and Y. Chan. 2011. Clinical characteristics and treatment outcomes of patients undergoing nail avulsion surgery for dystrophic nails. Hong Kong Med J. 17: 127–31.

Larsen, G.K., M. Haedersdal and E.L. Svejgaard. 2003. The prevalence of onychomycosis in patients with psoriasis and other skin diseases. Acta Derm Venereol. 83: 206–9.

Lavrijsen, A.P., K.J. Balmus, W.M. Nugteren-Huying, A.C. Roldaan, J.W. Van't Wout and B.H. Stricker. 1992. Hepatic injury associated with itraconazole. Lancet 340: 251–2.

Lecha, M., I. Effendy, M. Feuilhade De Chauvin, N. Di Chiacchio, R. Baran and E. Taskforce On Onychomycosis. 2005. Treatment options—development of consensus guidelines. J Eur Acad Dermatol Venereol. 19(Suppl 1): 25–33.

Lee, R.E., T.T. Liu, K.S. Barker, R.E. Lee and P.D. Rogers. 2005. Genome-wide expression profiling of the response to ciclopirox olamine in Candida albicans. J Antimicrob Chemother. 55: 655–62.

Levy, L.A. 1997. Epidemiology of onychomycosis in special-risk populations. J Am Podiatr Med Assoc. 1997 Dec; 87(12): 546–50.

Mahoney, J.M., J. Bennet and B. Olsen. 2003. The diagnosis of onychomycosis. Dermatol Clin. 21: 463–7.
Matricciani, L., K. Talbot and S. Jones. 2011. Safety and efficacy of tinea pedis and onychomycosis treatment in people with diabetes: a systematic review. J Foot Ankle Res. 4: 26.
Mercantini, R., R. Marsella and D. Moretto. 1996. Onychomycosis in Rome, Italy Mycopathologia 136. 25–32.
Monod, M., S. Capoccia, B. Lechenne, C. Zaugg, M. Holdom and O. Jousson. 2002. Secreted proteases from pathogenic fungi. Int J Med Microbiol. 292: 405–19.
Munoz, P., S. Moreno, J. Berenguer, J.C. Bernaldo De Quiros and E. Bouza. 1991. Fluconazole-related hepatotoxicity in patients with acquired immunodeficiency syndrome. Arch Intern Med. 151: 1020–1.
Murdan, S. 2008. Enhancing the nail permeability of topically applied drugs. Expert Opin Drug Deliv. 5: 1267–82.
Pandhi, D. and P. Verma. 2012. Nail avulsion: indications and methods (surgical nail avulsion). Indian J Dermatol Venereol Leprol. 78: 299–308.
Pharaon, M., M. Gari-Toussaint, A. Khemis, K. Zorzi, L. Petit, P. Martel, R. Baran, J.P. Ortonne, T. Passeron, J.P. Lacour and P. Bahadoran. 2014. Diagnosis and treatment monitoring of toenail onychomycosis by reflectance confocal microscopy: Prospective cohort study in 58 patients. J Am Acad Dermatol. 2014 Jul; 71(1): 56–61.
Piraccini, B.M., A. Sisti and A. Tosti. 2010. Long-term follow-up of toenail onychomycosis caused by dermatophytes after successful treatment with systemic antifungal agents. J Am Acad Dermatol. 62: 411–4.
Piraccini, B.M., R. Balestri, M. Starace and G. Rech. 2013. Nail digital dermoscopy (onychoscopy) in the diagnosis of onychomycosis. J Eur Acad Dermatol Venereol. 27: 509–13.
Pleacher, M.D. and W.W. Dexter. 2007. Cutaneous fungal and viral infections in athletes. Clin Sports Med. 26: 397–411.
Ploysangam, T. and A.W. Lucky. 1997. Childhood white superficial onychomycosis caused by Trichophyton rubrum: report of seven cases and review of the literature. J Am Acad Dermatol. 36: 29–32.
Polak, A. 1992. Preclinical data and mode of action of amorolfine. Dermatology 184(Suppl 1): 3–7.
Polak, A. 1993. Kinetics of amorolfine in human nails. Mycoses 36: 101–3.
Rich, P. 2001. Nail biopsy: indications and methods. Dermatol Surg. 27: 229–34.
Rich, P., B. Elewski, R.K. Scher and D. Pariser. 2013. Diagnosis, clinical implications, and complications of onychomycosis. Semin Cutan Med Surg. 32: S5–8.
Richert, B., N. Lateur, A. Theunis and J. Andre. 2009. New tools in nail disorders. Semin Cutan Med Surg. 28: 44–8.
Rippon, J.W. 1988. Medical Mycology: The Pathogenic Fungi and the Pathogenic Actinomycetes. Saunders, Philadelphia.
Rock, F.L., W. Mao, A. Yaremchuk, M. Tukalo, T. Crepin, H. Zhou, Y.K. Zhang, V. Hernandez, T. Akama, S.J. Baker, J.J. Plattner, L. Shapiro, S.A. Martinis, S.J. Benkovic, S. Cusack and M.R. Alley. 2007. An antifungal agent inhibits an aminoacyl-tRNA synthetase by trapping tRNA in the editing site. Science 316: 1759–61.
Rollman, O. 1982. Treatment of onychomycosis by partial nail avulsion and topical miconazole. Dermatologica 165: 54–61.
Rosenbach, M.L. and J.E. Schneider. The burden of onychomycosis in the Medicare population. Health Economic Research Inc., Sandoz Pharmaceuticals Corporation (on file).
Rothmund, G., E.C. Sattler, R. Kaestle, C. Fischer, C.J. Haas, H. Starz and J. Welzel. 2013. Confocal laser scanning microscopy as a new valuable tool in the diagnosis of onychomycosis—comparison of six diagnostic methods. Mycoses 56: 47–55.
Salgo, P.L.D., C.R.; A.K. Gupta, J.D. Mozena and S.W. Joseph. 2003. Onychomycosis disease management. Medical Crossfire: debates, peer exchange and insights in medicine 4: 1–17.
Saunte, D.M., J.B. Holgersen, M. Haedersdal, G. Strauss, M. Bitsch, O.L. Svendsen, M.C. Arendrup and E.L. Svejgaard. 2006. Prevalence of toe nail onychomycosis in diabetic patients. Acta Derm Venereol. 86: 425–8.
Scher, R.K. 1994. Onychomycosis is more than a cosmetic problem. Br J Dermatol. 130(Suppl 43): 15.
Scher, R.K. 1996. Onychomycosis: a significant medical disorder. J Am Acad Dermatol. 35: S2–5.
Scher, R.K. and R. Baran. 2003. Onychomycosis in clinical practice: factors contributing to recurrence. Br J Dermatol. 149(Suppl 65): 5–9.
Scher, R.K. and C.R. Daniel. 2005. Nails Diagnosis, Therapy, Surgery, 3rd ed. Elsevier Saunders, Oxford.

Scher, R.K., D. Breneman, P. Rich, R.C. Savin, D.S. Feingold, N. Konnikov, J.L. Shupack, S. Pinnell, N. Levine, N.J. Lowe, R. Aly, R.B. Odom, D.L. Greer, M.R. Morman, A.D. Bucko, E.H. Tschen, B.E. Elewski and E.B. Smith. 1998. Once-weekly fluconazole (150, 300, or 450 mg) in the treatment of distal subungual onychomycosis of the toenail. J Am Acad Dermatol. 38: S77–86.

Scher, R.K., P. Rich, D. Pariser and B. Elewski. 2013. The epidemiology, etiology, and pathophysiology of onychomycosis. Semin Cutan Med Surg. 32: S2–4.

Scher, R.K., N. Nakamura and A. Tavakkol. 2014. Luliconazole: a review of a new antifungal agent for the topical treatment of onychomycosis. Mycoses. 2014 Jul; 57(7): 389–93.

Seebacher, C., J.P. Bouchara and B. Mignon. 2008. Updates on the epidemiology of dermatophyte infections. Mycopathologia 166: 335–52.

Shear, N.H., V.V. Villars and C. Marsolais. 1991. Terbinafine: an oral and topical antifungal agent. Clin Dermatol. 9: 487–95.

Sigurgeirsson, B. and R. Baran. 2013. The prevalence of onychomycosis in the global population—A literature study. J Eur Acad Dermatol Venereol. 2014 Nov; 28(11): 1480–91.

Sigurgeirsson, B., K. Van Rossem, S. Malahias and K. Raterink. 2013. A phase II, randomized, double-blind, placebo-controlled, parallel group, dose-ranging study to investigate the efficacy and safety of 4 dose regimens of oral albaconazole in patients with distal subungual onychomycosis. J Am Acad Dermatol. 69: 416–25.

Singal, A. and D. Khanna. 2011. Onychomycosis: Diagnosis and management. Indian J Dermatol Venereol Leprol. 77: 659–72.

Smith, E.S., A.B. Fleischer, Jr. and S.R. Feldman. 2001. Demographics of aging and skin disease. Clin Geriatr Med. 17: 631–41, v.

Smith, M.B.M., M.R. 2008. Diagnostic histopathology. pp. 37–51. *In*: D.R. Hospenthal and M.G. Rinaldi (ed.). Diagnosis and Treatment of Human Mycoses. Humana Press, Totowa, NJ.

Sotiriou, E., T. Koussidou-Eremonti, G. Chaidemenos, Z. Apalla and D. Ioannides. 2010. Photodynamic therapy for distal and lateral subungual toenail onychomycosis caused by Trichophyton rubrum: Preliminary results of a single-centre open trial. Acta Derm Venereol. 90: 216–7.

Stevens, D.A. 1988. The new generation of antifungal drugs. Eur J Clin Microbiol Infect Dis. 7: 732–5.

Stokes, J.H. 1999. Letter to referring physician--1938. Int J Dermatol. 38: 827, 829, 840.

Thomas, J., G.A. Jacobson, C.K. Narkowicz, G.M. Peterson, H. Burnet and C. Sharpe. 2010. Toenail onychomycosis: an important global disease burden. J Clin Pharm Ther. 35: 497–519.

Tosti, A., B.M. Piraccini and M. Iorizzo. 2003. Management of onychomycosis in children. Dermatol Clin. 21: 507–9, vii.

Tosti, A., R. Hay and R. Arenas-Guzman. 2005. Patients at risk of onychomycosis—risk factor identification and active prevention. J Eur Acad Dermatol Venereol. 19 Suppl 1: 13–6.

Trial, C. 2013a. Pilot Study to Evaluate Plasma Treatment of Onychomycosis [Online]. MOE Medical Devices. Available: http://clinicaltrials.gov/ct2/show/NCT01819051 [Accessed 21 June 2014.

Trial, C. 2013b. Safety and Efficacy of Luliconazole Solution, 10% in Subjects With Mild to Moderate Onychomycosis (SOLUTION) [Online]. Topica Pharmaceuticals. Available: http://clinicaltrials.gov/ct2/show/NCT01431820 [Accessed 21 June 2014.

Uchida, K., T. Tanaka and H. Yamaguchi. 2003. Achievement of complete mycological cure by topical antifungal agent NND-502 in guinea pig model of tinea pedis. Microbiol Immunol. 47: 143–6.

Van Cauteren, H.C., W.; J. Vandenberghe et al. 1987. The toxicological properties of itraconazole. pp. 262–271. *In*: Fromtling, R.A. (ed.). Recent Trends in the Discovery, Development and Evaluation of Antifungal Agents. Prous Science Publishers, Barcelona.

Van Peer, A., R. Woestenborghs, J. Heykants, R. Gasparini and G. Gauwenbergh. 1989. The effects of food and dose on the oral systemic availability of itraconazole in healthy subjects. Eur J Clin Pharmacol. 36: 423–6.

Velez, A., M.J. Linares, J.C. Fenandez-Roldan and M. Casal. 1997. Study of onychomycosis in Cordoba, Spain: prevailing fungi and pattern of infection. Mycopathologia 137: 1–8.

Villars, V.V. and T.C. Jones. 1992. Special features of the clinical use of oral terbinafine in the treatment of fungal diseases. Br J Dermatol. 126(Suppl 39): 61–9.

Villasenor-Mora, C., A.G. Vega, M.E. Garay-Sevilla, J.A. Padilla-Medina and L.I. Arteaga-Murillo. 2013. Procedure to diagnose onychomycosis through changes in emissivity on infrared images. J Biomed Opt. 18: 116005.

Vural, E., H.L. Winfield, A.W. Shingleton, T.D. Horn and G. Shafirstein. 2008. The effects of laser irradiation on Trichophyton rubrum growth. Lasers Med Sci. 23: 349–53.

Wang, Y.R. and D. Margolis. 2006. The prevalence of diagnosed cutaneous manifestations during ambulatory diabetes visits in the United States, 1998–2002. Dermatology 212: 229–34.

Watanabe, S., H. Takahashi, T. Nishikawa, I. Takiuchi, N. Higashi, K. Nishimoto, S. Kagawa, H. Yamaguchi and H. Ogawa. 2007. Dose-finding comparative study of 2 weeks of luliconazole cream treatment for tinea pedis—comparison between three groups (1%, 0.5%, 0.1%) by a multi-center randomised double-blind study. Mycoses 50: 35–40.

Weinberg, J.M., E.K. Koestenblatt, W.D. Tutrone, H.R. Tishler and L. Najarian. 2003. Comparison of diagnostic methods in the evaluation of onychomycosis. J Am Acad Dermatol. 49: 193–7.

Welsh, O., L. Vera-Cabrera and E. Welsh. 2010. Onychomycosis. Clin Dermatol. 28: 151–9.

Wiederhold, N.P., A.W. Fothergill, D.I. Mccarthy and A. Tavakkol. 2014. Luliconazole demonstrates potent in vitro activity against dermatophytes recovered from patients with onychomycosis. Antimicrob Agents Chemother. 58: 3553–3555.

Yin, Z., J. Xu and D. Luo. 2012. A meta-analysis comparing long-term recurrences of toenail onychomycosis after successful treatment with terbinafine versus itraconazole. J Dermatolog Treat. 23: 449–52.

CHAPTER 3

Mycotic Keratitis: Current Perspectives

*Sabyasachi Bandyopadhyay** and *Mita Saha (Dutta Chowdhury)*

Introduction

Corneal opacity, mostly as a sequel of infectious keratitis is a leading cause of preventable monocular blindness worldwide after cataract and glaucoma (Upadhyay et al. 1991; Gilbert et al. 1995; Whitcher et al. 2001; Resnikoff et al. 2004). Mycotic keratitis or fungal corneal ulcer usually presents as a suppurative and ulcerative corneal infection. Fungi are ubiquitous organisms that are more frequently implicated as ocular pathogens in agrarian, tropical countries than in the developed world (McLeod 2009). This type of keratitis reportedly occurs much more frequently in developing countries like India than in developed countries like United States. This entity may account for 30% to 62% of culture positive infectious keratitis in different studies in tropical and subtropical countries and at least 70 genera of fungi have been isolated from corneal specimen (Srinibasan et al. 1997; Agarwal et al. 2001; Gopinathan et al. 2002; Bharathi et al. 2003; Basak et al. 2005; Bandyopadhyay et al. 2012). The filamentous fungi, *Aspergillus* and *Fusarium* contribute up to 70% of the cultures (Agarwal et al. 1994). The determination of regional etiology is important as the causative fungi differ from region to region and within the same region over time. The incidence of fungal keratitis in the United States is about 1500 cases per year (O'Day 1996). The majority of cases occur in the warmer southern and southwestern states with septate filamentous fungi like *Fusarium* and *Aspergillus* as in most other parts of the world (O'Day 1996). But, in the northern states, Candida is the most frequently isolated fungal organism.

Assistant Professor, Department of Ophthalmology, R.G. Kar Medical College and Hospital, 1, Khudiram Bose Sarani, Kolkata - 700 004.
* Corresponding author: sabyasachi.bandyopadhyay@yahoo.com

Aspergillus was the predominant isolate in eastern India (Basak et al. 2005; Bandyopadhyay et al. 2012), parts of south India, north India, Nepal and Bangladesh (Williams et al. 1987; Venugopal et al. 1989; Upadhyay et al. 1991; Despande and Koppikar 1999; Punia et al. 2014). But, *Fusarium* spp. were found to be more common in other studies in south India (Srinibasan et al. 1997; Leck et al. 2002; Bharathi et al. 2003), Paraguay, Florida, Hong Kong and Singapore (Liesegang and Forster 1980; Mino de Kasper et al. 1991; Wong et al. 1997; Houang et al. 2001). These differences in predominant fungal isolates could be attributed to different climatic conditions. *Candida albicans*, the yeast like fungus, was the predominant isolate mainly in United States (Thygeson and Ocumoto 1974; Rosa et al. 1994; Tanure et al. 2000), Denmark (Nielsen et al. 2014) and the second most common isolate (after *Aspergillus*) in one study in Nepal (Upadhyay et al. 1991) but not in other major studies in the tropics.

Fungi cannot penetrate the intact corneal epithelium and therefore do not enter the cornea through episcleral limbal vessels (Insan et al. 2013). Corneal trauma, however trivial it might be, frequently precedes infection in the warmer states of the United States and in the tropical countries. Associated contamination with organic vegetative matter presents an increased risk for filamentous fungal keratitis. In colder climates candida infections are predominant. Pre-existing corneal disease, local immunosuppression by prolonged use of corticosteroids and other systemic debilitating illnesses like diabetes mellitus, alcoholism and vitamin A deficiency are frequently associated with this type of infection.

The complications of fungal keratitis are devastating therefore early and correct diagnoses of the etiological agent as well as aggressive treatment are necessary. Mycotic keratitis should be suspected in every case with corneal lesion and must be ruled out before starting antibiotics or steroids. Direct microscopic examinations of the corneal scrapings by 10% KOH, Gram staining or Lactophenol cotton blue (LPCB) followed by culture in the Sabouraud glucose-neopeptone agar or blood agar usually confirm the diagnosis. Topical Natamycin (5%) is the first agent of choice against filamentous fungi like *Aspergillus* and *Fusarium* and can be combined with oral Ketoconazole in deep fungal stromal keratitis (Prajna et al. 2013). Topical Amphotericin B (0.15%) is the drug of choice for the candida keratitis and is also effective against filamentous fungi (Tanure et al. 2000; Thomas 2003).

Voriconazole (1%) is a new triazole derivative with potent broad spectrum antifungal activity and can be combined with Natamycin to yield promising results against all type of fungal keratitis (Sharma et al. 2013). Despite prompt and aggressive treatment some cases result in corneal perforations requiring urgent penetrating keratoplasty. In worst cases the development of endophthalmitis or panophthalmitis can lead to complete loss of vision in the affected eye. Recent molecular identification methods with prompt detection of DNA of fungal elements by polymerase chain reaction (PCR) for species-level diagnosis, development of newer antifungals and use of disease modifying agents against cytokines to reduce corneal inflammation and tissue destruction can prevent such disastrous consequences.

Epidemiology

Corneal infections due to filamentous fungi usually occur following trauma (commonly by a vegetative matter) in healthy young males engaged in agricultural or other outdoor activities (Thomas 2003). Fungal conidia can be directly implanted in the corneal stroma through the traumatizing agents or the abraded epithelium can permit invasion by exogenous fungi (Thomas 2003). Incidence of fungal keratitis is most common during monsoon time corresponding to paddy harvesting in tropical countries (Basak et al. 2005). Next to monsoon the incidence is more common in winter than summer. Winter season also corroborates with high agricultural activity. A dry or humid,windy climate along with a largely agrarian population creates favorable conditions for fungal ulceration in cornea (Houang et al. 2001). Less frequent predisposing factors include immunological incompetence, prior administration of corticosteroids or antibacterial agents, allergic conjunctivitis, and the use of hydrophilic contact lenses (Hagan et al. 1995; Houang et al. 2001). Keratitis due to *Candida albicans* and related fungi usually occurs in pre-existing ocular diseases like tear film dysfunction, insufficient lid closure, herpetic keratitis or abrasions caused by contaminated contact lenses and in systemic conditions like diabetes mellitus, immunosuppression, alcoholism, vitamin A deficiency and the use of corticosteroids (Thomas 1998; McLeod 2009). Unlike fungal endophthalmitis, fungal keratitis is not associated with systemic fungemia (Arffa 1991; McLeod 2009). This type of keratitis is also possible after penetrating keratoplasty, clear cornea (suture less) cataract surgery, photorefractive keratectomy or laser *in situ* keratomileusis (LASIK) surgery (Kuo et al. 2001; Garg et al. 2003).

Morphology of the Fungi Causing Keratitis

Fungi invading the cornea can be broadly classified as yeasts or molds. Yeasts are unicellular fungi and *Candida albicans* is the representative of such fungi as corneal pathogen. It has oval or round structure called blastoconidium. In culture, yeasts reproduce by budding. In tissue, the yeast may develop elongated buds which look like hyphal form and are termed pseudohyphae. Molds are filamentous fungi which possess true hyphae and grow by apical extension or branching. Filamentous fungi are classified as septate or nonseptate depending upon the presence or absence of the cross walls in the hyphae. All corneal pathogenic molds form septate hyphae.

The reproduction of fungi usually occurs by asexual method through the formation of various types of spores. The fungi in asexual reproductive stage are described to be in imperfect stage and were previously called Fungi Imperfecti. The fungal organisms causing corneal disease are usually in the asexual phase of their life cycle (McGinis 1980).

Corneal pathogenic fungi can be *Aspergillaceae* and *Nectriaceae* (Hyaline or light coloured fungi such as *Aspergillus* and *Fusarium*), *Pleosporaceae* or *Dematiaceae* (dark coloured fungi such as *Curvularia*) and yeasts (*Candida*). *Aspergillus* species are found in most parts of the world thriving in decaying vegetation and soil. *Aspergillus* has been found in more than one third of the cases of keratomycosis in large studies of tropical regions (Williams et al. 1987; Upadhyay et al. 1991; Despande and Koppikar

1999; Basak et al. 2005; Bandyopadhyay et al. 2012). *A. flavus* is the most common species of *Aspergillus* isolated but other species like *A. fumigatus*, *A. niger*, *A. tamarii*, *A. pseudotamarii*, *A. nomius*, *A. brasiliensis* and *A. tubingensis* are also detected (Kredics et al. 2007; Kredics et al. 2009; Manikandan et al. 2009; Manikandan et al. 2010; Baranyia et al. 2013). The *Aspergillus* species is diagnosed microscopically by the conidiophores with swollen terminal ends surrounded by the sterigmata which produces long chains of coccoid conidia that radiate from the terminal end. Hyphae of *Aspergillus* are septate and branch dichotomously (O'Day and Burd 1996).

Fusarium fungi are saprophytic or parasitic in nature and grow widely in decaying organic vegetation and different parts of wild and cultivated plants. *Fusarium* strains are responsible for at least one third of culture positive mycotic keratitis and *Fusarium solani* is the most common species detected (Mino de Kasper et al. 1991; Srinibasan et al. 1997; Leck et al. 2002; Bharathi et al. 2003). *Fusarium* species are diagnosed by their characteristic macroconidia and microconidia. They produce large banana shaped macroconidia on short lateral hyphae or conidiospores. Microconidia are less specific and formed by other fungi such as *Acremonium* (O'Day and Burd 1996). The closely related fungal species are indistinguishable morphologically and a species level diagnosis is achieved only by internal transcribed spacer (ITS) sequence-based molecular identification method (Kuo et al. 2012; Manikandan et al. 2013). The ITS region that contains the target gene (5.8S rRNA gene) is amplified by polymerase chain reaction (PCR) and semi nested PCR to detect fungal DNA (Kuo et al. 2012).

Dematiaceous fungi are saprophytic in the plant world and are characterized by the brown pigmentation of their colonies. *Curvularia*, *Bipolaris* and *Botryodiplodia* are the frequent isolates from fungal corneal ulcers from this group.

Yeast infection of the cornea is represented by *Candida albicans* which is distributed uniformly in the environment. It is found in the diseased skin, in the gastrointestinal tract, in the sputum, in the urine of catheterized patients and in the female genital tract (O'Day and Burd 1996). Candida strains form smooth creamy white colonies which grow well in the blood agar. The presence of budding yeasts in a corneal scrape is almost diagnostic for *Candida*.

Cultural Characteristics of Causal Fungi

Sabouraud's dextrose agar without cycloheximide inhibitor and brain-heart infusion broth are preferred media for culture of corneal pathogens. Cycloheximide is omitted as this inhibits the growth of saprophytic fungi. There may be mixed bacterial and fungal pathogens in corneal ulcers. An antibiotic like chloramphenicol is added to the brain-heart infusion broth to prevent growth of bacteria. Though fungal growth mostly occurs within 3–4 days, it is recommended to wait for at least three weeks before declaring it to be negative. As there is a chance of contamination by other saprophytic fungi which are also known to be corneal pathogens, utmost care is taken to inoculate the plates and to maintain them free of contamination.

Aspergillus flavus and *Aspergillus fumigatus* are the two most common species of Aspergillus isolates. *Aspergillus flavus* are genetically and phenotypically diverse with some types producing abundant conidia, large (L) sclerotia and variable amounts of aflatoxins, while other isolates producing abundant small (S) sclerotia, fewer conidia and high levels of aflatoxins (Varga et al. 2011). *A. flavus* is morphologically indistinguishable from *A. tamari* and *A. nomius* and molecular level identification is necessary for diagnosis of the exact species (Varga et al. 2011). Colonies of *A. fumigatus* are white at first and then turn velvet green due to the pigmentation of the conidia after production of spores (O'Day and Burd 1996). *A. niger* colonies are also white initially but become completely black following sporulation (Mcginis 1980). Morphology of *A. niger* sp. is very much similar to *A. tubingensis* and *A. brasiliensis* and here also a molecular level identification can clinch the diagnosis (Varga et al. 2007). *Fusarium* colonies are usually white in the early stages of development. Following maturation a yellow, red purple coloured pigment develops in the colonies which is best seen on the under surface of the colony. This is called reverse pigmentation (Nelson et al. 1983).

The colony of *Candida* organisms on Sabouraud agar is white to tan or opaque with a smooth round contour and soft consistency. The colonies reach a few millimeters in diameter after 48 hours of incubation at 30°C. It has a distinctive fruity odour which aids in the identification (O'Day and Burd 1996).

Pathogenesis

Fungi are ubiquitous in nature thriving mainly in hot and humid conditions rich in organic matter. They do not invade the cornea easily and require trauma, immunocompromisation or tissue devitalization for their penetration. The fungi most commonly present in the environment have been found as transient commensals in the conjunctival sac of some healthy individuals (Srinibasan et al. 1991). These commensals are believed to be virulent following trauma or administration of corticosteroids. Still this mechanism of infection is not as important as direct injury and the inoculation of external fungi (Thomas 2003). After inoculation the virulence of the organisms is dependent upon their ability to proliferate within corneal tissue overcoming the host defence and causing tissue damage. They secrete various toxic substances like proteases, hemolysins and exotoxins for their penetration (Abad and Foster 2000). Furthermore, in the presence of fungal elements neutrophils release a number of proteolytic enzymes including corneal matrix metalloproteinases causing additional tissue damage (Zhu et al. 1990). The complete ingestion of the fungal element by neutrophils or macrophages is difficult due to the large size of the hyphae of filamentous fungi or the pseudohyphae of the yeasts. As the fungus penetrates deeper into the corneal tissue there may be negative cultures in superficial scrapes (Abad and Foster 2000). Also this form of infection is resistant to the treatment with antifungals, the tissue penetration of which is poor (Abad and Foster 2000). It has been shown histopathologically that the invasion of the mycelia of the filamentous fungi occurs parallel to the corneal stromal lamellae or can be perpendicular in more virulent organisms (O'Day and Burd 1996). Though, the intact Descemet's membrane offers some resistance, the fungi have to penetrate it in order to enter into the anterior chamber (Abad and Foster 2000). The presence of fungal elements on the traumatized corneal epithelium and it's germination, toxin secretion and hyphal invasion of the tissues either initiates a host response that controls infection at the expense of extensive tissue damage and corneal scarring or the host is unable to control fungal growth which becomes unregulated causing corneal perforation (Tarabishy et al. 2008). The innate immunity response in experimental *Fusarium* keratitis in an immunocompetent murine model has been studied in details. The response included recognition of *Fusarium* by resident cells in the corneal stroma with production of IL-1α and IL-1R1/MyD88-dependent CXC chemokine. There is recruitment of neutrophils from limbal vessels to the central cornea and TLR4-dependent antifungal activity by neutrophils followed by tissue damage and corneal opacification either due to production of cytotoxic mediators by fungi or by products of neutrophil degranulation (Tarabishy et al. 2008). This has highlighted the importance of disease modifying agents in control of tissue damage as an adjunct to antifungal drugs.

Clinical Features

Fungal keratitis presents as suppurative, usually ulcerative corneal lesions. Time of presentation is more prolonged (about 5–10 days) and runs a more indolent course than bacterial corneal ulcers (Thomas 2003). Filamentous fungal keratitis can involve any part of the cornea and usually presents with firm (sometimes dry looking), raised slough [Figs. 1 and 2] and feathery extension lines from the edge of the ulcer into the normal cornea [Figs. 3 and 4]. Multifocal granular greywhite satellite lesions (stromal infiltrations), circular immune ring, Descemet's folds and mild iritis may also be present (Thomas 1998). An endothelial plaque or hypopyon (pus in the anterior chamber) can occur after the first week (Thomas 2003). These basic features can vary in severity in individual cases depending upon the causative fungal types and status of the host defence. *Fusarium* keratitis is usually very severe and may cause corneal perforation, deep extension or malignant glaucoma (Sharma et al. 1993; Vemuganti et al. 2002). *Aspergillus* infection on the other hand is less severe and more responsive to therapy than *Fusarium* keratitis. *Curvularia* species commonly present as superficial feathery infiltrations with slow progression to focal suppuration. There may be mixed bacterial

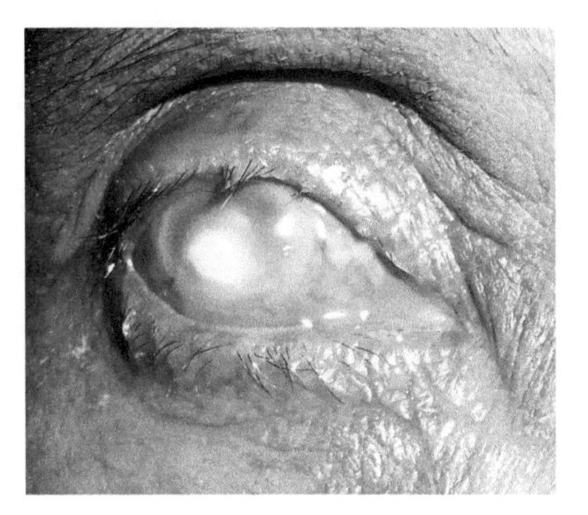

Figure 1. Fungal corneal ulcer of the right eye showing raised (dry looking) necrotic slough with hypopyon.

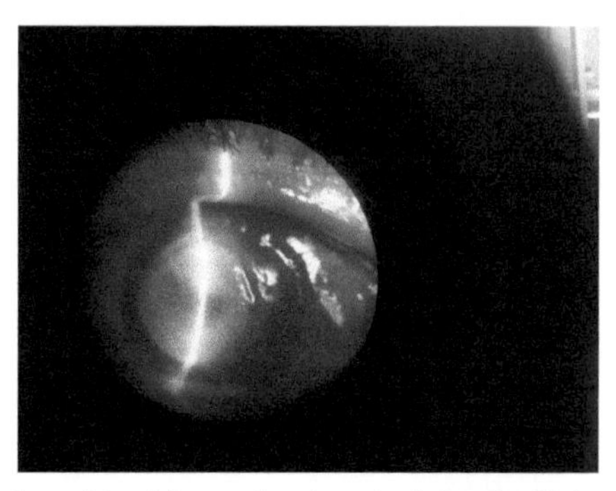

Figure 2. Fungal corneal ulcer of the right eye showing raised slough with hypopyon.

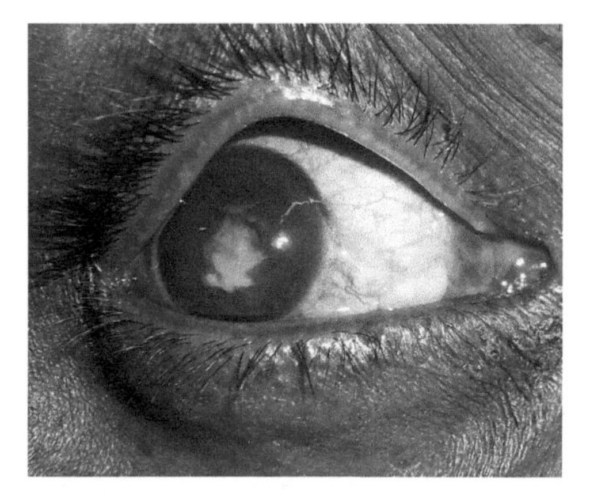

Figure 3. Fungal corneal ulcer of the right eye showing feathery (hyphate) margins with satellite lesions.

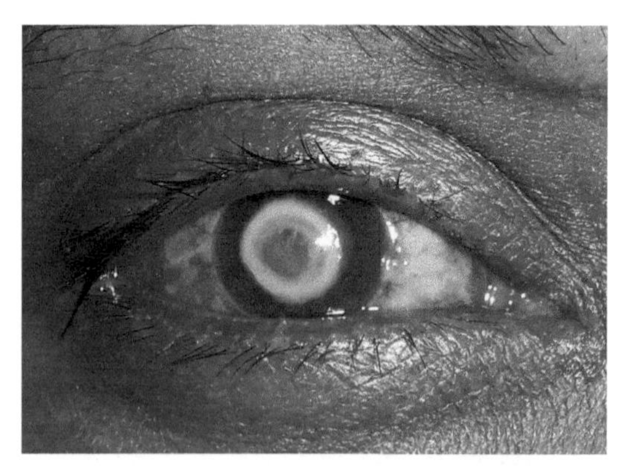

Figure 4. Fungal corneal ulcer of the right eye showing ring shaped ulcer with feathery margins.

and fungal infection necessitating treatment against both types of organisms. Chronic, severe filamentous fungal keratitis can mimic bacterial suppuration and may involve the whole cornea (Thomas 2003). Stromal keratitis by *Candida* species presents with a collar button configuration and is often more localized than filamentary mycotic keratitis. These ulcers are usually small with expanding infiltrate and are associated with chronic debilitating ocular conditions.

Laboratory Diagnosis

In suspected fungal infections of the cornea, thorough microbiological investigations should be performed. The material for microscopy or culture is obtained by scraping the base and edge of the ulcer several times with a No. 15 Bard Parker blade or sharpened Kimura platinum spatula (Abad and Foster 2000). This also reduces the fungal load and aids in the treatment but care should be taken not to perforate the cornea. It may be difficult to obtain fungal scrapes when the epithelial defect is small and fungal proliferation occurs in deep stroma. In that case corneal biopsy can be undertaken where partial thickness trephination is done similar to performing a lamellar keratoplasty (Alexandrakis et al. 2000). The depth and diameter of the trephination should be sufficient to obtain the offending agent. Corneal material can also be collected at the time of performing penetrating keratoplasty (Vemuganti et al. 2002). There are several methods for direct microscopic examinations of the corneal scrapings which allow rapid presumptive diagnosis of fungal keratitis (Rao 1989). A wet preparation of 10% potassium hydroxide (10% KOH) of corneal smear under the microscope can detect fungal hyphal elements or yeast form. Gram staining has a sensitivity of 45% to 73% and stains yeast cells and fungal hyphae equally well. It can also detect bacteria in case of mixed infection. Special fungal stains like lactophenol cotton blue (LPCB), Gomori methenamine silver (GMS), periodic acid Schiff (PAS) or calcofluor white(CW) can also be used to detect fungal elements in thinly spread out direct corneal smears with a sensitivity ranging from 80 to 90% (Thomas et al. 1991; Thomas 1998) [Table 1].

Corneal materials for culture are inoculated onto different solid agar plates (blood agar, Sabouraud glucose-neopeptone agar, brain-heart infusion agar, cystine tryptone agar) in the form of several 'C' streaks. Only growths in the 'C' streaks are considered significant (Thomas 2003). Some corneal material is also inoculated in liquid broth media (Sabouraud broth, brain-heart infusion broth, thioglycollate broth). The contaminant growth of bacteria can be prevented by addition of antibacterials (gentamicin, chloramphenicol). If mixed bacterial and fungal infections are suspected bacterial cultures are done separately. Aerobical incubations were done for all inoculated media (Srinibasan et al. 1997). The inoculated Sabouraud's dextrose agar media were incubated at 27°C and after daily examination they were discarded after 3 weeks, if there was no growth. The blood agar, chocolate agar, thioglycolate broth and brain-heart infusion broth after inoculation were incubated at 37°C, examined daily and discarded if no growth was seen in 7 days. The growth of fungi in culture is considered significant if it correlates with clinical presentation, if growth of the same fungus is detected on two or more solid media, if there is semi confluent growth at the site of inoculation on one solid medium, growth in the liquid media and growth

Table 1. Types of stain required for direct microscopic visualization of fungus with their special features and drawbacks.

1. Potassium hydroxide (KOH) wet mounts	i) KOH causes good digestion of thick corneal samples ii) Added use of ink, periodic acid-Schiff (PAS) or acridine orange (AO) provides good contrast of fungal structures against background iii) Inexpensive iv) Sensitivity of 75–90%	i) Artifacts are common ii) Swelling of corneal cells may be inadequate to provide transparent preparations iii) long viewing time (12–18 h) for ink-KOH mounts iv) Ultraviolet UV microscope needed when AO counterstain is used. Viewer should have protection against UV light
2. Gram staining	i) Yeast cells and fungal hyphae are stained equally well. Also, bacteria can be differentiated if there is mixed infection ii) Takes only 5 min to perform iii) Sensitivity of 45–73%	i) Sometimes irregular stain of fungal hyphae causes difficulty in diagnosis ii) Thick preparations stain poorly iii) False-positive artifacts are common iv) Crystal violet precipitates create confusion
3. Lactophenol cotton blue (LPCB)	i) All common ocular fungi and *Acanthamoeba* cysts are detected by this simple, inexpensive one-step method ii) The stain has long shelf-life and can be kept for years iii) Sensitivity of 70–80%	i) As there is no tissue digestion, thick corneal preparations may be unsuitable for staining ii) There may be insufficient contrast between fungi and background iii) Unusual fungi are difficult to detect
4. Gomori Methenamine silver (GMS) staining (modified)	i) Fungal cell walls and septa are clearly delineated against pale green background ii) *Acanthamoeba* cysts and *Pneumocystis carinii* are also detected, if present iii) Sensitivity of 89%	i) The procedure has multiple steps and takes about 60 min to perform ii) Excessive deposition of silver may obscure details of the fungi and the background iii) false-positive results may occur due to staining of cellular debris and melanin
5. Calcofluor white (CW)	i) Fungal hyphae as well as yeast cells are clearly delineated against dark background even in thick smears ii) *Acanthamoeba* cysts and *Pneumocystis carinii* can also be detected iii) Sensitivity of 80–90%	i) UV microscope is needed ii) All reagents are to be used fresh to avoid false-positive artifacts iii) Reagents and procedures require standardization

consistent with the direct microscopic observations as in KOH preparation or Gram stain (Srinibasan et al. 1997; Sharma and Athmanathan 2002).

Cultures from corneal biopsy specimens can yield better results than corneal scrapes particularly when there is active fungal growth in deep stroma with small epithelial defect. Direct microscopical examination of corneal biopsy materials also produces better results than corneal scrapes when stained by PAS, GMS or CW stains (Vemuganti et al. 2002).

Patients who are not willing to have a corneal biopsy can undergo an impression debridement procedure (Arora and Singhvi 1994). A cellulose acetate filter paper as used in conjunctival impression cytology is gently applied over the corneal ulcer to obtain a replica which is then stained and examined under microscope. The laboratory diagnosis of fungal keratitis is sometimes very difficult due to the very small sample obtained by scraping the corneal ulcer. Alternative methods for the identification of fungal element include immunofluorescence staining, electron microscopy, and confocal microscopy. Confocal microscopy can help in diagnosing early stages of fungal keratitis and in monitoring progress of the disease at the edges and in deep stroma (Winchester et al. 1997). It may be helpful in taking timely decision for keratoplasty and in determining the duration of medication (Vaddavalli et al. 2011).

Molecular diagnosis of fungal keratitis depends upon DNA detection of particular fungus by polymerase chain reaction (PCR) in corneal scrapes (Ferrer and Alio 2011; Zhao et al. 2014). PCR is the enzymatic exponential amplification of specific target region using short primers resulting in detectable amounts of amplified DNA from one or a few original sequences. In real time PCR, a less time consuming method, the PCR products are measured automatically during each cycle in a closed tube format using an integrated fluorimeter (Atkins and Clark 2004). The target gene (5.8S rRNA gene) in Internal transcribed spacer (ITS) region is amplified by PCR and semi nested PCR to detect fungal DNA (Kuo et al. 2012). A comparative study was done between a PCR based assay to amplify a part of the fungal 18S ribosomal RNA gene and conventional culture from corneal scrapes of 30 patients with presumed fungal keratitis (Gaudio et al. 2002). PCR and fungal cultures matched in 74% cases. When compared to culture the sensitivity of the PCR was 94% but specificity was only 50% (Gaudio et al. 2002). The suggested way to achieve species-level diagnosis is culture of the fungi combined with ITS sequence-based molecular identification (Tsui et al. 2011).

The Clinical and Laboratory Standards Institute [CLSI; formerly called the National Committee for Clinical Laboratory Standards (NCCLS)] has developed standardized methods for *in vitro* susceptibility testing of antifungal agents for filamentous fungi and yeasts (Badiee et al. 2012). These methods include variables in inoculum preparation size, duration of incubation, medium, temperature and minimum inhibitory concentration (MIC) endpoint determination. The Etest is an agar dilution-diffusion method that has been used alternatively to evaluate the susceptibility of fungi to antifungal drugs (Badiee et al. 2012). This method is faster and simpler than the CLSI method for routine use in laboratories and has comparable results to the CLSI method (Matar et al. 2003; Badiee et al. 2012). Lalitha et al. (2014) studied the minimum inhibitory concentration of fungal isolates to natamycin and voriconazole in 323 cases of fungal keratitis and observed that *Fusarium* isolates were least susceptible to voriconazole and *A. flavus* isolates were least susceptible to natamycin as compared to other filamentous fungi. Sun et al (2014) investigated the association between *in vitro* susceptibility to natamycin and voriconazole and clinical outcomes in mycotic keratitis. They observed that the susceptibility to natamycin was associated with increased infiltrate or scar size and increased odds of perforation. However, they found no association between susceptibility to voriconazole and disease outcome.

Treatment

Medical treatment

The currently available antifungal agents are mostly fungistatic (except amphotericin B which is fungicidal in high doses), therefore, prolonged course of therapy and an immunocompetent host are necessary to combat fungal infections of the cornea. Most antifungals except natamycin were developed for systemic mycoses and then subsequently used for fungal keratitis.

Polyenes

Polyene antibiotics (amphotericin B, natamycin) bind ergosterol to the fungal cell membrane and create pores that disrupt the homeostasis of the organism leading to death.

Amphotericin B is the drug of choice for *Candida* keratitis and is also effective in therapy of filamentous fungal keratitis (Tanure et al. 2000; Thomas 2003). It is prepared as 0.15% concentration in distilled water and used every 15 to 30 minutes for first 24 to 48 hours. Daily scraping of the ulcer aids in drug penetration and decreases the fungal load.

Natamycin, another polyene antibiotic, is used as 5% suspension. It is the first agent of choice against filamentous fungi like *Aspergillus* and *Fusarium* (Prajna et al. 2013). It is also effective against yeasts. The drug is water insoluble and penetrates ocular tissue poorly rendering it ineffective in stromal infections. Natamycin and amphotericin B can be combined to be used alternately on hourly basis against infections not responding to a single drug (McLeod 2009).

Fluorinated pyrimidines

Flucytosine is a pyrimidine antifungal agent and is converted into a thymidine analog within fungal cells that blocks fungal thymidine kinase. It is not used alone due to formation of resistance. The drug is synergistic with amphotericin B and used topically (1% solution) and systemically (150 mg/kg/day orally in four divided doses) in combination with amphotericin B against candidal keratitis (Abad and Foster 2000).

Imidazoles and Triazoles

Imidazoles (like clotrimazole, miconazole, ketoconazole) and triazoles (like fluconazole, itraconazole) inhibit sterol 14-α-demethylase, a microsomal P-450-

related enzyme thus impairing biosynthesis of ergosterol for cytoplasmic membranes. Ketoconazole when used topically or systemically (in adults, 200 mg oral tablet once daily for 14 days) achieves adequate level in cornea and is effective against both yeast and filamentous fungi. The drug can be used orally in combination with topical natamycin in deep fungal stromal keratitis. It has systemic side effects like transient elevation of liver enzymes, blocking of hepatic microsomal systems, decreased libido and menstrual irregularities.

Fluconazole is a triazole with fewer side effects than ketoconazole. It has adequate corneal penetration following topical administration and also achieves adequate therapeutic concentration in cornea when taken orally. It is effective when used topically in candidal keratitis and orally in combination with natamycin against filamentous keratitis. But, most of the filamentous fungi are resistant to fluconazole when used alone (Manikandan et al. 2013).

Newer triazole derivatives (voriconazole, posaconazole or ravuconazole) have potent broad spectrum antifungal activity. Voriconazole 1% solution is used for local instillation, every hour during the day and every 2 hours during the night. Recently, Prajna et al. (2013) published a report of multicentric randomized trial comparing topical natamycin with voriconazole in the treatment of filamentous fungal keratitis. Natamycin treatment was found to be associated with significantly better clinical and microbiological outcomes than voriconazole treatment for smear positive filamentous fungal keratitis, particularly against *Fusarium* cases. Sharma et al. (2013) compared the efficacy of topical versus intrastromal voriconazole as a combination to topical natamycin in recalcitrant fungal keratitis. They concluded that topical voriconazole was a useful adjunct to natamycin in fungal keratitis not responding to topical natamycin alone. Intrastromal voriconazole had no beneficial effect over topical therapy.

Other Antifungal Agents

Topical silver sulfadiazine which is usually used for antibiotic prophylaxis in burn patients can be used against fungal keratitis. In a study of 110 patients in India it was detected to be effective when used five times daily in 76.3% cases (Mohan et al. 1988a). In another study it was found more effective than miconazole (80% and 55% respectively) (Mohan et al. 1988b).

After the diagnosis of the particular fungal element as the offending agent, treatment is started with topical polyene antifungal medications. Initially eye drops are applied every 5 minutes for the first five doses followed by every hour during the day and every 2 hours during the night. Azoles can be added as oral preparations in case of deep stromal keratitis. If there is a favorable response night time application is decreased followed by a reduction in day time instillation. Table 2 summarizes the treatment modalities against commonly encountered filamentous fungi and yeasts.

Use of corticosteroids is not recommended in fungal keratitis as they aggravate the condition and enhance fungal viability. In case of fungal infection of recently done corneal grafting there is a chance of graft failure as steroids are to be avoided. Topical cyclosporine A which has anti-inflammatory as well as antifungal properties can be used alternatively in such cases.

Table 2. Treatment modalities of the frequently encountered filamentous hyaline (*Aspergillus* spp., *Fusarium* spp.) or Dematiacious (*Curvularia* spp.) fungi and yeasts (*Candida* spp.).

Type of Fungi isolated	Therapeutic agent	Route & Dosage-concentration
A) Hyaline fungi		
i) *Aspergillus* spp. ii) *Fusarium* spp.	1. Natamycin: It is the first agent of choice against filamentous fungi like *Aspergillus* and *Fusarium*.	1. Natamycin used topically as 5% suspension
	2. Amphotericin B: Second agent of choice against filamentous fungi. Can be used in combination with natamycin.	2. Amphotericin B: It is prepared as 0.15% concentration in distilled water and used every 15 to 30 minutes for first 24 to 48 hours. Natamycin and amphotericin B can be combined to use alternately on hourly basis against infections not responding to a single drug. Intra vitreal Amphotericin B (5 mcg/0.1 ml) is used in fungal endophthalmitis.
	3. Ketoconazole when used orally achieves adequate level in cornea and is effective against both filamentous fungi and yeast. The drug can be used orally in combination with topical natamycin in deep fungal stromal keratitis.	3. Ketoconazole: Orally 200 mg tablet once daily for 14 days in adults.
	4. Voriconazole 1%: Newer triazole derivative with potent broad spectrum antifungal activity. Effective in combination with natamycin in recalcitrant Fusarium keratitis not responding to natamycin alone.	4. Voriconazole: 1% solution is used for local instillation, every hour during the day and every 2 hours during the night.
	5. Fluconazole is effective when used topically in candidal keratitis and orally in combination with natamycin against filamentary keratitis.	5. Fluconazole: Topically as 1–2% solution. Orally as 200 mg/day.
B) Dematiacious fungi		
Curvularia spp.	1. Natamycin: used alone or combined with oral Ketoconazole or topical Amphotericin B against Dematicious fungi.	1.Natamycin used topically as 5% suspension.
	2. Ketoconazole: The drug can be used orally in combination with topical natamycin in deep fungal stromal keratitis.	2. Ketoconazole: Orally 200 mg tablet once daily for 14 days in adults.
	3. Voriconazole.	3. Voriconazole: 1% solution is used for local instillation.
C) *Candida* spp.	1. Amphotericin B is the drug of choice for the candida keratitis.	1. Amphotericin B: It is prepared as 0.15% concentration in distilled water and used every 15 to 30 minutes for first 24 to 48 hours. Intra vitreal Amphotericin B (5 mcg/0.1 ml) is used in Candida endophthalmitis.

Table 2. contd....

Table 2. contd.

Type of Fungi isolated	Therapeutic agent	Route & Dosage-concentration
	2. Natamycin: It is also effective against yeasts. 3. Flucytosine is synergistic with Amphotericin B and used topically and systemically in combination with Amphotericin B against candidal keratitis. 4. Voriconazole 1%: Potent broad spectrum antifungal activity. 5. Fluconazole is effective when used topically in candidal keratitis.	2. Natamycin used topically as 5% suspension. 3. Flucytosine is used topically (1% solution) and systemically (150 mg/kg/day orally in four divided doses). 4. Voriconazole: 1% solution is used for local instillation. 5. Fluconazole: Topically as 1–2% solution.

Vajpayee et al. (2014) studied the effect of corneal collagen cross-linking as an additional therapy in mycotic keratitis. But they did not find any advantage over medical management in cases with moderate mycotic keratitis.

The course of medical treatment is protracted and is usually maintained for about 12 weeks with careful monitoring of the response and subsequent tapering of the doses. Liver function tests are to be performed when ketoconazole or fluconazole is taken orally. The improvement of the disease process can be detected by diminution of pain, decrease in size of infiltrate, disappearance of satellite lesions and rounding out of the feathery margins of the ulcer in the region of healing fungal lesions. There may be toxicity of antifungal agents in the form of conjunctival chemosis and punctuate epithelial keratopathy.

Surgical treatment

In case of fungal keratitis not responding to medical treatment, surgical intervention should be considered. Fungal elements can pierce the Descemet's membrane and enter in to the anterior chamber. To prevent such disastrous consequences, penetrating keratoplasty should be undertaken to completely remove the fungal load. About 30% of fungal keratitis cases do not respond to antifungal therapy and require penetrating keratoplasty (Xie et al. 2001). It is necessary to make the graft margin free of infection so the decision of the surgery should be taken early before the ulcer involves whole of the cornea. When removing the infected corneal button there should be as much clear margin as is possible. Otherwise there will be a risk of endophthalmitis where the fungal infection invades the intraocular contents with very poor visual prognosis. Furthermore, any intraocular surgery during active fungal keratitis will increase the chance of intraocular infection. Lamellar keratoplasty can be considered when the infection is limited to the anterior stroma. However in case of doubt, penetrating keratoplasty should be preferred. Penetrating keratoplasty with a good anterior and posterior chamber wash, extracapsular lens extraction if the lens is involved, postoperative systemic antifungal therapy and full dilatation of pupils is usually recommended; a more invasive procedure like anterior vitrectomy may not be needed (Kuriakose and Thomas 1991).

Prognosis

The outcome of mycotic keratitis depends on the complex interplay of agent (invasiveness, toxigenicity and resistance to drugs), host (inflammatory response, hypersensitivity reactions) and predisposing factors (Thomas 2003). Rapid progression of mycotic keratitis in the early phases is thought to be mainly by agent factors like large fungal inoculum and deep penetration into the corneal stroma whereas progression in the later phases is due to a combination of agent and host factors and resistance to antifungals (Vemuganti et al. 2002).

Patients with deep stromal infections particularly with steroid use carry poor prognosis. A delay in diagnosis, inadequate treatment or failure to do penetrating keratoplasty in optimum time can lead to potential blinding consequences like corneal perforation, scleritis, endophthalmitis and panophthalmitis (Abad and Foster 2000).

Conclusions

Mycotic keratitis continues to be a significant cause of blindness particularly in the developing countries. Although, newer antifungal drugs like voriconazole, posaconazole or ravuconazole are giving promising results, therapeutic inadequacy still persists necessitating future developments. Currently Etest and CLSI methods are being used for *in vitro* susceptibility testing of antifungal agents with optimum results but have limited availability. Newer improved standardization methods need to be undertaken so that *in vitro* antifungal sensitivities correspond with *in vivo* clinical response. Additionally, these standardized methods should be widely available particularly in the developing countries.

Fundamental researches are now being undertaken to study the role of IL-1R1/MyD88-dependent CXC chemokines in corneal inflammation and tissue destruction. Further studies will have to focus on the role of C-type lectins and additional host and fungal virulence factors in disease pathogenesis. Future treatment modalities should incorporate regimens to control such cornea destroying factors apart from the usual antifungal measures to improve the final outcome.

References

Abad, J.C. and C.S. Foster. 2000. Fungal keratitis. pp. 906–914. *In*: Albert, Daniel, M. and Frederick A. Jakobiec (eds.). Principals and Practice of Ophthalmology, Vol. 2, 2nd ed. W.B. Saunders Company, Philadelphia, Pennsylvania.

Agarwal, P.K., P. Roy, A. Das, A. Banerjee, P.K. Maity and A.R. Banerjee. 2001. Efficacy of topical and systemic itraconazole as a broad-spectrum antifungal agent in mycotic corneal ulcer: A preliminary study. Indian J Ophthalmol. 49: 173–76.

Agarwal, V., J. Biswas, H.N. Madhaban, G. Mangat, M.K. Reddy, J.S. Saini, S. Sharma and M. Srinivasan. 1994. Current perspectives in infectious keratitis. Indian J Ophthalmol. 42: 171–91.

Alexandrakis, G., R. Haimovici, D. Miller and E.C. Alfonso. 2000. Corneal biopsy in the management of progressive microbial keratitis. Am J Ophthalmol. 129: 571–576.

Arffa, R.C. 1991. Infectious keratitis: fungal and parasitic. Grayson's diseases of the cornea, 3rd ed. St Louis, Mosby Year Book 1991: 199–223.

Arora, I. and S. Singhvi. 1994. Impression debridement of corneal lesions. Ophthalmology 101: 1935–40.

Atkins, S. and I. Clark. 2004. Fungal molecular diagnostics: a mini review. J. Appl. Genet. 45: 3–15.

Badiee, P., A. Alborzi, M. Moeini, P. Haddadi, S. Farshad, A. Japoni and M. Ziyaeyan. 2012. Antifungal susceptibility of the aspergillus species by Etest and CLSI reference methods. Arch Iran Med. 15(7): 429–432.

Bandyopadhyay, S., D. Das, K.K. Mondal, A.K. Ghanta, S.K. Purkait and R. Bhaskar. 2012. Epidemiology and laboratory diagnosis of fungal corneal ulcer in the Sunderban region of West Bengal, eastern India. Nep J Oph. 4: 29–36.

Baranyia, N., S. Kocsubé, A. Szekeres, A. Raghavan, V. Narendran, C. Vágvölgyi, K.P. Selvam, Singh Y.R. Babu, L. Kredics, J. Varga and P. Manikandan. 2013. Keratitis caused by Aspergillus pseudotamarii. Med Mycol Case Rep. 2: 91–94.

Basak, S.K., S. Basak, A. Mohanta and A. Bhowmick. 2005. Epidemiological and microbiological diagnosis of suppurative keratitis in Gangetic West Bengal, eastern India. Indian J Ophthalmol. 53: 17–22.

Bharathi, M.J., R. Ramakrishnan, S. Vasu, R. Meenakshi and R. Palaniappan. 2003. Epidemiological characteristics and laboratory diagnosis of fungal keratitis: a three-year study. Indian J Ophthalmol. 51: 315–21.

Despande, S.D. and G.V. Koppikar. 1999. A study of mycotic keratitis in Mumbai. Indian J Pathol Microbiol. 42: 81–87.

Ferrer, C. and J.L. Alio. 2011.Evaluation of molecular diagnosis in fungal keratitis. Ten years of experience. J .Ophthalmic. Inflamm. Infect. 1: 15–22.

Garg, P., S. Mahesh, A.K. Bansal, U. Gopinathan and G.N. Rao. 2003. Fungal infection of sutureless self-sealing incision for cataract surgery. *Ophthalmology* 110: 2173–7.

Gaudio, P.A., U. Gopinathan, V. Sangwan and T.E. Hughes. 2002. Polymerase chain reaction based detection of fungi in infected corneas. Br J Ophthalmol. 86: 755–760.

Gilbert, C.E., M. Wood, K. Wadel and A. Foster. 1995. Causes of childhood blindness in East Africa; results in 491 pupils attending 17 school for the blind in Malawi, Kenya and Uganda. Ophthalmic Epidemiol. 2: 77–84.

Gopinathan, U., P. Garg, M. Fernandes, S. Sharma, S. Atmanathan and G.N. Rao. 2002. The epidemiological features and laboratory results of fungal keratitis: A 10-year review at a referral eye care center in south India. Cornea 21: 555–59.

Hagan, M., E. Wright, M. Newman, P. Dolin and G. Johnson. 1995. Causes of suppurative keratitis in Ghana. Br J Ophthalmol. 79: 1024–1028.

Houang, E., D. Lam, D. Fan and D. Seal. 2001. Microbial keratitis in Hong Kong: relationship with climate, environment and contact lens-disinfection. Trans R Soc Trap Med Hyg. 95: 361–67.

Insan, N.G., Mane V. Vijay, B.L. Chaudhary, M.S. Danu, A. Yadav and V. Srivastava. 2013. A review of fungal keratitis: etiology and laboratory diagnosis. Int J Curr Microbiol App Sci. 2: 307–314.

Kredics, L., J. Varga, S. Kocsubé, I. Doćzi I., Robert A. Samson, R. Rajaraman, V. Narendran, M. Bhaskar, C. Vágvölgyi and P. Manikandan. 2007. Case of keratitis caused by Aspergillus tamari. J Clin Microbiol. 45: 3464–3467.

Kredics, L., J. Varga, S. Kocsubé, R. Revathi, R. Anita and I. Dóczi. 2009. Infectious keratitis caused by Aspergillus tubingensis Cornea. 28: 951–954.

Kuo, I.C., T.P. Margolis, V. Cevallos and D.G. Kwang. 2001. Aspergillus fumigatus keratitis after laser *in situ* keratomileusis. Cornea. 20: 342–344.

Kuo, M.T., H.C. Chang and C.K. Cheng. 2012. A highly sensitive method for molecular diagnosis of fungal keratitis: a dot hybridization assay. Ophthalmol. 119: 2434–42.

Kuriakose, T. and P.A. Thomas. 1991. Keratomycotic malignant glaucoma. Indian J Ophthalmol. 39: 118–121.

Lalitha, P., C.Q. Sun, N.V. Prajna, R. Karpagam, M. Geetha, K.S. O'Brien, V. Cevallos, S.D. McLeod, N.R. Acharya, T.M. Lietman and Mycotic Ulcer Treatment Trial Group. 2014. *In vitro* susceptibility of filamentous fungal isolates from a corneal ulcer clinical trial. Am J Ophthalmol. 157: 318–26.

Leck, A.K., P.A. Thomas, M. Hagan, J. Kaliamurthy, E. Ackuaku, John M. Newman M.J., F.S. Codjoe, J.A. Opintan, C.M. Kalavathy, V. Essuman, C.A.N. Jesudasan and G.J. Johnson. 2002. Aetiology of suppurative corneal ulcers in Ghana and south India, and epidemiology of fungal keratitis. Br J Ophthalmol. 86: 1211–15.

Liesegang, T.J. and R.K. Forstor. 1980. Spectrum of microbial keratitis in south Florida. Am J Ophthalmol. 90: 38–47.

Manikandan, P., J. Varga, S. Kocsubé, Robert A. Samson, R. Anita, R. Revathi, I. Doćzi, T.M. Ne´meth, V. Narendran, C. Vágvölgyi, C. Manoharan and L. Kredics. 2009. Mycotic keratitis due to *Aspergillus nomius*. J Clin Microbiol. 47: 3382–3385.

Manikandan, P., J. Varga, S. Kocsubé, R. Revathi, R. Anita, I. Dóczi, R.A. Samson and R. Rajaraman. 2010. Keratitis caused by the recently described new species Aspergillus brasiliensis: two case reports. J Med Case Rep. 4: 68.

Manikandan, P., J. Varga, S. Kocsubé, R. Anita, R. Revathi, T.M. Németh, V. Narendran, C. Vágvölgyi, S.K. Panneer, C.S. Shobana, Singh Y.R. Babu and L. Kredics. 2013. Epidemiology of Aspergillus keratitis at a tertiary care eye hospital in South India and antifungal susceptibilities of the causative agents. Mycoses 56: 26–33.

Matar, M.J., L. Ostrosky-Zeichner, V.L. Paetznick, J.R. Rodriguez, E. Chen and J.H. Rex. 2003. Correlation between E-Test, disk diffusion, and microdilution methods for antifungal susceptibility testing of fluconazole and voriconazole. Antimicrob Agents Chemother. 47(5): 1647–1651.

McGinis, M.R. 1980. Laboratory Handbook of Medical Mycology. Academic Press, New York.

McLeod, S.D. 2009. Fungal keratitis. pp. 271–273. *In*: Yanoff, M. and J.S. Duker (eds.). Ophthalmology, 3rd ed. Mosby, St. Louis.

Mino de Kasper, H., G. Zoulek, M.E. Paredes, R. Alborno, D. Medina, M. Centurian de Morinigo, M. Ortiz de fresco and F. Aguero. 1991. Mycotic keratitis in Paraguay. Mycoses 34: 251–54.

Mohan, M., S.K. Gupta, V.K. Kalra, R.B. Vajpayee and M.S. Sachdev. 1988a. Topical silver sulphadiazine: a new drug for ocular keratomycosis. Br J Ophthalmol. 72: 192–95.

Mohan, M., S.K. Gupta and R.B. Vajpayee. 1988b. Management of keratomycosis with 1% silver sulfadiazine: A prospective controlled clinical trial in 110 cases. pp. 495–498. *In*: Cavanagh, H.D. (ed.). The Cornea: Transactions of the World Congress on the Cornea III. Raven, New York.

Nelson, P.D., T.A. Toussoun and W.F.O. Marasas. 1983. Fusarium Species: An Illustrated Manual for Identification. The Pennsylvania State University Press, University Park.

Nielsen, S.E., E. Nielsen, H.O. Julian, J. Lindegaard, K. Højgaard, A. Ivarsen, J. Hjortdal and S. Heegaard. 2014. Incidence and clinical characteristics of fungal keratitis in a Danish population from 2000 to 2013. Acta Ophthalmol. 2014 May 19. doi: 10.1111/aos.12440 [Epub ahead of print].

O'Day, D.M. 1996. Fungal keratitis. pp. 1048–1061. *In*: Pepose, J.S., G.N. Holland and K.R. Wilhelmus (eds.). Ocular Infection and Immunity, Mosby Year Book, St. Louis.

O'Day, D.M. and E.M. Burd. 1996. Fungal keratitis and conjunctivitis: mycology. pp. 229–239. *In*: Smolin, G and R.A. Thoft (eds.). The Cornea: Scientific Foundations and Clinical Practice, 3rd edn. Little, Brown & Co, Boston.

Prajna, N.V., T. Krishnan, J. Mascarenhas, R. Rajaraman, L. Prajna, M. Srinivasan, A. Raghavan, M.D., C.E. Oldenburg, K.J. Ray, M.E. Zegans, S.D. McLeod, T.C. Porco, N.R. Acharya and T.M. Lietman. 2013. The mycotic ulcer treatment trial: a randomized trial comparing natamycin vs. voriconazole. JAMA Ophthalmol. 131: 422–429.

Punia, R.S., R. Kundu, J. Chander, S.K. Arya, U. Handa and H. Mohan. 2014. Spectrum of fungal keratitis: clinicopathologic study of 44 cases. Int J Ophthalmol. 7: 114–7.

Rao, N.A. 1989. A laboratory approach to rapid diagnosis of ocular infections and prospects for the future. Am J Ophthalmol. 107: 283–291.

Resnikoff, S., D. Pascolini, D. Etya'ale, I. Kocur, R. Pararajasegaram, G.P. Pokharel and S.P. Mariotti. 2004. Global data on visual impairment in the year 2002. WHO Bulletin. 2004. Available at http://whqlibdoc.who.int/bulletin/2004/Vol82-No11/bulletin_2004_82(11)_844-851.pdf?ua=1.

Rosa, R.H., D. Miller and E.C. Alfonso. 1994. The changing spectrum of fungal keratitis in South Florida. Ophthalmology 101: 1005–1013.

Sharma, N., J. Chacko, T. Velpandian, J.S. Titiyal, R. Sinha, G. Satpathy, R. Tandon and R.B. Vajpayee. 2013. Comparative evaluation of topical versus intrastromal voriconazole as an adjunct to natamycin in recalcitrant fungal keratitis. Ophthalmology 120: 677–81.

Sharma, S., Athmanathan. 2002. Diagnostic procedures in infectious keratitis. pp 232–253. *In*: Nema, H.V. and N. Nema (eds.). Diagnostic Procedures in Ophthalmology. Jaypee Brothers Medical Publishers, New Delhi.

Sharma, S., M. Srinivasan and C. George. 1993. The current status of Fusarium species in mycotic keratitis in south India. Indian J Med Microbiol. 11: 140–147.

Srinibasan, M., C.A. Gonzales, C. George, V. Cevallos, J.M. Mascarenhas, B. Asokan, J. Wilkins, G. Smolin and J.P. Whitcher. 1997. Epidemiology and aetiological diagnosis of corneal ulceration in Madurai, south India. Br J Ophthalmol. 81: 965–71.

Srinivasan, R., R. Kanungo and J.L. Goyal. 1991. Spectrum of oculomycosis in South India. Acta Ophthalmol. 69: 744–749.

Sun, C.Q., P. Lalitha, NV. Prajna, R. Karpagam, M. Geetha, K.S. O'Brien, C.E. Oldenburg, K.J. Ray, S.D. McLeod, N.R. Acharya, T.M. Lietman and Mycotic Ulcer Treatment Trial Group. 2014. Association between in vitro susceptibility to natamycin and voriconazole and clinical outcomes in fungal keratitis. Ophthalmology 121: 1495–500.

Tanure, M.A., E.J. Cohen, S. Grewal, C.J. Rapuano and P.R. Laibson. 2000. Spectrum of fungal keratitis at Wills Eye Hospital, Philadelphia, Pennsylvania. Cornea 19: 307–312.

Tarabishy, A.B., B. Aldabagh, Y. Sun, Y. Imamura, P.K. Mukherjee, J.H. Lass, M.A. Ghannoum and E. Pearlman. 2008. MyD88 Regulation of fusarium keratitis is dependent on TLR4 and IL-1R1 but not TLR21. J Immunol. 181: 593–600.

Thomas, P.A. 1998. Tropical ophthalmomycoses. pp. 121–142. *In*: Seal, D.V., A.J. Bron and J. Hay (eds.). Ocular Infection: Investigation and Treatment in Practice. Martin Dunitz, London.

Thomas, P.A. 2003. Fungal infections of the cornea. Eye 17: 852–862.

Thomas, P.A., T. Kuriakose, M.P. Kirupashanker and V.S. Maharajan. 1991. Use of Lactophenol cotton blue mounts of corneal scrapings as an aid to the diagnosis of mycotic keratitis. Diagn Microbiol Infect Dis. 14: 219–224.

Thygeson, P. and M. Okumoto. 1974. Keratomycosis: a preventable disease. Trans. AM Acad Ophthalmol Otolaryngol. 78: 433–439.

Tsui, C.K.M., J. Woodhall, W. Chen, C.A. Lévesque, A. Lau, C.D. Schoen, C. Baschien, M.J. Najafzadeh and G. Sybren de Hoog. 2011. Molecular techniques for pathogen identification and fungus detection in the environment. IMA Fungus 2: 177–189.

Upadhyay, M.P., P.C. Karmacharya, S. Koirala, N.R. Tuladhar, L.E. Bryan, G. Smolin and J.P. Whitcher. 1991. Epidemiological characteristics, predisposing factors and etiologic diagnosis of corneal ulceration in Nepal. Am J Ophthalmol. 111: 92–99.

Vaddavalli, P.K., P. Garg, S. Sharma, V.S. Sangwan, G.N. Rao and R. Thomas. 2011. Role of confocal microscopy in the diagnosis of fungal and acanthamoeba keratitis. *Ophthalmology* 118: 29–35.

Vajpayee, R.B., S.N. Shafi, P.K. Maharana, N. Sharma and V. Jhanji. 2014. Evaluation of corneal collagen cross-linking as an additional therapy in mycotic keratitis. Clin Experiment Ophthalmol. 2014 Jul 28. doi: 10.1111/ceo.12399 [Epub ahead of print].

Varga, J., S. Kocsubé', B. Tóth, J.C. Frisvad, G. Perrone, A. Susca, M. Meijer and R.A. Samson. 2007. Aspergillus brasiliensis sp. nov., a biseriate black Aspergillus species with world-wide distribution. Int J Syst Evol Micr. 57:1925–1932.

Varga, J., J.C. Frisvad and R.A. Samson. 2011. Two new aflatoxin producing species, and an overview of Aspergillus section Flavi. Stud Mycol. 69: 57–80.

Vemuganti, G.K., P. Garg, U. Gopinathan, T.J. Naduvilath, R.K. John, Buddi R. and G.N. Rao. 2002. Evaluation of agent and host factors in progression of mycotic keratitis: a histologic and microbiologic study of 167 corneal buttons. Ophthalmology 109: 1538–1546.

Venugopal, P.L., T.L. Venugopal, A. Gomathi, E.S. Ramkrishna and S. Ilavarasi. 1989. Mycotic keratitis in Madras. Indian J Pathol Microbiol. 32: 190–97.

Whitcher, J.P., M. Srinibasan and M.P. Upadhyay. 2001. Corneal blindness: a global perspective. Bull World Health Organ. 79: 214–21.

Williams, G., F. Billson, R. Husain, S.A. Howlader, N. Islam and K. McCellan. 1987. Microbiological diagnosis of suppurative keratitis in Bangladesh. Br J Ophthalmol. 71: 315–21.

Winchester, K., W.D. Mathers and J.E. Sutphin. 1997. Diagnosis of Aspergillus keratitis *in vivo* with confocal microscopy. Cornea 16: 27–31.

Wong, T.Y., K.S. Fong and D.T.H. Tan. 1997. Clinical and microbiological spectrum of fungal keratitis in Singapore: a 5-year retrospective study. Int Ophthalmol. 21: 127–30.

Xie, L., X. Dong and W. Shi. 2001. Treatment of fungal keratitis by penetrating keratoplasty. Br J Ophthalmol. 85: 1070–1074.
Zhao, G., H. Zhai, Q. Yuan, S. Sun, T. Liu and L. Xie. 2014. Rapid and sensitive diagnosis of fungal keratitis with direct PCR without template DNA extraction. Clin Microbiol Infect. 2014 Jan 29. doi: 10.1111/1469-0691.12571 [Epub ahead of print].
Zhu, W.S., K. Wojdyla, K. Donlon, P.A. Thomas and H.I. Eberle. 1990. Extracellular proteases of Aspergillus flavus. Diagn Microbiol Infect Dis. 13: 491–497.

SECTION II

Emerging Mycoses Caused by Opportunistic Fungal Pathogens

CHAPTER 4

Incidence of *Candida* Species in Urinary Tract Infections and Their Control by Using Bioactive Compounds Occurring in Medicinal Plants

Vaibhav Tiwari,[1] *Mamie Hui*[2] and *Mahendra Rai*[1,*]

Introduction

There has been an abrupt increase in opportunistic fungal infections during the past two decades (Sahiner et al. 2011; Khan and Ahmad 2012; Suzuki et al. 2014). Moreover, the incidences of these infections are predominantly likely to cause infection in the urinary tract. The urinary tract consists of the kidneys, ureters, bladder and urethra. Urinary tract infection (UTI) is caused by various pathogenic organisms for example bacteria, yeast, fungi and parasites (Sobel and Vasquez 1999; de Marie 2000; Chen et al. 2002; Ghotaslou et al. 2010; Behzadi et al. 2010; Al-Mathkurthy and Abdul-Gaffar 2011). These infections are mainly caused by impairments in host defense mechanisms as a consequence of viral infections, especially the human immunodeficiency virus, hematological disorders such as different types of leukemia, organ transplants, and more intensive and aggressive medical practices. Many clinical procedures and

[1] Department of Biotechnology, S.G.B. Amravati University, Amravati – 444 602, Maharashtra, India.
[2] Department of Microbiology, Faculty of Medicine, The Chinese University of Hong Kong.
* Corresponding author: mkrai123@rediffmail.com

treatments, such as surgery, the use of catheters, injections, radiation, chemotherapy, antibiotics and steroids, are risk factors for fungal infections (Soll 2000; Kojic and Darouiche 2004; Tiwari et al. 2009; Behiryet al. 2010; Fisher 2011; Sobel et al. 2011; Buonsenso and Cataladi 2012; Suzuki et al. 2014). There are specific terminologies that confine the urinary tract infection to the major structural segment involved such as urethritis (urethral infection), cystitis (bladder infection), ureter infection and pyelonephritis (kidney infection). The various risk factors associated with the urinary tract infections have been given in Table 1. However, infections in urinary tract are common with the enormous preponderance of *Candida* species as compared to the other medically important fungi such as *Aspergillus* and *Cryptococcus* species causing common clinical syndromes of urethritis, cystitis, and pyelonephritis.

Table 1. Route and risk factors involved in urinary tract infections from *Candida* species.

S.N.		Route	Risk factors
1.	**Renal Candidiasis**	Hematogenous (anterograde)	Prolonged neutropenia, intravascular drug use, burns, recent surgery, systemic infection
2.	**Lower UTI *Candida* infections**	Ascending (retrograde)	Catheterization, extremes of age, diabetes mellitus, obstruction/stasis, recent antibacterial therapy, recent bacterial UTI, urinary stent, nephrostomy tube, renal transplantation
3.	**Pyelonephritic *Candida* infections**	Ascending	Diabetes, obstruction/stasis, instrumentation, Post-operation, nephrostomy tube, ureteral stent, nephrolithiasis

Urinary Tract Infections (UTI)

UTI are more common in women as compared to men (Rifkind and Frey 1972; Tatfeng et al. 2003; Abdulhadi et al. 2008; Di Paola et al. 2011; Kim et al. 2012; Suzuki et al. 2014). This is because, in females, the urethra is much closer to the anus and is shorter than in males (urethra length is approximately 25–50 mm (1–2 inches) long in females, versus about 20 cm (8 inches) in males. Also, women lack the bacteriostatic properties of prostatic secretions. UTI causes problems ranging from dysuria (pain/burning while urinating) to organ damage and even death. Candiduria is one of the most common causes of lower urinary tract infection by yeast (Nayman et al. 2011). The frequency of infection is mostly associated with patients with catheters and other urinary tract manipulations (Kojic and Darouiche 2004; Revankar and Sobel 2010) but the infection can also occur in neonates and children of different age groups (Rao and Ali 2005; Abdulhadi et al. 2008). Acute pyelonephritis is one of the common infections in childhood. However, febrile UTI's had the highest incidence during the first year of life in both the sexes; whereas non-febrile UTI's predominantly occur in girls older than three years (Montini et al. 2011). Thus, UTI can lead to renal scars and if undiagnosed leads to permanent renal damage causing hypertension or end stage renal diseases.

Micro-organisms can reach the urinary tract by haematogenous or lymphatic spread, but there is abundant clinical and experimental evidence to show that the ascent of micro-organisms from the urethra is the most common pathway that leads to a UTI, especially organisms of enteric origin (i.e., *Escherichia coli* and other Enterobacteriaceae). This provides a rational account for the greater frequency of UTIs in women than in men and for the increased risk of infection following bladder catheterization or instrumentation. A single insertion of a catheter into the urinary bladder in ambulatory patients results in urinary infection in 1–2% of cases. Indwelling catheters with open-drainage systems result in bacteriuria in almost 100% of cases within 3–4 days. Bacteria such as *Escherichia coli, Klebsiella, Proteus, Staphylococcus, Streptococcus, Enterococcus* and *Pseudomonas* are more common to cause UTI's (Taneja et al. 2010).

Besides this, there are incidences of yeast and fungal infections in the urinary tract. These types of infections are commonly known as mycoses. By far *Candida* species are the most prominent cause of UTI's (Revankar and Sobel 2010). *Candida* causes a clinical UTI via the haematogenous route, but is also an infrequent cause of an ascending infection if an indwelling catheter is present, or following antibiotic therapy.

The Genus *Candida*

Candida belongs to the class Ascomycetes and the family Saccharomycetaceae. *Candida* is a diploid asexual and dimorphic fungus and depending upon environmental conditions can exist as unicellular yeast (blastospores and chlamydospores) as well as in different filamentous forms (hypha, pseudo hyphae) (McCullough et al. 1996; Molero et al. 1998).

Of the *Candida* species afflicting humans, *Candida albicans* is by far the most common (Jha et al. 2006; Tiwari and Rai 2008; Johnson 2009). This yeast can live as a harmless commensal in many different body locations, and is carried in almost half of the population. However, in response a change in the host environment, *C. albicans* can convert from a benign commensal into a disease causing pathogen, causing infections in the oral, gastrointestinal and genital tracts, suggesting its dimorphic nature (Kojic and Darouiche 2004).

Currently, over 40 of more than the 200 known *Candida* species have been associated in human infection, although for many years it has been recognized that only a few cause invasive infection on a regular basis (Jones 1990). These are *Candida albicans, Candida glabrata, Candida tropicalis, Candida parapsilosis, Candida krusei, Candida guilliermondii, Candida lusitaniae, Candida utilis* and *Candida kefyr* (Odds 1988; Hazen 1995; Hazen et al. 1999; Hajjeh et al. 2004; Sim et al. 2005; Pfaller and Diekema 2007; Johnson 2009; Kothavade et al. 2010; Vidigal et al. 2011). Different *Candida* species according to their pathogenesis have been listed in Table 2.

Among these listed species *C. albicans* is the most infectious. The fascinating feature of *C. albicans* is its ability to grow in two different ways; reproduction by budding, forming an ellipsoid bud, and in hyphal form, which can periodically fragment and give rise to new mycelia, or yeast-like forms (Molero et al. 1998). Switching between the two phenotypes can be due to several environmental factors such as pH or temperature or different compounds such as N-acetylglucosamine or proline. This ability to switch between the yeast and the hyphal mode of growth has been involved in its pathogenicity (Cutler 1991; Leberer et al. 1997; Pukkila-Worley et al. 2009; Pastuer et al. 2011).

Table 2. Different *Candida* species associated with human infection.

Common infectious species	Less common infectious species	Rare infectious species
Candida albicans	*Candida dubliniensis*	*Candida blankii*
Candida glabrata	*Candida famata*	*Candida bracarensis*
Candida tropicalis	*Candida inconspicua*	*Candida catenulate*
Candida parapsilosis	*Candida lipolytica*	*Candida chiropterorum*
Candida krusei	*Candida metapsilosis*	*Candida ciferri*
Candida guilliermondii	*Candida norvegensis*	*Candida eremophila*
Candida lusitaniae	*Candida orthopsilosis*	*Candida fabianii*
Candida kefyr	*Candida pelliculosa*	*Candida fermentati*
Candida utilis	*Candida rugosa*	*Candida freyschussii*
	Candida zeylanoides	*Candida haemulonii*
		Candida intermedia
		Candida lambica
		Candida magnolia
		Candida membranaefaciens
		Candida nivariensis
		Candida palmioleophila
		Candida pararugosa
		Candida pseudohaemulonii
		Candida pseudorugosa
		Candida pintolopesii
		Candida pulcherrima
		Candida thermophila
		Candida valida
		Candida viswanathii

But, in addition to the pathogenicity of *C. albicans,* the incidence of non-*albicans* species such as *C. tropicalis, C. parapsilosis, C. glabrata* has also been increased with alarming frequency (Krcmery and Barnes 2002; Roilides et al. 2003; Dixon et al. 2004; Sim et al. 2005; Nosek et al. 2009; Kothavade et al. 2010; Vidigal et al. 2011).

UTI Caused by *Candida* species

The infection caused by *Candida* species can be defined in terms of two broad categories, superficial mucocutaneous and systemic invasive, which involves the spread of *Candida* to the blood stream (candidemia) and to the major organs. Systemic candidemia is often fatal. Superficial infections affect the various mucous membrane surfaces of the body such as in oral and vaginal thrush. The incidence of vulvovaginal candidiasis (thrush) has increased approximately 2 fold in the last decade (Holland et al. 2003; Tiwari et al. 2009; Weichert et al. 2012). Approximately 75% of all women experience a clinically significant episode of vulvovaginalcadidiasis (VVC) at least once during the reproductive period. It is proposed that the infection is due to the minor changes in epithelial conditions, such as pH, altered glucose/glycogen concentration or changes in epithelial integrity. During pregnancy the risk of vaginal thrush increases, possibly due to changes in hormone production, leading to increased glycogen content in the vagina (Holland et al. 2003).

However, the source of *Candida* infections has been the subject of considerable debate. The prevalence is more in patients with impaired immune system suggesting a role for depressed cell mediated immunity in candidiasis, diabetes mellitus, AIDS and cancer patients and iatrogenic factors like antibiotic use, indwelling devices, intravenous drug use and hyper-alimentation fluids (Minari et al. 2001; Manfredi et al. 2002). The infection of *C. albicans* is mostly common in UTI's but the probability of infections of non-*albicans* species has also been increased with an alarming rate (Blinkhorn et al. 1989; Jacobs 1996; Jacobs et al. 1996; Hazen et al. 1999; Osmanagaoglu et al. 2000; François et al. 2001).

It is apparent that differences emerge when specific groups of patients or geographical locations are examined (Pfaller and Diekema 2007). In patients with hematologic malignancy and consequently much reduced immune function, some of the less pathogenic *Candida* species become more prevalent (Holzschu et al. 1979; Hazen et al. 1999; François et al. 2001; Minari et al. 2001).

In recent years, a number of ascomycetous yeast species that were formerly considered to be food yeasts or harmless commensals have emerged as significant pathogens in patients with compromised immune function (Table 2).

It is difficult to establish their true pervasiveness because many studies on epidemiology list the most prevalent *Candida* species and then combine the remaining uncommon or single isolations as *Candida* species. In some cases, this may be due to the fact that they were not definitively identified. These are known as cryptic species (Johnson 2009). *C. dubliniensis* was previously misidentified as *C. albicans* (Sullivan et al. 1995) and was also isolated and identified from HIV-infected and non HIV-infected individuals (Jabra-Rizk et al. 2000). Similarly, *C. parapsilosis* groups I, II, and III have been given formal species status as *C. parapsilosis*, *Candida orthopsilosis*, and *Candida metapsilosis*, respectively (Tavanti et al. 2005). The species *C. glabrata*, based on phenotypic identification characteristics, has been found to comprehend two new species, *Candida nivariensis* and *Candida bracarensis*. Thus, it is possible that infections due to these species are increasing together with the increase in numbers of other less common infectious *Candida* species. Hence, there is a pressing need for studies based on epidemiology and molecular characterization of isolates of these species because it may be difficult or even impossible to discriminate them from closely related species based on phenotypic characteristics alone (Johnson 2009; Tiwari et al. 2009).

Role of Biofilms in *Candida* Infections

Biofilms are highly structured, hydrated microbial communities containing sessile cells embedded in a self-produced extra-celllular polymeric matrix (containing polysaccharides, DNA and other components) (Donlan and Costertan 2002; Jin et al. 2004). In comparison to their free floating cells in suspension, sessile cells are frequently much more resistant to antimicrobial agents and this increased resistance has a considerable impact on the treatment of biofilm-related infections (Mah and O'Toole 2001; Kumamoto 2002; Bendel 2003; Lewis 2008).

Candida species are capable of causing a variety of superficial and deep seated mycoses. All the opportunists are liable to attack immunocompromised hosts or those debilitated in some other way (Coenye et al. 2011). *C. albicans* is the principal pathogen and considered to be the most virulent strain among pathogenic fungi. The most recent surveys have shown *Candida* to be the third or fourth most commonly isolated pathogen (Shin et al. 2002).

C. albicans is capable of invading virtually every site on the body, including deep tissues and organs, superficial sites such as skin, nails and mucosa. Superficial infections, such as acute infections of the oral cavity or vagina, are some of the most frequently encountered infections (Sandven 2000; Ramage et al. 2001). Moreover, biomaterials such as stents, shunts, prostheses (voice, heart valve, knee, etc.), implants (lens, breast, denture, etc.) endotracheal tubes, pacemakers and various types of catheters to name a few, have all been shown to support colonization and biofilm formation by *Candida* (Goldmann and Pier 1993; Cardinal et al. 1996; Gilbert et al. 1996; Niazi et al. 1996; Fukasawa et al. 1997; Radford et al. 1999; Soll 2000; Douglas 2003; Starakis and Mazokopakis 2009; Mohandas and Ballal 2011; Hwang et al. 2012).

Mechanism Involved in Biofilm Resistance

Several mechanisms are thought to be involved in biofilm antimicrobial resistance including:

- Slow penetration of the antimicrobial agent into the biofilm
- Changes in the chemical microenvironment within the biofilm, leading to zones of slow or no growth
- Adaptive stress responses, and
- Presence of a small population of extremely resistant "persister" cells.

The development of a biofilm occurs in several distinct phases (Chandra et al. 2001; Seneviratne et al. 2008). Adhesion between cell-surface components and another surface is mediated by reversible hydrophobic and electrostatic forces and microbial attachment is the result of a balance between attraction and repulsion. Adhesion to abiotic surfaces is primarily mediated by hydrophobic interactions, whereas microbial adherence to biological surfaces is controlled by adhesins, e.g., lectins. Adherence is not limited to one single species as most biofilms in nature are polymicrobial (Lee and King 1983; Dunne 2002; Stewart and Franklin 2008).

Biofilm Formation by *Candida albicans*

In human body, *Candida* species occur as commensals on the skin as well as in the oral cavity, the gastrointestinal tract, the urogenital tract and the vagina. Virulence factors of *C. albicans* include proteases, adhesins and the morphological conversion from budding yeast to a filamentous form. The increasing use of indwelling medical devices in conjunction with an ageing/increasingly immunocompromised population has resulted in a surge of hospital acquired *Candida* spp. infections, *C. albicans* ranking high among nosocomial pathogens. *Candida* infections are frequently associated with the formation of biofilms on implantable medical devices (Kumamoto 2002). These devices readily support biofilm formation and are responsible for a considerable percentage of clinical candidiasis cases. Several experimental parameters such as the nature of the surface material (Hawser and Douglas 1994), the growth medium (Jin et al. 2004; Krom et al. 2007) and conditions of incubation (Millsap et al. 1999; Gallardo-Moreno et al. 2004) influence *C. albicans* biofilm formation and structure.

Raz-Pastuer and colleagues (2011) carried out Scanning electron microscope (SEM) observations to analyze the capability of *C. albicans*, *C. tropicalis* and *C. parapsilosis* to adhere to human skin model. The skin sections were inoculated with low and high concentration of the yeasts and followed for one and six days; they were then viewed by SEM. The electron microscopy observations revealed that all three yeasts tested adhered to the skin but *C. albicans* covered the entire skin model to a higher extent than *C. tropicalis* or *C. parapsilosis*.

Thus, the basic structural features and properties of *C. albicans* biofilms have been established. Sequencing of the *C. albicans* genome is the key in defining the biofilm phenotype of *C. albicans* with a view to identifying possible targets for novel, biofilm-specific antifungal agents. However, little is currently known about the chemical composition of the matrix material and it needs further evaluation.

Role of Bioactive Compounds against *Candida* species

There has been an abrupt increase in opportunistic fungal infections during the past two decades (Marine et al. 2010; Behzadi et al. 2010; Nayman et al. 2011). *Candida* species are the most prominent causal agents among these infections (Marine et al. 2010; Sahiner et al. 2011; Kim etal. 2012). Candidiasis is the most common fungal infection found all over the world. *Candida* species isolates are the fourth most common cause of blood stream infections among hospital patients. The incidence of such infections is increasing because of a rising number of immunocompromised patients, widespread use of broad-spectrum antibiotics and invasive devices or procedures (Moran et al. 2003; Runyoro et al. 2006; Di Paola et al. 2011; Kim et al. 2012; Weichert et al. 2012).

Systemic fungal infections are most difficult to diagnose and treat and have a mortality rate of over 40% (Kumar et al. 2011). *Candida albicans* was found to be responsible for more than half of the candidiasis cases as compared to other *Candida* species (Fisher 2011; Kim et al. 2012). However, the incidence of other non-*albicans* species such as *Candida tropicalis*, *Candida parapsilopsis* and *Candida glabrata* is also increasing very briskly (Marine et al. 2010; Ishida et al. 2011).

The most generalized therapies to treat fungal infections are based on disrupting the fungal membrane homeostasis (Sobel et al. 2011). The most frequently used groups of fungal agents used for treating fungal infections include the polyenes (e.g., Amphotericin B), which interrupt membrane function by forming a direct association with fungal sterols and the azoles (e.g., flucanazole, itraconazole and posaconazole), which hamper sterol biosynthesis (Marine et al. 2010; Ishida et al. 2011). Also, the treatment of invasive *Candida* species is often complicated due to limited number of effective antifungal agents, high toxicity, low tolerability or a narrow spectrum of antifungal drugs and as well as the increase in the number of azole-resistant strains (Ishida et al. 2011). Hence, there has been a steep rise in the reoccurrence of these fungal infections with more severity (Runyoro et al. 2006; Khan and Ahmad 2012).

Thus, the problems associated with the management of *Candida* infections necessitate the discovery of new antifungal agents in order to broaden the spectrum activity against *Candida* and combat strains depicting resistance to antifungal agents (Runyoro et al. 2006; Ishida et al. 2011).

Medicinal plants have formed a major part of human medicine since the dawn of human civilization. Hence, plants form the backbone of traditional medicinal systems in India. Our ancient Ayurvedic books "Charak Samhita" and "Sushrat Samhita" also show the use of plant extracts to treat microbial infections (Kumar et al. 2011). Due to increased prevalence of drug resistant microbes there is an immense need to search for new effective drugs having natural or synthetic origin. Plant extracts and their products are clinically safer as compared to the commercially available antibiotics (Kim et al. 2012). Thus, plant derived natural products could offer a potential lead to new compounds, which could act on fungal infections (Kumar et al. 2011).

Hammer et al. (1998) studied the *in vitro* activity of 24 essential oils and in particular tea tree oil against *Candida albicans* isolates and non-*albicans Candida* isolates (*C. glabrata, C. gulliermondii, C. tropicalis, C. parapsilosis* and *C. stellatoidea*) using broth microdilution method. Among the 24 essential oils no oil inhibited *C. albicans* at the lowest concentration (0.03%). All essential oils except cedar wood, sweet almond and evening primrose inhibited *C. albicans* at concentration of ≤2.0%. The bioactivity of tea tree oil was also checked against 81 C. *albicans* isolates and 33 non-*albicans Candida* isolates, it was observed that tea tree oil inhibited 90% of the *albicans* and non-*albicans* isolates at a minimum inhibitory concentration (MIC) of 0.25% for *C. albicans* and 0.5% for non-*albicans Candida* species. In the study, three intra-vaginal tea tree oil products were also tested for determination of MIC and minimal fungicidal concentration (MFC) compared to non-formulated tea tree oil. The results depicted that the products possessed MIC and MFC similar to non-formulated tea tree oil.

The antifungal activity of plant extract of *Echinophora platyloba* against *C. albicans* was reported by Avijgan et al. (2006). The authors used the ethanolic extract of *Echinophora platyloba* at different concentrations (1, 2, 4, 8, 16, 32, 64, 128, and 256 mg/ml) against *C. albicans*. The activity of the plant extract was tested using the agar dilution method and the results were noted 21 days after the incubation period which ensured sufficient growth for *C. albicans*. The results of the above study revealed that the plant extract at a minimum concentration of 2mg/ml effectively inhibited the growth of *C. albicans*. Thus, *Echinophora platyloba* extract can be used as an effective agent against *C. albicans*.

Similarly, Khan et al. (2009) demonstrated antimicrobial activities of the crude ethanolic extracts of five plants (*Acacia nilotica, Cinnamum zeylanicum*, *Synzium aromaticum*, *Terminalia arjuna* and *Eucalyptus globules*) against multidrug resistant (MDR) strains of *Escherichia coli, Klebsiella pneumoniae* and *Candida albicans.* ATCC strains of *Streptococcus mutans, Staphylococcus aureus, Enterococcus faecalis, Streptococcus bovis, Pseudimonas aeruginosa, Salmonella typhimurium, Escherichia coli, Klebsiella pneumoniae* and *Candida albicans* were also tested in the study. It was observed that *A. nilotica, C. zeylanicum* and *S. aromaticum* acquired the most potent activity against all the microorganisms studied. Whereas the MDR strains exhibited strong resistance to the extracts of *Terminulla arjuna* and *Eucalyptus globulus*. Among the extracts the most potent antimicrobial plant was *A. nilotica* (MIC range 9.75–313 μg/ml), whereas other crude plant extracts were found to exhibit higher MIC values than *A. nilotica*. Hence, it was concluded that *A. nilotica, C. zeylanicum* and *S. aromaticum* can be used against multidrug resistant microbes causing nosocomial and community acquired infections.

Hassan et al. (2012) conducted a study to determine antimicrobial activity of stem extracts (*Mangifera indica, Jacaranda mimosifolia, Azhardicta indica, Salvadora persica and Acacia nilotica*) popularly used in folk medicine to treat dental plague and caries in human. Determination of antimicrobial activity was done by an agar well diffusion method and determination of MIC and the aqueous, ethanolic, and hexane extracts were assayed for antimicrobial activities. The antimicrobial activities of stems of the different plant extracts indicated that ethanol extracts showed higher anti-microbial activity. The MIC results of *M. indica* showed significant MIC against *Aspergillus niger, Penicillium notatum* and *Candida albicans* at 5.0, 10, and 15mg/ml while, no results were observed for *J. mimosifolia* and *S. persica* and *A. nilotica* and *A. indica* have high potency for *Aspergillus niger, Penicillium notatum* but no results were found for *Candida albicans*. The study thus justifies that ethanolic plant extracts show medicinal use as a dental plague remedy.

Even though several effective antifungal agents are available for oral candida infections but their failure is uncommon because most of the times *C. albicans* exhibit resistance to the drug during therapy. Thus, Doddanna et al. (2013) performed a study to evaluate the antimicrobial effects of plant extracts of Tea leaves, Onion leaves, Onion bulb (*Allium cepa*), Mint leaves, Aloe and Curry leaves on *Candida albicans*. An additional objective of the study was to identify an alternative, inexpensive, simple, and effective method of preventing and controlling *Candida albicans*. For the study ethanolic extracts of plants were used and the stock solutions (10 mg/ml) of plant extracts were inoculated on petri plates containing species of *Candida albicans* and incubated at 25 ± 2°C for 72 h. The results of the above study revealed that alcoholic curry leaves showed the maximum zone of inhibition against *Candida albicans* followed by aqueous tea leaves. The other plant extracts like alcoholic onion leaves, alcoholic tea leaves, alcoholic onion bulb, alcoholic aloe vera, and alcoholic mint leaves also inhibited the growth of *Candida albicans* but to a lesser extent. Thus, the present study rendered an alternative medicine for the treatment of oral candidasis.

Al-Judaibi and Al-Yousef (2014) also investigated the *in vitro* antimicrobial activity of ethanolic extracts of medicinal plants including *Rhamnus globosa, Ocimum basilicum, Tecoma stans* and *Coleus forskohlii* against *Candida* species. The authors performed antifungal assays, MIC, MFC and Time kill determination studies for the evaluation of antimicrobial activity. The result of the above study showed that high inhibition in growth of *Candida* was observed after treatment with *R. globosa* and *O. basilicum. C. tropicalis* was shown to be a sensitive strain with an inhibition of 29, 28, 35, 25 and 27 mm after treatment with *R. globosa, R. globosa* "leaf with thorns," *O. basilicum, Tecoma stans* and *Coleus forskohlii*, respectively. Thus, the results confirmed the fungicidal effect of *O. basilicum* and *R. globosa* with a 20 and 30% reduction in CFU compared with the starting inoculums in the time-kill studies.

Higher incidence of *Candida* infection and increased levels of *Candida* spp. have been identified in denture wearer and the existing antifungal agents like Nystatin and Fluconazole have been found to be toxic on long term application, also chances of development of drug resistance is high. Use of medicinal plants to inhibit growth and development of fungal infection is found to be a remedy. Thus, Bhat et al. (2014) evaluated the antifungal activity of some of the medicinal essential oils (*Origanum vulgare*, *Pogostemon patchouli*, *Ocimum sanctum* and Neem) against oral isolates of *Candida* obtained from denture wearing patients. The oral isolates of *Candida* species were collected after conducting a survey of complete denture wearers for at least more than a year. The presence of *Candida* was confirmed with the help of germ tube test, other biochemical tests and chlamydospore formation tests. Moreover, the essential oils of different herbs were extracted using hydro distillation methods and the antifungal activity of these oils was tested and compared with antifungal activity of Nystatin and fluconazole using modified Kirby-Baeur method. Among the *Candida* species, *Candida albicans* was more prevalent followed by *C. tropicalis* and *C. glabrata* in the denture wearers. The results of the study depicted that among the tested essential oils, *Origanum vulgare* (stored and fresh) and clove oil conferred positive results against all isolates. The fresh samples gave better results than the stored and all three oils showed more antifungal activity than Nystatin and fluconazole. Thus, *Origanum vulgare* and clove oil are potent antifungal agents against oral species of *Candida* and can be either used separately or their synergistic activity could be explored against denture stomatitis.

Sumathi and Gowthami (2014) reported the antimicrobial activity of aqueous and solvent extracts of *Carica papaya* against *Candida albicans*. The antimicrobial assay of plant extract against clinical isolates was performed by Agar Well Diffusion method. Among the leaf, stem and root extracts, only the leaf extracts showed inhibitory effect against *Candida albicans,* whereas stem and root extracts were ineffective. Thus, among the leaf, stem and root extracts, the leaf extract was found to exhibit more antimicrobial activity than the stem and root extracts.

Thus, the use of natural products for the control of fungal diseases is an exciting alternative to synthetic fungicides due to their lower negative effect and reduced cost also. India is considered to be rich in drug plants, which are mainly used in preventive and curative medicine. Hence, screening of medicinal plants for antimicrobial agents has gained much impetus.

Conclusions and Future Perspectives

In the context of the increasing number of immunocompromised patients, combined with the advances in medical technology, fungi have emerged as a major cause of infectious diseases, with *Candida albicans* being the major pathogen. *Candida* spp. are known to form biofilms upon contact with various surfaces. Biofilm-associated fungal infections are particularly serious because sessile *Candida* cells are comparatively resilient to a wide spectrum of antifungal drugs. Additionally, the few antimycotics that are active against microbial biofilms repeatedly result in only fractional killing of the biofilm cells, leaving a subpopulation of the biofilm cells alive, so called persisters. Since they start growing over again when the antimycotic pressure drops, persisters are considered as one of the most important reasons for the recurrence of biofilm-associated infections. In spite of these there have been reports of various bioactive agents that can act as potent antifungal agents. Advanced research based on these bioactive compounds can lead to the formation of novel antifungal agents for combating these pathogens. Natural products synthesized from medicinal plants are among the safest sources of new medications and antifungal drugs. Thus, there is a necessity to search for novel and effective anti-infective agents especially from plants for the treatment of infectious and non-infectious diseases.

References

Abdulhadi, S.K., A.H. Yashua and A. Uba. 2008. Organisms causing urinary tract infection in paediatric patients at Murtala Muhammad Specialist Hospital, Kano, Nigeria. Int J Biomed Health Sci. 4(4): 165–167.

Al-Judaibi, A. and F. Al-Yousef. 2014. Antifungal effect of ethanol plant extract on *Candida* sp. American J Agricult Biol Sci. 9(3): 277–283.

Al-Mathkurthy, H.J. and S.N. Abdul-Gaffar. 2011. Urinary tract infections caused by *staphylococcus aureus* DNA and in comparison to the *Candida albicans* DNA. N Am J Med Sci. 3(12): 565–569.

Avijgan, M., M. Hafizi, M. Saadat and M.A. Nilforoushzadeh. 2006. Antifungal effect of *Echinophora Platyloba's* extract against *Candida albicans*. Iranian J Pharm Res. 4: 285–289.

Behiry, I.K., S.K.E. Hedeki and M. Mahfouz. 2010. *Candida* infection associated with urinary catheter in critically ill patients: identification, antifungal susceptibility and risk factors. Res J Med Med Sci. 5(1): 79–86.

Behzadi, P., E. Behzadi, H. Yazdanbod, R. Aghapour, Akbari, M. Cheshmeh and D. Salehiah Omran. 2010. Urinary tract infections associated with *Candida albicans*. Maedica. (Buchar). 5(4): 277–279.

Bendel, C.M. 2003. Colonization and epithelial adhesion in the pathogenesis of neonatal candidiasis. Seminars Perinatol. 27(5): 351–424.

Bhat, V., S.M. Sharma, V. Shetty, C.S. Shastry, V. Rao, S.M. Shenoy, S. Saha and S. Balaji. 2014. Screening of selected plant essential oils for their antifungal activity against *Candida* species isolated from denture stomatitis patients. Nitte Univ J Health Sci. 4(1): 46–51.

Blinkhorn, R.J., D. Adelstein and P.J. Spagnuolo. 1989. Emergence of a new opportunistic pathogen, *Candida lusitaniae*. J Clin Microbiol. 27(2): 236–240.

Buonsenso, D. and L. Cataladi. 2012. Urinary tract infections in children: a review. Minerva Pediatr. 64(2): 145–157.

Cardinal, E., E.M. Braunstein, W.N. Capello and D.A. Heck. 1996. *Candida albicans* infection of prosthetic joints. Orthopedics 19: 247–251.

Chandra, J., D.M. Kuhn, P.K. Mukherjee, L.L. Hoyer, T. McCormick and M.A. Ghannoum. 2001. Biofilm formation by the antifungal pathogen *Candida albicans*: development, architecture, and drug resistance. J Bacteriol. 183: 5385–5394.

Chen, S.C.A., C.L. Halliday and W. Meyer. 2002. A review of nucleic acid based diagnostic tests for systemic mycoses with an emphasis on polymerase chain reaction-based assays. Med Mycol. 40: 333–357.

Chic, O.I. and T.T. Amom. 2013. Phytochemical and antimicrobial evaluation of leaf-extracts of *Pterocarpus santalinoides*. European J Med Plants 4(1): 105–115.

Cutler, J.E. 1991. Putative virulence factors of *Candida albicans*. Annu. Rev. Microbiol. 45: 187–218.

de Marie, S. 2000. New developments in the diagnosis and management of invasive fungal infections. Haematologica 85: 88–93.

Di Paola, G., A. Mogovorich, G. Fiorini, M.G. Cuttano, F. Manassero and C. Selli. 2011. *Candida* bezoars with urinary tract infections in two women without immunocompromising conditions. Scientific World J. 1168–1172.

Dixon, T.C., W.J. Steinbach, D.K. Benjamin, L.W. Williams, Jr. and L.A. Myers. 2004. Disseminated *Candida tropicalis* in a patient with chronic mucocutaneous candidiasis. South Med J. 97: 788–790.

Doddanna, S.J., S. Patel, M.A. Sundarrao and R.S. Veerabhadrappa. 2013. Antimicrobial activity of plant extracts on *Candida albicans*: An *in vitro* study. Ind J Dental Res. 24(4): 401–405.

Donlan, R.M. and J.W. Costerton. 2002. Biofilms: survival mechanisms of clinically relevant microorganisms. Clin Microbiol Rev. 15: 167–93.

Douglas, J.L. 2003. *Candida* biofilms and their in infection. Trends Microbiol. 11(1): 30–36.

Dunne, W.M. 2002. Bacterial adhesion: seen any good biofilms lately. Clin Microbiol Rev. 15: 155–166.

Fisher, J.F. 2011. *Candida* urinary tract infections- epidemiology, pathogenesis, diagnosis and treatment: executive summary. Clin Infec Dis. 429–432.

François, F., T. Noël, R. Pépin, A. Brulfert, C. Chastin, A. Favel and J. Villard. 2001. Alternative identification test relying upon sexual reproductive abilities of *Candida lusitaniae* strains isolated from hospitalized patients. J Clin Microbiol. 39(11): 3906–3914.

Fukasawa, N. and K. Shirakura. 1997. *Candida* arthritis after total knee arthroplasty—a case of successful treatment without prosthesis removal. Acta Orthop Scand. 68: 306–307.

Gallardo-Moreno, A.M., M.L. Gonzalez-Martin, C. Perez-Giraldo, J.M. Bruque and A.C. Gomez-Garcia. 2004. The measurement temperature: an important factor relating physicochemical and adhesive properties of yeast cells to biomaterials. J Colloid Interf Sci. 271: 351–358.

Ghotaslou, R., A. Yaghoubi and S. Sharify. 2010. Urinary Tract Infections in hospitalized Patients during 2006 to 2009 in Madani Heart Center Tabriz, Iran. J Cardiovasc Thorac Res. 2(1): 39–42.

Gilbert, H.M., E.D. Peters, S.J. Lang and B.J. Hartman. 1996. Successful treatment of fungal prosthetic valve endocarditis: case report and review. Clin Infect Dis. 22: 348–354.

Goldmann, D.A. and G.B. Pier. 1993. Pathogenesis of infections related to intravascular catheterization. Clin Microbiol Rev. 6: 176–192.

Hajjeh, R.A., A.N. Sofair, L.H. Harrison, G.M. Lyon, B.A. Arthington- Skaggs, S.A. Mirza, M. Phelan, J. Morgan, W. Lee-Yang, M.A. Ciblak, L.E. Benjamin, L.T. Sanza, S. Huie, S.F. Yeo, M.E. Brandt and D.W. Warnock. 2004. Incidence of bloodstream infections due to *Candida* species and *in vitro* susceptibilities of isolates collected from 1998 to 2000 in a population based active surveillance program. J Clin Microbiol. 42: 1519–1527.

Hammer, K.A., C.F. Carson and T.V. Riley. 1998. Antimicrobial activity and phytochemical analysis of ethanolic extracts of twelve medicinal plants against oral microorganisms. J Antimicro Chemo. 42: 591–595.

Hassan, A., Q. Syed, I. Amjad and A. Hassan. 2012. Antimicrobial evaluation of some dental remedial plant extracts from Pakistan. Med Plant Res. 2(3): 11–17.

Hawser, S.P. and L.J. Douglas. 1994. Biofilm formation by *Candida* species on the surface of catheter materials *in vitro*. Infect. Immun. 62: 915–921.

Hazen, K.C. 1995. New and emerging yeast pathogens. Clin Microbiol Rev. 8: 462–478.

Hazen, K.C., G.W. Theisz and S.A. Howell. 1999. Chronic urinary tract infections due to *Candida utilis*. J Clin Microbiol. 37(3): 824–827.

Holland, J., M.L. Young, O. Lee and S.C.-A. Chen. 2007. Vulvovaginal carriage of yeasts other than *Candida albicans*. Sex Transm Infec. 79: 249–250.

Holzschu, D.L., H.L. Presley, M. Miranda and H.J. Phaff. 1979. Identification of *Candida lusitaniae* as an opportunistic yeast in humans. J Clin Microbiol. 10(2): 202–205.

Hwang, B.H., J.Y. Yoon, C.H. Nam, K.A. Jung, S.C. Lee, C.D. Han and S.H. Moon. 2012. Fungal periprosthetic joint infection after primary total knee replacement. Bone Joint Surg Br. 94(B): 656–659.

Ishida, K., J.C.F. Rodrigues, S. Cammerer, J.A. Urbina, I. Gilbert, W. de Souza and S. Rozental. 2011. Synthetic arylquinuclidine derivatives exhibit antifungal activity against *Candida albicans*, *Candida tropicalis* and *Candida parapsilopsis*. Ann Clin Microbiol Antimicrobial. 10(3): 1–10.

Jabra-Rizk, M.A., W.A. Falkler, Jr., W.G. Merz, A.A.M.A. Baqui, J.I. Kelley and T.F. Meiller. 2000. Reptrospective identification and characterization of *Candida dubliniensis* isolates among *Candida albicans* clinical laboratory isolates from human immunodeficiency virus (HIV)-infected and non HIV-infected individuals. J Clin Microbiol. 38: 2423–2426.

Jacobs, L.G. 1996. Fungal urinary tract infections in the elderly: treatment guidelines. Drugs Aging 8: 89–96.

Jacobs, L.G., E.A. Skidmore, K. Freeman, D. Lipschultz and N. Fox. 1996. Oral fluconazole compared with bladder irrigation with amphotericin B for treatment of fungal urinary tract infections in elderly patients. Clin Infec Dis. 22: 30–35.

Jha, B.K., S. Dey, M.D. Tamang, P.G. Shivananda and K.N. Brahmadatan. 2006. Characterization of *Candida* species isolated from cases of lower respiratory infection. Kathmandu Univ Med J. 4(3): 290–294.
Jin, Y., L.P. Samaranayake, Y. Samaranayake and H.K. Yip. 2004. Biofilm formation of *Candida albicans* is variably affected by saliva and dietary sugars. Arch. Oral Biol. 49: 789–798.
Johnson, E.M. 2009. Rare and emerging *Candida* species. Curr. Fung. Infec. Rep. 3: 152–159.
Jones, J.M. 1990. Laboratory diagnosis of invasive candidiasis. Clin Microbiol Rev. 3: 32–45.
Khan, M.S. and I. Ahmad. 2012. Biofilm inhibition by *Cymbopogon citratus* and *Syzgium aromaticum* essential oils in the strains of *Candida albicans*. J Ethnopharmacol. 140(2): 416–423.
Khan, R., B. Islam, M. Akram, S. Shakil, A. Ahmad, S.M. Ali, M. Siddiqui and A.U. Khan. 2009. Antimicrobial activity of five herbal extracts against Multi Drug Resistant (MDR) strains of bacteria and fungus of clinical origin. Molecules 14: 586–597.
Kim, J.Y., Y. Yi and Y. Lim. 2012. Antifungal activity of coptidis rhizome against *Candida* species. J Med Plant Res. 6(12): 2295–2298.
Kojic, E.M. and R.O. Darouiche. 2004. *Candida* infections of medical devices. Clin Microbiol Rev. 17(2): 255–267.
Kothavade, R.J., M.M. Kura, A.G. Valand and M.H. Panthaki. 2010. *Candida tropicalis*: its prevalence, pathogenicity and increasing resistance to fluconazole. J Med Microbiol. 59: 873–880.
Krcmery, V. and A.J. Barnes. 2002. Non-*albicans Candida* spp. causing fungaemia: pathogenicity and antifungal resistance. J Hosp Infect. 50: 243–260.
Krom, B.P., J.B. Cohen, G.E. McElhaney Feser and R.L. Cihlar. 2007. Optimized candidal biofilm microtiter assay. J Microbiol Meth. 68: 421–423.
Kumamoto, C.A. 2002. *Candida* biofilms. Curr Opin Microbiol. 5: 608–611.
Kumar, A., V. Bhatii, A. Kumar, S. Patil, V. Bhatia and A. Kumar. 2011. Screening of various plant extracts for antifungal activity against *Candida* species. World J Sci Technol. 1(10): 43–47.
Leberer, E., K. Ziegelbauer, A. Schmidt, D. Harcus, D. Dignard, J. Ash, L. Johnson and D.Y. Thomas. 1997. Virulence and hyphal formation of *Candida albicans* require the Ste20p-like protein kinase CaCla4p. Curr. Biol. 7: 539–546.
Lee, J.C. and R.D. King. 1983. Characterization of *Candida albicans* adherence to human vaginal epithelial cells *in vitro*. Infec Immun. 41(3): 1024–1030.
Lewis, K. 2008. Multidrug tolerance of biofilms and persister cells. Curr Top Microbiol. 322: 107–131.
Mah, T.F. and G.A. O'Toole. 2001. Mechanisms of biofilm resistance to antimicrobial agents. Trends Microbiol. 9: 34–39.
Manfredi, R., L. Calza and F. Chiodo. 2002. Dual *Candida albicans* and *Cryptococcus neoformans* fungaemia in an AIDS presenter: a unique disease association in the highly active antiretroviral therapy (HAART) era. J Med Microbiol. 51: 1135–1137.
Marine, M., F.J. Pastor and J. Guarro. 2010. Efficacy of posaconazole in a murine disseminated infection by *Candida tropicalis*. Antimicrob. Agents Chemother. 54(1): 530–532.
McCullough, M.J., B.C. Ross and P.C. Reade. 1996. *Candida albicans*: a review of its history, taxonomy, epidemiology, virulence attributes, and methods of strain differentiation. Int. J Oral Maxillofac Surg. 25(2): 136–144.
Millsap, K.W., R. Bos, H.J. Busscher and H.C. Van der Mei. 1999. Surface aggregation of *Candida albicans* on glass in the absence and presence of adhering *Streptococcus gordonii* in a parallel-plate flow chamber: a surface thermodynamical analysis based on acid base interactions. J Colloid Interf Sci. 212: 495–502.
Minari, A., R. Hachem and I. Raad. 2001. *Candida lusitaniae*: A cause of breakthrough fungemia in cancer patients. Clin Infec Dis. 32: 186–190.
Mohandas, V. and M. Ballal. 2011. Distribution of *Candida* species in different clinical samples and their virulence: biofilm formation, proteinase and phospholipase production. J Global Infect Dis. 3(1): 4–8.
Molero, G., R. Díez-Orejas, F. Navarro-García, L. Monteoliva, J. Pla, C. Gil, M. Sánchez-Pérez and C. Nombela. 1998. *Candida albicans*: genetics, dimorphism and pathogenicity. Int Microbiol. 1: 95–106.
Montini, G., K. Tullus and I. Hewitt. 2011. Febrile urinary tract infections in children. N Engl J Med. 365: 239–250.
Moran, G.P., D.J. Sullivan and D.C. Coleman. 2003. Emergence of non-*Candida albicans* species as pathogens. pp. 37–53. *In*: Calderone, R.A. (ed.). Candida and Candidiasis, 1st edition. American Society Microbiol., Washington D.C.
Nayman, A.S., I. Ozgunes, O.T. Ertem, N. Erben, K.E. Doyuk, M. Tozun and G. Usler. 2011. Evaluation of risk factors in patients with candiduria. Mikrobiyol Bull. 45(2): 318–324.

Niazi, Z.B., C.A. Salzberg and M. Montecalvo. 1996. *Candida albicans* infection of bilateral polyurethane-coated silicone gel breast implants. Ann Plast Surg. 37: 91–93.

Nosek, J., Z. Holesova, P. Kosa, A. Gacser and L. Tomaska. 2009. Biology and genetics of the pathogenic yeast *Candida parapsilosis*. Curr Genet. 55: 497–509.

Odds, F.C. 1988. *Candida* and Candidosis, Odds, F.C. and B. Tindall. 2nd edn. London. Bailliere Tindall, London, UK.

Osmanagaoglu, O., N. Altinlar, S.C. Sacilik, C. Cokmus and A. Akin. 2000. Identification of different *Candida* species isolated in various hospitals in Ankara by fungichrom test kit and their differentiation by SDS-PAGE. Turk. J Med Sci. 30: 355–358.

Pastuer, A.R., Y. Ullmann and I. Berdicevsky. 2011. The pathogenesis of *Candida* infections in a human skin model: scanning Electron Microscope observations. ISRN Dermatol. doi:10.5402/2011/150642.

Pfaller, M.A. and D.J. Diekema. 2007. Epidemiology of invasive candidiasis: a persistent public health problem. Clin Microbiol Rev. 20: 133–163.

Pukkila-Worley, R., A.Y. Peleg, E. Tampakakis and E. Mylonakis. 2009. *Candida albicans* hyphal formation and virulence assessed using a *Caenorhabditiselegans* infection model. Eukar Cell 8(11): 1750–1758.

Radford, D.R., S.J. Challacombe and J.D. Walter. 1999. Denture plaque and adherence of *Candida albicans* to denture-base materials *in vivo* and *in vitro*. Crit Rev Oral Biol Med. 10: 99–116.

Ramage, G., K. VandeWalle, B.L. Wickes and J.L. López–Ribot. 2001. Characteristics of biofilm formation by *Candida albicans*. Rev Iberoam Micol. 18: 163–170.

Rao, S. and U. Ali. 2005. Systemic fungal infections in neonates. J. Postgrad. Med. 51: 27–29.

Raz-Pasteur, A., Y. Ullmann and I. Berdicevsky. 2011. The pathogenesis of *Candida* infections in a human skin model: scanning electron microscope observations. ISRN Dermatol. doi:10.5402/2011/150642.

Revankar, S.G. and J.D. Sobel. 2010. Yeast infections of the lower urinary tract: Recommendations for diagnosis and treatment. Curr Fungal Infect Rep. 4: 175–178.

Rifkind, A. and J.A. Frey. 1972. Influence of gonadectomy on *Candida albicans* urinary tract infection in CFW mice. Infec. Immunity 5(3): 332–336.

Roilides, E., E. Farmaki, J. Evdoridou, A. Francesconi, M. Kasai, J. Filioti, M. Tsivitanidou, D. Sofianou, G. Kremenopoulos and T.J. Walsh. 2003. *Candida tropicalis* in a neonatal intensive care unit: epidemiologic and molecular analysis of an outbreak of infection with an uncommon neonatal pathogen. J Clin Microbiol. 41: 735–741.

Sahiner, F., K. Ergunay, M. Ozyurt, N. Ardic, T. Hozbul and T. Haznedagrolu. 2011. Phenotypic and genotypic identification of *Candida* strains isolated as noscomial pathogens. Mikrobiyol Bull. 45(3): 478–488.

Sandven, P. 2000. Epidemiology of candidemia. Rev Iberoam Micol. 17: 73–81.

Seneviratne, C.J., L. Jin and L.P. Samaranayake. 2008. Biofilm lifestyle of *Candida*: a mini-review. Oral Dis. 14: 582–590.

Shin, J.H., S.J. Kee, M.G. Shin, S.H. Kim, D.H. Shin, S.K. Lee, S.P. Suh and D.W. Ryang. 2002. Biofilm production by isolates of *Candida* species recovered from nonneutropenic patients: comparison of bloodstream isolates with isolates from other sources. J Clin Microbiol. 40(4): 1244–1248.

Sim, J.P., B.C. Kho, H.S. Liu, R. Yung and J.C. Chan. 2005. *Candida tropicalis* arthritis of the knee in a patient with acute lymphoblastic leukaemia: successful treatment with caspofungin. Hong Kong Med J. 11: 120–123.

Sobel, J.D. and J.A. Vasquez. 1999. Fungal infections of the urinary tract. World J Urol. 17: 410–414.

Sobel, J.D., J.F. Fisher, C.A. Kauffman and C.A. Newman. 2011. *Candida* urinary tract infections-epidemiology. Clin Infec Dis. 433–436.

Soll, D.R. 2000. The ins and outs of DNA fingerprinting of infectious fungi. Clin Microbiol Rev. 13: 332–370.

Starakis, I. and E.E. Mazokopakis. 2009. Prosthetic valve endocarditis: diagnostic approach and treatment options. Cardiovasc. Haemato. Disorders-Drugs Targets 9: 249–260.

Stewart, P.S. and M.J. Franklin. 2008. Physiological heterogeneity in biofilms. Nat Rev Microbiol. 6: 199–210.

Sullivan, D.J., T.J. Westerneng, K.A. Haynes, D.E. Bennett and D.C. Coleman. 1995. *Candida dubliniensis* sp. Nov.: phenotypic and molecular characterization of a novel species associated with oral candidosis in HIV infected individuals. Microbiology 141: 1507–1521.

Sumathi, R. and M. Gowthami. 2014. Phytochemical analysis and *in vitro* Antimicrobial activity of Aqueous and Solvent extracts of *Carica papaya* against clinical Pathogens. Internat. J Adv Res Biol Sci. 1(1): 73–77.

Suzuki, M., M. Hiramatsu, M. Fukazawa, M. Matsumoto, K. Honda, Y. Suzuki and Y. Kawabe. 2014. Effect of SGLT2 inhibitors in a murine model of urinary tract infection with *Candida albicans*. *Diabetes*, Obesity Metabol. doi: 10.1111/dom.12259.

Taneja, N., S.S. Chatterjee, M. Singh, S. Singh and M. Sharma. 2010. Pediatric urinary tract infections in a tertiary care center from north India. Ind J Med Res. 131: 101–105.

Tatfeng, Y.M., M.I. Agba, G.O. Nwobu and D.E. Agbonlahor. 2003. *Candida albicans* in urinary tract or in seminal sac. Online J Health Allied Sci. 4: 5.

Tavanti, A., A.D. Davidson, N.A.R. Gow, M.C.J. Maiden and F.C. Odds. 2005. *Candida orthopsilosis* and *Candida metapsilosis* spp. Nov. to replace *Candida parapsilosis* groups II and III. J. Clin Microbiol. 43: 284–292.

Tiwari, V.V. and M.K. Rai. 2008. Incidence of *Candida albicans* infection in cerebrospinal fluid—A First Report from Vidarbha, Central India. Curr Trends Biotechnol Pharm. 3(1): 648–652.

Tiwari, V.V., M.N. Dudhane and M.K. Rai. 2009. Molecular tools for identification and differentiation of different human pathogenic *Candida* species. pp. 349–370. *In*: Gherbawy, Y., R.L. Mach and M.K. Rai (eds.). Current Advances in Molecular Mycology. Nova Science Publishers, New York.

Vidigal, P. Gonçalves, Santos, S. Aparecida, M.A. Fernandez, P. Bonfim, H.V. Martinez and T.I.E. Svidzinski. 2011. Candiduria by *Candida tropicalis* evolves to fatal candidemia. Med Case Stud. 2(2): 22–25.

Weichert, S., K. Reinshagen, K. Zahn, G. Geginat, A. Dietz, A.K. Killan, H. Schroten and T. Tenenbaum. 2012. Candidiasis caused by *Candida kefyr* in a neonate: case report. BMC Infect Dis. 12(1): 61.

CHAPTER 5

Otorhinolaryngology-Related Fungal Diseases: A Convenient Classification for Better Clinical Practice

Ahmed Ragab

Introduction

Fungi show similar latitudinal diversity gradients as other organisms with approximately 1.5 to 5 million different species in our globe, but only about 300 are known to cause diseases (Tedersoo et al. 2014). They are eukaryotic structures that have a cell wall and a cell membrane. The cell wall is made up of a poly-N-acetylglucosamine. The cell membrane contains ergosterol and chitin in contrast to mammalian cells which contain cholesterol. Clinicians find it more facile to treat bacterial related diseases rather than fungal related ones. Generally speaking fungal diseases are very difficult to treat because fungi are chemically and genetically more similar to animals than to other organisms (Hawksworth 2001). Specifically the forms of fungal related otorhinolaryngological (ORL) diseases have no characteristic presentation, with the immune status of the host playing a noteworthy etiopathological role. As a result accurate diagnosis and targeted treatment cannot be established without an optimum classification which was overlooked in Otorhinolaryngology (ORL) specialty.

Department of Otorhinolaryngology, Faculty of Medicine, Menoufia University, Shebin El-Kom, 32155, Egypt.
E-mail: ahmedragab2000@hotmail.com

Are Fungi Normally Present in ORL Organs or they are a Mark of Disease?

Knowledge of the composition of the normal flora at certain sites is useful for predicting which organisms may be involved in an infection adjacent to that site. Also it can assist in the selection of a logical antimicrobial therapy, even before the exact microbial etiology of the infection is known. The trachea, bronchi, esophagus and stomach are not normally colonized by indigenous flora. However, a limited number of transient organisms may be present at these sites from time to time. At other sites like the oral cavity and pharynx (with its subsites), vestibule of the nose and paranasal sinuses the densities of microorganisms are much higher. The mean colony densities of microorganisms in the nose and paranasal cavities are $10–10^4/10–10^5$ per ml and the mean for oral subunits is $10^7–10^{10}/10^8–10^{10}$ per ml.

The relative and total counts of organisms can be affected by various factors, such as age, diet, anatomic variations, illness, hospitalization, and antimicrobial therapy. However, established sets of flora, with predictable patterns, remain stable throughout life, despite being subject to perturbing factors.

Ear canal fungal flora

The external auditory canal is the only skin-lined cul-de-sac in the human body, so it traps moisture readily, and is an excellent environment for the growth of bacteria and fungi. Cerumen serves as a primary microbial barrier due to the presence of mucin, zinc-alpha-2-glycoprotein, serpins, apolipoprotein D, prolactin-inducible protein and cathepsin D (Feig et al. 2013; Karaca et al. 2013).

Stroman et al. (2001) performed a study to isolate and characterize bacteria and fungi from the healthy ears and to obtain susceptibility profiles on each bacterial isolate. One hundred forty-eight cerumen specimens from one hundred sixty-four yielded 314 organisms, including 23 (7.0%) fungi. One hundred forty-seven canal specimens yielded 310 organisms, including 8 (2.5%) fungi. They isolated 99% Gram-positive bacteria from cerumen and 96% from canal, and 63% of them were staphylococci. *Candida parapsilosis* and *Penicillium* species were the most common cultured fungi (Stroman et al. 2001).

Oral and pharynx fungal flora

Oral flora includes a miscellaneous group of microorganisms, e.g., bacteria, fungi, mycoplasma, protozoa, and possibly a viral flora which may persist from time to time. Bacteria are the predominant group of microorganisms. There are probably 350 different cultivable species and a further proportion of uncultivable flora, which are identified using molecular techniques (Listgarten 2000). This, and the fact that oral cavity has a wide range of sites with different environmental conditions, makes studying of microbiology very complex and difficult (Patil et al. 2013). Noteworthy information is that many microorganisms commonly isolated from neighboring ecosystems such as the skin are not found in the mouth. So, this indicates the unique and selective properties of the oral cavity with regard to microbial colonization.

The formation of the normal oropharyngeal flora is initiated at birth. Certain organisms such as lactobacilli and anaerobic streptococci, which establish themselves at an early date, reach high numbers within a few days. *Actinomyces, Fusobacterium*, and *Nocardia* are acquired by the age of 6 months. Following that time, *Prevotella* and *Porphyromonas* spp., *Leptotrichia, Propionibacterium*, and *Candida* also become part of the oropharyngeal flora (Socransky and Manganiello 1971). *Fusobacterium* populations attain high numbers after dentition and reach maximal numbers at the age of 1 year.

After adulthood, species of *Gemella*, *Granulicatella*, *Streptococcus*, and *Veillonella* were commonly detected at most sites (Socransky and Manganiello 1971). *Streptococcusmitis* is the most commonly found species at essentially all sites and in all subjects. Some species are location specific at one or multiple locations, while other species are individual specific but they are different from the disease states.

For fungi, chronic colonization of mucosal surfaces, including the oral cavity and oropharynx are common, especially for *Candida* spp., and carriage rates of *Candida* among healthy adults ranges from 30 to 70%. Because *Candida* can colonize a variety of microbiota-containing host niches (both biotic tissues and abiotic indwelling medical devices), its pathogenic potential from within the microbiome is related to its ability to adapt, survive, and grow in constantly changing environments (Aas et al. 2005). Members of the agglutinin-like sequence family of adhesions mediate *C. albicans* allow aggregation with other bacteria and yeasts. At the same time, farnesol represses hyphal growth and early biofilm formation and has similar effects on other bacterial pathogens, inhibiting biofilm growth and formation (Gary 2013). In addition to cross-kingdom adhesion and quorum sensing, metabolic changes in the environment (pH, metabolic substrate production, inhibitor production, nutrient sensing/sequestering, etc.) and indirect activity on the host response are mechanisms by which *Candida* and other fungi interact with bacteria in the microbiome to regulate fungal levels and host responses to fungal colonization (Gary 2013).

Most individuals are asymptomatically colonized with this organism. However, when environmental conditions permit the outgrowth of *C. albicans*, colonization can lead to infection and invasion of host tissues. Candida infections may remain localized to mucosal sites (e.g., oral or vaginal candidiasis) or may spread hematogenously leading to candidemia or deep-seated mycoses of other tissues, which can lead to death. Two environmental variables that can lead to increased levels of *C. albicans* in the host are immunosuppression and antibiotic treatment, the latter of which decreases bacterial colonization resistance against *Candida*.

Nose and paranasal sinuses fungal flora

Fungi are ubiquitous saprophytes that reproduce by the formation of spores that are able to enter the respiratory tract by means of inhalation. The retention and clearance of fungal spores after inhalation depends on many factors, e.g., the physical and chemical properties of both the spores and the mucous surfaces. The nature of the structures with which the particles interact at the site of deposition (type of mucociliary layer at air fluid interface, the aqueous phase, free cells, epithelial cells or denteritic cells residing near the basal aspect of epithelium) are other effective factors. Fungal spores after inhalation are submerged in the aqueous lining layers of the airways (Ragab 2010). So during respiration many spores will stick more to the mucus gel layer than to the skin of the vestibule that contains long chain fatty acids that inhibit fungal colonization. This can be phagocytosed by macrophages in contrast to nano particles (NP) in which other inflammatory cells present on airway surfaces, e.g., eosinophils may substantially contribute to its uptake (Geiser et al. 2014).

Using total nasal lavage technique of collection as indicated by Ponikau et al. (1999) a 100% culture rate from the total nasal lavages of the healthy subjects has been found (Braun et al. 2003; Ragab et al. 2006). The type of genus and species mostly reflects the outside indoor and outdoor fungi, e.g., *Aspergillus* sp. and *Pencillium* sp. Also as the nose represents the first mucus lined station of the respiratory tract, more fungi were recovered from the nasal lining than from the lower airway (Ragab et al. 2006).

Two fungal forms can be inhaled; fungal spores and fungal fragments. Three sizes of fungal fragments can be present. Airborne particles were there in three distinct size fractions according to their aerodynamic size: (i) >2.25 μm (spores); (ii) 1.05–2.25 μm (mixture of spores and fragments); and (iii) <1.0 μm (submicrometer-sized fragments). The fraction able to enter the airways through the nose or mouth is called the inhalable fraction (Madsen et al. 2009).

Smaller-sized fungal fragments may be important for several reasons. Smaller-sized fragments have longer lifetimes in the air compared to larger spores and can penetrate deeply into the alveolar region when inhaled. Fungal fragments have been shown to contain fungal antigens, mycotoxins, and (1→3)-β-D-glucan (Madsen et al. 2013). The small size, large quantities, and biological properties of fungal fragments suggest that these particles may potentially contribute to the adverse health effects and raises the need for further characterizations of fungal fragments in fungal diseases.

Can We Develop a Classification for Mycotic ORL Diseases?

In otorhinolaryngology (ORL), fungal related diseases have no specific classification to follow in clinical practice. The only classification existed was related to paranasal sinuses fungal related diseases. The classification that exists classify the diseases into invasive and non-invasive fungal rhinosinusitis. Six forms of fungal rhinosinusitis disorders are currently recognized. Three of the disorders are tissue invasive, whereas the other three are non-invasive (Ragab 2010).

In fungal related ORL (FRORL) infections: The term fungal infections should be replaced with fungal diseases. The clinical nomenclatures used for the mycoses are based on the (1) site of the infection, (2) route of acquisition of the pathogen, and (3) type of virulence exhibited by the fungus. Generally mycoses are classified as superficial, cutaneous, subcutaneous, or systemic (deep) infections depending on the degree of tissue involvement and the host response to the pathogen. Infecting fungi may be either exogenous or endogenous. Endogenous infection involves colonization by a member of the normal flora or reactivation of a previous infection. Routes of entry for exogenous fungi include airborne, cutaneous or percutaneous. Primary pathogens can establish infections in normal hosts. Opportunistic pathogens cause disease in individuals with compromised host defense mechanisms (Odds et al. 1992).

For FRORL diseases the situation is difficult with a large area of mucosal surfaces are exposed to various types of fungi, fungal spores and fungal particles.

So general classification (Fig. 1) based on tissue invasiveness is appropriate for clinical practice including the invasive and non-invasive forms. The superficial disease is the non-invasive form. It can be in two forms according to the body reaction, allergic non-invasive fungal category and fungal balls. The invasive can be (mucocutaneous) mycoses or deep mycoses. The deep mycoses depend on the immunological host responses that include the endemic mycoses and the opportunistic mycoses. According to the onset and duration of the disease; opportunistic mycosis can be acute or chronic invasive disease.

Fungal related ORL diseases depend on three aspects to produce the selected category of mycological disease including; host immunological state, ORL part involved with the disease (Table 1) and type of fungal genus and species. The ORL area involved with the disease depends upon the organ involved whether the ear, nose and paranasal sinuses, mouth, pharynx and larynx. The host factor is an integral part of the disease process. The host immunity can be classified as immune-compromised and immune-competent hosts. Lastly the types of the fungal genus and species are corner stone in identification. Tissue identification is the basic to prove FRORL diagnosis (Fig. 2).

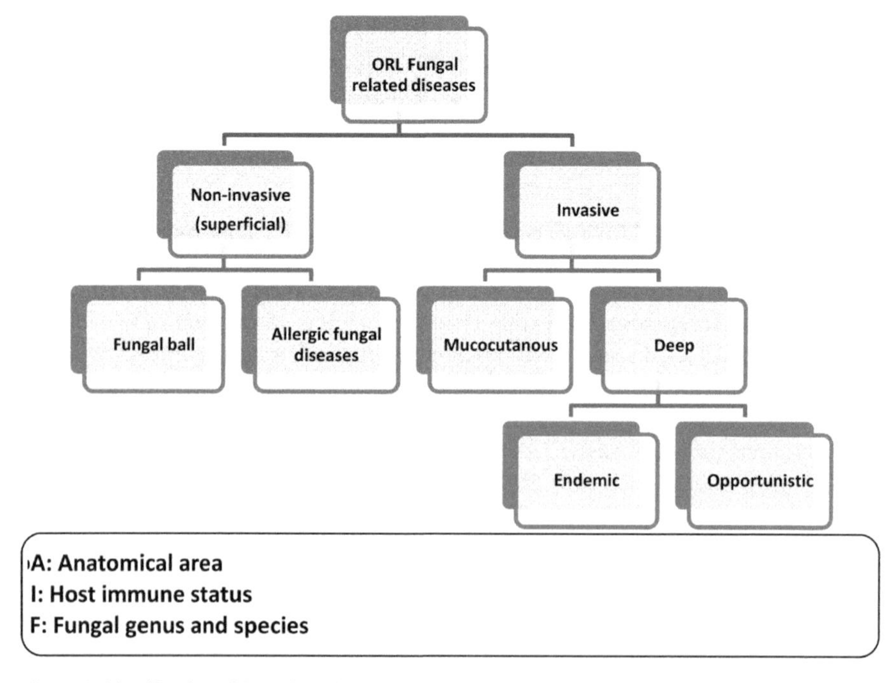

Figure 1. Classification of ORL fungal related diseases.

Table 1. Application of the ORL fungal related diseases classification according to the otorhinolaryngological anatomical area involved with the disease.

Area	**Non-invasive**		**Invasive**		
	Superficial	Allergic fungal	Mucocutaneous mycoses	Deep mycoses	
				Opportunistic Mycoses	Primary (Endemic mycoses)
Ear	Otomycosis	Otitis media with effusion of fungal etiology	Erosive otitis externa of fungal etiology	Necrotizing (malignant) otitis externa of fungal etiology	Histoplasmosis Paracoccidioidomycosis Coccidioidomycosis
Nose & PNS	Fungal balls	Allergic fungal rhinosinusitis	Chronic granulomatous	- Acute invasive - Chronic invasive	Histoplasmosis Paracoccidioidomycosis Coccidioidomycosis
Oral cavity and Pharynx	- Pseudo-membranous candidiasis (thrush): - Erythematous candidiasis: - Angular cheilitis: - Median rhomboid glossitis (MRG)		Chronic hyperplastic candidiasis	- Acute invasive - Chronic invasive	
Larynx	Candidiasis		Chronic invasive	- Acute invasive - Chronic invasive	Histoplasmosis Paracoccidioidomycosis Coccidioidomycosis

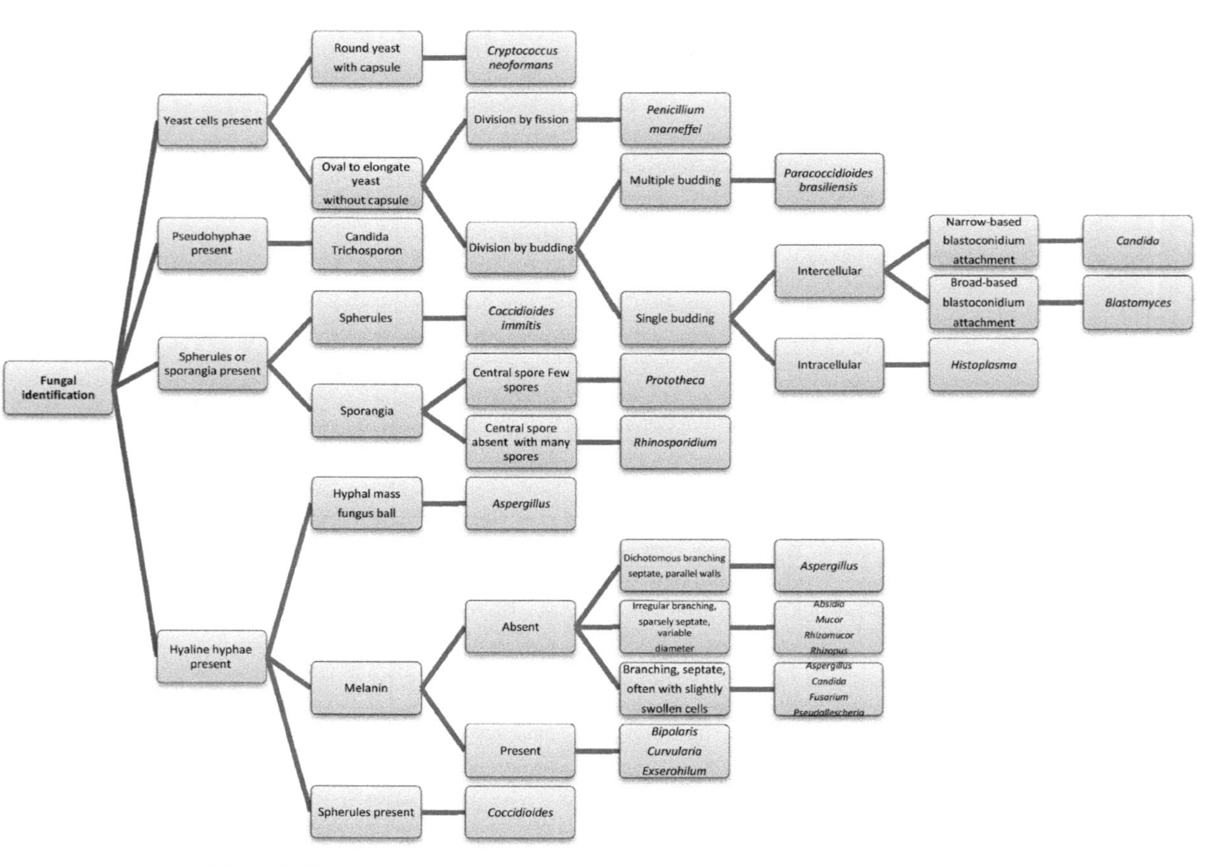

Figure 2. Flow chart for identification of fungal genera in histopathological examination.

Superficial non-Invasive mycoses

The superficial non-invasive mycoses are those diseases that are limited to the very superficial surfaces of the mucocutaneous lining. The immunity of the body can determine the type of the body reaction as fungal balls or allergic fungal disease.

In the ear: According to the body reaction the otomycosis is the non-invasive form of fungal diseases that affect the external canal when there are no associated symptoms and signs of inflammation. In the middle ear eosinophilic fungal OME is the form of non-invasive diseases in the middle ear.

Ear diseases with superficial non-invasive mycoses

Otomycosis

The most frequent species were *Aspergillus flavus* (42.4%), *A. niger* (35.9%), *A. fumigatus* (12.5%). *Aspergillus* spp. was the most commonly isolated fungi in the immunocompetent group and *Candida albicans* in the immunocompromised group (Viswanatha et al. 2012). Bilateral involvement was more common in the immunocompromised group. In its superficial form it is characterized by a scant, odorless secretion as well as mild discomfort and pruritis.

If the disease progress to a moderate clinical stage the erythema increases and is joined by edema (particularly in the thicker pain exacerbated by tragal pressure or movement of the auricle. In the severe stage, pain is intense, the lumen of the canal obstructed, and extra-canal signs such as auricular cellulitis, parotitis, or adenopathy are likely (Ho et al. 2006).

Otitis media with effusion of fungal etiology

It was clarified by use of the methenamine silver staining method that fungi were present in the middle ear fluid in 100% of the studied cases of eosinophilic otitis media or OME (Murakami et al. 2012). Whether fungi are also present in the middle ear cavity of normal persons is unknown, but the possibility that they may contribute as a cause of both diseases cannot be excluded. In such diseases the eosinophil granule proteins can be released from the degenerated eosinophils to cause epithelial injury of the middle ear. Whether or not fungi induce the eosinophilia in the middle ear cannot be proved or excluded.

Nose and paranasal sinuses diseases with superficial non-invasive mycoses

Fungal Balls (FB)

FBs are non-invasive chronic fungal rhinosinusitis in immunocompetent patients with regional characteristics. In fungal balls fungal hyphae compressed into a mat or ball lying within the sinus lumen extrinsic to the mucosa. In this disease fungi are usually confined to one sinus, in which the maxillary sinus is the most common location of infection (87.8%). Fungal sinus balls are usually caused by fungi that are abundant in the air and biologically apt to implantation and these include species of *Aspergillus*.

The most common fungi vary according to country and geographical region, and while *Aspergillus fumigatus* is the most frequent agent in Europe and South Asia (Nomura et al. 2013), *Aspergillus flavus* is more frequent in India and Sudan (Ragab 2010).

Allergic Fungal Rhinosinusitis (AFRS)

AFRS is considered to be a non-tissue invasive fungal process with an allergic/hypersensitivity response to the presence of extra-mucosal fungi within the sinus cavity. This is possibly analogous to allergic broncho-pulmonary aspergillosis (ABPA). In the diagnosis of AFRS, the detection of fungi in allergic mucin is considered the hallmark of the disease, although hyphae are sparse in sinus content. Patients with AFS uniformly show five characteristics as described by Bent and Kuhn: gross production of eosinophilic mucin-containing non-invasive fungal hyphae, nasal polyposis, characteristic radiographic findings, immunocompetence and allergy to fungi (Bent and Kuhn 1996).

Mouth and pharynx ear diseases with superficial non-invasive mycoses

Candidiasis

Three forms of candidiasis that affect the mouth and pharynx can be considered as superficial non-invasive Mycoses:

Pseudomembranous candidiasis (thrush) is the characteristic pattern of superficial disease. It presents as white curd-like pseudomembranes, which can be taken away, leaving behind an erythematous mucosa. It is usually detected in immunocompromised patients, irradiated patients, elderly people, and patients with xerostomia or diabetes mellitus. Additionally it can be seen in nursing infants (due to the low pH in the oral cavity and the lack of a stable microbial ecosystem capable of inhibiting development of the fungus). The lesions appear especially in the oropharyngeal region, cheek mucosa and lateral surfaces of the tongue. It can be seen in the form of whitish-yellow plaques (masses of hyphae, yeasts, cell detritus and desquamated epithelial cells) of a soft and gelatinous consistency that exhibit a centrifugal growth pattern. The whitish-yellow plaques detach upon rasping, leaving an erythematous zone. Symptoms are sparse including burning, itching sensation and odynophagia.

Acute erythematous candidiasis is the most common appearance in both immune depressed and immunocompetent individuals, and tends to be visible as a complication of broad spectrum antibiotic treatment. By physical examination there are erythematous and atrophic regions located anywhere within the oral cavity, but particularly on the palate and tongue. The filiform papillae disappear, and the dorsal surface of the tongue appears smooth. The condition is frequently asymptomatic or is associated with a mild burning and itching sensation (Coronado-Castellote and Jiménez-Soriano 2013; Rautemaa and Ramage 2011).

Angle cheilitis is mostly manifested in patients with vitamin and iron deficiencies, staphylococcal infections, as well as in totally or partially edentulous individuals with vertical dimension loss. The consequence is the formation of fissures or cracks at the lip commissures, with the accumulation of saliva that supports the generation of a humid environment and colonization by Candida. Over time, the fissures and erosions can produce crusts and pain. Such patients had mixed infections with bacteria, e.g., *Staphylococcus aureus* or certain streptococci (Rautemaa and Ramage 2011).

Median rhomboid glossitis (MRG) has an effect on smokers and diabetics with incidence in less than 1% of the population. The posterior region of the back of the tongue, on the midline anterior to the lingual "V", and rarely the paramedial zone are the most commonly affected sites. Rounded or rhomboid plaque of firm consistency and an intense red or pink color, secondary to atrophy or the absence of filiform papillae, with limits that are clearly differentiated from the rest of the tongue are the clinical signs of the disease. It is usually asymptomatic with two dissimilar clinical presentations: an atrophic or macular and non-elevated (flat) form, or a mamillated, hyperplastic or exophytic lesion that can appear elevated (up to 2–5 mm). In some cases a fissured or lobulate lesion can be observed (Coronado-Castellote and Jiménez-Soriano 2013).

In the larynx

Isolated fungal infections of the larynx are rare, accounting for approximately 40 reported cases in literature. Larynx can be involved primarily by inoculation of sputum, or secondarily due to predisposing conditions. Isolated laryngeal candidiasis is infrequently recognized and poorly documented (Vrabec 1993). It has been described in patients with predisposing factors, such as underlying chronic diseases, granulocytopenia, broad spectrum antimicrobial therapy, steroid administration, diabetes mellitus, and mechanical, chemical, or thermal injury to the mucosal barrier.

Most cases are usually mild and typically show Punctuated white patches on the laryngeal mucosa (Neuenschwander et al. 2001; Tashijan and Peacock 1984; Walsh and Grey 1987).

Candida and *Aspergillus* are possible causative agents (Vrabec 1993). Laryngeal fungal infections can mimic laryngopharyngeal reflux (LPR) disease, granulomatous diseases, leukoplakia, and carcinoma (Chandran et al. 1988). In a 10-year retrospective analysis of 54 patients, inhaled steroids were the single most common predisposing factor, as seen in 89% of patients. Conditions such as LPR produced mucosal injury through retrograde acid transit via the esophagus, while prolonged antibiotic therapy can reduce the bacterial burden at the larynx, thereby predisposing the patient to fungal laryngitis (Henry et al. 2005; Neuenschwander et al. 2001).

Clinical presentation of fungal laryngitis is variable and can include hoarseness, dysphagia, dysphonia, odynophagia, stridor, and respiratory distress. A high index of suspicion is necessary to make a diagnosis of isolated laryngeal candidiasis. *Candida albicans* usually causes superficial infections that are characterized by epithelial hyperplasia as a result of direct mucosal invasion. Most cases of muco-cutaneous candidiasis are superficial, but they can be invasive when they present with systemic manifestations of fever, chills, and malaise (Nunes et al. 2008).

Video laryngoscopy typically reveals edema, erythema, hyperkeratosis, adherent white plaques, shallow ulcerations, and white or grey pseudomembrane formations (Zhang 2000). Diffuse mucosal edema and erythema despite sufficient LPR therapy should prompt consideration of fungal laryngitis (Henry et al. 2005).

Mucocutaneous mycoses

Mucocutaneous mycoses are diseases of the mucocutaneous covering. These diseases may elicit a host response and become symptomatic.

In the ear

Erosive external otitis (EEO) of fungal etiology: is an erosive process of the EAC that occurs in healthy patients. The clinical presentation of EEO is mild, with little or no aural discomfort. On examination, granulation tissue extending laterally from the annulus can be seen. A more benign, nonspecific pattern of inflammation, as opposed to the intense polymorphonuclear infiltration and necrosis can be observed in malignant otitis externa (MEO) (Nguyen et al. 2010).

In the oral cavity and pharynx

Chronic hyperplastic candidiasis can manifest in nodular form or as whitish plaques that cannot be attributed to any other disorder, do not detach upon rasping, and are typically located on the cheek mucosa and tongue, and especially bilaterally at both lip retro-commissures. In this form of the disease the Candida hyphae are not only found at epithelial surface level but also invade deeper levels where epithelial dysplasia can be observed—with the associated risk of malignization (Rautemaa and Ramage 2011).

In the nose and paranasal sinuses

Chronic granulomatous fungal rhinosinusitis

A chronic granulomatous form has been reported in Sudan, and the non-granulomatous forms occur mainly in diabetic patients with type 2 diabetes. On intranasal examination severe nasal congestion and polypoid mucosa may be noted. There may be soft tissue mass that can be either mucosally covered or ulcerated with overlying debris or dried secretions. Proptosis is the most common presentation of the granulomatous form in Sudan. *Aspergillus* species are the most common organisms involved and *A. jlavus* is the most common cause in Sudan. Three variants occur: proliferative (granulomatous pseudo tubercles in a fibrous tissue stroma), exudative-necrotizing (with prominent foci of necrosis) and a mixed form. The granulomatous reaction is composed of eosinophilic material surrounded by fungus, giant cells and palisading nuclei and non-granulomatous characterized by tissue necrosis with little inflammatory infiltrates and dense hyphal accumulation like fungus balls (Ragab 2010).

Deep mycoses

Deep mycoses involve the deeper layers of the mucus membrane (skin in the ears), including muscle, connective tissue and bone. The host immune system recognizes the fungi, resulting in variable tissue destruction. Two forms can exist according to the immune status of the host, the endemic and the opportunistic mycoses (De Pauw et al. 2008).

The endemic mycoses

The endemic mycoses are fungal infections caused by the classic dimorphic fungal pathogens *Histoplasma capsulatum*, *Blastomyces dermatitidis*, *Coccidioides immitis*, *Coccidioides posadasii*, *Paracoccidioides brasiliensis* and *Penicillium marneffei*. These fungi exhibit thermal dimorphism (exist as yeasts or spherules at 37°C and molds at 25°C) and are generally confined to geographic regions where they occupy specific environmental or ecologic niches. The endemic mycoses are often referred to as systemic mycoses, because these organisms are true pathogens and can cause infection in healthy individuals. All of these agents produce a primary infection in the lung, with subsequent dissemination to other organs and tissues. In ORL the most susceptible area is the larynx. Endemic mycoses include histoplasmosis, blastomycosis, coccidioidomycosis, paracoccidioidomycosis, sporotrichosis, and infection due to *P. marneffei* (De Pauw et al. 2008).

A. Proven Endemic Mycosis (De Pauw et al. 2008; Marchetti et al. 2012)

In a host with an illness consistent with an endemic mycosis, one of the following:

Recovery in culture from a specimen obtained from the affected site or from blood.

Histopathological or direct microscopic demonstration of appropriate morphologic forms with a truly distinctive appearance characteristic of dimorphic fungi, such as *Coccidioides* species spherules, *B. dermatitidis* thick-walled broadbased budding yeasts, *P. brasiliensis* multiple budding yeast cells, and in the case of histoplasmosis, the presence of characteristic intracellular yeast forms in a phagocyte in a peripheral blood smear or in tissue macrophages.

For Coccidioidomycosis, demonstration of coccidioidal antibody in CSF, or a 2-dilution rise measured in 2 consecutive blood samples tested concurrently in the setting of an ongoing infectious disease process. For paracoccidioidomycosis, demonstration in 2 consecutive serum samples of a precipitin band to paracoccidioidin concurrently in the setting of an ongoing infectious disease process.

B. Probable Endemic Mycosis (De Pauw et al. 2008; Marchetti et al. 2012)

Presence of a host factor, plus a clinical picture consistent with endemic mycosis and mycological evidence, such as a positive Histoplasma antigen test result from urine, blood, or CSF.

Opportunistic Invasive Fungal Diseases (IFD)

Examples are Candidiasis, Cryptococcosis, Aspergillosis, Zygomycosis and a whole lot more. The list could go on and on because we see organisms, that we have never thought would cause disease, cause disease in patients here and cause substantial infection that often leads to death. The infection may be acute or chronic. Revised definitions of invasive fungal disease from (EORTC/MSG) Consensus Group identified the disease to be proven, probable or possibly diagnosed (De Pauw et al. 2008; Marchetti et al. 2012):

- Proven (IFRORL) Disease: Cases of the IFD can be diagnosed when demonstrations of fungal elements in diseased tissue using histopathological examination exist.
- Probable (IFRORL) Disease: Cases of probable IFD require that a host factor, clinical features, and mycological evidence be present as listed below.
- Possible (IFRORL) Disease: include only those cases with the appropriate host factors and with sufficient clinical evidence consistent with IFD but for which there was no mycological support.

Criteria for Probable Invasive Fungal Disease Except for Endemic Mycoses (De Pauw et al. 2008; Marchetti et al. 2012)

a. Host factors

- Recent history of neutropenia (less than 0.5 × 1.9 neutrophils/L [less than 500 neutrophils/mm^3] for more than 10 days) temporally related to the onset of fungal disease.
- Receipt of an allogeneic stem cell transplant.
- Prolonged use of corticosteroids (excluding among patients with allergic bronchopulmonary aspergillosis) at a mean minimum dose of 0.3 mg/kg/day of prednisone equivalent for more than 3 weeks.
- Treatment with other recognized T cell immunosuppressants, such as cyclosporine, TNF-a blockers, specific monoclonal antibodies (such as alemtuzumab), or nucleoside analogues during the past 90 days.
- Inherited severe immunodeficiency (such as chronic granulomatous disease or severe combined immunodeficiency).

b. Clinical criteria

- Lower respiratory tract fungal disease: The presence of 1 of the following 3 signs on CT: Dense, well-circumscribed lesions(s) with or without a halo sign, Air-crescent sign and Cavity.
- Tracheobronchitis: Tracheobronchial ulceration, nodule, pseudomembrane, plaque, or eschar seen on bronchoscopic analysis.
- Sinonasal infection: Imaging showing sinusitis plus at least 1 of the following 3 signs: Acute localized pain (including pain radiating to the eye), nasal ulcer with black eschar and extension from the paranasal sinus across bony barriers, including into the orbit.

- Ear infection: necrotizing otitis externa:
  The diagnostic criteria of the disease can be divided into two categories: obligatory and occasional. The obligatory criteria are: pain, edema, exudate, granulations, microabscess (when operated), positive bone scan or failure of local treatment often more than 1 week. The occasional criteria are diabetes, cranial nerve involvement, positive radiograph, debilitating condition and old age. All of the obligatory criteria must be present in order to establish the diagnosis. The presence of occasional criteria alone does not establish it. The importance of Tc99 scan in detecting osteomyelitis is stressed. When bone scan is not available, a trial of 1–3 weeks of local treatment is suggested. Failure to respond to such treatment may assist in making the diagnosis of MEO (Cohen and Friedman 1987).
- CNS infection: 1 of the following 2 signs: Focal lesions on imaging and meningeal enhancement on MRI or CT.
- Disseminated candidiasis: At least 1 of the following 2 entities after an episode of candidemia within the previous 2 weeks: Small, target-like abscesses (bull's-eye lesions) in liver or spleen and progressive retinal exudates on ophthalmologic examination.

c. Mycological criteria

1. Direct test (cytology, direct microscopy, or culture): Mold in sputum, bronchoalveolar lavage fluid, bronchial brush, or sinus aspirate samples, indicated by 1 of the following:
   - Presence of fungal elements indicating a mold.
   - Recovery by culture of a mold (e.g., *Aspergillus*, *Fusarium*, Zygomycetes, or *Scedosporium* species).
2. Indirect tests (detection of antigen or cell-wall constituents)
   - Aspergillosis: Galactomannan antigen detected in plasma, serum, bronchoalveolar lavage fluid, or CSF.
   - Invasive fungal disease other than cryptococcosis and zygomycosis: β-d-glucan detected in serum.

Conclusions

The nomenclature and classifications are an essential part of scientific methodology in health care. This is the first time an applied classification of FRORL diseases has been presented in our medical field. The classification and the nomenclature in this sector of healthcare activity enables accurate diagnosis and treatment of such fungal disease categories. This classification is essential as it provides a common language for reporting and monitoring diseases. It facilitates the collection and storage of data for analysis and evidence-based decision-making.

References

Aas, J.A., B.J. Paster, L.N. Stokes, I. Olsen and F.E. Dewhirst. 2005. Defining the normal bacterial flora of the oral cavity. J Clin Microbiol. 43: 5721–5732.

Akçay, S., B. Akman, H. Ozdemir, F.O. Eyüboğlu, O. Karacan and N. Ozdemir. 2013. External auditory canal microbiology and hearing aid use Am J Otolaryngol. 34: 278–281.

Bent, J.P. and F.A. Kuhn. 1996. Antifungal activity against allergic fungal sinusitis organisms. Laryngoscope 106: 1331–1334.

Braun, H., W. Buzina, K. Freudenschuss, A. Beham and H. Stammberger. 2003. Eosinophilic fungal rhinosinusitis: A common disorder in Europe? Laryngoscope 113: 264–269.

Chandran, S.K., K.M. Lyons, V. Divi, M. Geyer and R.T. Sataloff. 2009. Fungal laryngitis. Ear Nose Throat J. 88: 1026–1027.

Cohen, D. and P. Friedman. 1987. The diagnostic criteria of malignant external otitis. J Laryngol Otol. 101: 216–221.

Coronado-Castellote, L. and J. Jiménez-Soriano. 2013. Clinical and microbiological diagnosis of oral candidiasis. J Clin Exp Dent. 5: e279–e286.

De Pauw, B., T.J. Walsh, J.P. Donnelly, D.A. Stevens, J.E. Edwards, T. Calandra et al. 2008. Revised definitions of invasive fungal disease from the European Organization for Research and Treatment of Cancer/Invasive Fungal Infections Cooperative Group and the National Institute of Allergy and Infectious Diseases Mycoses Study Group (EORTC/MSG) Consensus Group. Clin Infect Dis. 146: 1813–1821.

Feig, M.A., E. Hammer, U. Völker and N. Jehmlich. 2013. In-depth proteomic analysis of the human cerumen-a potential novel diagnostically relevant biofluid. J Proteomics 83: 119–129.

Geiser, M., C. Wigge, M.L. Conrad , S. Eigeldinger-Berthou, L. Künzi, H. Garn, H. Renz and M.A. Mall. 2014. Nanoparticle uptake by airway phagocytes after fungal spore challenge in murine allergic asthma and chronic bronchitis. BMC Pulm Med. 14: 116.

Hawksworth, D.L. 2001. The magnitude of fungal diversity: the 1.5 million species estimate revisited. Mycol Res. 105: 1422–1432.

Henry, L.R., M.D. Packer and J. Brennan. 2005. Airway-obstructing laryngeal candidiasis in an immunocompetent host. Otolaryngol. Head Neck Surg. 133: 808–810.

Ho, T., J.T. Vrabec, D. Yoo and N.J. Coker. 2006. Otomycosis: clinical features and treatment implications. Otolaryngol. Head Neck Surg. 135: 787–791.

Huffnagle, G.B. and M.C. Noverr. 2013. The emerging world of the fungal microbiome. Trends Microbiol. 21: 334–341.

Listgarten, M.A. 2000. The structure of dental plaque. Periodontology 5: 52–65.

Madsen, A.M., V. Schlünssen, T. Olsen, T. Sigsgaard and H. Avci. 2009. Airborne fungal and bacterial components in PM1 dust from biofuel plants. Ann Occup Hyg. 53: 749–757.

Madsen, A.M., K. Tendal, T. Thilsing, M.W. Frederiksen, J. Baelum and J.V. Hansen. 2013. Fungi, β-glucan, and bacteria in nasal lavage of greenhouse workers and their relation to occupational exposure. Ann Occup Hyg. 57: 1030–1040.

Marchetti, O., F. Lamoth, M. Mikulska, C. Viscoli, P. Verweij and S. Bretagne. 2012. European Conference on Infections in Leukemia (ECIL) Laboratory Working Groups. ECIL recommendations for the use of biological markers for the diagnosis of invasive fungal diseases in leukemic patients and hematopoietic SCT recipients. Bone Marrow Transplant 47: 846–854.

Murakami, A., T. Tutumi and K. Watanabe. 2012. Middle ear effusion and fungi. Ann Otol Rhinol Laryngol. 121: 609–614.

Neuenschwander, M.C., A. Cooney, J.R. Speigel, K.M. Lyons and R.T. Sataloff. 2001. Laryngeal candidiasis. Ear Nose Throat J. 80: 138–139.

Nguyen, L.T., J.P. Harris and Q.T. Nguyen. 2010. Erosive External Otitis: A Novel Distinct Clinical Entity of the External Auditory Canal in Nonimmunosuppressed Individuals. Otol Neurotol. 31(9): 1409–1411.

Nomura, K., D. Asaka, T. Nakayama, T. Okushi, Y. Matsuwaki, T. Yoshimura, M. Yoshikawa, N. Otori, T. Kobayashi and H. Moriyama. 2013. Sinus fungus ball in the Japanese population: clinical and imaging characteristics of 104 cases. Int J Otolaryngol. 73: 16–40.

Nunes, F.P., T. Bishop, M.L. Prasad, J.M. Madison and D.Y. Kim. 2008. Laryngeal candidiasis mimicking malignancy. Laryngoscope 118: 1957–1959.

Odds, F.C., T. Arai, A.F. DiSalvo, E.G.V. Evans, R.J. Hay, H.S. Randhawa, M.G. Rinaldi and T.J. Walsh. 1992. Spectrum of mycoses. J Med Vet Mycol. 30: 1.

Patil, S., R.S. Rao, D.S. Sanketh and N. Amrutha. 2013. Microbial flora in oral diseases. J Contemp Dent Pract. 14: 1202–1208.
Ponikau, J.U., D.A. Sherris, E.B. Kern, H.A. Homburger, F. Frigas, T.A. Gaffey and G.D. Roberts. 1999. The diagnosis and incidence of allergic fungal sinusitis. Mayo Clin Proc. 74: 877–884.
Ragab, A. 2010. Current trends in fungal sinusitis. pp. 193–231. *In*: Rai, Mahendra (ed.). Progress in Mycology, 1st edition. Springer-Verlag, India.
Ragab, A. and P. Clement. 2007. The role of fungi in the airway of chronic rhinosinusitis patients. Curr Opin Allergy Clin Immunol. 7: 17–24.
Ragab, A., P. Clement, W. Vincken, N. Nolard and F. Simones. 2006. Fungal cultures of different parts of the upper and lower airways in chronic rhinosinusitis. Rhinology 44: 19–25.
Rautemaa, R. and G.E. Ramage. 2011. Oral candidosis: clinical challenges of a biofilm disease. Crit Rev Microbiol. 37: 328–336.
Reponen, T., S. Seo, F. Grimsley, T. Lee, C. Crawford and S.A. Grinshpun. 2007. Fungal fragments in moldy houses: a field study in homes in New Orleans and Southern Ohio. Atmos Environ. 41: 8140–8149.
Socransky, S.S. and S.D. Manganiello. 1971. The oral microflora of man from birth to senility. J Periodontol. 42: 485–496.
Stroman, D.W., P.S. Roland, J. Dohar et al. 2001. Microbiology of normal external auditory canal. Laryngoscope 111: 2054–2059.
Tashijan, L.S. and J.E. Peacock. 1984. Laryngeal candidiasis. Report of seven cases and review of the literature. Arch Otolaryngol. 110: 806–809.
Tedersoo, L., M. Bahram, S. Põlme et al. 2014. Fungal biogeography. Global diversity and geography of soil fungi. Science 346: 1256688.
Viswanatha, B., D. Sumatha and M.S. Vijayashree. 2012. Otomycosis in immunocompetent and immunocompromised patients: comparative study and literature review. Ear Nose Throat J. 91: 114–121.
Vrabec, D.P. 1993. Fungal infections of the larynx. Otolaryngol. Clin North Am. 26: 1091–1114.
Walsh, T.J. and W.C. Gray. 1987. *Candida* epiglottitis in immunocompromised patients. Chest. 91: 482–485.
Zhang, S., T.L. Farmer, M.A. Frable and C.N. Powers. 2000. Adult herpetic laryngitis with concurrent candidal infection: a case report and literature review. Arch. Otolaryngol. Head Neck Surg. 126: 672–674.

CHAPTER 6

Fungal Infection in Renal Transplant Patients

Salwa S. Sheikh,[1,*] *Abdul Razack A. Amir*[2] and *Samir S. Amr*[3]

Introduction

Since the first successful human kidney transplant performed in 1954, solid organ transplantation (SOT) has increased worldwide. With time, as immunosuppressive therapy and graft survival improved, the risk of infectious complications and malignancies also increased. In particular invasive fungal infection (IFI) is a major complication and an important cause of morbidity and mortality among SOT patients. The most common causes of IFIs in SOT are Candidiasis followed by Aspergillosis, and Cryptococcosis except in lung transplant recipients where *Aspergillus* is the most common cause of IFI. Fungal infections in renal transplant recipients had been reported since the early days of the introduction of kidney transplantation as a mode of therapy for renal failure. The estimated risk of developing IFI in renal transplant recipients varies from 1.4–9.4%, commonly between 3–5%, depending on different reports from various transplant centers (Peterson et al. 1982; Nampoory et al. 1996; Patel and Paya 1997; Stitt 2003; Badiee et al. 2005; Einollahi et al. 2008; Low and Rotstein 2011). The risk of opportunistic infections in kidney transplant patients is determined by the interaction between epidemiologic exposures to offending organisms, whether it is within the hospital or in the community, or recipient's net-state of immunosuppression. These patients may acquire fungal infection via one of the two main routes.

[1] Pathology Services Division, Johns Hopkins Aramco Healthcare, JHAH, Dhahran, Building 62, Rm: 296A Dhahran, Saudi Arabia.
[2] Internal Medicine Division, Nephrology Services, Johns Hopkins Aramco Healthcare, JHAH, Dhahran, Saudi Arabia.
[3] Department of Pathology and Laboratory Medicine, King Fahad Specialist Hospital, Dammam, Saudi Arabia.
* Corresponding author: sheikhss@aramco.com.sa; salwa.sheikh@aramco.com

1. Infections caused by geographically limited endemic systemic mycosis are usually infrequent and lead to either disseminated primary infection or reactivation infection. Examples of such fungi include histoplasmosis, coccidioidomycosis, paracoccidioidomycosis, and blastomycosis.
2. Infections caused by opportunistic fungal species that only rarely cause invasive fungal infection in normal host and are usually nosocomial acquired within the hospital environment. Examples of such organisms include *Candida* species, *Aspergillus* species, Mucorales species (Zygomycetes), and *Cryptococcus neoformans* (Tolkoff-Rubin and Rubin 1992; Badiee et al. 2005; Badiee and Alborzi 2011).

Risk of Infection in Transplant

The risk of infection in renal transplant recipients is determined by a balanced relationship between two factors—Epidemiologic exposures and Net state of immunosuppression.

1. **Epidemiological Exposures**

   Epidemiological exposures vary based upon the nature of immune deficits with most transplant patients having multiple deficits. For example, transplant recipients with neutropenia are more prone to develop bacterial and fungal infections, while those with T cell immune deficits are more commonly infected with viral and intracellular organisms such as *Mycobacterium tuberculosis (M. Tb)*. In addition latent infections are often reactivated in the setting of immunosuppression. In fact, *Strongyloides stercoralis* may be reactivated many years after transplantation (Fishman and Davis 2008; Snydman et al. 2009; Fishman 2014).
   Epidemiological exposures of importance may be divided into four categories:

   a) Donor-Derived infections:
   Infection derived from donor tissue and activated in the recipient is among the most important and least appreciated exposures in transplantation. Some of these are latent while others are active infections that are not diagnosed at the time of transplantation. Three types of infections need special attention. First are bloodstream infections where donors may have active infection at the time of procurement. Second, some bacterial and fungal organisms selectively adhere to the vascular or urinary anastomotic sites leading to leaks or mycotic aneurysms. Examples of such organisms include *Candida* species, and bacteria such as *Salmonella*, *Staphylococcus*, *Streptococcus pneumonae*, and *Escherichia coli*. Second are the viral infections such as Cytomegalovirus (CMV) and Epstein Barr virus (EBV) that are often associated with particular syndromes and illnesses in immunocompromised population. The greatest risk is in seronegative recipients receiving infected grafts from seropositive donors with latent infection. Third is the late latent infection such as tuberculosis and

histoplasmosis that can be reactivated many years after transplantation. Donor screening before transplantation is limited by the available testing technology and the time available within which organs from deceased donors must be used. As a result, some active infections may go undetected and be transmitted to the recipient. Fungal infections that can be transmitted from donor include *Candida* often related to central venous catheter, *Aspergillus*, *Cryptococcus neoformans*, and endemic fungi such as *Histoplasma capsulatum*, *Coccidioides immitis*, and *Blastomyces dermatidis*. Certain organisms when identified in donors before transplantation should be considered contraindication to organ donation. These include Cryptococcal fungal infection among several other viral, bacterial, and parasitic infections (Fishman and Davis 2008; Snydman et al. 2009; Morris et al. 2010; Fishman et al. 2012). Donor-derived fungal infections will be discussed further in detail later in the chapter.

b) Recipient-Derived infections: These are often latent infections that activate later in the setting of immunosuppression and require detailed history, including history of vaccination status, and dietary habits. Notable among these infections are viruses such as herpes simplex, varicella zoster, hepatitis B (HBV), hepatitis C (HCV), human immunodeficiency virus (HIV), *M. tuberculosis*, *S. stercolaris,* and fungal infections such as coccidioidomycosis, and histoplasmosis.
c) Community acquired pathogens: Community exposures are often related to water-borne, food borne, respiratory viral, fungal, and parasitic organisms. A number of potential organisms in the community may lead to infection in transplant recipients. Viral pneumonia due to common respiratory viruses as well as atypical respiratory pathogens predisposes these patients to bacterial or fungal superinfections. Recent or remote exposure to endemic fungal organisms such as *H. capsulatum*, *B. dermatidis*, and *Coccidioides immitis* may lead to localized pulmonary or systemic infection.
d) Nosocomial infections: Nosocomial infections in transplant patients are of increasing significance as organisms with significant antimicrobial resistance are present in most hospitals that include methicillin-resistant staphylococcus, vancomycin-resistant enterococci, and fluconazole-resistant *Candida*. Other fungal infections that can be acquired in hospitals are *Aspergillus* and *non-albicans Candida* species (Pappas et al. 2010). In addition, misuse of antibiotics has led to increased *Clostridium difficile* infection rate. Outbreaks of *Legionella* in hospitals are often seen in association with contaminated water supplies and ventilation systems.

2. Net State of Immunosuppression

The net state of immunosuppression in transplant recipients is determined by several factors (Fishman and Davis 2008; Fishman 2014):

a) Immunosuppressive therapy including dose, duration, and sequence of agents. Specific immunosuppressive agents are associated with increased risk of infection by certain specific organisms, mainly viral and bacterial.
b) Underlying or comorbid diseases.

c) Prolonged instrumentation and invasive devices such as vascular access devices, catheters, surgical drains, and airway intubation.
d) Prolonged use of broad spectrum antimicrobials.
e) Presence of devitalized tissue or fluid collection in or around transplanted organ.
f) Infection with immunomodulating viruses such as CMV, Epstein Barr virus (EBV), HBV, HCV, or HIV.
g) Other host factors including renal or hepatic dysfunction, graft dysfunction, and metabolic dysfunctions such as diabetes, uremia, and malnutrition.

Timing of Infection after Transplantation

Specific infections vary with a predictable pattern depending on the time after transplantation. This in turn depends on multiple factors including changing risk factors over time such as immunosuppression, surgical procedures or hospitalization, latent or new infections, and graft rejection.

1. ***First four weeks after transplantation*** including perioperative period reflects surgical and technical complications. Three types of infection occur during this period. *First* is the emergence of recipient carried infections that are inadequately treated and not eradicated. These are activated in a setting of immunosuppression, and stress. Examples include pretransplant pneumonia, vascular access infection, and colonization of recipient with resistant organisms that may infect catheters or surgical drains. *Second* type of infection is donor derived and may be nosocomially derived as a systemic infection or contamination during organ procurement. Nosocomial infection could also be acquired through infected lines and may be caused by staphylococcus aureus, resistant gram negative bacilli, or *Candida* species. *Third* and most common source of infection is related to surgical procedure and includes wound infection, aspiration pneumonia, urinary tract infection, infection from vascular access or catheter, infection of fluid collection, and leak of vascular or urinary anastomosis. Unfortunately most of these cases show only subtle signs of infection during this period due to the immunosuppression. Fungal organisms that commonly infect at this time are *Candida* and *Aspergillus* species.
2. ***One to six months after transplantation*** reflect intense immunosuppression. Usually this is the time of opportunistic infections such as *Aspergillus*, Pneumocystis, Listeria, and Toxoplasma among others. Viral activation may also occur during this time. In addition infection due to surgical and technical complications as described above can also be seen during this time.
3. ***More than 6–12 months after transplantation*** reflect mostly community acquired infections. Majority of these patients (70–80%) have reduced immunosuppression with satisfactory allograft function. These patients exhibit similar behaviour as the general community with community acquired respiratory virus posing a major risk. The second subgroup of transplant recipients has chronic viral infections which in absence of effective therapy may lead to end-organ damage or malignancy. A third group consists of transplant recipients with allograft

dysfunction and severe immunosuppression. These patients are susceptible to chronic viral infections and opportunistic infections. Invasive fungal organisms that can infect at this stage include *Aspergillus*, *Zygomycetes,* and *Dermaticeae* (Fishman and Davis 2008).

Invasive Fungal Infection in Renal Transplant; General Considerations and Commonly Encountered Pathogens

IFIs are on the rise in hematopoietic stem cell and SOT patients, including renal transplant patients, often leading to substantial morbidity and mortality (2) (Low and Rotstein 2011). The annual number of deaths with IFI increased from 1557 to 6534 over a period of 17 years accounting for approximately 320% rise (17) (McNeil et al. 2001). The 1 year cumulative incidence of IFI varies in SOT recipients depending on the organ transplanted and is highest for small bowel (11.6%) and lowest for kidney transplant recipients (1.3%) (Pappas et al. 2010; Fishman 2014).

IFIs are a significant and often lethal complication in renal transplant patients. Common predisposing factors to fungal infection include immunosuppression, indwelling catheters, use of broad spectrum antibiotics, technical/surgical complications with wound and fluid collections, disruption of epithelial surfaces, and other risk factors like diabetes and prolonged dialysis before renal transplant (Abbott et al. 2001; John et al. 2003). The Transplant-Associated Infections Surveillance Network (TRANSNET), a consortium of 23 transplant centers in United States, reported *Candida* (53%) to be the most common cause of IFI in SOT followed by *Aspergillus* (19%), *cryptococcus* (8%), non-*Aspergillus* molds (8%), endemic fungi (5%), and *Zygomycetes* (2%). Data from 15 centers shows the mortality in SOT patients at 12 months is approximately 40% for Aspergillosis, 34% for Candidiasis, and 34% for Cryptococcosis (Pappas et al. 2010; Spellberg et al. 2005).

With improvement in anti-rejection therapy and enhanced immunosuppression, significant improvement in graft survival has been seen since the early 90s. The current rate of graft survival is reported to be more than 90% in most cases. Likewise, the rate of fungal infection has also been reported to increase (Badiee and Alborzi 2011).

IFIs by geographically limited endemic fungi are uncommon. Opportunistic fungal infections, on the other hand, are the most commonly encountered fungal infections and characteristically lead to disseminated infection. These infections are either primary or secondary. The primary IFIs usually involve the lungs mostly caused by *C. neoformans*, *Aspergillus*, and less commonly Mucoraceae. The secondary infection is either of lungs or mucocutaneous surfaces that are damaged or disrupted by other processes or through infected lines and catheters by *Candida*, *Aspergillus*, *Torulopsis (Candida) glabrata*, and Mucoraceae (Badiee et al. 2005). In case of Mucoraceae, few metabolic factors have been identified as risk factors that include systemic acidosis often predisposing patients to invasive pulmonary and rhinocerebral Mucormycosis, deferoxamine therapy in dialysis patients, and renal failure. Among kidney transplant recipients Mucormycosis comprises about 2–6% of cases, however, it is associated with the highest 2 year mortality rate and longest duration of hospitalization (Abbott et al.

2001; Kuy and Cronin 2013). In addition, use of voriconazole or capsofungin increases the risk of Zygomycosis by 4.4 times (Van Cutsem and Boelaert 1989; Spellberg et al. 2005; Veroux et al. 2006; Singh et al. 2009; Pappas et al. 2010; Kuy et al. 2013; Chung et al. 2013). Combined use of rituximab and plasmapheresis pre-transplant also increases post-transplant infection in renal transplant recipients. It is important to emphasize that identification of not only the primary site of IFI but also the other involved sites is of utmost significance as it has major implications on treatment and prognosis. For example the survival rate of patients with localized invasive pulmonary *Aspergillus* is 85% whereas it is less than 5% with central nervous system (CNS) involvement (Badiee et al. 2005).

Most IFIs in renal transplant patients manifest as either superficial (cutaneous/ subcutaneous) or systemic forms. Cutaneous/superficial infections can pose a significant problem and may involve the cornea and external auditory canal (Badiee and Alborzi 2011). Majority of systemic infections occur in patients with poor renal function and intensive immunosuppression. Colonization with a fungal species specially *Candida* is a risk factor for infection. *Candida* colonization is reported in up to 45% of kidney transplant recipients with almost all cases developing IFI. Another mean of acquiring these infections is via inhaling fungal spores. These invasive infections are most often associated with intravascular invasion or infection of sinuses, CNS, and orbit (Peterson et al. 1982).

There are several new emerging epidemiological trends that have been noted recently. There is increasing incidence of infection by *non-albicans Candida* species. Although *C. albicans* still remains the most common *Candida* species infecting renal transplant recipients, *C. tropicalis, C. parapsilosis,* and *C. glabrata* are now more frequently isolated. The second observation, as reported by SENTRY Antimicrobial Surveillance Program, is the increase in resistance of *Candida* species to azole treatment not only by *C. albicans* but also *non-albicans* species. In addition, with the increasing use of voriconazole for prophylaxis and treatment of *Aspergillus*, there is increase in incidence of Zygomycosis (Veroux et al. 2006; Fishman et al. 2012; Fishman 2014).

Most IFIs occur late after transplant with majority occurring more than 90 days after transplantation. The median time to onset is 103 days for invasive Candidiasis, 184 days for invasive Aspergillosis, 312 days for Zygomycosis, 343 days for endemic fungal infections, 467 days for non-*Aspergillus* molds, and 575 days for *Cryptococcus*. IFIs may occur even 3 years after transplantation. Occurrences of early IFI within the first 90 days post-transplantation are mostly caused by invasive Candidiasis, and invasive Aspergillosis (Pappas et al. 2010; Fishman 2014).

Symptoms of IFI can be subtle and nonspecific particularly at early stages and are therefore difficult to diagnose requiring vigilant clinical and diagnostic workup of suspected cases for early diagnosis and proper treatment.

In an autopsy study on renal transplant recipients who died of various causes, Rifkind et al. 1967, reported IFIs in 45% (23 of 51) patients. These infections were mostly caused by *Candida* (12 cases), followed by *Aspergillus* in five, *Nocardia* in two, *Histoplasma* in one case, and combinations of these fungi in three (Rifkind et al. 1967).

Several studies from various countries elaborated on various types of fungal infections in their renal transplant recipients. There had been some variation in the fungal species infecting renal transplant recipients from one country to another. In a large study from USA, TRANSNET, on SOT totaling 16,808 cases, including 8,672 renal transplants, there were a total of 1208 IFIs identified among 1063 solid organ transplant recipients. Out of these, 332 occurred in renal transplant patients. Candidiasis was responsible for the highest number of cases accounting for 49% of IFI, followed by Cryptococcosis (15%), Aspergillosis (14%), Endemic mycoses (10%), and Zygomycosis (2%) (16).

In one study of 471 patients from Australia, IFI developed in 2.1% (10 cases) of renal transplant recipients. *C. neoformans* was responsible for 50% of episodes (5 cases) followed by *A. fumigatus* (3 cases), and *Pseudallescheria boydii* (3 cases) and a single case of mucormycosis. The most commonly involved sites are the lungs followed by meninges and skin (28) (Ezzatzadegan et al. 2012). A study of 512 renal transplant recipients from Kuwait revealed 18 (3.5%) instances of IFI. These included candidiasis (8 cases), aspergillosis (5 cases), cryptococcosis (3 cases) and zygomycosis (2 cases) (Nampoory et al. 1996). Another study from Tunisia reported 11 cases of IFIs among 321 (3.4%) renal transplant recipients, including four cases of pneumocystosis, two cases of candidiasis, two cases of aspergillosis, two cases of cryptococcosis and only one case of mucormycosis (Trabelsi et al. 2013). A large study of 2410 recipients from Iran, showed 21 patients (0.87%) with IFI in renal transplant recipents, with mucormycosis being the most common pathogen responsible for 52% of the infections (11 cases), followed by disseminated candidiasis, aspergillosis, nocardiasis, and histoplasmosis (Einollahi et al. 2008). Two studies from different parts of India report the incidence of IFI in transplant recipients to be 6.1% and 6.9%. One of these studies showed *Candida* to be the most common infecting organism followed by *Aspergillus*, *Cryptococcus, Pneumocystis*, and Mucormycosis. The second study showed *Cryptococcus* to be the most common organism followed by *Candida,* mucormycosis, *Aspergillus*, and mixed *Aspergillus* with cryptococcosis (Chugh et al. 1993; Jayakumar et al. 1998).

The incidence of IFI in 296 kidney graft recipients admitted to a medical center in Turkey between 1986 and 1999 was found to be 4%. IFI was diagnosed in 18% of 28 recipients who were transplanted in India and 8% of 12 recipients that were transplanted in Russia, however it was encountered in only 2% of recipients transplanted at the same center. The most common etiologic agent was *A. fumigatus* but C*andida* spp., *Rhizopus* spp. and *C. neoformans* were also encountered. In 2 patients, 2 different pathogens were isolated at the same time (Altiparmak et al. 2002).

These variations could be related to prevalence of certain fungi in the environment of a given geographic location or to nosocomial infections in the center where the study was conducted. Another factor could be related to commercial transplants. We present herein a detailed review of some common invasive fungal infections.

Candidiasis

Candida is the most common systemic/invasive fungal infection encountered in renal transplant patients accounting for over half of all IFIs in this population, and has associated infection-related mortality rate of 1.5% (16, 29, 33, 34). In a large prospective study, invasive candidiasis had a 12-month cumulative incidence of 1.9%, the highest of all IFIs. It is seen more frequently in small intestine, pancreas, liver, kidney, heart and lung transplant recipients in descending order and occurs earlier than other invasive mycoses, generally within the first 3 months after transplantation. The most common sites of infection are bloodstream infection, intra-abdominal and urinary tract infection (Silveira et al. 2013).

Mucocutaneous infections are most commonly seen in patients on a high dose steroid treatment, during broad spectrum antibiotic therapy, and with diabetes. Oral *Candida* (OC) carriage and infection have been reported to be associated with a greater risk for systemic infection in transplant recipients. In a study on the prevalence of oropharyngeal candidiasis, the oral carrier status, *Candida* titers, and the involved species in 90 kidney and heart transplant subjects with 72 matched healthy controls; it was found that seven transplant patients had infection with *C. albicans,* and none among the controls. The transplant group had significantly higher OC titers than the control group. The most frequent species combination in transplant subjects was *C. albicans* and *C. glabrata*. The later was isolated from 13.5% of transplant carriers, however none of the controls was positive (Dongari-Bagtzoglou et al. 2009).

The prevalence and risk of OC was looked at comparing 500 renal transplant recipients and 501 healthy controls in Spain. The prevalence of OC is found to be 7.4% in the transplanted subjects, compared with 4.19% in healthy controls ($P < 0.03$) and the most frequent type of OC in both transplant and healthy groups is denture stomatitis (López-Pintor et al. 2013).

In Saudi Arabia, a study was conducted to determine and compare the prevalence of oral candidal colonization and OC in 58 renal transplant recipients and 52 healthy control subjects. Prevalence of oral fungal colonization was not significantly higher in transplant recipients than in healthy subjects (74.1% vs. 59.6%, respectively), but the density of growth was significantly higher in the transplant subjects. OC was diagnosed in 15.5% of transplant subjects but in none of the healthy controls. In addition, transplant subjects who used a chewing stick (Miswak: Salvadora persica) for oral hygiene had a significantly lower prevalence of OC compared with other transplant recipients (Al-Mohaya et al. 2002).

The same researchers reviewed the prevalence of intra-oral lesions in the same group of 58 medically stable Saudi renal transplant patients, in comparison with age and sex-matched 52 healthy control subjects. Gingival overgrowth, erythematous candidiasis and hairy leukoplakia were diagnosed in renal transplant patients with prevalence of 74.1%, 15.5%, and 8.6%, respectively. *C. albicans* was isolated from five lesions (55.6%), while *C. dubliniesis* and *C. famata* each was isolated from two lesions (22.2%). None of the healthy control subjects had OC (Al-Mohaya et al. 2009).

An earlier study from London University Hospital of 159 renal transplant recipients and 160 control patients, showed the most common lesion to be cyclosporin-induced gingival hyperplasia. OC was observed in 9.4% of renal transplant recipients compared with 2.5% of the controls and 3.8% of renal transplant recipients exhibited EC which was not seen in the controls (King et al. 1994).

A study of from Iranon 120 renal transplant subjects evaluated fungal colonization by mouth, vagina, urine, and rectal swabs. Fifty four kidney recipients (45%) had *Candida* colonization in different sites of their bodies. Fungal infections presented in 13 of 120 recipients (10.8%). Five recipients had IFIs; 3 had fungal pneumonitis and 2 severe esophagitis (Badiee et al. 2005).

In Poland, 185 patients who were below the age of 25, and who underwent kidney or liver transplant were evaluated with control subjects for the frequency of OC. *Candida* spp. colonies were isolated in the oral mucosa in 34% graft recipients and 27% control subjects (Olczak-Kowalczyk et al. 2010).

Aggressive management with eradication of *Candida* is required when it is seen in association with catheters, vascular access, and surgical drains requiring removal of the "foreign body" and systemic antifungal therapy.

Candiduria is a unique problem in renal transplant patients which is often asymptomatic. In a study of 1738 recipients of kidney transplant, 192 patients had 276 episodes of candiduria. *C. glabrata*, was the most common pathogen identified and was recovered in 51% of patients. Most candiduria patients are asymptomatic. Independent predictors of candiduria are female gender, intensive care unit admission, and antibiotic use during the month preceding candiduria, presence of an indwelling bladder catheter, diabetes, neurogenic bladder, and malnutrition. A variety of regimens are used for treatment including removal of the indwelling bladder catheter. Candiduria may be cleared with treatment in two-thirds of the patients, however treatment is not associated with an improved survival rate (Safdar et al. 2005).

One case from Saudi Arabia described a patient who received cadaveric renal transplant and developed urinary tract infection, following allograft rejection and treatment by antilymphocyte globulin, prednisone and cyclosporine. It was caused by *Hansenula anomala*, which waslater reclassified as *Candida pelliculosa.* The patient had spontaneous resolution of his candiduria without any need for antifungal treatment (Qadri et al. 1988).

Obstructing fungal balls can occur at uretropelvic junction particularly in patients with poor bladder function leading to obstructive uropathy, pyelonephritis, and systemic infection. Fungemia with *Candida* in the bloodstream carries more than 50% risk of visceral invasion (Fishman and Davis 2008).

Renal transplantation can be one of the etiological factors for candidemia. Intravascular lines are noted to be the most significant etiological factor, observed in 88% of candidemia cases. Other causes include admission to intensive care (51%), corticosteroid therapy (12%), chemotherapy (11%), multiple antibiotics (74%), parenteral nutrition (35%), renal transplant (5%), and neutropenia (3%) (43) (Schelenz and Gransden 2003). The most frequently isolated spp. are reported to be *C. albicans* followed by *C. glabrata, C. tropicalis,* and *C. parapsilosis*. Data was extracted from Prospective Antifungal Therapy Alliance (PATH) database, which is a comprehensive registry that collects information regarding IFIs. This data was obtained to review the epidemiology and outcome of candidemia in patients from USA and Canada. Candidemia was diagnosed in 2019 out of 4010 patients representing 50.3%. The incidence of *non-candida albicans* (54.4%) was higher than *C. albicans* (45.6%). The overall crude 12-week mortality rate is 35.2% (Horn et al. 2009).

Candidemia can vary in different types of SOT. In a study from Turkey on bloodstream infections in kidney (556 cases), liver (307 cases) and heart (64 cases) transplants, candidemia was encountered in only one patient with renal transplant (0.18%), and in seven patients with liver transplant (2.3%). None of the heart transplant patients had candidemia (Yesilkaya et al. 2013).

A nationwide surveillance study from Spain on bloodstream infections among transplant recipients, encountered 6 cases of candidemia (5%) and 6 cases of non-candida fungemia (5%) among 1400 renal transplant recipients (Moreno et al. 2007).

Rare cases of unusual presentations have been reported that include disseminated intravascular coagulation (DIC) with associated purpura fulminans of the skin associated with *C. tropicalis, trnaisent fungemia,* and right knee artheritis in a kidney and pancreas transplant recipient due to *C. zeylanoides,* and psoas abscess in renal transplant recipient with prolonged pyuria and candiduria (Silverman et al. 1986; Bisbe et al. 1987; Ozgur et al. 2014).

Esophageal Candidiasis (EC) is another *Candida* infection that can affect renal transplant recipients. Gupta et al. (1994) conducted a retrospective study of 265 live related renal allograft recipients to investigate *Candida* esophagitis, and compared three groups: patients given azathioprine and prednisolone (group I), those given cyclosporine, azathioprine, and prednisolone (group II), and those given cyclosporine and prednisolone (group III). The overall incidence of EC was 10.5%. Group II patients had a significantly higher incidence (28.6%) than those in group I (10.4%) and group III (3.8%). Dysphagia (57.1%) was the most common presenting symptom of EC. 21.4% of patients were asymptomatic and oral thrush was present in 42.9%. The entire esophageal mucosa was affected in six (46.1%) patients in group II and one (20%) in group III. Treatment failure occurred in seven (25%). Three patients died of disseminated candidiasis (Gupta et al. 1994).

Aspergillosis

Aspergillus as an IFI following renal transplant was recognized back in the late 1960s and early 1970s after the introduction of renal transplantation for treatment of end stage kidney disease (ESKD). The frequency of *Aspergillus* infection in renal transplant recipients is between 0.5–2.2% with a mortality rate of more than 88%. It can occur as a primary or secondary infection (Einollahi et al. 2008; Morris et al. 2010; Fishman et al. 2012). Fatal systemic aspergillosis was reported in 1967 in one study in five patients out of 23 who died of IFI (Rifkind et al. 1967). In 1972, Burton et al. (1972) reported four patients who had received kidney transplants and became infected with *A. fumigatus* while receiving immunosuppressive therapy. Three of them had invasive pulmonary infection while the fourth had a disseminated disease. All four were effectively treated with amphotericin B in low, widely spaced doses (Burton et al. 1972).

In 1969, a series of 65 patients from United Kingdom was reported who received cadaveric renal transplants. 33 (51%) died within six months after transplantation. Out of these, 26 died of sepsis, including 8 due to IFI caused by *A. fumigatus* (6 patients) and *Candida* (2 patients) (Pletka et al. 1969). Hill et al. (1967) analyzed the causes of death in 60 patients out of 123 who received SOT. There were 107 recipients of renal transplants, 49 (46%) of them died including 7 who died of *Aspergillus* infection (Hill et al. 1967).

In a study of 3215 organ transplant patients from Korea (2954 kidney and 261 liver recipients), nine patients (0.003%) developed invasive pulmonary aspergillosis (7 kidney and 2 liver recipients) (Ju et al. 2009). This low incidence of invasive aspergillosis in renal transplant recipients was also observed in a study from Cleveland Clinic of 2046 solid organ transplant recipients, including kidney (733 cases), lung (188 cases), heart (686 cases) and liver (439 cases). There were a total of 33 cases of invasive aspergillosis, mostly in lung transplant recipients (24 patients; 12.8%). There were only 3 renal transplant recipients (0.4%) who developed invasive aspergillosis (Minari et al. 2002).

In an effort to understand the predisposing factors for infection in a renal transplant population by invasive aspergillosis, Gustafson et al. (1983) studied 9 patients who developed invasive aspergillosis among a total of 148 who received cadaveric renal transplants. Despite an extensive search, no common environmental source of contamination was found. The administration of high-dose corticosteroids posed a significant risk (Gustafson et al. 1983).

In a comprehensive review of *Aspergillus* infection in transplant recipients, Singh and Patterson (2005) discussed the epidemiological aspects of invasive aspergillosis in SOT and hematopoietic stem cell transplants (HSCT). *Aspergillus* infections have been reported in 2 to 26% of HSCT recipients and in 1 to 15% of SOT. In kidney transplant recipients, the incidence rate ranged between 0% and 4% (mean 0.7%), which was the least among solid organ recipients. Despite this overall low incidence compared with other organ transplants, invasive aspergillosis is a significant contributor to morbidity in renal transplant patients. The national registry of U.S. Renal Data System documented that between 1994 to 1997, the estimated hospitalizations due to fungal infections by *Aspergillus* was 12% (Singh and Paterson 2005).

Although invasive aspergillosis is relatively uncommon among renal transplant recipients, occasional outbreaks in hospitals caring for such patients had been reported. Panackal et al. (2003) reported a cluster of invasive aspergillosis among kidney transplant recipients at a California hospital from January to February 2001, when construction was ongoing. Four cases occurred among 40 kidney transplant recipients hospitalized during the study period. Factors associated with an increased risk of invasive aspergillosis included prolonged hemodialysis, lengthy corticosteroid treatment post-transplant, and use of sirolimus alone or with mycophenolate. After the study period, there were three additional cases; two *Aspergillus* isolates recovered from these patients had indistinguishable profiles by DNA fingerprinting, suggesting common-source exposure. The authors recommended that measures to prevent invasive aspergillosis in these patients should be taken during hospital construction (Panackal et al. 2003).

In a recent study of 27 patients with invasive aspergillosis in 1762 SOT patients including 9 in renal transplant recipients from Spain, Hoyo et al. (2014) found that invasive aspergillosis appears late in renal transplant recipients and is often associated with co-morbid conditions such as chronic lung disease and chronic heart failure (Hoyo et al. 2014).

Aspergillus infection in renal transplant recipient is considered a medical emergency. In a systemic review of the literature on case-fatality rate (CFR) associated with invasive aspergillosis in 1941 cases due to a variety of underlying conditions, including 21 recipients of renal transplantation (1.1%), CFR was found to be 58%, indicating that invasive aspergillosis is a highly lethal opportunistic infection. Interestingly, patients with bone marrow transplants had a CFR of 86.7%, a rate higher than that for any other group. On the other hand, renal transplant recipients had a CFR of 63% (Lin et al. 2001).

More than 90% of cases show the portal of entry to be lungs or sinuses, followed by skin in most of the remaining cases. In a review of invasive aspergillosis, Denning classified invasive *Aspergillus* infections into infection associated with tissue damage, surgery, or a foreign body, and infection predominantly in immunocompromised hosts. In that review, the incidence of invasive aspergillosis varied according to the underlying conditions. Heart and lung or lung transplantations have the highest frequency that ranged from 19% to 26%. Heart alone and renal transplantation has a lower rate, ranging from 0.5% to 10% (Denning 1998).

Most infections are caused by *A. fumigatus* and *A. flavum* species with only occasional cases of amphotericin-resistant *A. terreus* reported. The hallmark of invasive Aspergillosis is vascular invasion resulting in tissue infarction, hemorrhage, and systemic dissemination. During early post-transplantation period, *Aspergillus* is the main cause of CNS infection whereas a year or more after transplant, *Zygomycetes* becomes the main culprit.

In a review of infections of the central nervous system (CNS) in transplant recipients, Singh and Husain pointed out that aspergillosis is by far the most frequently occurring etiology of brain abscesses in organ transplant recipients. *Aspergillus* brain abscesses are notable for their occurrence in the early post-transplant period in these patients. Thirty-one percent (8/26) of the CNS lesions occurring within 30 days of liver transplants were infectious, with *Aspergillus* accounting for 75% (6/8) of these lesions (Singh and Husain 2000). An earlier study of the clinical characteristics and neuropathological findings of *Aspergillus* infection of CNS in 22 organ transplant recipients out of 218 patients who had complete necropsy carried out, including 13 patients who had liver transplant, 6 kidney transplants, 2 heart transplants and one who had cluster transplants, showed *Aspergillus* spp. invasion of the blood vessels with subsequent ischemic or hemorrhagic infarcts, and solitary or multiple abscesses were the predominant neuropathological findings. Concomitant diabetes mellitus was noted in 59% of the patients and bacterial or other severe infections in 86%. CNS aspergillosis was preceded by organ rejection and the need for intense immunosuppression and re-transplantation in the majority of the patients (Torre-Cisneros et al. 1993).

In addition to involvement of the lungs and the brain in invasive aspergillosis in renal transplant recipients, other sites had been reported infrequently including skin, bone, joints, vertebrae, paraspinal area, eye, and CNS (Tack et al. 1982; Alvarez et al. 1995; Schelenza and Goldsmith 2003; Park et al. 2004a, 2004b; Ersoy et al. 2007; Horn et al. 2009; Gamaletsou et al. 2014). A report from Canada documented a case of cutaneous aspergillosis due to *A. fumigatus* that developed about 3 weeks after renal transplantation at the site of recently evacuated hematoma surrounding an arteriovenous fistula on the left wrist (Langolis et al. 1980). Other rare sites of invasive aspergillosis in renal transplant recipients include the prostate (Shirwany et al. 1998), the testis (Singer et al. 1998), and the ureter (Vuruskan et al. 2005).

The treatment of choice is voriconazole with surgical debridement being an essential requirement to clear the invasive infection.

Cryptococcosis

In 1970, Libero Ajello, a leading mycologist, described cryptococcosis as the sleeping giant among the systemic mycoses; eight years later Kaufman and Bulmer stated that this mycosis was not dormant but it was truly the awakening giant among the systemic mycosis in terms of morbidity and mortality and since 1987, after the beginning of the AIDS epidemic, it is the systemic mycosis which presents the highest rates of morbidity and mortality. Cryptococcosis includes two different diseases, the one produced by *C. neoformans* which has a wide geographic distribution and behaves as an opportunistic infection, especially in HIV-positive patients and the other due to *C. gattii* which is more frequent in tropical and subtropical areas and produces severe infections in patients who do not present risk factors for opportunistic mycoses. Both are very serious infections and the CNS is often affected (Negroni 2012). Cryptococcosis is considered to be the third most common IFI in SOT recipients with incidence ranging from 0.3–5%. In renal transplant, Cryptococcosis is an extremely rare infection and is seen in 0.2–1.2% of cases with an associated mortality rate of 72% (Einollahi et al. 2008). Cryptococcosis should be suspected in transplant recipients presenting with headaches, fever, failure to thrive, altered consciousness, and skin lesions especially presenting 6 months after transplantation.

In a review of the incidence of cryptococcosis in various groups of SOT recipients during the tacrolimus era at the University of Pittsburgh Medical Center (1990–2000), 55 cases of cryptococcosis were diagnosed in 5377 recipients. 9 cases were seen in 2120 kidney recipients, an incidence of 4.2 per 1000 patients. Interestingly, heart transplant recipients had a much higher incidence of cryptococcosis (34.9% per 1000 patients). There was one case of cryptococcal pneumonia, two cases of cryptococcal meningitis and seven cases with disseminated disease (Vilchez et al. 2002).

From the same center, another study came out on clinical manifestations and the correlation with a positive serum cryptococcal antigen in 55 consecutive SOT recipients diagnosed with cryptococcosis. These included heart (Snydman et al. 2009), lung (Nampoory et al. 1996), liver (Ezzatzadegan et al. 2012), kidney (Badiee et al. 2005), and small bowel (Stitt 2003) recipients. While there were no significant differences in the manifestations of cryptococcosis in heart and lung recipients, kidney recipients had disseminated disease as the most common presentation (Vilchez et al. 2003).

Approximately 8% of IFIs in SOT recipients are due to cryptococcosis. Cryptococcosis is typically a late-occurring infection; the median time to onset usually ranges from 16 to 21 months post-transplantation. It usually manifests as CNS disease (meningitis) or pneumonia, but can affect multiple sites, including skin and soft tissues, prostate, liver, kidney, bones, and joints (Baddley et al. 2013). The incidence of cryptococcal meningitis is more in heart and intestine transplant recipients and the mortality is about 50%, observed mostly in liver transplant patients (Wu et al. 2002).

A review of 235 cases of proven infection due to *Cryptococcus* species in USA collected at the Prospective Antifungal Therapy (PATH) Alliance, showed all cases to be due to *C. neoformans.* There were 52 cases in SOT recipients, over half of them (27 recipients) had kidney transplants; 107 in patients infected with HIV; and 76 with neither HIV nor organ transplantation. Vast majority of Cryptococcosis patients manifested with CNS disease and meningitis. SOT recipients receiving calcineurin inhibitors (CNIs) are less likely to have CNS involvement in cryptococcal infection (40.1% versus 66.7%). Overall, 12-week mortality for patients with cryptococcal infection in the PATH Alliance registry is 22.6% (21.2% for SOT, 15.9% for HIV-infected patients, and 32.9% for patients with risk factors other than HIV infection or organ transplantation (Davis et al. 2009).

The impact of calcineurin-inhibitor agents on mortality in cryptococcal infection in SOT recipients was evaluated by Singh et al. (2007). They evaluated 111 organ transplant recipients, that included 56 kidney transplant recipients with cryptococcosis. Fifty-four percent of patients had pulmonary infection, 52.2% had CNS, and 8.1% had skin, soft-tissue, or osteo-articular infections. Sixty-one percent of the patients had disseminated cryptococcosis, and, in 32.4% of the patients, the infection was limited to the lungs. The overall mortality rate at 90 days was 14%. Patients receiving a calcineurin-inhibitor–based regimen (tacrolimus or cyclosporine A) were significantly less likely to have CNS infection (48% vs. 80%) and were more likely to have cryptococcosis limited to the lungs (36.6% vs. 6.6%). CNS infection was present in 47.3% of the patients receiving tacrolimus recipients, 50% of patients receiving cyclosporine A, and 80% of the patients receiving azathioprine or mycophenolate mofetil without a calcineurin-inhibitor agent (Singh et al. 2007). The type of primary immunosuppressive agent used in transplantation influences the predominant clinical manifestation of cryptococcosis. Patients receiving tacrolimus are significantly less likely to have central nervous system involvement (78% versus 11%) and more likely to have skin, soft-tissue, and osteoarticular involvement (66% versus 21%) than patients receiving non-tacrolimus-based immunosuppression (Husain et al. 2001).

The outcome of cryptococcosis of CNS varies with host immune function and the overall acuity of illness is worse among non-immunosuppressed patients with Cryptococcosis as compared to immunocompromised patients (Hong Nguyen et al. 2010).

Cryptococcus is known to reside in pigeon excreta (guano). Three reported cases of renal transplant recipients who developed *C. meningitis* lived in rural areas and were exposed to birds' excreta. All 3 patients had negative cryptococcal titers prior to transplant surgery and after treatment two patients were alive and well with excellent allograft function and the third patient had marginal renal function (Kapoor et al. 1999).

Review of 29 cases of cryptococcosis in renal transplant recipients from China showed *C. neoformans* var. *grubii* VNI strains to be the most common agent with an average time to infection after kidney transplantation being 5.16+/–3.97 years (Yang et al. 2014).

Pulmonary cryptococcosis had been recognized in renal transplant recipients as early as 1975 when it was reported in 15 out of 193 renal transplant recipients (7.8%) who developed pulmonary fungal infection. Agents responsible for infection included *Nocardia asteroides* in 8 cases, *A. flavus* in 5 cases, *C. neoformans* in 4 patients and *C. albicans* in 2 cases. Two cases had mixed mycotic infections. Ten out of these 15 patients died (Mills et al. 1975).

Cutaneous involvement by cryptococcosis can be primary or part of disseminated disease. Primary cutaneous cryptococcosis is an extremely rare entity of the disease, defined as the occurrence of lesions associated with a skin portal of entry for *C. neoformans* without systemic involvement. They appear as tubercle, nodule, or abscess at the site of penetration and rarely with satellite lymphangitis and adenopathy. Skin lesions are observed in 6% of AIDS-associated disseminated cryptococcosis and in 10% to 15% of those associated with sarcoidosis, organ transplantation or treatment with high doses of corticosteroids.

SOT recipients treated with tacrolimus have a higher frequency of skin, soft tissue, and osteoarticular lesions (Biancheri et al. 2012). In addition to skin and subcutaneous tissue involvement, cryptococcosis can cause cellulitis and necrotizing fasciitis (Carlson et al. 1987; Yoneda et al. 2014). Other organs that have been rarely reported to be involved include parathyroid (Thalla et al. 2009), prostate (Siddiqui et al. 2005), and joints (Bruno et al. 2002).

Mucormycosis (Zygomycosis)

Mucormycosis is an extremely rare and potentially rapidly fatal complication after kidney transplantation. The incidence is 0.2%–1.2% (Nampoory et al. 1996). It is a rare opportunistic infection that is ubiquitious in the environment. Mucormycosis refers to a spectrum of disease presentations caused by fungi of class *Zygomycetes*, order *Mucorales*. These fungi are commonly found on decaying vegetations and in soil. They grow rapidly and release large numbers of spores that can become airborne (Ahmadpour et al. 2009; Godara et al. 2011). It rarely causes disease in immunocompetent hosts except in settings of uncontrolled diabetes mellitus, iron overload, and heavy exposure such as in natural disaster. Recipients of SOT are at higher risk due to multiple risk factors including diabetes mellitus (*de novo* or post-transplant diabetes mellitus), immunosuppression (including induction therapy and treatment of rejection episodes), neutropenia within 60 days prior to the onset of infection, HCV infection, prior exposure to capsofungin or variconazole, renal failure, prolonged ICU stay, use of ureteral stents during renal transplant, and age more than 40 years (Godara et al. 2011; Hamdi et al. 2014). Pulmonary disease is often reported to be more in patients who received azathioprine compared to those on mycophenolate mofetil. On the contrary, use of tacrolimus is associated with 4-fold reduction in the risk of developing mucormycosis (Singh et al. 2009). The most common forms of presentation are rhinocerebral, pulmonary, GI, cutaneous, and disseminated (Einollahi et al. 2011; Godara et al. 2011). Dissemination is defined as an infection of 2 non-contagious sites. It is a serious complication and has a grave prognosis. The risk of dissemination is related to the primary site of infection and the type of transplant. Mucormycosis after transplantation has a poor prognosis particularly in patients with pulmonary involvement. The overall mortality ranges from 38%–56.5% with marked increase in mortality reaching upto 100% in disseminated disease. The species of mucormycosis also has an effect on the outcome with the highest treatment success rate with *Rhizopus* (68%), followed by *Mucor* (59%) and *Mycocladus* (50%). Despite early diagnosis and treatment, graft loss and mortality rate remain high in patients with disseminated disease.

Dematiaceous Fungi (Phaeohyphomycosis)

In a comprehensive review of melanized fungi in human diseases in 2010, Revankar and Sutton reported over 150 species and 70 genera of dematiaceous fungi implicated in human and animal diseases. With the exponential increase in the number of patients who are immunologically compromised, including SOT patients, more species are added to an already large list (Revankar and Sutton 2010). The most common agents of phaeohyphomycosis are the genus *Alternaria, Exophiala, Cladosporium, Bipolaris, Wangiella, Phoma, Phialophora, Colletotrichum, Curvularia* and *Rhinocladiella (Ramichloridium)*, which are normally found on the soil and organic vegetable material. These fungi can cause subcutaneous haeohyphomycosis including mycotic cysts (Isa-Isa et al. 2012). Surgical excision of the lesions in combination with itraconazole for at least 6 months is considered an effective treatment for subcutaneous phaeohyphomycosis (Marufuji Ogawa et al. 2009). We shall list down some of the species that have been associated with infection in renal transplant recipients.

Alternaria

Alternaria is a ubiquitous dematiaceous (phaeoid) melanin producing fungus, which is isolated from plants, soil, food, and indoor air environment. It includes 50 species, among these, *Alternaria alternata* is the most common one isolated from human infections. Cutaneous and subcutaneous phaeohyphomycosis in immunosuppressed individuals is the most common presentation (Gilaberte et al. 2005; Vermeirea et al. 2010). Other species of *Alternaria* have been implicated in cutaneous infections such as *A. tenuissima* and *A. infectoria* (Romano et al. 1997; Mesa et al. 1999; Salido-Vallejoa et al. 2014).

Exophiala

It is a dematiaceous fungus widely distributed in water, soil, plants, and decaying wood. It encompasses several species; the commonest ones are *Exophiala castellanii*, *Exophiala jeanselmei*, and *Exophiala moniliae.* It can result in subcutaneous infections such as mycetoma and chromoblastomycosis. These infections are usually acquired via traumatic implantation in immunosuppressed patients (Ronan et al. 1993; Sabbaga et al. 1994; Lief et al. 2010).

A review of all reported cases from 1966–2009 of *Exophiala* infection in SOT, 13 out of a total 28 cases (46%) were renal transplant recipients. All cases of Exophiala infection in SOT patients presented as infection of the skin and soft tissues, mostly as slowly growing subcutaneous nodules on the extremities. No systemic infections were reported. In most cases the *Exophiala* isolate is *E. jeanselmei*. Other species are also reported including *E. Spinifera*, *E. mansoni,* and *E. oligosperma* (Dutriaux et al. 2005; Gonzalez-Lopez et al. 2007; Lief et al. 2010).

Rhinocladiella

Rhinocladiella mackenziei, also formerly known as *Rhamichloridium mackenziei* is a rare neurotropic fungus that is mostly limited to the Middle East, with most cases reported from countries in the Arabian Peninsula. This infection is life threatening and often fatal. Campbell and Al Hedaithy in a review of five cases of brain abscesses due to *R. mackenziei* reported two cases to be in kidney transplant patients from Qatar and Oman (Campbell and Al-Hedaithy 1993). A report of two cases from Saudi Arabia and review of additional ten cases with various underlying co-morbid conditions showed the high mortality rate of this invasive fungal infection with majority of patients dying of disease, some despite aggressive antifungal chemotherapy with or without neurosurgical drainage or resection (Kanj et al. 2001).

On the contrary, another case of renal transplant was reported from Saudi Arabia, where the patient developed cerebral abscess secondary to *R. mackenziei*. The infection progressed despite surgical evacuation and therapy with liposomal amphotericin B, itraconazole, and 5-flucytosine. The patient was subsequently treated with the investigational triazole posaconazole oral suspension that resulted inprogressive clinical and radiologic improvement and the patient was alive for four years following this treatment (Al-Abdely et al. 2005).

Cladophialophora

Cladophialophora bantiana is another neurotropic dematiaceous fungus, known by many other names including *Xylohypha bantiana'*, '*Xylohypha emmonsii'*, and '*Cladosporium trichoides*. It is considered to be the most common and dangerous neurotropic fungus which causes brain abscess. Few cases of brain abscess in renal transplant recipients due to *C. bantiana* have been reported (Gupta et al. 1997; Ajantha and Kulkarni 2011).

In a review of 101 reported cases of CNS phaeohyphomycosis in literature between 1966 and 2002, there were six cases in renal transplant recipients. The causative agents included *C. bantiana* in 2 cases; *R. mackenziei* in 2; *Fonsecaea pedrosoi* in one and *Chaetomium atrobrunneum* in one case (Revankar et al. 2004).

Histoplasmosis

Histoplasmosis, caused by *H. capsulatum*, is a common infection endemic in many regions of America, Asia, India and Africa, with sporadic cases also occurring throughout the world. It is acquired by the inhalation of the organism's mold form, microconidia, which transforms to the yeast-form in tissues. Most infections are not clinically recognized, but are rather identified as incidental radiographic or pathological findings. Histoplasmosis affects primarily the lungs, but can also affect the skin, brain, GI tract and can be disseminated, particularly in immunocompromised patients (Wheat 2003).

In the largest study of histoplasmosis in SOT recipients, Assi et al. (2013) reported 152 cases from 24 medical centers in USA. 78 cases (51%) were among renal transplant recipients. The median time from transplant to diagnosis was 27 months, but 34% were diagnosed in the first year after transplant. Twenty-eight percent of patients had severe disease (requiring intensive care unit admission) and 81% had disseminated disease. Although late cases occur, the first year is the period of highest risk for histoplasmosis. There is 10% infection related mortality with more than 70% succumbing during the first month of diagnosis (Davies et al. 1979; Assi et al. 2013).

Two outbreaks of histoplasmosis in renal transplant recipients were reported in 1983. Ten episodes of histoplasmosis were documented in eight renal allograft recipients out of a total of 379 recipients (2.1%). In five patients, there were associated cytomegalovirus infections. Prolonged fever was the predominant clinical finding; and dissemination was observed in seven of nine patients, including three with meningitis. Treatment with amphotericin B resulted in prompt clinical improvement in all patients, but relapse occurred in two patients one year following the therapy (Wheat et al. 1983).

Histoplasmosis is endemic in the Ohio and the Mississippi river valleys. It is also endemic in several other areas such as Latin America, India, and Bangladesh. The incidence of Histoplasmosis is low (1.1%). Almost all these patients present with either disseminated histoplasmosis or pulmonary disease. Rarely may the patients develop hemophagocytic syndrome. The duration of immunosuppression prior to the diagnosis of infection ranges from 84 days to 14 years (Peddi et al. 1996; Cuellar-Rodriguez et al. 2009; Lobo et al. 2009; Rappo et al. 2010; Nieto-Ríos et al. 2014).

Involvement of the CNS had been recognized clinically in 5% to 10% of patients with progressive disseminated histoplasmosis. The risk of developing CNS histoplasmosis is increased in individuals with impaired cellular immunity including renal transplant recipients, but not all patients with this condition are immuno-compromised. CNS involvement can occur in conjunction with progressive disseminated histoplasmosis (PDH) or as an apparently isolated manifestation. About 50% of patients with CNS histoplasmosis have PDH (Saccente 2008).

H. capsulatum can involve the skin and subcutaneous fat in renal transplant recipients leading to panniculitis and cellulitis that may progress to a fatal outcome (McGuinn et al. 2005; Dufresne et al. 2013). Gastrointestinal involvement by histoplasmosis usually occurs as part of PDH and may cause perforation (Brett et al. 1988).

The incidence of Histoplasma infection is 1 case per 1000 person-year and the prognosis is good but requires protracted therapy. Patients with latent infection will most likely not develop post-transplant histoplasmosis when prophylaxis is used.

Donor-Derived Fungal Infection

Fungal infections can be donor-derived and transmitted with the allograft. Donor-derived fungal infections (DDFI), although rare, are associated with serious complications. The characteristics of and risks posed by these infections are often poorly understood and they have a unique spectrum of illnesses. DDFIs are most commonly encountered in Kidney transplant recipients, among all SOTs accounting for approximately 91%. Common risk factors include transplant tourism/commercial transplants, near drowning events, and immunosuppression of donors including donors who themselves were transplant recipients. Overall graft loss is 83% and mortality 17% in DDFI with a majority of patients presenting with vascular complications related to graft vasculature (65%), allograft dysfunction (43%), and unexplained febrile illness (39%) (Tolkoff-Rubin and Rubin 1992). In many DDFI cases contaminated preservation fluid is the proposed source of infection. *Candida* is the most common organism infecting the preservation fluid and the contamination can occur prior to or at the time of organ procurement (Tolkoff-Rubin and Rubin 1992; Mai et al. 2006; Matignon et al. 2008; Albano et al. 2009; Canaud et al. 2009; Veroux et al. 2009). Early recognition, surgical intervention and antifungal therapy are crucial for graft as well as recipient survival.

Donor-derived Candidiasis is seen in 1 in 1000 kidney transplant recipients (Albano et al. 2009). The infection is often manifested as candidemia, perirenal hematoma, abscess, fungal ball, infected urinoma, and vascular complications such as anastamosis rupture and mycotic aneurysm, the latter considered to be the most serious complication (Laouad et al. 2005; Mai et al. 2006; Albano et al. 2009; Singh et al. 2012). Contaminated preservation fluid has also been shown to transmit aspergillus and mucormycosis (Battaglia et al. 2004; Cerutti et al. 2006).

In a review of 23 donor-derived filamentous infections, Aspergillus was found to be the most common fungal infection seen in 71% of renal transplant recipients (Gomez and Singh 2013).

Cryptococcosis is seen in 0.3–5% with majority secondary to infection reactivation in recipients. Donors with crypococcosis of any site may transmit infection to the transplant recipient. Similarly disseminated infection may occur from infection transmitted by unrecognized meningoencephalitis in a donor (Lyon et al. 2011; Singh et al. 2012).

Histoplasmosis is present in the soil and is endemic in certain areas reaching up to 75% rate of infection in some places. It occurs in 0.1–0.5% of transplant patients with 5% of these infections being donor-derived (Davies et al. 1979; Wheat et al. 1983; Peddi et al. 1996; Freifeld et al. 2005; Cuellar-Rodriguez et al. 2009; Singh et al. 2012).

Coccidioidomycosis occurs in 1.5–8.7% of transplant cases in endemic areas with only few DDFI reported. Usually these infections are seen during the first year after transplantation (Blair and Logan 2001; Singh et al. 2012).

The precise risk of DDFI from donors with near-drowning experience is not known, however, near-drowning victims may acquire unusual fungi from contaminated water that may be transmitted with the grafted organs such as *Apohysomyces elegans* (mucormycosis species) and *Scedosporium apiospermum* (Van der Vliet et al. 1980; Alexander et al. 2010).

IFIs in Commercial Transplantation

Medical tourism refers to patients travelling across national borders to receive healthcare. Due to increased demand and limited cadaveric and living related organ supply for transplantation worldwide, transplant tourism or commercial transplantation is gaining popularity. It accounts for approximately 5–10% of kidney transplants per year worldwide (Akoh 2012). Commercial kidney transplants are associated with higher incidence of surgical complications, acute rejection, and infections leading to significant morbidity and mortality. A literature review by Shoham et al. (2012) identified 19 cases of IFIs in commercial kidney transplant recipients with the most common organism being *Aspergillus* (63%), followed by Zygomycosis (26%), and other fungi (5%). IFI was present at the graft site in 35%, with an overall mortality of 59%. These fungal infections, although have been associated with higher graft loss and mortality, it is of note to mention that the rate of these life threatening opportunistic infections has decreased in recent years (Qunibi 1997; Shoham et al. 2010).

Rare cases are reported with recipients developing multiple bacterial, viral, parasitic, and fungal infections after transplant. More than one fungal organism can be isolated in occasional cases of this deep mycosis (Tomazie et al. 2007).

Fungal arteritis is also a rare complication of commercially transplanted kidney and is associated with significant morbidity and mortality. Less than 20 cases are reported in literature of fungal arteritis in SOT mostly caused by *Candida, Aspergillus, Mucor*, and rarely *Trichosporon*. Fungal arteritis typically involves graft arteries on SOT and the infection of these graft vessels with extension to recipient iliac vessels invariably leads to rupture of arteries requiring surgical excision of the graft and the iliac vessels with appropriate graft to avoid future recurrent rupture secondary to infected and or disrupted anastomosis (Chkhotua et al. 2001; Laouad et al. 2005; Mai et al. 2006; Fadhil et al. 2011).

Another major area of concern is the lack of communication between the transplant team/center and centers subsequently taking care of these transplant recipients that negatively impacts the care of these critical patients (Kennedy et al. 2005).

More than half of the renal transplants occurring in tropical countries including majority of commercial transplant centers develop serious infections with 20–40% infection associated mortality (Levy and Bia 1995; Jha and Chugh 2002; Tomazie et al. 2007; Turner 2009). Multiple factors lead to this predisposition including poor hygienic conditions, hot humid climate, and limited diagnostic utilities. In addition, the ultimate purpose of most of these commercial transplant centers is predominantly financial gain rather than patient well-being.

Diagnosis

Early detection of systemic fungal infection regardless of the techniques or test methodologies used is difficult due to the limitations of each method. Therefore it is recommended to use a combination of various diagnostic tests in transplant patients suspected to have fungal infection. Current methods of IFI detection include clinical signs and symptoms, imaging, biopsies with histopathologic examination with appropriate ancillary studies such as special/immunohistochemical stains, traditional culture growth, antigen and antibody based assays, and molecular techniques including polymerase chain reaction (PCR). The standard criteria for diagnosis are identification of fungus by histological evaluation, positive culture from sterile body fluids (such as cerebrospinal fluid, blood, and pleural/pericardial fluid) and tissue biopsy samples.

Renal transplant recipients, in general show very few if any clinical manifestations. Most patients present with a rise in baseline Creatinine that could be seen in rejection as well as opportunistic infections. Sensitive imaging techniques are required to identify various lesions such as mass, cyst, cavity, fungal ball, or an abscess.

New assays based on antigen and antibody detection and metabolite detection are on board that can be useful. The mean antibody index is not shown to be significant between normal, probable, and confirmed cases of IFI (Cray et al. 2009). Antigen detection assays are more sensitive and include enzyme immune assays, radioimmunoassay, and latex agglutination assays. Tests based on *Candida mannan* antigen can detect invasive Candidiasis in 80% of cases whereas those based on galactomannan molecules of fungal wall, in particular of *Aspergillus*, has variable sensitivity and specificity in different populations. It is shown to have 100% sensitivity and 90.8% specificity in bronchoalveolar lavage samples in SOT patients (Badiee et al. 2005; Clancy et al. 2007; Badiee and Alborzi 2011).

Molecular assays based on amplification and hybridization techniques are on the rise in clinical practice as these can detect IFI before even the appearance of clinical signs and symptoms in these immunocompromised patients. Nested PCR has sensitivity of 92.8% and specificity of 94% in detecting these fungal infections. In addition, real-time PCR is a quantitative assay that can quantitate Candida and Aspergillus deoxyribonucleic acid with accuracy (Badiee et al. 2005; Badiee and Alborzi 2011).

Blood cultures are not the method of choice as only 45–75% of autopsy-proven cases of systemic Candidiasis can show growth (Thaler et al. 1988; Goodrich et al. 1991). The growth of mold in sputum should be a possible indicator of pulmonary infection. However as yeasts are the normal flora of gastrointestinal tract and may be seen as contaminants in sputum, it is recommended to have 3 smear-positive pure sputum cultures or positive lung biopsy proven disease.

Tissue biopsy, if feasible, is gold standard for histological detection of fungus. Special stains that best show fungi include Grocott-Gomori silver stain, acridine Orange, periodic acid-Schiff reaction, lectins, and calcofluor white. If enough sample is available potassium hydroxide wet mount smears are very sensitive means of early detection. The detection is even better if potassium hydroxide is combined with calcofluor white and specimens examined under fluorescence microscopy. Direct ink/ India ink preparation is useful in detection of Cryptococcus in cerebrospinal fluid.

The current European Organization for Research and Treatment of Cancer/Mycosis Study Group (EORTC/MSG) criteria for diagnosis of IFI involves a combination of host factors, histopathology, imaging, and microbiological cultures or markers to pinpoint this challenging and difficult diagnosis.

Treatment of Fungal Infection in Renal Transplant Patients

Fungal infections among solid organ transplant recipients are associated with a high mortality rate. The one year cumulative incidence of invasive fungal infections (IFI) varies depending on the type of the solid organ transplanted, being higher among intestinal transplant (11.6%) and lower among renal transplant (1.3%) recipients. Candidiasis, invasive aspergillus, and cryptococcosis are the leading causative agents (Pappas et al. 2010). Early detection and treatment can reduce overall morbidity and mortality among these immunocompromised populations.

The treatment of different fungal organisms varies substantially from species to another. Several factors affect the type and the duration of the antifungal therapy, these include: the anatomic location of the infection, the patients' underlying co-morbid medical conditions, his immune status, the presence of risk factors for infection, and sensitivity to given antifungal agent.

Antifungal classes used to treat various fungal infections include, polyenes, azoles, and echinocandins. Azoles such as Fluconazole, itraconazole, voriconazole, and posaconazole work primarily by inhibiting the conversion of lanosterol to ergosterol, a vital component of the cellular membrane of fungi. Azoles interact with several other medications that also utilize cytochrome P450 enzymes including calcineurin inhibitors and Sirolimus; and therefore alternative antifungal agents, such as echinocandins, may be preferred in such cases.

Echinocandins include caspofungin, anidulafungin, and micafungin are inhibitors of the synthesis of 1,3-beta-D-glucan, which is an integral component of the fungal cell wall. Diminishing the concentration of the glucan in the cellular wall leads to changes in the permeability of the membrane and eventually cell lysis. The echinocandins are rapidly fungicidal against most *Candida* species and fungistatic against *Aspergillus* species. They are not active at clinically relevant concentrations against *Cryptococcus neoformans*, or *Fusarium* species (Denning 2003). Amphotericin B is a polyene antifungal agent that disrupts fungal cell wall synthesis leading to the formation of pores that allow leakage of cellular components. The original Amphotericin B deoxycholate is associated with significant nephrotoxicity and therefore several lipid-based less nephrotoxic derivatives were created, including liposomal amphotericin B and Amphotericin B lipid complex and Amphotericin B colloidal dispersion.

Prophylactic Therapy for Solid Organ Transplant Recipients

Patient undergoing liver or pancreatic transplant are at higher risk of developing invasive candidiasis (Collins et al. 1994; Karchmer et al. 1994; Hadley et al. 1995), and therefore they should receive prophylactic antifungal therapy with liposomal amphotericin B at 1 mg/kg/day, or fluconazole 400 mg daily during the early postoperative period (Kung et al. 1995; Tollemar et al. 1995). The risk of invasive candidiasis following other solid organs transplantation is low and therefore does not require the use of systemic prophylaxis.

Treatment of Candidiasis

The susceptibility of *Candida* species to different antifungal agents can be predicted; however, occasionally individual isolates do not follow the general pattern (Vazquez and Sobel 2011). The following is the recommended treatment guideline by the Infectious Diseases Society of America (IDSA) for the treatment of various *candida* infections (Rex et al. 2000; Pappas et al. 2009).

Acute Hematogenously disseminated candidiasis

Candida is now considered the fourth most common bloodstream isolate and the most common of developing invasive fungal infection in critically ill nonneutropenic patients. Several factors increasing the risk of invasive candidiasis include the prolonged use of antibacterial antibiotics, central venous catheters, hyperalimentation, abdominal surgery involving transaction of the gut wall, and prolonged ICU stay.

Candidemia due to *C. parapsilosis* appears to be associated with line related infection and carries relatively lower mortality rate than other *Candida* species (Anaissie et al. 1998; Coleman et al. 1998; Kossoff et al. 1998; Levy et al. 1998). In such cases removal of all existing central venous catheters should be completed.

The choice of medical therapy would vary depending on the clinical status of the patient and the suspected fungal species. In stable patients most experts would initiate therapy with fluconazole at ≥6 mg/kg/day. In the clinically unstable patient infected with an isolate of unknown species, the choice of using fluconazole versus amphotericin B at ≥0.7 mg/kg/day is justifiable (Edwards et al. 1997). The susceptibility to various antifungal medications can be predicted once the isolate is identified.

For example, *C. albicans, C. tropicalis*, and *C. parapsilosis* can be treated with either fluconazole at 6 mg/kg/day or amphotericin B at 0.6 mg/kg/day. *C. glabrata* on the other hand has reduced susceptibility to both azoles and amphotericin B, and therefore most authorities recommend amphotericin B at ≥0.7 mg/kg/day as initial therapy or Fluconazole at 12 mg/kg/day particularly in less-critically ill patients. Higher dose of Amphotericin B (1.0 mg/kg/day) is preferred for treating cases infected with *C. krusei.* Many but not all isolates of *C. lusitaniae* are resistant to amphotericin B; thus, fluconazole at 6 mg/kg/day is the preferred therapy for this species. For candidemia, therapy should be continued for 2 weeks after the last positive blood culture and resolution of signs and symptoms of infection.

Urinary Candidiasis

Candiduria should be treated in symptomatic patients, neutropenic patients, and patients with renal allografts. Urinary tract instrumentation, recent antibiotic therapy, and advanced age are the most common risk factors for developing candiduria (Hamory and Wenzel 1978). Changing the Foley catheter alone in candiduric individuals rarely results in elimination of the infection (<20%). Discontinuation of the catheter will result in eradication of candiduria in almost 40% of patients (Sobel et al. 2000).

However, if complete removal of urinary stents and Foley catheters is not possible, then placement of new instruments may be beneficial.

Therapy with fluconazole at 200 mg/day for 7–14 days or amphotericin B at doses ranging (0.3–1.0 mg/kg/d for 1–7 days) have been used (Jacobs et al. 1996). In the absence of renal insufficiency, oral flucytosine at 25 mg/kg/q.i.d. may be beneficial in eradicating candiduria, especially those caused by non-*albicans Candida* species. Resistance to flucytosine occurs rapidly when this drug is used alone (Francis and Walsh 1992). Bladder irrigation with amphotericin B (50–200 µg/mL) can transiently clear funguria (Leu and Huang 1995). Relapse is frequent even after successful use of local or systemic antifungal therapy. In immunocompromised patients with persistent candiduria further evaluation with imaging studies is recommended.

Oropharyngeal Candidiasis

Topical therapy with Clotrimazole troches (one troche 5 times daily) or Nystatin suspension (100,000 U/mL [4–6 mL swish-and-swallow q.i.d.] for 7–14 days) can be used as an initial therapy for treatment of oropharyngeal candidiasis. Most patients will respond initially to topical therapy (Pons et al. 1993; Sangeorzan et al. 1994; Finlay et al. 1996). Oral Fluconazole (100 mg/d for 7–14 days orally) is considered to be superior to topical therapy and has less reported symptomatic relapses. Ketoconazole and itraconazole capsules are less effective than fluconazole because of variable absorption (De Wit et al. 1989). However, Itraconazole solution (200 mg/d for 7–14 days orally) is as efficacious as fluconazole (Cartledge et al. 1997; Graybill et al. 1998; Phillips et al. 1998). Fluconazole-refractory oropharyngeal candidiasis will respond to oral itraconazole solution (≥200 mg/d orally). In Itraconazole refractory cases, Amphotericin B oral suspension (1 mL q.i.d. of the 100 mg/mL suspension) may be used. Intravenous amphotericin B (0.3 mg/kg/d) may be used as a last resort in patients with refractory disease.

Esophageal Candidiasis

Systemic therapy is required for effective treatment of esophageal candidiasis. A 14–21 day course of either fluconazole (100 mg/d orally) or itraconazole solution (200 mg/d orally) is usually effective. Fluconazole-refractory esophageal candidiasis should be treated with itraconazole solution (≥200 mg/d orally) and if this fails then a trial of intravenous amphotericin B (0.3–0.7 mg/kg/day) may be used.

Genital Candidiasis

Two forms of vaginal candida infection are reported. Uncomplicated vaginitis which accounts for more than 90% of the cases and responds readily to short course of oral or topical treatment, and the complicated type which requires longer antimycotic therapy for at least 7 days (Sobel et al. 1998).

Treatment of Invasive Aspergillosis

Invasive aspergillosis (IA) has a high mortality rate and therefore, rapid institution of therapy in suspected cases while awaiting for tests results is considered life saving. Treatment of IA in kidney transplant recipients appears to be no different than non transplant patients, however extra cautions should be exercised to avoid medication related nephrotoxicity and potential interactions with immunosuppressant drugs. In the years before the availability of lipid-associated amphotericin B and itraconazole, conventional amphotericin B in doses of 0.5–1.0 mg/kg was often successful in salvaging the patient, if all immunosuppression was stopped and the transplanted kidney sacrificed. Even this strategy was often unsuccessful if the patient had severe bilateral pulmonary involvement or disseminated disease. A preferable alternative with less nephrotoxicity is a lipid-associated amphotericin B preparation, at a daily dose of 4–5 mg/kg (White et al. 1997; Ringdén et al. 1991). Recently, Voriconazole is considered the drug of choice for invasive aspergillosis when compared to amphotericin B (Herbrecht et al. 2002). However concern exists regarding mounting azole resistance (Howard et al. 2009) Amphotericin B may be the drug of choice when used empirically to treat a suspected case of mucormycosis as voriconazole is ineffective for *Zygomycetes* infection. Treatment with voriconazole is initiated with a loading dose of 6 mg/kg IV every 12 hr for 2 doses, followed by 4 mg/kg every 12 hr. It is important to notice that voriconazole interacts with both corticosteroids and cyclosporine and therefore dose adjustment is necessary. A 50–60% reduction in the dose of calcineurin-inhibitor agents may be necessary with the concurrent use of voriconazole (Saad et al. 2006). The use of sirolimus is contraindicated in patients receiving voriconazole.

Caspofungin is also approved for treatment of invasive aspergillosis when species are resistant to other therapies (Maertens et al. 2004). The currently recommended dosage regimen of caspofungin in adults consists of a single loading dose of 70-mg on day 1, followed by 50 mg/day thereafter. Higher doses of caspofungin (70 mg/day) are used in salvage combination therapy of invasive aspergillosis (Maertens et al. 2006). Caspofungin may reduce tacrolimus concentrations by up to 20% and may increase cyclosporine A plasma concentrations by 35% (Sable et al. 2002).

Combination of voriconazole and caspofungin might be considered preferable therapy for subsets of solid organ transplant recipients. In a prospective multicenter study by Singh et al; the combination therapy of voriconazole and caspofungin as primary therapy for invasive aspergillosis was independently associated with an improved 90-day survival in multivariate analysis. Combination therapy was associated with a trend towards lower mortality (HR 0.58, 95% CI: 0.30–1.14) when controlled for CMV infection and renal failure. When 90-day mortality was analyzed in subgroups of patients, combination therapy was independently associated with reduced mortality in patients with renal failure, and in those with *A. fumigatus* infection, even when adjusted for other factors predictive of mortality in the study population (Singh 2006). However initial combination therapy is usually not indicated and should generally be reserved for treatment failures. Concomitant therapy with azole antifungals and amphotericin may not be synergistic and could result in diminished efficacy. The optimal duration of therapy for invasive aspergillosis depends upon the response to therapy, and the patient's underlying immune status. Treatment is usually continued for 12 weeks however, the precise duration of therapy will be determined by clinical response. A reasonable course would be to continue therapy until all clinical and radiographic abnormalities have been resolved, and repeat cultures are negative.

The level of immunosuppression in renal transplant patients with invasive aspergillosis should be tapered and the corticosteroid dose must be reduced or discontinued.

Finally, the use of interferon-gamma (IFN-γ) to treat invasive fungal infections after combination antifungal therapy failed to eradicate the infection has been reported (Estrada et al. 2012). Current immunosuppressive therapy blunts cell-mediated immunity, thereby predisposing organ transplant recipients to invasive fungal infections. IFN-γ has the potential to augment this defect in immunity, eradicate invasive fungal disease, and thus far has not been associated with allograft rejection (Armstrong-James et al. 2010). Guidelines of the IDSA suggest a role for IFN-γ as adjunctive antifungal therapy for invasive aspergillosis in immunocompromised non neutropenic host.

Allergic Bronchopulmonary Aspergillosis (ABPA)

This is an exaggerated hypersensitivity immune response to fungus *aspergillus*. Therapeutic options aim to control episodes of acute inflammation and limit progressive lung injury. Monotherapy with oral corticosteroids is the mainstay of treatment. However, such therapy is associated with significant immune modulation and metabolic disorders. Acute flare of ABPA is treated with prednisone at 1 mg/kg/day for 14 days, followed by conversion to an everyday regimen that will be tapered over a six months period. Inhaled steroids do not have documented efficacy in preventing acute episodes of ABPA. The concomitant use of oral itraconazole twice daily for four months along with corticosteroids helps to reduce fungal causes of bronchial inflammation. This would allow rapid resolution of pulmonary infiltrates and gradual tapering corticosteroid (Salez et al. 1999; Stevens et al. 2000; Wark et al. 2003a, 2003b).

Chronic Necrotizing Pulmonary Aspergillosis

Treatment consists of therapy with voriconazole, or caspofungin, or amphotericin B or amphotericin lipid formulation. A prolonged course of therapy with the goal of radiographic resolution is needed. In addition, reduction or elimination of immunosuppression should be attempted, if possible. Surgical resection may be considered when localized disease fails to respond to antifungal therapy.

Aspergilloma

Surgery is indicated for persistent, or a life-threatening hemoptysis, for lesions in the proximity of great vessels or pericardium, for single cavitary lung lesion which progresses despite adequate treatment, for lesions invading the pericardium, bone, or invading the thoracic tissue (Walsh et al. 2008). Surgical resection for aspergilloma is curative but may not be possible in patients with limited pulmonary reserve. The use of oral itraconazole may provide partial or complete resolution of aspergillomas in 60% of patients. Successful intracavitary instillation of amphotericin treatment alone or in combination with other drugs, has been reported in limited number of cases. A cohort of fifteen patients with active inoperable pulmonary aspergilloma underwent percutaneous intracavitary injection of a special therapeutic paste of glycerin and amphotericin B. In 12 cases the aspergilloma regressed within 3 months and as a result the serology became negative. In three cases, there was no change in the cavity, but hemoptysis did not recur (Giron et al. 1993). Bronchial artery embolization may be used for life-threatening hemoptysis in patients with limited pulmonary reserve to undergo surgery (Mal et al. 1999).

Treatment of Cryptococcal Disease

According to the updated 2010 infectious diseases society of America guidelines, transplant recipients with documented severe cryptococcal infection of the central nervous system (CNS) should receive an induction therapy with the combination of intravenous liposomal amphotericin B at a dose of 3–4 mg/kg per day in addition to flucytosine 100 mg/kg per day given in 4 divided doses for at least 2 weeks. This is followed by a maintenance course of oral fluconazole of 400–800 mg per day for 8 weeks and then a daily dose of 200–400 mg for 6–12 months. Severe non-CNS infection or disseminated diseases are treated similar to CNS infection. However, for a mild-to-moderate non-CNS infection including pulmonary infiltrates, fluconazole 400 mg per day for 6–12 months should be adequate. Immunosuppressive therapy should be gradually tapered with consideration of lowering the corticosteroid dose first (Perfect 2010).

Treatment of Mucormycosis

A prospective study among solid organ transplant recipients showed that renal failure, diabetes mellitus, iron overload and prior use of voriconazole and/or caspofungin were associated with a higher risk of infection with mucormycosis (Singh and Sun 2008; Singh et al. 2009). Interestingly, the use of the calcineurin inhibitor tacrolimus in this study was associated with a 4-fold reduction in the risk of mucormycosis.

Immediate initiation of antifungal therapy in a documented case of mucormycosis is favorable and would increase the patient's chance of survival (Chamilos et al. 2008). Liposomal amphotericin B is the drug of choice and the dose should be at least 5 mg/kg/day, as suggested by uncontrolled retrospective study (Sun et al. 2009). Higher dose of liposomal amphotericin B (10 mg/kg/day) for the initial treatment of mucormycosis was also successful but with high drug related renal toxicity. Combination therapy of amphotericin B and caspofungin has been described as successful in a limited number of diabetic patients with rhinocerebral mucormycosis (Reed et al. 2008). The duration of acute phase treatment is not well defined. However, treatment should continue for a minimum of 6–8 weeks or until complete resolution of the clinical symptoms.

Conclusion

In summary SOT recipients are susceptible to various opportunistic infections and with the improvement of immunosuppressive therapy and graft survival with time, the risk of infectious complications has increased. In particular IFI is becoming a major complication and is an important cause of morbidity and mortality among these transplant recipients. This chapter represents a comprehensive review with an outline of incidence, timings, risk factors, prevention and treatment strategies of most common post renal transplant fungal infections. In addition there is a commentary on Invasive Mycosis among commercial organ transplantation with its unique risks and concerns.

References

Abbott, K., I. Hypolite, R. Poropatich, P. Hshieh, D. Cruess, C. Hawkes et al. 2001. Hospitalizations for fungal infections after renal transplantation in the United States. Transplant Inf Dis. 3(4): 203–211.

Ahmadpour, P., M. Lessan-Pezeshki, M. Ghaidiani, F. Pour-Reza-Gholi, F. Samadian, J. Aslani et al. 2009. Mucormycosis after living donor kidney transplantation: a multicenter retrospective study. Int J Nephrol Urol. 1(1): 39–44.

Ajantha, G.S. and R.D. Kulkarni. 2011. *Cladophialophora bantiana*, the neutropic fungus—a mini review. J Clin Diag Res. (Suppl-1) 5(6): 1301–1306.

Akoh, J.A. 2012. Key issues in transplant tourism. World J Transplant. 2(1): 9–18.

Al-Abdely, H.M., A.M. Alkhunaizi, J.A. Al-Tawfiq, M. Hassounah, M.G. Rinaldi and D.A. Sutton. 2005. Successful therapy of cerebral phaeohyphomycosis due to *Ramichloridium mackenziei* with the new trizole posaconazole. Med. Mycol. 43(1): 91–95.

Alangaden, G.J., R. Thyagarajan, S.A. Gruber, K. Morawski, J. Garnick, J.M. El-Amm et al. 2006. Infectious complications after kidney transplantation: current epidemiology and associated risk factors. Clin Transplant. 20(4): 401–409.

Albano, L., S. Bretagne, M.-F. Mamzer-Bruneel, I. Kacso, M. Desnos-Ollivier, P. Guerrini et al. 2009. Evidence that graft-site candidiasis after kidney transplantation is acquired during organ recovery: a multicenter study in France. Clin Inf Dis. 48(2): 194–202.

Alexander, B., W. Schell, A. Siston, C. Rao, W. Bower, S. Balajee et al. 2010. Fatal Apophysomyces elegans infection transmitted by deceased donor renal allografts. American J Transplant. 10(9): 2161–2167.

Al-Mohaya, M.A., A. Darwazeh and W. Al-Khudair. 2002. Oral fungal colonization and oral candidiasis in renal transplant patients: the relationship to Miswak use. Oral Surgery, Oral Medicine, Oral Pathology, Oral Radiology, and Endodontology 93(4): 455–460.

Al-Mohaya, M.A., A.M.G. Darwazeh, S. Bin-Salih and W. Al-Khudair. 2009. Oral lesions in Saudi renal transplant patients. Saudi J Kidney Dis Transplant. 20(1): 20–29.

Altiparmak, M.R., S. Apaydin, S. Trablus, K. Serdengecti, R. Ataman, R. Ozturk et al. 2002. Systemic fungal infections after renal transplantation. Scandinavian J Inf Dis. 34(4): 284–288.

Alvarez, L., E. Calvo and C. Abril. 1995. Articular aspergillosis. A case report. Clin Inf Dis. 20(2): 457–460.

Anaissie, E.J., J.H. Rex, Ö. Uzun and S. Vartivarian. 1998. Predictors of adverse outcome in cancer patients with candidemia. Am J Med. 104: 238–45.

Armstrong-James, D., I.A. Teo, S. Shrivastava et al. 2010. Exogenous interferon-γ immunotherapy for invasive fungal infections in kidney transplant patients. Am J Transplant. 10(8): 1796–1803.

Assi, M., S. Martin, L.J. Wheat, C. Hage, A. Freifeld, R. Avery et al. 2013. Histoplasmosis after solid organ transplant. Clin Inf Dis. 57(11): 1542–1549.

Baddley, J.W., G.N. Forrest and the AST Infectious Diseases Community of Practice. 2013. Cryptococcosis in solid organ transplantation. American J Transplant. 13(4): 242–249.

Badiee, P. and A. Alborzi. 2011. Invasive fungal infections in renal transplant recipients. Expt Clin Transplant 9(6): 355–62.

Badiee, P., P. Kordbacheh, A. Alborzi, F. Zeini, H. Mirhendy and M. Mahmoody. 2005. Fungal infections in solid organ recipients. Expt Clin Transplant. 3(2): 385–389.

Battaglia, M., P. Ditonno, O. Selvaggio, L. Garofalo, S. Palazzo, A. Schena et al. 2004. Kidney transplants from infected donors: our experience. Transplant Proc. 36(3): 491–492.

Biancheri, D., J. Kanitakis, A.L. Bienvenu, S. Picot, E. Morelon, M. Faure et al. 2012. Cutaneous cryptococcosis in solid organ transplant recipients: epidemiologic, clinical, diagnostic and therapeutic features. European J Dermatol. 22(5): 651–657.

Bisbe, J., J. Vilardell, M. Vails, A. Moren, M. Brancos and J. Andreu. 1987. Transient fungemia and candida arthritis due to *Candida zeylanoides.* European J Clin Microbiol. 6(6): 668–669.

Blair, J.E. and J.L. Logan. 2001. Coccidioidomycosis in solid organ transplantation. Clin Inf Dis. 33(9): 1536–1544.

Brett, M.T., J.T.C. Kwan and M.R. Bending. 1988. Cecal perforation in a renal transplant patient with disseminated histoplasmosis. J Clin Pathol. 41(9): 992–995.

Bruno, K.M., L. Farhoomand, B.S. Linman, C.N. Pappas and F.J. Landry. 2002. Cryptococcal arthritis, tendinitis, tenosynovitis, and carpal tunnel syndrome: Report of a case and review of the literature. Arth Care Res. 47(1): 104–108.

Burton, J.R., J.B. Zachery, R. Bessin, H.K. Rathbun, W.B. Greenough 3rd., S. Sterioff et al. 1972. Aspergillosis in four renal transplant recipients. Diagnosis and effective treatment with amphotericin B. Ann Inter Med. 77(3): 383–388.

Campbell, C.K. and S.S. Al-Hedaithy. 1993. Phaeohyphomycosis of the brain caused by *Rhamichloridium mackenziei* sp. nov. in Middle Eastern countries. J Med Vet Mycol. 31(3): 325–332.

Canaud, G., M.-O. Timsit, J. Zuber, M.-E. Bougnoux, A. Méjean, E. Thervet et al. 2009. Early conservative intervention for candida contamination of preservative fluid without allograft nephrectomy. Nephrol Dial Transplant. 24(4): 1325–1327.

Carlson, K.C., M. Mehlmauer, S. Evans and P. Chandrasoma. 1987. Cryptococcal cellulitis in renal transplant recipients. J American Acad Dermatol. 17(3): 469–472.

Cartledge, J.D., J. Midgely and B.G. Gazzard. 1997. Itraconazole solution: higher serum drug concentrations and better clinical response rates than the capsule formulation in acquired immunodeficiency syndrome patients with candidosis. J Clin Pathol. 50*:* 477–80.

Cerutti, E., C. Stratta, R. Romagnoli, R. Serra, M. Lepore, F. Fop et al. 2006. Bacterial-and fungal-positive cultures in organ donors: clinical impact in liver transplantation. Liver Transplant. 12(8): 1253–1259.

Chamilos, G., R.E. Lewis and D.P. Kontoyiannis. 2008. Delaying amphotericin B-based frontline therapy significantly increases mortality among patients with hematologic malignancy who have zygomycosis. Clin Inf Dis. 47(4): 503–509.

Chkhotua, A., A. Yussim, A. Tovar, M. Weinberger, V. Sobolev, N. Bar-Nathan et al. 2001. Mucormycosis of the renal allograft: case report and review of the literature. Transplant Int. 14(6): 438–441.

Chugh, K.S., V. Sakhuja, S. Jain, P. Talwar, M. Minz, K. Joshi et al. 1993. High mortality in systemic fungal infections following renal transplantation in third-world countries. Nephrol Dial Transplant. 8(2): 168–172.

Chung, B., J. Yun, S. Ha, J. Kim, I. Moon, B. Choi et al. 2013. Combined use of rituximab and plasmapheresis pre-transplant increases post-transplant infections in renal transplant recipients with basiliximab induction therapy. Transplant Inf Dis. 15(6): 559–568.

Clancy, C.J., R.A. Jaber, H.L. Leather, J.R. Wingard, B. Staley, L.J. Wheat et al. 2007. Bronchoalveolar lavage galactomannan in diagnosis of invasive pulmonary aspergillosis among solid-organ transplant recipients. J Clin Microbial. 45(6): 1759–1765.

Coleman, D.C., M.G. Rinaldi, K.A. Haynes et al. *1998.* Importance of *Candida* species other than *Candida albicans* as opportunistic pathogen. Med Mycol. 36*:* 156–65.

Collins, L.A., M.H. Samore, M.S. Roberts et al. 1994. Risk factors for invasive fungal infections complicating orthotopic liver transplantation. J Infect Dis. 170*:* 644–52.

Cray, C., T. Watson and K.L. Arheart. 2009. Serosurvey and diagnostic application of antibody titers to Aspergillus in avian species. Avian Dis. 53(4): 491–494.

Cuellar-Rodriguez, J., R. Avery, M. Lard, M. Budev, S. Gordon, N. Shrestha et al. 2009. Histoplasmosis in solid organ transplant recipients: 10 years of experience at a large transplant center in an endemic area. Clin Inf Dis. 49(5): 710–716.

Davies, S.F., G.A. Sarosi, P.K. Peterson, M. Khan, R.J. Howard, R.L. Simmons et al. 1979. Disseminated histoplasmosis in renal transplant recipients. The American J Surg. 137(5): 686–691.

Davis, J.A., D.L. Horn, K.A. Marr and J.A. Fishman. 2009. Central nervous system involvement in cryptococcal infection in individuals after solid organ transplantation or with AIDS. Transplant Inf Dis. 11(5): 432–437.

De Wit, S., D. Weerts, H. Goossens and N. Clumeck. 1989. Comparison of fluconazole and ketoconazole for oropharyngeal candidiasis in AIDS. Lancet 1: 746–8.

Denning, D.W. 1998. Invasive aspergillosis. Clin Inf Dis. 26(4): 781–803.

Denning, D.W. 2003. Echinocandin antifungal drugs. Lancet 362: 1142–51.

Dongari-Bagtzoglou, A., P. Dwivedi, E. Ioannidou, M. Shaqman, D. Hull and J. Burleson. 2009. Oral *Candida* infection and colonization in solid organ transplant recipients. Oral Microbiol Immunol. 24(3): 249–254.

Dufresne, S.F., R.E. LeBlanc, S.X. Zhang, K.A. Marr and D. Neofytos. 2013. Histoplasmosis and subcutaneous nodules in a kidney transplant recipient: erythema nodosum versus fungal panniculitis. Transplant Inf Dis. 15(2): E58–E63.

Dutriaux, C., I. Saint-Cyr, N. Desbois, D. Cales-Quist, A. Diedhou and A.M. Boisseau-Garsaud. 2005. Subcutaneous phaeohyphomycosis due to exophiala spinifera in a renal transplant recipient. Annales de Dermatologie et de Venereologie 132(3): 259–262.

Edwards, J.E., Jr., G.P. Bodey, R.A. Bowden et al. 1997. International conference for the development of a consensus on the management and prevention of severe candidal infections. Clin Infect Dis. 25: 43–59.

Einollahi, B., M. Lessan-Pezeshki, V. Pourfarziani, E. Nemati, M. Nafar, F. Pour-Reza-Gholi et al. 2008. Invasive fungal infections following renal transplantation: a review of 2410 recipients. Ann Transplant. 13(4): 55–58.

Einollahi, B., M. Lessan-Pezeshki, J. Aslani, E. Nemati, Z. Rostami, M.J. Hosseini, M.H. Ghaidani, P. Ahmadpour, H. Shahnazian, F. Pour-Reza-Gholi, M.R. Ghanji, A. Hossein and N. Nouri-Majalan. 2011. Two decades of experience in Mucormycosis after kidney transplantation. Ann Transplant. 16(3): 44–48.

Ersoy, A., I. Akdag, H. Akalin, B. Sarisozen and B. Ener. 2007. *Aspergillosis* osteomyelitis and joint infection in a renal transplant recipient. Transplant Proc. 39(5): 1662–1663.

Estrada, C., A.G. Desai, L.M. Chirch et al. 2012. Invasive Aspergillosis in a Renal Transplant Recipient Successfully Treated with Interferon-Gamma. Case reports in Transplant. 493758: 5.

Ezzatzadegan, S., S. Chen and J.R. Chapman. 2012. Invasive fungal infections after renal transplantation. Int J Organ Transplant Med. 3(1): 18–25.

Fadhil, R., H. Al-Thani, Y. Al-Maslamani and O. Ali. 2011. Trichosporon fungal arteritis causing rupture of vascular anastamosis after commercial kidney transplantation: A case report and review of literature. Transplant Proc. 43(2): 657–659.

Finlay, P.M., M.D. Richardson and A.G. Robertson. 1996. A comparative study of the efficacy of fluconazole and amphotericin B in the treatment of oropharyngeal candidosis in patients undergoing radiotherapy for head and neck tumours. Br J Oral Maxillofac Surg. 34: 23–5.

Fishman, J. 2014. Infection in the solid organ transplant recipient. UptoDate.

Fishman, J. 2014. Infection in kidney transplant recipients. *In*: Morris, P.J. and S.J. Knechtle (eds.). Kidney Transplantation. Principles and Practice. Elsevier Saunders, China, pp. 491–510.

Fishman, J.A., M.A. Greenwald and P.A. Grossi. 2012. Transmission of infection with human allografts: essential considerations in donor screening. Clin Inf Dis. 55(5): 720–727.

Francis, P. and T.J. Walsh. 1992. Evolving role of flucytosine in immunocompromised patients: new insights into safety, pharmacokinetics, and antifungal therapy. Clin Infect Dis. 15: 1003–18.

Freifeld, A., P. Iwen, B. Lesiak, R. Gilroy, R. Stevens and A. Kalil. 2005. Histoplasmosis in solid organ transplant recipients at a large Midwestern university transplant center. Transplant Inf Dis. 7(3-4): 109–115.

Gamaletsou, M.N., B. Rammaert, M.A. Bueno, B. Moriyama, N.V. Sipas, D.P. Kontoviannis et al. 2014. Aspergillus osteomyelitis: epidemiology, clinical manifestations, management, and outcome. J Inf. 68(5): 478–493.

Gilaberte, M., R. Bartralot, J.M. Torres, F.S. Reus, V. Rodriguez, A. Alomar et al. 2005. Cutaneous alternariosis in transplant recipients: clinicopathologic review of 9 cases. J American Acad Pathol. 52(4): 653–659.

Giron, J.M., C.G. Poey, P.P. Fajadet, G.B. Balagner, J.A. Assoun, G.R. Richardi et al. 1993. Inoperable pulmonary aspergilloma: percutaneous CT-guided injection with glycerin and amphotericin B paste in 15 cases. Radiology 188(3): 825–7.

Godara, S.M., V.B. Kute, K.R. Goplani, M.R. Gumber, D.N. Gera, P.R. Shah et al. 2011. Mucormycosis in renal transplant recipients: predictors and outcome. Saudi J Kidney Dis Transplant. 22(4): 751–756.

Gomez, C.A. and N. Singh. 2013. Donor-derived filamentous fungal infections in solid organ transplant recipients. Curr Opin Inf Dis. 26(4): 309–316.

Gonzalez-Lopez, M.A., R. Salesa, M.C. Gonzalez-Vela, H. Fernandez- Llaca, J.F. Val-Bernal and J. Cano. 2007. Subcutaneous phaeohyphomycosis caused by *Exophiala oligosperma* in a renal transplant recipient. British J Dermatol. 156 (4): 762–764.

Goodrich, J.M., E.C. Reed, M. Mori, L.D. Fisher, S. Skerrett, P.S. Dandliker et al. 1991. Clinical features and analysis of risk factors for invasive candidal infection after marrow transplantation. J Inf Dis. 164(4): 731–740.

Graybill, J.R., J. Vazquez, R.O. Darouiche et al. 1998. Randomized trial of itraconazole oral solution for oropharyngeal candidiasis in HIV/AIDS patient. Am J Med. 104: 33–9.

Gupta, K.L., A.K. Ghosh, R. Kochhar, V. Jha, A. Chakrabarti and V. Sakhuia. 1994. Esophageal candidiasis after renal transplantation: Comparative study in patients on different immunosuppressive protocols. American J Gastroenterol. 89(7): 1062–1065.

Gupta, S.K., K.S. Manjunath-Prasad, B.S. Sharma, V.K. Khosla, V.K. Kak, M. Minz et al. 1997. Brain abscess in renal transplant recipients: Report of three cases. Surgical Neurol. 48(3): 284–287.

Gustafson, T.L., W. Schaffner, G.B. Lavely, C.W. Stratton, H.K. Johnson and R.H. Hutcheson, Jr. 1983. Invasive aspergillosis in renal transplant recipients: correlation with corticosteroid therapy. J Inf Dis. 148(2): 230–238.

Hadley, S., M.H. Samore, W.D. Lewis, R.L. Jenkins, A.W. Karchmer and S.M. Hammer. 1995. Major infectious complications after orthotopic liver transplantation and comparison of outcomes in patients receiving cyclosporine or FK506 as primary immunosuppression. Transplantation 59: 851–9.

Hamdi, T., V. Karthikeyan and G. Alangaden. 2014. Mucormycosis in a renal transplant recipient: Case report and comprehensive review of literature. Int J Nephrol. Article ID 950643, http://dx.doi.org/10.1155/2014/950643.

Hamory, B.H. and R.P. Wenzel. 1978. Hospital-associated candiduria: predisposing factors and review of the literature. J Urol. 120: 444–8.

Herbrecht, R., D.W. Denning, T.F. Patterson, J.E. Bennett, R.E. Greene, J.W. Oestmann et al. 2002. Voriconazole versus amphotericin B for primary therapy of invasive aspergillosis.

Hill, R.B. Jr., B.E. Dahrling 2nd, T.E. Starzl and D. Rifkind. 1967. Death after transplantation; an analysis of sixty patients. American J Med. 42(3): 327–334.

Hong Nguyen, M., S. Husain, C.J. Clancy, J.E. Peacock, C.C. Hung, D.P. Kontoyiannis et al. 2010. Outcomes of central nervous system cryptococcosis vary with host immune function: Results from a multi-center, prospective study. J Inf. 61(5): 419–426.

Horn, D., S. Sae-Tia and D. Neofytos. 2009. *Aspergillus osteomyelitis*: review of 12 cases identified by the Prospective Antifungal Therapy Alliance registry. Diag Microbiol Inf Dis. 63(4): 384–387.

Horn, D.L., D. Neofytos, E.J. Anaissie, J.A. Fishman, W.J. Steinbach, A.J. Olyaei et al. 2009. Epidemiology and outcomes of candidemia in 2019 patients: data from the prospective antifungal alliance registry. Clin Inf Dis. 48(12): 1695–1703.

Howard, S.J., D. Cerar, M.J. Anderson et al. 2009. Frequency and evolution of azole resistance in Aspergillus fumigatus associated with treatment failure. Emerg Inf Dis. 15(7): 1068–1076.

Hoyo, I., G. Sanclemente, J.P. de la Bellacasa, F. Cofan, M.J. Ricart, M. Cardona et al. 2014. Epidemiology, clinical characteristics, and outcome of invasive aspergillosis in renal transplant patients. Transplant Inf. Dis. doi: 10.1111/tid.12301 [Epub ahead of print].

Husain, S., M.M. Wagener and N. Singh. 2001. Cryptococcus neoformans infection in organ transplant recipients: variables influencing clinical characteristics and outcome. Emer Inf Dis. 7(3): 375–381.

Isa-Isa, R., C. García, M. Isa and R. Arenas. 2012. Subcutaneous phaeohyphomycosis (mycotic cyst). Clin Dermatol. 30(4): 425–431.

Jacobs, L.G., E.A. Skidmore, K. Freeman, D. Lipschultz and N. Fox. 1996. Oral fluconazole compared with bladder irrigation with amphotericin B for treatment of fungal urinary tract infections in elderly patients. Clin Infect Dis. 22: 30–5.

Jayakumar, M., F. Gopalakrishnan, R. Vijayakumar, S. Rajendran and M.A. Muthusethupathi. 1998. Systemic fungal infections in renal transplant recipients at Chennai, India. Transplant Proc. 30(7): 3135.

Jha, V. and K.S. Chugh. 2002. Posttransplant infections in the tropical countries. ArtOrg. 26(9): 770–777.

John, G.T., V. Shankar, G. Talaulikar, M.S. Mathews, M.A. Abraham, P.P. Thomas et al. 2003. Epidemiology of systemic mycoses among renal-transplant recipients in India. Transplant 75(9): 1544–1551.

Ju, M.K., D.J. Joo, S.J. Kim, H.K. Chang, M.S. Kim, S.I. Kim et al. 2009. Invasive pulmonary Aspergillosis after solid organ transplantation: Diagnosis and treatment based on 28 years of transplantation experience. Transplant Proc. 41(1): 375–378.

Kanj, S.S., S.S. Amr and G.D. Roberts. 2001. *Ramichloridium mackenziei* brain abscess: report of two cases and review of the literature. Med Mycol. 39(1): 97–102.

Kapoor, A., S.M. Flechner, K. O'Malley, D. Paolone, T.M. File, Jr. and A.F. Cutrona. 1999. Cryptococcal meningitis in renal transplant patients associated with environmental exposure. Transplant Inf Dis. 1(3): 213–217.

Karchmer, A.W., M.H. Samore, S. Hadley, L.A. Collins, R.L. Jenkins and W.D. Lewis. 1994. Fungal infections complicating orthotopic liver transplantation. Trans Am Clin Climatol Assoc. 106*:* 38–48.

Kennedy, S.E., Y. Shen, J.A. Charlesworth, J.D. Mackie, J.D. Mahony, J. Kelly et al. 2005. Outcome of overseas commercial kidney transplantation: an Australian perspective. Med J Australia 182(5): 224–227.

King, G.N., C.M. Healy, M.T. Glover, J.T.C. Kwan, D.M. Williams, I.M. Leigh and M.H. Thornhill. 1994. Prevalence and risk factors associated with leukoplakia, hairy leukoplakia, erythematous candidiasis, and gingival hyperplasia in renal transplant recipients. Oral Surgery, Oral Medicine and Oral Pathology 78(6): 718–26.

Kossoff, E.H., E.S. Buescher and M.G. Karlowicz. 1998. Candidemia in a neonatal intensive care unit: trends during fifteen years and clinical features of 111 cases. Pediatr Inf Dis *J.* 17*:* 504–8.

Kung, N., N. Fisher, B. Gunson, M. Hastings and D. Mutimer. 1995. Fluconazole prophylaxis for high-risk liver transplant recipients. Lancet 349*:* 1234–5.

Kuy, S., C. He and D.C. Cronin. 2013. Renal Mucormycosis: A Rare and Potentially Lethal Complication of Kidney Transplantation. Case Rep Transplant. 2013.

Langolis, R.P., K.M. Flegel, J.L. Meakins, D.D. Morehouse, H.G. Robson and R.D. Guttmann. 1980. Cutaneous aspergillosis with fatal dissemination in renal transplant recipient. Can Med Assoc J. 122(6): 673–676.

Laouad, I., M. Buchler, C. Noel, T. Sadek, H. Maazouz, P. Westeel and Y. Lebranchu. 2005. Renal Artery Aneurysm Secondary to *Candida albicans* in Four Kidney Allograft Recipients. Transplant Proc. 37(6): 2834–2836.

Leu, H.S. and C.T. Huang. 1995. Clearance of funguria with short-course antifungal regimens: a prospective, randomized, controlled study. Clin Infect Dis. 20*:* 1152–7.

Levy, E. and M.J. Bia. 1995. Isolated renal mucormycosis: case report and review. J American Soc Nephrol. 5(12): 2014–2019.

Levy, I., L.G. Rubin, S. Vasishtha, V. Tucci and S.K. Sood. 1998. Emergence of *Candida parapsilosis* as the predominant species causing candidemia in children. Clin Infect Dis. 26*:* 1086–8.

Lief, M.H., D. Caplivski, E.J. Bottone, S. Lerner, C. Vidal and S. Huprikar. 2010. Exophiala jeanselmei infection in solid organ transplant recipients: report of two cases and review of the literature. Transplant Inf Dis. 13(1): 73–79.

Lin, S.-J., J. Schranz and S.M. Teutsch. 2001. Aspergillosis case-fatality rate: systematic review of the literature. Clin Inf Dis. 32(3): 358–366.

Lobo, V., A. Joshi, P. Khatavkar and M.K. Kale. 2009. Pulmonary histoplasmosis in renal allograft recipient. Ind J Nephrol. 24(2): 120–123.

López-Pintor, R.M., G. Hernández, L. de Arriba and A. de Andrés. 2013. Oral candidiasis in patients with renal transplants. Medicina Oral, Patologia Oral y Cirugia Bucal 18(3): e381–387.

Low, C.Y. and C. Rotstein. 2011. Emerging fungal infections in immunocompromised patients. F1000 medicine reports.

Lyon, M., D. Kaul, A. Ehsan, S. Covington, S. Taranto, K. Taylor et al. 2011. Infectious disease transmission from organ donors with meningitis or encephalitis. Am J Transplant. 11(suppl 2): 200.

Maertens, J., I. Raad, G. Petrikkos, M. Boogaerts, D. Selleslag, F.B. Petersen et al. 2004. Efficacy and safety of caspofungin for treatment of invasive aspergillosis in patients refractory to or intolerant of conventional antifungal therapy. Clin. Infect. Dis. 39(11): 1563–71.

Maertens, J., A. Glasmacher, R. Herbrecht et al. 2006. Multicenter, noncomparative study of caspofungin in combination with other antifungals as salvage therapy in adults with invasive aspergillosis. Cancer 107: 2888–97.

Mai, H., L. Champion, N. Ouali, A. Hertig, M.-N. Peraldi, D. Glotz et al. 2006. *Candida albicans* arteritis transmitted by conservative liquid after renal transplantation: a report of four cases and review of the literature. Transplantation 82(9): 1163–1167.
Mal, H., I. Rullon, F. Mellot, O. Brugière, C. Sleiman, Y. Menu et al. 1999. Immediate and long-term results of bronchial artery embolization for life-threatening hemoptysis. Chest 115(4): 996–1001.
Marufuji Ogawa, M.M., N.Z. Galante, P. Godoy, O. Fischman-Gompertz, F. Martelli A.L. Colombo et al. 2009. Treatment of subcutaneous phaeohyphomycosis and prospective follow-up of 17 kidney transplant. J American Acad Dermatol. 61(12): 977–985.
Matignon, M., F. Botterel, V. Audard, B. Dunogue, K. Dahan, P. Lang et al. 2008. Outcome of renal transplantation in eight patients with Candida sp. contamination of preservation fluid. American J Transplant. 8(3): 697–700.
McGuinn, M.L., M.E. Lawrence, L. Proia and J. Segreti. 2005. Progressive disseminated histoplasmosis presenting as cellulitis in a renal transplant recipient. Transplant Proc. 37(10): 4313–4314.
McNeil, M.M., S.L. Nash, R.A. Hajjeh, M.A. Phelan, L.A. Conn, B.D. Plikaytis et al. 2001. Trends in mortality due to invasive mycotic diseases in the United States, 1980–1997. Clin Inf Dis. 33(5): 641–647.
Mesa, A., J. Henao, M. Gil and G. Durango. 1999. Phaeohyphomycosis in kidney transplant patients. Clin Transplant. 13(3): 273–276.
Mills, S.A., H.F. Seigler and W.G. Wolfe. 1975. The incidence and management of pulmonary mycosis in renal allograft patients. Ann Surg. 182(5): 617–26.
Minari, A., R. Husni, R.K. Avery, D.L. Longworth, M. DeCamp, M. Bertin et al. 2002. The incidence of invasive aspergillosis among solid organ transplant recipients and implications for prophylaxis in lung transplants. Transplant Inf Dis. 4(4): 195–200.
Moreno, A., C. Cervera, J. Gavalda, M. Rovira, I. De La Cámara, I. Jarque et al. 2007. Bloodstream infections among transplant recipients: results of a nationwide surveillance in Spain. American J Transplant. 7(11): 2579–2586.
Morris, M., S. Fisher and M. Ison. 2010. Infections transmitted by transplantation. Inf Dis Clin North America 24: 497–504.
Nampoory, M., Z. Khan, K. Johny, J. Constandi, R. Gupta, I. Al-Muzairi et al. 1996. Invasive fungal infections in renal transplant recipients. J Infection 33(2): 95–101.
Negroni, R. 2012. Cryptococcosis. Clin Dermatol. 30(6): 599–609.
Nieto-Ríos, J.F., L.M. Serna-Higuita, C.E. Guzman-Luna, C. Ocampo-Kohn, A. Aristizabal-Alzate, I. Ramírez et al. 2014. Histoplasmosis in renal transplant patients in an endemic area at a reference hospital in Medellin, Columbia. Transplantation Proc. 46(9): 3004–3009.
Olczak-Kowalczyk, D., J. Pawlowska, B. Garczewska, E. Smirska, R. Grend, M. Syczewska et al. 2010. Oral candidiasis in immunosuppressed children and young adults after liver or kidney transplantation. Pediatric Dentistry 32(3): 189–194.
Ozcan, D., A.T. Gulec and M. Haberal. 2008. Multiple subcutaneous nodules leading to the diagnosis of pulmonary aspergillosis in a renal transplant recipient. Clin Transplant. 22(1): 120–123.
Ozgur, N., N. Seyahi, U. Sili, M. Oruc, B. Mete, R. Ataman et al. 2014. Candidal psoas abscess following persistent pyuria in a renal transplant recipient. Int Urol Nephrol. 46(1): 269–273.
Panackal, A.A., A. Dahlman, K.T. Keil, C.L. Peterson, L. Mascola, S. Mirza et al. 2003. Outbreak of invasive aspergillosis among renal transplant recipients. Transplantation 75(7): 1050–1053.
Pappas, P.G., C.A. Kauffman, D.R. Andes et al. 2009. Clinical practice guidelines for the management candidiasis': 2009 update by the Infectious Diseases Society of America. Clin Inf Dis. 48(5): 503–535.
Pappas, P.G., B.D. Alexander, D.R. Andes, S. Hadley, C.A. Kauffman, A. Freifeld et al. 2010. Invasive fungal infections among organ transplant recipients: results of the Transplant-Associated Infection Surveillance Network (TRANSNET). Clin Inf Dis. 50(8): 1101–1111.
Park, S.B., M.J. Kang, E.A. Whang, S.Y. Han, H.C. Kim and K.K. Park. 2004a. A case of primary cutaneous aspergillosis in a renal transplant recipient. Transplantation Proc. 36(7): 2156–2157.
Park, S.B., M.J. Kang, E.A. Whang, S.Y. Han and H.C. Kim. 2004b. A case of fungal sepsis due to Aspergillus spondylitis followed by cytomegalovirus infection in a renal transplant recipient. Transplantation Proc. 36(7): 2154–2155.
Patel, R. and C.V. Paya. 1997. Infections in solid-organ transplant recipients. Clin Microbiol Rev. 10(1): 86–124.
Peddi, V., S. Hariharan and M. First. 1996. Disseminated histoplasmosis in renal allograft recipients. Clin Transplant. 10(2): 160–165.

Perfect, J.R. 2010. Cryptococcal Disease: 2010 Update by the Infectious Diseases Society of America. Clin Inf Dis. 50: 291–322.

Peterson, P., R. Ferguson, D. Fryd, H. Balfour, Jr., J. Rynasiewicz and R. Simmons. 1982. Infectious diseases in hospitalized renal transplant recipients: a prospective study of a complex and evolving problem. Medicine 61(6): 360–372.

Phillips, P., K. De Beule, G. Frechette et al. 1998. A double-blind comparison of itraconazole oral solution and fluconazole capsules for the treatment of oropharyngeal candidiasis in patients with AIDS. Clin Infect Dis. 26: 1368–73.

Pletka, P., J.R. Kenyon, M. Snell, S.L. Cohen, K. Owen, J.F. Mowbray et al. 1969. Cadaveric renal transplantation. An analysis of 65 cases. Lancet. 1(7584): 1–6.

Pons, V., D. Greenspan and M. Debruin. 1993. Multicenter Study Group. Therapy for oropharyngeal candidiasis in HIV-infected patients: a randomized, prospective multicenter study of oral fluconazole versus clotrimazole troches. J Acquir Immune Defic Syndr. 6: 1311–6.

Qadri, S.M.H., F. Al Dayel, M.J. Strampfer and B.A. Cunha. 1988. Urinary tract infection caused by *Hansenula anomala*. Mycopathologica 104(1): 99–101.

Qunibi, W. 1997. Commercially motivated renal transplantation: results in 540 patients transplanted in India. Clin Transplant. 11(6): 536–544.

Rappo, U., J.R. Beitler, J.R. Faulhaber, B. Firoz, J.S. Henning, K.M. Thomas et al. 2010. Expanding the horizons of histoplasmosis: disseminated histoplasmosis in a renal transplant patient after a trip to Bangladesh. Transplant Inf Dis. 12(2): 155–160.

Reed, C., R. Bryant, A.S. Ibrahim et al. 2008. Combination polyene-caspofungin treatment of rhino-orbital-cerebral mucormycosis. Clin Inf Dis. 47(3): 364–371.

Revankar, S.G. and D.A. Sutton. 2010. Melanized fungi in human disease. Clin Microbiol Rev. 23(4): 884–928.

Revankar, S.G., D.A. Sutton and M.G. Rinaldi. 2004. Primary central nervous system phaeohyphomycosis: A review of 101 cases. Clin Inf Dis. 38(2): 206–216.

Rex, J.H., T.J. Walsh, J.D. Sobel et al. 2000. Practice guidelines for the treatment of candidiasis. Infectious Diseases Society of America. Clin Infect Dis. 30(40): 662–678.

Rifkind, D., T.L. Marchioro, S.A. Schneck and R.B. Hill. 1967. Systemic fungal infections complicating renal transplantation and immunosuppressive therapy. Clinical, microbiologic, neurologic and pathologic features. American J Med. 43(1): 28–38.

Ringdén, O., F. Meunier, J. Tollemar, P. Ricci, S. Tura, E. Kuse et al. 1991. Efficacy of amphotericin B encapsulated in liposomes (AmBisome) in the treatment of invasive fungal infections in immunocompromised patients. J Antimicrob Chemother 28(suppl B): 73–82.

Romano, C., L. Valenti, C. Miracco, C. Alessandrini, E. Paccagnini, E. Faggi et al. 1997. Two cases of cutaneous phaeohyphomycosis by Alternaria alternate and Alternaria tenuissima. Mycopathol. 137(1): 65–74.

Ronan, S.G., I. Uzoaru, V. Nadimpalli, J. Guitart and J.R. Manaligod. 1993. Primary cutaneous phaeohyphomycosis: report of seven cases. J Cut Pathol. 20(3): 223–228.

Saad, A.H., D.D. DePestel and P.L. Carver. 2006. Factors influencing the magnitude and clinical significance of drug interactions between azole antifungals and select immunosuppressants. Pharmacotherapy 26: 1730–44.

Sabbaga, E., L.M. Tedesco-Marchesi, C. Lacaz, L.C. Cuce, A. Salebian, E.M. Heins-Vaccari et al. 1994. Subcutaneous phaeohyphomycose due to Exophiala jeanselmie. Report of 3 cases in patients with kidney transplant. Revista do Instituto de Medicina Tropical de Sao Paulo 36(2): 175–183.

Sable, C.A., B.Y. Nguyen, J.A. Chodakewitz and M.J. DiNubile. 2002. Safety and tolerability of caspofungin acetate in the treatment of fungal infections'. Transplant Infect Dis. 4: 25–30.

Saccente, M. 2008. Central nervous system histoplasmosis. Curr Treat Opt Neurol. 10(3): 161–167.

Safdar, N., W.R. Slattery, V. Knasinski, R.F. Gangnon, Z. Li, J.D. Pirsch and D. Andes. 2005. Predictors and outcomes of candiduria in renal transplant recipients. Clin Inf Dis. 40(10): 1413–1421.

Salez, F., A. Brichet, S. Desurmont, J.M. Grosbois, B. Wallaert and A.B. Tonnel. 1999. Effects of itraconazole therapy in allergic bronchopulmonary aspergillosis. Chest. 116(6): 1665–8.

Salido-Vallejoa, R., M.J. Linares-Sicilia, G. Garnacho-Saucedoa, M. Sánchez-Frías, F. Solís-Cuesta, J. Genée et al. 2014. Subcutaneous phaeohyphomycosis due to Alternaria infectoria in a renal transplant patient: Surgical treatment with no long-term relapse. Revista Iberoamericanade Micología 31(2): 149–151.

Sangeorzan, J.A., S.F. Bradley, X. He et al. 1994. Epidemiology of oral candidiasis in HIV-infected patients: colonization, infection, treatment, and emergence of fluconazole resistance. Am J Med. 97: 339–46.

Schelenz, S. and W.R. Gransden. 2003. Candidaemia in a London teaching hospital: analysis of 128 cases over a 7-year period. Mycoses 46(9-10): 390–396.

Schelenza, S. and D.J.A. Goldsmith. 2003. Aspergillus endophthalmitis: an unusual complication of disseminated infection in renal transplant patients. J Inf. 47(4): 336–343.

Sever, M.Ş., R. Kazancioglu, A. Yildiz, A. Türkmen, T. Ecder, S.M. Kayacan et al. 2001. Outcome of living unrelated (commercial) renal transplantation. Kidney Int. 60(4): 1477–1483.

Shirwany, A., S.J. Sargent, R.R. Dmochowski and M.S. Bronze. 1998. Urinary tract Aspergillosis in a renal transplant recipient. Clin Inf Dis. 27(5): 1336.

Shoham, S. and K.A. Marr. 2012. Invasive fungal infections in solid organ transplant recipients. Future Microbiol. 7(5): 639–655.

Shoham, S., F. Hinestrosa, J. Moore, S. O'Donnell, M. Ruiz and J. Light. 2010. Invasive filamentous fungal infections associated with renal transplant tourism. Transplant Inf Dis. 12(4): 371–374.

Siddiqui, T.J., T. Zamani and J.P. Parada. 2005. Primary cryptococcal prostatitis and correlation with serum prostate specific antigen in renal transplant recipient. J Inf. 51(3): 153–156.

Silveira, F.P., S. Kusne and the AST Infectious Diseases Community of Practice. 2013. Candida infections in solid organ transplantation. American J Transplant. 13(s4): 220–227.

Silverman, R.A., A.R. Rhodes and P.H. Dennehy. 1986. Disseminated intravascular coagulation and purpura fulminans in a patient with *Candida* sepsis. Biopsy of purpura fulminans as an aid to diagnosis of systemic *Candida* infection. The American J Med. 80(4): 679–684.

Singer, A.J., I.B. Kubak and K.H. Anders. 1998. Aspergillosis of the testis in a renal transplant recipient. Urology 51(1): 119–121.

Singh, N. 2006. Combination of voriconazole and caspofungin as primary therapy for invasive aspergillosis in solid organ transplant recipients: a prospective, multicenter, observational study. Transplant. 81: 320–326.

Singh, N. and S. Husain. 2000. Infections of the central nervous system in transplant recipients. Transplant Inf Dis. 2(3): 101–111.

Singh, N. and D.L. Paterson. 2005. *Aspergillus* infections in transplant recipients. Clin Microbiol Rev. 18(1): 44–69.

Singh, N. and H.-Y. Sun. 2008. Iron overload and unique susceptibility of liver transplant recipients to disseminated disease due to opportunistic pathogens. Liver Transplant. 14(9): 1249–1255.

Singh, N., B.D. Alexander, O. Lortholary, F. Dromer, K.L. Gupta, G.T. John et al. 2007. Cryptococcus neoformans in organ transplant recipients: Impact of calcineurin-inhibitor agents on mortality. J Inf Dis. 195(5): 756–764.

Singh, N., J.M. Aguado, H. Bonatti, G. Forrest, K. Gupta, N. Safdar et al. 2009. Zygomycosis in solid organ transplant recipients: A prospective study to asses risks for disease and outcome. J Inf Dis. 200(6): 1002–1011.

Singh, N., S. Huprikar, S. Burdette, M. Morris, J. Blair and L. Wheat. 2012. Donor-derived fungal infections in organ transplant recipients: guidelines of the American Society of Transplantation, Infectious Diseases Community of Practice. American J Transplant. 12(9): 2414–2428.

Snydman, D.R., A.C. Roxby, G.S. Gottlieb and A.P. Limaye. 2009. Strongyloidiasis in transplant patients. Clin Inf Dis. 49(9): 1411–1423.

Sobel, J.D., S. Faro, R.W. Force et al. 1998. Vulvovaginal candidiasis: epidemiological, diagnostic, and therapeutic considerations. Am J Obstet Gynecol. 178: 203–11.

Sobel, J.D., C.A. Kauffman, D. McKinsey et al. 2000. Candiduria: a randomized, double-blind study of treatment with fluconazole and placebo. Clin Infect Dis. 30: 19–24.

Spellberg, B.J., Edwards and A. Ibrahim. 2005. Novel perspectives on mucormycosis: pathophysiology, presentation, and management. Clin Microbiol Rev. 18(3): 556–569.

Stevens, D.A., H.J. Schwartz, J.Y. Lee, B.L. Moskovitz, D.C. Jerome, A. Catanzaro et al. 2000. A randomized trial of itraconazole in allergic bronchopulmonary aspergillosis'. N Engl J Med. 342(11): 756–62.

Stitt, N. 2003. Infection in the Transplant Recipient. Organ Transplant (Medscape Online).

Sun, H.-Y., J.M. Aguado, H. Bonatti et al. 2009. Pulmonary zygomycosis in solid organ transplant recipients in the current era. Am J Transplant. 9(9): 2166–2171.

Tack, K.J., F.S. Rhames, B. Brown and R.C. Thompson, Jr. 1982. Aspergillus osteomyelitis. Report of four cases and review of the literature. American J Med. 73(2): 295–300.

Thaler, M., B. Pastakia, T.H. Shawker, T. O'leary and P.A. Pizzo. 1988. Hepatic candidiasis in cancer patients: the evolving picture of the syndrome. Ann Inter Med. 108(1): 88–100.
Thalla, R., D. Kim, K.K. Venkat and R. Parasuraman. 2009. Sequestration of active *Cryptococcus neoformans* infection in the parathyroid gland despite prolonged therapy in renal transplant recipient. Transplant Inf Dis. 11(4): 349–352.
Tolkoff-Rubin, N.E. and R.H. Rubin. 1992. Opportunistic fungal and bacterial infection in the renal transplant recipient. J American Soc Nephrology 2(12): S264.
Tollemar, J., K. Hockerstedt, B.G. Ericzon, H. Jalanko and O. Ringden. 1995. Liposomal amphotericin B prevents invasive fungal infections in liver transplant recipients: a randomized, placebo-controlled study. Transplant. 59: 45–50.
Tomazie, J., M. PIRS, T. Matos, D. Ferluga and J. Lindic. 2007. Multiple infections after commercial renal transplantation in India. Nephrol Dial Transplant. 22(3): 972–973.
Torre-Cisneros J., O.L. Lopez, S. Kusne, A.J. Martinez, T.E. Starzl, R.L. Simmons et al. 1993. CNS aspergillosis in organ transplantation. A clinicopathological study. J Neurol Neurosurgery, and Psychiatry 56(2): 188–193.
Trabelsi, H., S. Néji, H. Sellami, S. Yaich, F. Cheikhrouhou, R. Guidara et al. 2013. Invasive fungal infections in renal transplant recipients: about 11 cases. J Med Mycol. 23(4): 255–260.
Turner, L. 2009. Commercial organ transplantation in the Philippines. Cambridge Quarterly of Healthcare Ethics. 18(02): 192–196.
Van Cutsem, J. and J.R. Boelaert. 1989. Effects of deferoxamine, feroxamine and iron on experimental mucormycosis (zygomycosis). Kidney International 36(6): 1061–1068.
Van der Vliet, J., G. Tidow, G. Kootstra, H. Van Saene, R. Krom, M. Slooff et al. 1980. Transplantation of contaminated organs. British J Surgery 67(8): 596–598.
Vazquez, J. and J. Sobel. 2011. Candidiasis. pp. 167–206. *In*: Kaufman, C., P. Pappas, J. Sobel and W. Dismukes (eds.). Essentilas of Clinical Mycology, 2nd edition. Oxford University Press, Oxford.
Vermeirea, S.E.M., H. de Jongea, K. Lagroub and D.R.J. Kuypersa. 2010. Cutaneous phaeohyphomycosis in renal allograft recipients: report of 2 cases and review of the literature. Diag Microbiol Inf Dis. 68(2): 177–180.
Veroux, M., M. Macarone, P. Fiamingo, D. Cappello, M. Gagliano, M. Di Mare et al. 2006. Caspofungin in the treatment of azole-refractory esophageal candidiasis in kidney transplant recipients. Transplantation Proc. 38(4): 1037–1039.
Veroux, M., D. Corona, G. Giuffrida, M. Gagliano, T. Tallarita, A. Giaquinta et al. 2009. Acute renal failure due to ureteral obstruction in a kidney transplant recipient with *Candida albicans* contamination of preservation fluid. Transplant Inf Dis. 11(3): 266–268.
Vilchez, R.A., J. Fung and S. Kunse. 2002. Cryptococcosis in organ transplant recipients: An overview. American J Transplant. 2(7): 575–580.
Vilchez, R., R. Shapiro, K. McCurry, R. Kormos, K. Abu-Elmagd, J. Fung and S. Kusne. 2003. Longitudinal study in cryptococcosis in adult solid-organ transplant recipients. Transplant Int. 16(5): 36–40.
Vuruskan, H., A. Ersoy, N.K. Girgin, M. Ozturk, G. Filiz, I. Yavascaoglu et al. 2005. An unusual cause of ureteral obstruction in a renal transplant recipient: ureteric aspergilloma. Transplant Proc. 37(5): 2115–2117.
Walsh, T.J., E.J. Anaissie, D.W. Denning, R. Herbrecht, D.P. Kontoyiannis, K.A. Marr et al. 2008. Treatment of aspergillosis: clinical practice guidelines of the infectious diseases society of America. CID 46: 327–60.
Wark, P.A., P.G. Gibson and A.J. Wilson. 2003a. Azoles for allergic bronchopulmonary aspergillosis associated with asthma. Cochrane. Database. Syst Rev. (3): CD001108.
Wark, P.A., M.J. Hensley, N. Saltos, M.J. Boyle, R.C. Toneguzzi, G.D. Epid et al. 2003b. Anti-inflammatory effect of itraconazole in stable allergic bronchopulmonary aspergillosis: a randomized controlled trial. J Allergy Clin Immunol. 111(5): 952–7.
Wheat, J.L. 2003. Current diagnosis of histoplasmosis. Tre Microbiol. 11(10): 488–494.
Wheat, L.J., E.J. Smith, B. Sathapatayavongs, B. Batteiger, R.S. Filo, S.B. Leapman et al. 1983. Histoplasmosis in renal allograft recipients: two large urban outbreaks. Archiv Inter Med. 143(4): 703–707.
White, M.H., E.J. Anaissie, S. Kusne, J.R. Wingard, J.W. Hiemenz, A. Cantor et al. 1997. Amphotericin B colloidal dispersion vs. amphotericin B as therapy for invasive aspergillosis. Clin Infect Dis. 24(4): 635–42.

Wu, G., R.A. Vilchez, B. Eidelman, J. Fung, R. Kormos and S. Kusne. 2002. Cryptococcal meningitis. An analysis among 5521 consecutive organ transplant recipients. Transplant Inf Dis. 4(4): 183–188.
Yang, Y.L., M. Chen, J.L. Gu, F.Y. Zhu, X.G. Xu, C. Zhang et al. 2014. Cryptococcosis in kidney transplant recipients in a Chinese university hospital and a review of published cases. Int J Inf Dis. 26: 154–61.
Yesilkaya, A., O.K. Azap, M.H. Demirkaya, M.A. Ok, H. Arslan and Akdur. 2013. Bloodstream infections among solid organ transplant recipients: Eight years' experience from a Turkish university hospital. Balkan Med J. 30(3): 282–286.
Yoneda, T., Y. Itami, A. Hirayama, T. Saka, K. Yoshida and K. Fujimoto. 2014. Cryptococcal necrotizing fasciitis in a patient after renal transplantation—A case report. Transplantation Proc. 46(2): 620–622.

CHAPTER 7

Clinical Importance of the Genus *Curvularia*

Krisztina Krizsán,[1] *Tamás Papp,*[1] *Palanisamy Manikandan,*[2,3] *Coimbatore Subramanian Shobana,*[4] *Muthusamy Chandrasekaran,*[5] *Csaba Vágvölgyi*[1,5] and *László Kredics*[1,*]

Introduction

Members of the genus *Curvularia* including species formerly classified as *Bipolaris* spp. in the *Cochliobolus* Group 2 are melanin-producing dematiaceous fungi. Among them *C. australiensis, C. geniculata, C. hawaiiensis, C. lunata* and *C. spicifera* are the species most frequently isolated from clinical samples. These fungi may cause local infections, primarily affecting the respiratory system, the cornea and the skin, in immunocompetent persons and invasive diseases in immunocompromised patients. Infections of the subcutaneous tissue, bone, endocardium and central nervous system have also been described.

[1] University of Szeged, Faculty of Science and Informatics, Department of Microbiology, Közép fasor 52, H-6726 Szeged, Hungary.

[2] Aravind Eye Hospital and Postgraduate Institute of Ophthalmology, Department of Microbiology, Avinashi road, Coimbatore 641 014, Tamilnadu, India.

[3] Department of Medical Laboratory Sciences, College of Applied Medical Sciences, Majmaah University, Al-Majmaah 11952, Kingdom of Saudi Arabia.

[4] Department of Microbiology, PSG College of Arts & Science, Avinashi road, Coimbatore 641 014, Tamilnadu, India.

[5] Botany and Microbiology Department, King Saud University, PB No. 2455, Riyadh 11451, Kingdom of Saudi Arabia.

[*] Corresponding author: kredics@bio.u-szeged.hu

The taxonomy of *Bipolaris*, *Curvularia* and related genera is confusing due to the nomenclatural conflicts of the teleomorphic genus *Cochliobolus* and their anamorphs in *Curvularia* and *Bipolaris,* and as a result of the frequent misidentifications of the species. Conventional identification of these fungi is based on the examination of the conidial morphology (i.e., comparison of the size and the number of the septa). Sequencing of the nuclear ribosomal ITS region has been most frequently used for molecular identification. However, applicability of this region to discern *Curvularia* spp. can be questionable. Accurate diagnosis of infection by these fungi is essential for the selection of the appropriate antifungal therapy. Although limited data is available on antifungal susceptibility of these fungi, significant differences could be observed in their sensitivity to certain antifungal agents.

The clinical importance of the genus *Curvularia* and the former genus *Bipolaris* were previously discussed in the literature separately from each other (Liu and Gray 2011; Schuetz 2011). This chapter provides an extensive overview about the most recent taxonomical status of the genus *Curvularia* (including former *Bipolaris*), about the developments in the field of molecular identification of *Curvularia* species, as well as the epidemiology, clinical manifestation and antimicrobial susceptibilities of the species most frequently isolated from human mycoses.

Taxonomic Status and Phylogeny of *Bipolaris* and *Curvularia*

The anamorph genera *Bipolaris* and *Curvularia* comprise dematiaceous filamentous fungi, among which several species are plant pathogens, mainly of graminaceous hosts. They belong to the family Pleosporaceae (Pleosporales, Ascomycota) and their teleomorph states, if known, are classified into the genus *Cochliobolus*.

Cochliobolus and related genera are notorious for their problematic taxonomies and their confused nomenclatures (Shimizu et al. 1998; Berbee et al. 1999; Zhang and Berbee 2001; Dela Paz et al. 2006; Manamgoda et al. 2012). Asexual states of *Cochliobolus* species together with those of two related genera, *Setosphaeria* and *Pyrenophora* were originally described in the genera *Helminthosporium* (characterized by melanized, transversely septate conidia) and *Curvularia* (characterized by melanized, transversely septate conidia where one of the cells is swollen and larger than the others) (Berbee et al. 1999). Today, anamorphs of *Setosphaeria* and *Pyrenophora* are classified in the genera *Exserohilum* and *Drechslera*, respectively (Schell 2002). *Exserohilum* can be distinguished from *Drechslera* and *Bipolaris* by the presence of a protuberant hilum on the conidia (Schell 2002). *Bipolaris* and *Drechslera* can be discerned based on the germination of their conidia. While conidia of *Bipolaris* primarily germinate from the two end cells with only a slightly protruding hilum, those of *Drechslera* can germinate from any cell. As conidia of *Bipolaris* and *Drechslera* are rather similar, they were frequently confused in the past and some earlier case reports even attributed phaeohyphomycoses in fact caused by *Bipolaris/Curvularia* to *Drechslera* (McGinnis et al. 1986). Moreover, Subramanian and Jain (1966) found that conidia of *Bipolaris* may also germinate in an amphigenous manner and placed the *Bipolaris* species into the genus *Drechslera*, which also increased the nomenclatural confusions (Dela Paz et al. 2006). Later, molecular phylogenetic studies analysed the relationships among the main groups of Pleosporaceae and clearly proved that *Bipolaris* and *Curvularia* cannot be merged in *Drechslera* (Berbee et al. 1999; Zhang and Berbee 2001; Zhang et al. 2009).

Unity of the genus *Cochliobolus* was already being questioned in the early studies and it was split into *Cochliobolus* and *Pseudocochliobolus* by certain authors on the basis of their sexual morphology (Tsuda et al. 1977). *Pseudocochliobolus* was characterized by ascospores less coiling than in *Cochliobolus* and the presence of a cylindrical stroma below the ascomata (Berbee et al. 1999; Manamgoda et al. 2012). Among the anamorphs of *Pseudocochliobolus*, both *Curvularia* and *Bipolaris* spp. were found. However, *Pseudocochliobolus* characters proved to be highly variable and they were found also in certain *Cochliobolus* species. Therefore, *Pseudocochliobolus* was later treated as a synonym of *Cochliobolus* (Alcorn 1983; Sivanesan 1987).

In the past 15 years, molecular phylogenetic studies greatly changed the taxonomy of *Cochliobolus*. In phylogenies inferred from ITS (internal transcribed spacer of the nuclear ribosomal RNA gene cluster), *gpd* (glyceraldehyde-3-phosphate dehydrogenase) and *Brn1* (1,3,8-trihydroxynaphthalene reductase) sequences, the genera *Bipolaris* and *Curvularia* proved to be polyphyletic and *Cochliobolus* was split into two clearly separated clades (Shimizu et al. 1998; Berbee et al. 1999; Dela Paz et al. 2006), among which one was connected with only *Bipolaris* anamorphs (Group 1), while the other was associated with both *Bipolaris* and *Curvularia* species (Group 2). Fungi in Group 1 have large, gently curving conidia (typical *Bipolaris*-type), while fungi in Group 2 have short and straight conidia or curved conidia (atypical *Bipolaris*-type) where one of the cells is large and swollen (*Curvularia*-type) (Fig. 1). All human pathogenic species including those formerly described as *Bipolaris australiensis*, *B. hawaiiensis* and *B. spicifera* were found in the *Cochliobolus* Group 2 in these phylogenies.

Recently a phylogeny inferred from the combined analysis of ITS, *gpd*, LSU (nuclear ribosomal large subunit RNA) and EF1-α (translation elongation factor 1-α) data sets was used to reinforce these results and to resolve the nomenclatural conflicts in this fungal group (Manamgoda et al. 2012). *Bipolaris*, *Cochliobolus* and *Curvularia* were separated into two monophyletic and well-defined genera: *Bipolaris* was maintained for the species in *Cochliobolus* Group 1, while the *Bipolaris* and

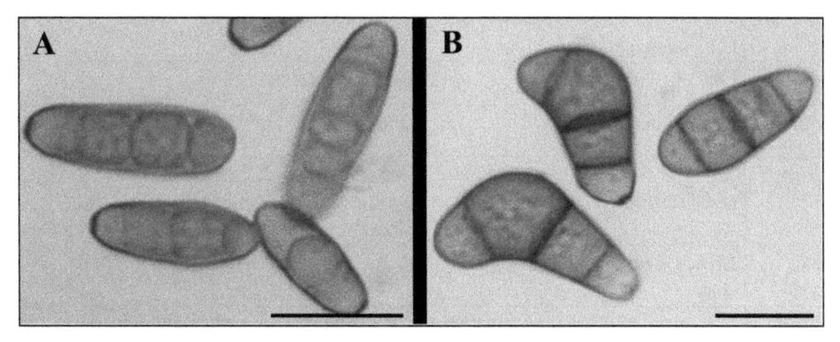

Figure 1. The conidial morphology of *Curvularia hawaiiensis* CBS 173.57 with atypical (A) and *Curvularia* sp. SZMC 13062 with typical (B) conidial morphology. Scale bars represent 10 μm.

Curvularia species of Group 2 were merged into *Curvularia*. They also provided clear generic descriptions for the re-evaluated genera, *Bipolaris* and *Curvularia* and the species involved in their study were transferred to the appropriate genus. Thus, all clinically relevant species reviewed in the present study were placed in *Curvularia*. Most recent works have also adopted this taxonomy and nomenclature (Paredes et al. 2013).

As a consequence of the above described taxonomic history, *Curvularia australiensis*, *C. hawaiiensis* and *C. spicifera* have been referred in earlier case reports as *Bipolaris*, *Drechslera* (*Bipolaris*) or *Drechslera* species.

Identification of *Curvularia* species

The clinically most relevant *Curvularia* species are *C. australiensis*, *C. hawaiiensis*, *C. spicifera, C. geniculata, C. lunata* and *C. pallescens* (Da Cunha et al. 2012; Alex et al. 2013). *C. australiensis*, *C. hawaiiensis* and *C. spicifera* have atypical, straight, short conidia and were originally classified as *Bipolaris* species (Fig. 1A). However, they have recently been transferred to the genus *Curvularia* based on molecular phylogenetic studies (Manamgoda et al. 2012), therefore the use of *Curvularia* generic name is suggested in the case of medically important *Bipolaris* species. *C. geniculata, C. lunata* and *C. pallescens* have typically curved, *Curvularia*-type conidia (Fig. 1B).

Differentiation of the opportunistic human pathogenic dematiaceous fungi primarily relies upon morphological characteristics (McGinnis et al. 1986; Guarro et al. 1999; Moore et al. 2001; Filizzola et al. 2003; Sun et al. 2003; Alexander and Pfaller 2006; Dyer et al. 2008; Revankar and Sutton 2010; Alex et al. 2013; Da Cunha et al. 2013). The basic diagnostic procedure of melanized fungi consists of the direct microscopic examination of the discriminative characteristics, mainly the pigmentation, septation, morphology and size of the hyphae, conidiophores and conidia (Table 1) (Filizzola et al. 2003; Kobayashi et al. 2008; Revankar and Sutton 2010; Da Cunha

Table 1. The main morphological characteristics of *Curvularia* species with typical and atypical conidia. The most important discriminative features are the shape and septation of the conidia.

	***Curvularia* spp. with atypical conidia**	***Curvularia* spp. with typical conidia**
Species	*C. australiensis, C. hawaiiensis, C. spicifera*	*C. geniculata, C. lunata, C. pallescens*
Germination	One or two distal cells	One or two distal cells
Basal germ tube	Semiaxial	Semiaxial
Conidium shape	Straight, narrowing toward the ends, uniformly brownish	Obovoid, curved; the penultimate cell is swollen, darker, with thicker wall than the others
Conidiophore morphology	Brown, unbranched, zigzagged shape	Brown, unbranched, zigzagged shape
Number of septa	3–6 (eusepta or distosepta)	3–4 (eusepta)
Hilum morphology	Slightly protruding	Slightly protruding

et al. 2013). For primary histopathological examination, the most frequently used techniques are Gross examination, 10% or 20% potassium hydroxide preparation, Gram, Giemsa and/or fluorescent calcofluor white stains; hyphal elements can be visualized by Fontana-Masson, Gömöri methenamine silver, hematoxylin and eosin and periodic acid-Schiff staining (Wilhelmus and Jones 2001; Thomas 2003; Srinivasan 2004; Alexander and Pfaller 2006; Shukla et al. 2008; Revankar and Sutton 2010; Isa-Isa et al. 2012; Moody et al. 2012). Parallel with the histological staining, fungal samples are suggested to culture. Commonly used fungal media, such as Sabouraud dextrose agar, Czapek Dox agar, brain heart infusion broth, blood agar or potato dextrose agar can be applied for this purpose (Wilhelmus and Jones 2001; Filizzola et al. 2003; Shukla et al. 2008). Confocal microscopy (Srinivasan 2004; Shukla et al. 2008) and scanning electron microscopy (Hosokawa et al. 2003) can also be useful approaches to identify melanized fungi.

However, morphology-based identification is often difficult and may be uncertain due to the high variability of the discriminative characters, which may depend on the isolate, the age of the colony and the culture conditions (Nakada et al. 1994; Manamgoda et al. 2012). For example, the diverse conidial morphology with many intraspecific size variants can be a typical source of uncertain identifications. Moreover, these fungi frequently produce conidial size reduction and decreased sporulation during the culturing and strain maintenance, which make the identification even more difficult (Hosokawa et al. 2003; Sun et al. 2003; Kumar et al. 2005; Pimentel et al. 2005; Da Cunha et al. 2012; Alex et al. 2013). Anyway, morphological identification requires great mycological expertise. Despite these frequently occurring problems, conventional morphology- and histology-based methods are currently the first line discriminative processes for the identification of *Curvularia* species, albeit the micromorphology-based differentiation is accurate only at the generic level (Filizzola et al. 2003; Pimentel et al. 2005; Toul et al. 2006).

Induction of the teleomorph state and examination of sexual morphology have been used as alternative species delimitation methods (Nakada et al. 1994; Hosokawa et al. 2003; Sun et al. 2003). This approach also has its difficulties due to the possibility of the loss of maternal factors and the long cultivation time. Therefore, this method has not become widespread in practice.

Specific antigen- or host response-based tests could be useful if identification using traditional methods is difficult. However, these methods also have limitations, such as possible cross-reactions or the lack of standardization (Shukla et al. 2008). An exoantigen-based differentiation of clinically important *Bipolaris/Curvularia, Drechslera* and *Exserohilum* species demonstrated that this method is suitable for genus level identification of these fungi (Pasarell et al. 1990).

Molecular methods provide great assistance in the course of differentiation of closely related species. These techniques are generally culture independent, sensitive and faster than the culture-based methods (Shin et al. 2003; Kobayashi et al. 2008; Pfaller et al. 2012). Currently, PCR amplification and sequencing of the internal transcribed spacer (ITS) region of the nuclear ribosomal RNA gene cluster is the most commonly used procedure for the identification of ascomycetous species. ITS has become the main molecular marker sequence for the identification of fungi because it is usually able to resolve inter- and sometimes intraspecific differences, it is present in all fungal genera in several copies which makes the amplification efficient, and it is uploaded in the highest numbers among the different markers to sequence databases (Ferrer et al. 2001; Schoch et al. 2012; Alex et al. 2013). Accordingly, ITS sequencing is widely used to confirm the morphology-based identification of *Curvularia* species (Fryen et al. 1999; Buzina et al. 2003; Pimentel et al. 2005; Toul et al. 2006; Yanagihara et al. 2010; Alex et al. 2013; Da Cunha et al. 2012, 2013). However, some doubts have also been raised about the use of ITS in the past few years, mainly for the following considerations: (i) only a few correctly evaluated reference sequences can be found in NCBI GenBank for *Bipolaris* and *Curvularia* species, especially for the clinically relevant members of this group; (ii) a number of the deposited sequences are incorrectly identified; (iii) the ITS was found to be not variable enough to discern certain closely related species, such as in the case of *C. australiensis* and *C. spicifera* (Balajee et al. 2009; Begerow et al. 2010; Revankar and Sutton 2010; Da Cunha et al. 2013). The unreliable use of ITS sequences was mentioned in a case report of sinusitis, in which the *Cochliobolus australiensis* and *Co. spicifera* proved to be indistinguishable based on BLAST similarity search with the ITS sequences (Toul et al. 2006). Similarly, in the case of a cutaneous *Curvularia* infection, the BLAST search of this sequence resulted in low similarity percentages (under the 97% threshold) for several different *Curvularia* species (Yanagihara et al. 2010).

To increase the sensitivity of the ITS-based identification, restriction fragment length polymorphism (RFLP) is also used. Digestion of the ITS with the restriction enzymes *Bam*HI and *Hae*III resulted in banding patterns specific to *C. hawaiiensis*, which could be distinguished from *C. australiensis* and *C. spicifera* (Nakada et al. 1994; Fryen et al. 1999). In another study, the enzyme *Rsa*1 produced only genus-specific patterns (Goh et al. 1998). Drawbacks of the ITS RFLP are that closely related species are generally indistinguishable and standardization is lacking.

There are some examples for the application of denaturing gradient gel electrophoresis (DGGE), temperature gradient gel electrophoresis (TGGE) (Kumar et al. 2005; Shukla et al. 2008) and single stranded conformation polymorphism (SSCP) (Kumar and Shukla 2005) using the ITS region to identify certain *Curvularia* species. These approaches might be useful, but lack standardization and there is not enough data available especially for the medically important *Curvularia* species.

In conclusion, the ITS region-based methods are applicable for identification of *Curvularia* spp. at the generic, rather than the species level. Previously, a similar problem was reported in the diagnosis of certain *Aspergillus* and *Fusarium* species concerning the applicability of the ITS region (Balajee et al. 2009; Wang et al. 2011). In those cases, the ITS region proved to be useful for the identification of "sections" or species complexes. For species level identification of aspergilli, sequencing of the β-tubulin gene (*tub*) while for the identification of *Fusarium* species, a multilocus approach involving EF1-α, *tub*, the calmodulin gene (*cmd*), LSU, the nuclear ribosomal small subunit RNA gene (SSU), the nuclear ribosomal intergenic spacer region (IGS) and the second largest subunit of RNA polymerase II (*rpb2*) are suggested.

If the closely related species could not be differentiated either by morphological characters or the ITS region, the use of alternative DNA markers is necessary. Among the melanized fungi in the genera *Ochroconis*, *Fonsecaea*, *Phoma* and *Aspergillus*, the *rpb2*, *tub*, actin (*act*) and chitin synthase 1 (*chs1*) proved to be useful instead of the ITS region (Santos et al. 2013). To identify members of the genus *Curvularia*, the second most frequently applied molecular marker is the LSU (Kobayashi et al. 2008; Singh et al. 2008; Yanagihara et al. 2010). In the case of a disseminated infection where the morphology-based method failed to identify the causal agent, the isolate could be identified by LSU-PCR followed by a BLAST similarity search with sequences in the GenBank (Kobayashi et al. 2008). However, it must be noticed that, besides the best hit, another six closely related species could be found above 98% identity. Moreover, during LSU sequence analysis of a *Curvularia* strain causing cutaneous infection, BLAST similarity search gave 100% identity with the LSU sequences of three different species (*Co. sativus, Co. nodulosus* and *Co. heterostrophus*) (Yanagihara et al. 2010). Similarly, LSU-RFLP analysis was effective only at genus level determination of the species *Bipolaris* and *Curvularia* (Goh et al. 1998). Therefore, it can be concluded that the application of LSU besides or instead of the ITS region does not improve the identification and there is a need to find more reliable, new DNA markers. The same observation was made during the multilocus sequence analysis of *Fusarium* species, where ITS, LSU and SSU were the least informative among the six tested DNA loci, while the IGS region proved to be phylogenetically the most informative (Wang et al. 2011). Based on previous phylogenetic studies, the 1,3,8-trihydroxynaphthalene reductase (*brn1*) (Sun et al. 2003) or the glyceraldehyde-3-phosphate dehydrogenase (*gpd*) genes (Manamgoda et al. 2012; Da Cunha et al. 2013) might be more reliable and sensitive molecular markers than the ITS or the LSU.

In conclusion, the most promising method for the accurate identification of *Curvularia* species could be found among the specific DNA motif-based molecular methods, which are culture-independent, rapid and cost effective, but the selection of the most suitable molecular marker is still in process.

Epidemiology and Clinical Manifestation of *Curvularia* Infections

Infections caused by *Curvularia* spp. are also referred to as "curvulariosis", which is a type of phaeohyphomycoses. Phaeohyphomycosis can be defined as a heterogeneous group of fungal infections caused by dematiaceous fungi (Naggie and Perfect 2009). In infected tissues these fungi may appear in the form of yeast-like cells, hyphae, or a combination of both (James et al. 2006). *Curvularia* species were reported to cause infections in animals (Bridges 1957; Muller et al. 1975; Kaplan et al. 1975; Kwochka et al. 1984; Qualls et al. 1985; Clark et al. 1986; Beale and Pinson 1990; Elad et al. 1991; Waurzyniak et al. 1992; Penrith et al. 1994; Herraez et al. 2001; Swift et al. 2006; Myers et al. 2009; Ben-Shlomo et al. 2010; Giri et al. 2011) and more importantly they are on the growing list of emerging fungal pathogens in humans. Diseases caused by *Curvularia* species in humans include eye infections, infections of the respiratory tract, skin and nail infections, infections complicating peritoneal dialysis as well as deep and disseminated infections. Furthermore, a single case of a urinary tract infection has also been reported in the literature (Robson and Craver 1994).

Eye Infections

Mycotic keratitis is the most common human disease caused by *Curvularia* species. Although *Curvularia* can cause various ocular diseases (Eghtedari and Pakshir 2006; Kaushik et al. 2001; Patil et al. 2011; Sodhi and Mehta 2003; Smith et al. 2007), the number of clinical cases involving mycotic keratitis and endophthalmitis has increased in the last two decades which can be attributed to the increasing attention given by microbiologists to these fungi that were previously considered nonpathogenic. A total of 58 cases of *Curvularia* eye infections could be recovered from the available literature and the case details are summarized in Table 2. Of these 58 cases, 39 are reported from India. The first report obtained for *Curvularia* infection was keratitis caused by *C. lunata* in a 36-year-old male from USA and the patient was healed after therapy with amphotericin B (Anderson et al. 1959).

Since this initial report, isolation of *Curvularia* species from eye infections has been reported from several countries (Table 3), including Bangladesh (Williams et al. 1987; Dunlop et al. 1994), China (Wang et al. 2000; He et al. 2011), Nepal (Amatya et al. 2012), Saudi Arabia (Alkatan et al. 2012), Vietnam (Nhung et al. 2012), Singapore (Wong et al. 1997), Thailand (Boonpasart et al. 2002; Sirikul et al. 2008), India (Poria et al. 1985; Anandi et al. 1988; Sundaram et al. 1989; Venugopal et al. 1989; Kotigadde et al. 1992; Hemashettar et al. 1992; Chander and Sharma 1994; Srinivasan et al. 1997; Rahman et al. 1998; Deshpande and Koppikar 1999; Kunimoto et al. 2000; Vajpayee et al. 2000; Leck et al. 2002; Gopinathan et al. 2002 and 2009; Bharathi et al. 2002, 2003, 2006 and 2009; Prajna et al. 2004; Chowdhary and Singh 2005; Singh et al. 2006; Saha and Das 2006; Chander et al. 2008; Sharma et al. 2008; Sherwal and Verma 2008; Kindo et al. 2009; Usha et al. 2009; Ragini et al. 2010; Parmjeet and Pushpa 2011; Kumar et al. 2011; Nath et al. 2011; Sengupta et al. 2011; Rautaraya et al. 2011; Tewari et al. 2012; Bandyopadhyay et al. 2012; Deorukhkar et al. 2012; Gajjar et al. 2013; Somabhai Katara et al. 2013; Krishna et al. 2013; Anusuya Devi et al. 2013), Ghana (Hagan et al. 1995; Leck et al. 2002), Brasil (Höfling-Lima et al. 2005), Paraguay (Miño de Kaspar et al. 1991; Laspina et al. 2004) and USA (Berger et al. 1991; Chung et al. 2000; Wilhelmus and Jones 2001; Marangon et al. 2004; Alfonso et al. 2006; Iyer et al. 2006).

Nityananda et al. (1964) reported a case of keratitis caused by *C. lunata* in a 38-year-old female who was subsequently treated with atropine eye drops along with bacitracin-polymyxin ointment. Agrawal et al. (1982) also reported *C. lunata* keratitis in a 35-year-old male farmer, where the infection could not be managed as the fungus had migrated into the deeper layers of the eye. Wanmali et al. (2014) reported *Curvularia* keratitis in a 17-year-old male following trauma with a stick. In all of the above cases the infection occurred following trauma with a foreign body blown into the eye (Anderson et al. 1959), injury from paddy (Nityananda et al. 1964) and magnetic iron particle (Agrawal et al. 1982). Dorey et al. (1997) reported a case of *Curvularia* keratitis contracted after swimming in Dead Sea, which was successfully treated with natamycin and gentamycin. An unusual case for *Curvularia* keratitis in a 50-year-old male was reported by Tuli and Yoo (2003) after laser *in situ* keratomileusis from a feline source (the patient's pet cat), he was treated with natamycin. Recurrence of infection in a case of *Curvularia* keratitis of a 38-year-old female without any history

Table 2. Case reports selected from the literature about eye infections (keratitis, endophthalmitis) caused by *Curvularia* species.

Country	Age/Sex	Clinical Diagnosis	Etiology	Therapy	Outcome/vision	Reference
USA	36/M	Keratitis	*C. lunata*	AMB	Healed	Anderson et al. 1959
Sri Lanka	35/F	Keratitis	*C. lunata*	Debridement	Healed	Nityananda et al. 1962
Sri Lanka, Colombo	38/M	Keratitis	*C. geniculata*	Bacitracin, polymyxin ointment	Healed	Nityananda et al. 1964
USA	ND	Keratitis	*C. lunata*	ND	ND	Georg 1964
USA, Florida	57/F	Keratitis	*C. lunata*	NTM	TPK	Wind and Polack 1970
India, Lucknow	35/M	Keratitis	*C. lunata* (wakker) var. Aeria (Batista, Lima and vasconcleos) Ellis	TPK advised	Enucleation advised	Agrawal et al. 1982
Argentina, Rosario	66/M	Keratitis	*C. lunata* var. *aeria*	KTZ, NTM	Improved/ND	Luque et al. 1986
South Africa, Pretoria	78/F	Keratitis	*C. brachyspora (Boedjin)*	AMB	ND	Marcus et al. 1992
UK, Surrey	53/M	Keratitis	*C. lunata*	NTM	Healed/6/12[e]	Dorey et al. 1997
Taiwan, Tainan	1. 64/M 2. 41/F	1. Keratitis 2. Keratitis	1. *Curvularia* sp.[a] 2. *Curvularia* sp.[a]	1. ND 2. TPK	1. ND 2. TPK	Fu-Chin et al. 1998
Korea, Seoul	58/F	1. Keratitis	*Curvularia* sp.	AMB	Healed	Kim et al. 1999
Spain, Reus	38/F	Keratitis	*C. senegalensis*	NTM, ITZ	TPK/ND	Guarro et al. 1999
India, Chandigarh	40/M	Keratitis/ Endophthalmitis	*C. lunata*	NTM, AMB	Cornea perforated/ ND	Kaushik et al. 2001
India, Vellore	45/M	Keratitis	*Curvularia* sp.[a]	KTZ	ND	Koshy and Ebenezer 2002
India, Madurai	60/F	Keratitis	*C. lunata*	NTM	Healed/ND	Prajna et al. 2002
India, Vellore	45/M	Keratitis	1 (Bilateral)	KTZ	Healed with opacity/ NA	Gopalakrishnan et al. 2003
India, New Delhi	1. 30/M 2. 35/F 3. 50/M 4. 45/M	1. Dacryocystitis 2. Dacryocystitis 3. Stromal abscess 4. Keratitis	1. *Curvularia* sp. 2. *Curvularia* sp. 3. *Curvularia* sp. 4. *Curvularia* sp.	1. FLZ 2. FLZ 3. FLZ 4. NTM, FLZ	1. Healed/ND 2. Healed/ND 3. Healed/ND 4. Healed with opacity	Sodhi and Kaur 2000

Table 2. contd....

Table 2. contd.

Country	Age/Sex	Clinical Diagnosis	Etiology	Therapy	Outcome/vision	Reference
USA, Florida	50/M	Keratitis	*Curvularia* sp.	NTM	Healed/20/60[e]	Tuli and Yoo 2003
India, Shahdara	28/F	Keratitis	*C. spicifera*[b]	NTM, Nystatin C, ITZ	Healed with opacity/ ND	Saha and Das 2005
Iran, Shiraz	35/F	Cojunctival cyst	*C. spicifera*[b]	ND	Healed/ND	Eghtedari and Pakshir 2006
India, Hyderabad	54/F	Endophthalmitis	*C. lunata*	AMB, NTM, ITZ	Improved/20/30[e]	Pathengay et al. 2006
Australia, Brisbane	57/M	Optic atrophy	*C. lunata* mucocoele	AMB, liposaomal AMB, VRZ	Improved/6/12[e]	Smith et al. 2007
Australia, Northern Territory	29/M	Keratitis	*C. australiensis*[c]	NTM	Healed/ND	Durkin et al. 2008
India, Kashmir	50/M	Keratitis/ Endophthalmitis	*C. hawaiiensis*[d]	NTM, ITZ	Eviscerated	Bashir et al. 2008
India, New Delhi	39/M	Keratitis	*Curvularia* sp.	NTM, FLZ	Healed/ND	Gupta et al. 2010
Brazil, Londrina	52/M	Endophthalmitis	*C. lunata*	ITZ, VRZ	Healed/20/30[e]	Berbel et al. 2011
USA, North Carolina	65/F	Endophthalmitis	*Curvularia* sp.	AMB, VRZ, PSZ	Enucleation	Ehlers et al. 2011
India, Sangli	38/M	Corneal abscess	*C. spicifera*[b]	NTM	Healed/ND	Patil et al. 2011
USA, Florida	1. 55/M 2. 23/M	1. Traumatic Endophthalmitis 2. Endophthalmitis	1. *Curvularia* sp. 2. *Curvularia* sp.	1. AMB 2. AMB, VRZ	1. TPK/20/200 2. lost follow up/20/400[e]	Pathengay et al. 2012
USA, Washington	74/F	Endophthalmitis	*C. lunata*	VRZ, AMB	Enucleated	Alex et al. 2013
Colombia, Medellín	NA	Endophthalmitis	*C. lunata*	AMB	Responded	Jaramillo and Varon 2013
India, Nagpur	17/M	Corneal ulcer	*Curvularia* sp.	NTM	TPK/ND	Sunay et al. 2014

M: male, F: female, TPK: therapeutic penetrating keratoplasty, AMB: amphotericin B, FLZ: fluconazole; ITZ: itraconazole, KTZ: ketoconazole, NTM: natamycin, PSZ: posaconazole, VRZ: voriconazole; [a]reported as *Bipolaris* sp., [b]reported as *B. spicifera*, [c]reported as *B. australiensis*, [c]reported as *B. australiensis*, [e]visual acuity, ND: no data available.

Table 3. Epidemiological data available in the literature about the involvement of *Curvularia* species in fungal keratitis.

Country	Period of study	Number of fungal isolates	Number of *Curvularia* isolates (%)	Reference
Bangladesh, Chittagong	11 months	51	8 (15.6)	Dunlop et al. 1994
Bangladesh, Chittagong	1991	39	1 (2.5)	Williams et al. 1987
Paraguay	April 1988–April 1989	26	1 (3.8)	Mino de Kaspar et al. 1991
Brasil	1975–2003	293	1 (0.4)	Höfling-Lima et al. 2005
China	October 2004–September 2009	139	1 (0.7)	He et al. 2011
China	January 1975–June 1997	615	NA (13.1)	Wang et al. 2000
Ghana	June 1999–May 2001	109	1 (0.9)	Leck et al. 2002
Ghana, Accra	NA	65	2 (3.0)	Hagan et al. 1995
Nepal	January 2004–December 2008	150	15 (10)	Amatya et al. 2012
Paraguay	1988–2001	209	15 (7.1)	Laspina et al. 2004
Saudi Arabia	January 2006 and December 2009	26	1 (3.8)	Alkatan et al. 2012
Singapore	January 1991–December 1995	29	1 (3.4)	Wong et al. 1997
Thailand (central)	January 1988–December 2000	34	7 (20.0)	Boonpasart et al. 2002
Thailand	January 2001–December 2004	49	2 (1.5)	Sirikul et al. 2008
USA, Florida	January 2004–December 2005	110	29 (26.4)	Alfonso et al. 2006
USA, Florida	January 1980–January 2002	421	34 (8.1)	Marangon et al. 2004
USA, Texas	1970–1999	ND	32	Wilhelmus and Jones 2001

Table 3. contd....

Table 3. contd.

Country	Period of study	Number of fungal isolates	Number of *Curvularia* isolates (%)	Reference
USA, Florida	January 1999 to June 2006	59	9 (15.3)	Iyer et al. 2006
Vietnam	2008	351	9 (2.6)	Nhung et al. 2012
India, Ahmedabad	July 2007–June 2008	31	5 (16.1)	Tewari et al. 2012
India, Amristar	NA	65	7 (9.7)	Parmjeet and Pushpa 2011
India, Assam	April 2007–March 2009	184	13 (7.7)	Nath et al. 2011
India, Baroda	June 2009–May 2012	73	4 (5.4)	Gajjar et al. 2013
India, Bhubaneswar	July 2006–December 2009	215	10 (4.7)	Rautaraya et al. 2011
India, Chandigarh (north)	6 years	61	5 (8.2)	Chander and Sharma 1994
India, Chandigarh (north)	January 1999–December 2003	34	4 (11.8)	Chander et al. 2008
India, Chennai (south)	July 2006–May 2008	20	2 (10)	Kindo et al. 2009
India, Coimbatore (south, children)	February 1997–January 2004	37	2 (5.4)	Singh et al. 2006
India, Gujarat	September 2003–June 2005	57	6 (10.5)	Kumar et al. 2011
India, Gujarat (west)	September 2006–February 2008	26	4 (15.3)	Somabhai Katara et al. 2013
India, Hyderabad (south)	February 1991–June 2001	1648	106 (6.4)	Gopinathan et al. 2009
India, Hyderabad (south)	January 1991–December 2000	1360	54 (4.0)	Gopinathan et al. 2002
India, Hyderabad	February 1991–June 1995	21	3 (4.1)	Kunimoto et al. 2000
India, Jamnagar	1985	37	2 (5.4)	Poria et al. 1985

India, Karnataka	October 1985–September 1988	67	2 (2.9)	Kotigadde et al. 1992
India, Karnataka	2012	36	4 (11.1)	Krishna et al. 2013
India, Karnataka	December 2009–February 2011	38	2 (5.2)	Anusuya Devi et al. 2013
India, Madurai (south)	January 1994–March 1994	155	6 (3.9)	Srinivasan et al. 1997
India, Madurai	December 2002–June 2003	100	03 (3.0)	Prajna et al. 2004
India, Madurai	NA	63	5 (7.9)	Rahman et al. 1998
India, Madras (south)	1980–1982	68	4 (5.8)	Sundaram et al. 1989
India, Madras	NA	322	1 (0.3)	Venugopal et al. 1989
India, Maharashtra	December 2004–December 2009	311	10 (3.2)	Deorukhkar 2012
India, Mumbai	NA	40	4 (10)	Usha et al. 2009
India, Mumbai (west)	1988–1996	387	10 (2.5)	Deshpande and Koppikar 1999
India, New Delhi	January 2000–December 2004	77	2 (2.6)	Saha and Das 2006
India, New Delhi	NA	13	2 (15.3)	Vajpayee et al. 2000
India, New Delhi	January 1999–June 2001	191	63 (33.0)	Chowdhary and Singh 2005
India, New Delhi	NA	39	7 (17.9)	Sherwal and Verma 2008
India, Pondicherry	January 2009–August 2008	373	84 (22.5)	Sengupta et al. 2011
India, Tiruchirapalli (south)	June 1999–May 2001	353	36 (10.2)	Leck et al. 2002
India, Tirunelveli (south)	September 1999–March 2001	554	35 (6.3)	Bharathi et al. 2002

Table 3. contd....

Table 3. contd.

Country	**Period of study**	**Number of fungal isolates**	**Number of *Curvularia* isolates (%)**	**Reference**
India, Tirunelveli	September 1999–August 2002	1100	55 (5.0)	Bharathi et al. 2003
India, Tirunelveli	September 1999–September 2002	1226	81 (6.6)	Bharathi et al. 2006
India, Tirunelveli	September 1999–September 2002	1226	81 (6.6)	Bharathi et al. 2009
India, Ujjain	April 2006–November 2007	37	6 (2.9)	Sharma et al. 2008
India, Varanasi	January 2004–December 2008	36	3 (8.3)	Ragini et al. 2010
India, West Bengal	February 2007–January 2011	399	41 (10.3)	Bandyopadhyay et al. 2012

ND: no data available

of trauma even after the treatment with itraconazole has been reported by Guarro et al. (1999). A rare case of simultaneous bilateral fungal keratitis caused by *Aspergillus* sp. and *C. lunata* in a 60-year-old Indian female which was resolved with natamycin was reported by Prajna et al. (2002).

Sodhi and Mehta (2003) reported two cases of *Curvularia* eye infection followed by dacryocystitis and one case each of corneal abscess without any history of trauma and corneal ulcer after trauma with a finger nail, which were treated successfully with oral fluconazole. A curious case of *Acanthamoeba* keratitis with *Curvularia* coinfection was reported in a 39-year-old Indian male and was treated with natamycin and fluconazole (Gupta et al. 2010). Among all these cases with *Curvularia* keratitis there was development of painful red eye and the formation of white plaque at the site of trauma (Anderson et al. 1959), brown- to orange-coloured discoid pellicle (Nityananda et al. 1964), hypopyon (Agrawal et al. 1982) and hard plaque (Dorey et al. 1997).

While analyzing the endophthalmitis cases where *Curvularia* was involved, it was noted that most of the cases developed the infection after cataract surgery (Pathengay et al. 2006; Berbel et al. 2011; Ehlers et al. 2011; Alex et al. 2013). There were two cases of *Curvularia* endophthalmitis followed by trauma: one while hammering a nail and other while cutting grass as reported by Pathengay et al. (2012). In these two cases, patients had pain and redness in the eye. The infection was resolved with itraconazole and voriconazole (Berbel et al. 2011), amphotericin B and voriconazole followed by flucytosine (Ehlers et al. 2011), or natamycin, itraconazole and amphotericin B (Pathengay et al. 2006). Kaushik et al. (2001) reported a postoperative *C. lunata* endophthalmitis with secondary keratitis in a 40-year-old male and the infection was resolved with amphotericin B. Smith et al. (2007) reported the first case of optic neuropathy caused by *C. lunata* in a 57-year-old male, which was resolved with amphotericin B. Endogenous endophthalmitis caused by *C. hawaiiensis* has also been reported in a patient with acquired immunodeficiency syndrome (Pavan and Margo 1993). Further cases of *Curvularia* endophthalmitis are known from retrospective studies (Sheyman et al. 2013; Rachitskaya et al. 2014; Minckler et al. 2014; Small et al. 2014).

In two cases of *Curvularia* eye infections in immunocompetent hosts, the infection was localised to the lacrimal sac only and presented in the form of acute to chronic dacryocystitis (Sodhi and Kaur 2000). Both cases responded well to oral fluconazole treatment.

Case studies reported the involvement of seven species from the genus *Curvularia*: *C. lunata* (Anderson et al. 1959; Nityananda et al. 1962; Georg 1964; Wind and Polack 1970; Agrawal et al. 1982; Luque et al. 1986; Berger et al. 1991; Dorey et al. 1997; Kaushik et al. 2001; Smith et al. 2007; Prajna et al. 2002; Pathengay et al. 2006; Berbel et al. 2011; Alex et al. 2013; Jaramillo and Varon 2013), *C. geniculata* (Nityananda et al. 1964), *C. senegalensis* (Guarro et al. 1999), *C. spicifera* (Hemashettar et al. 1992; Saha and Das 2005; Eghtedari and Pakshir 2006; Patil et al. 2011), *C. australiensis* (Durkin et al. 2008), *C. brachyspora* (Marcus et al. 1992) and *C. hawaiiensis* (Anandi et al. 1988; Bashir et al. 2009). The role of *Curvularia* enzymes and other toxins in fungal keratitis have not yet been studied (Wilhelmus and Jones 2001). An experimental study (Agrawal et al. 1982) from India applied a rabbit model of *Curvularia* keratitis and reported the severity of corneal infection that was worsened by topical corticosteroids in the absence of antifungal therapy. The authors also observed that the poor growth of *C. lunata* var. *aeria* at 37°C was not affecting the pathogenic potential of the strain as there was a severe infection in the mycotic keratitis patient at the site where this fungus was isolated.

Infections of the Respiratory Tract

Case reports selected from the literature about *Curvularia* species causing respiratory tract infections are summarized in Table 4. The likely source of these infections is inhalation from the atmosphere. After mycotic keratitis, the second most common human infection caused by *Curvularia* species is sinusitis.

Fungal sinusitis has four different forms, two of which, the acute/fulminant sinusitis and the chronic sinusitis are invasive, while the other two, mycetoma/fungus ball and allergic fungal sinusitis are non-invasive (Willard et al. 2003). Allergic fungal sinusitis (AFS) is related to IgE-mediated hypersensitivity to fungal antigens. Most of the sinusitis cases caused by *Curvularia* species reported in the literature belong to the AFS category (Table 4). Invasive *Curvularia* sinusitis cases were also reported, although they were less frequent (Travis et al. 1991; Ismail et al. 1993; Ebright et al. 1999; Castelnuovo et al. 2004; Ambrosetti et al. 2006; Shetty et al. 2006; Peden et al. 2008; Viola and Sutton 2010). The sphenoid and posterior ethmoid sinuses are most often involved, followed by the anterior ethmoid sinus, frontal sinus and maxillary sinus (Aribandi and Bazan 2007). Pansinusitis may also occur (Harpster et al. 1985; Rinaldi et al. 1987).

General symptoms of *Curvularia* sinusitis are facial pain, headache, nasal stuffiness, nasal obstruction, difficulty breathing through the nose, mucosal thickening, purulent discharge and mass in the sinus(es), however, other symptoms like the deviation of the nasal septum, the presence of intranasal polips and bony erosions were also frequently reported. In certain cases the infection led to obstruction of the lacrimal duct with enlargement of the lacrimal sac, painful swelling of the eyelid as well as decreased or lost visual acuity. Predisposing conditions included asthma, allergy, previous surgery and the usage of nasal topical corticosteroids.

Curvularia sinusitis cases were reported from the United States, mainly the southern part (Loveless et al. 1981; Sobol et al. 1984; Rolston et al. 1985; Adam et al. 1986; Harpster et al. 1985; Rinaldi et al. 1987; Frenkel et al. 1987; Jay et al. 1988; Pratt and Burnett 1988; Gourley et al. 1990; Maskin et al. 1989; Killingsworth and Wetmore 1990; Travis et al. 1991; Pingree et al. 1992; El-Zaatari et al. 1996; Klapper et al. 1997; Quraishi and Ramadan 1997; Ebright et al. 1999; Schroeder et al. 2002; Willard et al. 2003; Safdar 2003; Parva et al. 2005; Shetty et al. 2006; Peden et al. 2008; Viola and Sutton 2010) and Hawaii (Antoine et al. 1989), as well as Argentina (Alvarez et al. 2011), Chile (Cruz et al. 2013), Qatar (Taj-Aldeen et al. 2004), China (Ruan et al. 2005; Wan et al. 2006), India (Koshi et al. 1987), Japan (Taguchi et al. 2004), South Africa (Young et al. 1978), Australia (Robson et al. 1989; Rao et al. 1989) and European countries including Spain (Del Palacio et al. 1997), Germany (Fryen

Table 4. Case reports selected from the literature about respiratory tract infections (sinusitis, bronchopulmonary mycosis) caused by *Curvularia* species.

Country	Age/Sex	Clinical diagnosis	Source of isolation	Etiology	Therapy	Outcome	Reference
South Africa	15/M	Sinusitis	Sinus material	*C. hawaiiensis*[a]	Surgery, AMB		Young et al. 1973
USA	47/M	Invasive nasal septum infection	Septum material	*Curvularia* sp. and *Alternaria* sp.	Surgical excision of the nasal septum	Cured	Loveless et al. 1981
USA	16/F	Sinusitis	Sinus material	*C. lunata*	Surgery, intravenous and topical AMB	Cured	Berry et al. 1984
USA	10/M	Sinusitis	Sinus material	*C. spicifera*[b]	Surgery	ND	Sobol et al. 1984
USA	12/F	Sinusitis	Sinus material	*C. spicifera*[c]	Surgery	ND	Sobol et al. 1984
USA	36/M	Sinusitis	Sinus material	*C. spicifera*[c]	Surgery	ND	Sobol et al. 1984
USA	19/F	Sinusitis	Sinus material	*C. spicifera*[c]	Surgery, AMB, KTZ	Cured	Rolston et al. 1985
USA	32/M	Sinusitis	Sinus material	*C. spicifera*	Surgery, AMB	Cured	Adam et al. 1986
USA	28/M	Sinusitis	Sinus material	*C. australiensis*[d]	Surgery, AMB	Cured	Harpster et al. 1985
USA	7/F	Sinusitis	Sinus material	*C. spicifera*[b]	Surgery, AMB	Recurrence	Frenkel et al. 1987
USA	17/M	Sinusitis	Sinus material	*C. spicifera*[b]	Surgery, AMB	ND	Jay et al. 1988
USA	18/M	Sinusitis	Sinus material	*C. spicifera*[b]	Surgery	ND	Jay et al. 1988
USA	20/F	Paranasal sinusitis	Brownish-yellow material from the sinus	*C. lunata*	Caldwell-Luc procedure of the left maxillary sinus	Cured	Rinaldi et al. 1987
USA	23/M	Paranasal sinusitis	Intraoperative cultures from five different sinuses	*C. lunata*	Bilateral ethmoidectomies, Caldwell-Luc antrostomies, and nasal polypectomies	Cured	Rinaldi et al. 1987

Table 4. contd....

Table 4. contd.

Country	Age/Sex	Clinical diagnosis	Source of isolation	Etiology	Therapy	Outcome	Reference
USA	11/F	Paranasal sinusitis	Material from the sinus	*C. lunata*	Caldwell-Luc procedure with transantral ethmoidectomy and partial inferior turbinectomy, topical nasal steroids and chlorpheniramine	Cured	Rinaldi et al. 1987
USA	26/M	Refractory chronic pansinusitis	Intraoperative potassium hydroxide smears	*C. lunata*	Left ethmoidectomy with left sphenoid and frontal sinusotomy, postoperative cephradine	Cured	Rinaldi et al. 1987
USA	30/M	Paranasal sinusitis	Thick inspissated mucous in the sinus	*C. lunata*	Left polypectomy, ethmoidectomy, and Caldwell-Luc procedure, with a left nasal antral window	Cured	Rinaldi et al. 1987
USA	21/F	Sinusitis	Sinus material	*C. spicifera*[c]	Surgery	Cured	Pratt and Burnett 1988
USA (Hawaii)	27/M	Sinusitis	Sinus material	*Curvularia* sp.[e]	Surgery, AMB	ND	Antoine and Raterink 1989
USA	16/F	Allergic fungal sinusitis	Thick peanut butter-like material from the sinuses	*C. spicifera*[b]	External ethmoidectomy, sphenoidotomy, Caldwell-Luc procedure	Cured	Gourley et al. 1990
USA	20/M	Allergic fungal sinusitis	Thick, tenacious material from sinuses	*C. spicifera*[b]	drainage of a left orbital subperiosteal abscess, trephination of the left frontal sinus, sinus obliteration, ethmoidectomy, and left external sphenoethmoidectomy, Caldwell-Luc procedure	Cured	Gourley et al. 1990
USA	40/F	Allergic fungal sinusitis	Maxillary sinus material	*C. spicifera*[b]	endoscopic surgical drainage	ND	Gourley et al. 1990

USA	30/F	Sinusitis	Sinus material	*C. spicifera*[b]	Surgery, AMB	ND	Gourley et al. 1990
USA	21/F	Sinusitis	Sinus material	*C. spicifera*[b]	Surgery	ND	Gourley et al. 1990
USA	12/F	Sinusitis	Sinus material	*C. spicifera*[b]	Surgery, AMB	ND	Gourley et al. 1990
USA	40/F	Sinusitis	Sinus material	*C. spicifera*[b]	Surgery	ND	Gourley et al. 1990
Spain	36/M	Allergic chronic sinusitis	Sinus biopsy	*C. australiensis*[d]	Endoscopic surgery	ND	del Palacio et al. 1997
USA	26/M	Allergic fungal sinusitis with orbital involvement	Biopsy specimens of ethmoid and maxillary sinuses	*C. spicifera*[c]	Surgeries	Cured after recurrence	Klapper et al. 1997
Germany	20/F	Allergic fungal sinusitis	Biopsies	*C. hawaiiensis*[f]	Endonasal microsurgical pansinus operations, local AMB, oral FLZ, systemic ITZ	Cured	Fryen et al. 1999
USA	46/F	Invasive sinusitis and cerebritis of the frontal lobes, skull base	Biopsies from the ethmoid sinuses	*C. clavata*	Surgery, AMB, ITZ, FLZ	Cured	Ebright et al. 1999
USA	35/M	Allergic fungal sinusitis	Right nasal specimen	*Curvularia* sp.	endoscopic microdebridement	Cured	Willard et al. 2003
USA	31/M	Allergic fungal sinusitis	Surgical specimen from the right nostril	*C. lunata*	Biopsy, endoscopic microdebridement, and lachrymal duct probing	Cured	Willard et al. 2003
Austria	19/M	Fungus balls of the sinuses and from chronic rhinosinusitis with massive polyposis	Mucus masses, fungus balls from endoscopic sinus surgery	*C. spicifera*[b]	endoscopic sinus surgery, postoperative ITZ, steroids	Cured	Buzina et al. 2003
Italy	39/F	Ethmoido-sphenoidal invasive fungal rhinosinusitis	Pus	*C. hawaiiensis*[f]	surgical drainage, intravenous AMB, oral ITZ	Cured	Castelnuovo et al. 2004

Table 4. contd....

Table 4. contd.

Country	Age/Sex	Clinical diagnosis	Source of isolation	Etiology	Therapy	Outcome	Reference
USA	59/M	Extensive paranasal sinus disease	Ethmoid sinus mucosa	*C. lunata*	Endoscopic surgery, terbinafine, oral itraconazole	Cured	Safdar 2003
Japan	70/M	Allergic fungal sinusitis	Contents of the paranasal sinuses	*C. spicifera*[b]	Drainage surgery	Cured	Taguchi et al. 2004
Qatar	16/M	Allergic rhinosinusitis	Nasal polyps and fungal mucin	*C. hawaiiensis*[f], *C. lunata*	Endoscopic sinus surgery	Cured	Taj-Aldeen et al. 2004
Qatar	19/M	Allergic rhinosinusitis	Nasal polyps and fungal mucin	*C. hawaiiensis*[f]	Endoscopic sinus surgery	Cured	Taj-Aldeen et al. 2004
Qatar	21/M	Allergic rhinosinusitis	Nasal polyps and fungal mucin	*Curvularia* sp.[e]	Endoscopic sinus surgery	Lost to follow up	Taj-Aldeen et al. 2004
USA	48/M	Rhinosinusitis	Sinus material	*Curvularia* sp.	Surgical debridement, irrigation, antifungal therapy	ND	Parva et al. 2005
Iran	20/M	Allergic fungal sinusitis	Maxillary antrostomy	*Curvularia* sp.[e]	Clearance of all affected sinuses with endoscopic surgery, local and systemic corticosteroids	Cured	Ashraf et al. 2005
France	50/M	Rhinosinusitis	Sinus masses	*Curvularia* sp.[e]	Endoscopic sinus surgery	Cured	Toul et al. 2006
France	50/M	Chronic and granulomatous invasive fungal sinusitis	Biopsy of mass involving the left sinuses with erosion of the orbit	*C. spicifera*[b]	Surgical resection	Cured	Ambrosetti et al. 2006
USA	48/M	*Curvularia* infection in the left turbinate followed by chronic craniofacial infection	Sinus samples, incisional biopsies of the facial dermis	*Curvularia* sp.	Surgical debridement, intravenous AMB, oral ITZ	Lost to follow up	Shetty et al. 2006
Italy	45/M	Eosinophilic fungal rhinosinusitis	Thick and viscous mucus	*C. inaequalis*	Polypectomy and clearance of the affected sinuses, oral ITZ	Cured	Posteraro et al. 2010
USA	52/M	Invasive sinusitis, orbital cellulitis	Sinus washing, excised tissue	*Curvularia* sp. and *Fusarium solani*	Surgery, AMB, VRZ	ND	Peden et al. 2008

USA	36/F	Allergic fungal sinusitis complicated with central nervous system invasion	Operative cultures	*C. australiensis*[d]	Emergent resection of the frontal lobe mass and ablation of the sinuses, AMB deoxycholate, vancomycin, ceftriaxone, and metronidazole	Died	Viola and Sutton 2010
Argentina	17/M	Allergic fungal rhinosinusitis	Sticky material removed during surgey	*Curvularia* sp.	Surgery, AMB, ITZ	Cured	Alvarez et al. 2011
Chile	56/M	Allergic rhinosinusitis	Mucin	*C. inaequalis*	Endoscopic surgical toilet, systemic ITZ, steroids	Cured	Cruz et al. 2013
Italy	16/M	Allergic fungal rhinosinusitis	Nasal sinus drainage, bioptical specimens	*C. lunata*	Surgery, postoperative ITZ, steroid	Cured	Cavanna et al. 2014
USA	23/M	Allergic bronchopulmonary mycosis	Surgical specimens	*Curvularia* sp.[g]	Lung resection	ND	Dolan et al. 1970
USA	40/M	Allergic bronchopulmonary mycosis	Surgical specimen	*Curvularia* sp.[g]	Lung resection	ND	Dolan et al. 1970
Australia	33/F	Allergic bronchopulmonary mycosis	Brown inspissated material from the resected bronchus, sputum specimens, sputum plugs	*C. lunata*	Segmental resection of the apical segment of the right lower lobe	Recovery	Mcaleer et al. 1981
Australia	36/F	Allergic bronchopulmonary mycosis	Sputum samples	*C. hawaiiensis*[a]	Bronchoscopies, prednisone	ND	Mcaleer et al. 1981
Australia	35/M	Allergic bronchopulmonary mycosis	Sputum samples	*C. lunata*, *C. hawaiiensis*[a]	Prednisone, potassium iodide	Recovery	Mcaleer et al. 1981
Australia	28/M	Allergic bronchopulmonary mycosis	Mucus plugs	*Curvularia* sp.	ND	ND	Matthiesson 1981
USA	30/M	Allergic bronchopulmonary mycosis	Sputum	*Curvularia* sp.[g]	Steroids, exacerbations on followup	ND	Hendrick et al. 1982
USA	21/M	Allergic bronchopulmonary mycosis	Sputum	*Curvularia* sp.[g]	Steroids	Remission	Halloran et al. 1983
USA	36/M	Allergic bronchopulmonary disease	Bronchial washing, sputum	*C. lunata*	Steroids, MCZ, KTZ	Improved	Halwig et al. 1985

Table 4. contd....

Table 4. contd.

Country	Age/Sex	Clinical diagnosis	Source of isolation	Etiology	Therapy	Outcome	Reference
USA	58/M	Allergic bronchopulmonary disease	Bronchial secretions	*Curvularia* sp.[e]	Steroids	Improved	Adam et al. 1986
USA	48/F	Allergic bronchopulmonary mycosis following allergic fungal sinusitis and brain abscess	Lung specimen, sinus aspirate	*C. lunata*	AMB, KTZ, TBF, resection of the right upper lobe	ND	Travis et al. 1991
USA	16/M	Allergic bronchopulmonary mycosis, allergic fungal sinusitis	Lung biopsy and sinus specimen	*C. senegalensis*	Right lateral rhinotomy, medial maxillectomy, right frontal ethmoidectomy, sphenoid sinusotomy, dacryocystorhinostomy	ND	Travis et al. 1991
USA	10/F	Allergic bronchopulmonary mycosis	Sputum plugs, bronchial aspirate	*C. senegalensis*	Steroids	Remission for 5 years	Mroueh and Spock 1992
USA	41/M	Allergic bronchopulmonary mycosis	Sputum plugs, bronchial washings, endobronchial biopsies	*C. hawaiiensis*[f]	Steroids, AMB lipid complex, ITZ	Recovery	Saenz et al. 2001
USA	48/M	Allergic bronchopulmonary mycosis	Bronchoalveolar lavage	*C. spicifera*[b]	VRZ, albuterol nebulizer treatments	Recovery	Hamilton et al. 2006
USA	40/M	Back pain associated with endobronchial mucus impaction	Transbronchial brushings, bronchoalveolar lavage and multiple transbronchial biopsies	*C. australiensis*[d] (ITS)	Oral ITZ, switched to oral VRZ, bronchoscopy with suctioning, oral prednisone	Recovery	Dyer et al. 2008
India	6/F	Allergic bronchopulmonary mycosis	Serial sputum and bronchoalveolar lavage specimens	*C. hawaiiensis*[f] (ITS)	Oral ITZ, intravenous liposomal AMB, local instillation with VRZ, oral prednisone	Recovery	Chowdhary et al. 2011

M: male, F: female, 5FC: 5-fluorocytosine, AMB: amphotericin B, FLZ: fluconazole, ITZ: itraconazole, KTZ: ketoconazole, MCZ: miconazole, TBF: terbinafine, VRZ: voriconazole; [a]reported as *Drechslera hawaiiensis*; [b]reported as *Bipolaris spicifera*; [c]reported as *Drechslera spicifera*; [d]reported as *Bipolaris australiensis*; [e]reported as *Bipolaris* sp.; [f]reported as *Bipolaris hawaiiensis*; [g]reported as *Helminthosporium* sp.; ND: no data available et al. 1999), Austria (Buzina et al. 2003), Italy (Castelnuovo et al. 2004; Posteraro et al. 2010) and France (Ambrosetti et al. 2006; Toul et al. 2006).

In the processed set of case reports (Table 4), the age of sinusitis patients was between 7 (Frenkel et al. 1987) and 70 (Taguchi et al. 2004) with an average of 29.4, with males more frequently affected than females (66% and 34%, respectively). Where the causal agent was identified to the species level, *C. lunata* (Berry et al. 1984; Brummund et al. 1986; Rinaldi et al. 1987; MacMillan et al. 1987; Nishioka et al. 1987; Bartynski et al. 1990; Travis et al. 1991; Ismail et al. 1993; Schroeder et al. 2002; Willard et al. 2003; Safdar 2003; Taj-Aldeen et al. 2004; Cavanna et al. 2014), *C. spicifera* (Young et al. 1978; Sobol et al. 1984; Adam et al. 1986; Rolston et al. 1985; Frenkel et al. 1987; Jay et al. 1988; Pratt and Burnett 1988; Gourley et al. 1990; Rao et al. 1989; Klapper et al. 1997; Buzina et al. 2003; Taguchi et al. 2004; Ruan et al. 2005; Wan et al. 2006; Ambrosetti et al. 2006), *C. australiensis* (Harpster et al. 1985; Del Palacio et al. 1997; Viola and Sutton 2010), *C. hawaiiensis* (Koshi et al. 1987; Robson et al. 1989; Maskin et al. 1989; Fryen et al. 1999; Castelnuovo et al. 2004; Taj-Aldeen et al. 2004), *C. clavata* (Ebright et al. 1999), *C. senegaliensis* (Travis et al. 1991) and *C. inaequalis* (Posteraro et al. 2010; Cruz et al. 2013) were reported. ITS-based molecular identification has also been carried out for some strains of *C. hawaiiensis* (Fryen et al. 1999), *C. spicifera* (Buzina et al. 2003), *C. inaequalis* (Posteraro et al. 2010) and *C. lunata* (Cavanna et al. 2014). A coinfection both by bacteria, e.g., *Staphylococcus aureus* and *Staphylococcus epidermidis* (Gourley et al. 1990) and by other fungi including *Alternaria* sp. (Loveless et al. 1981) or *Fusarium solani* (Peden et al. 2008) is possible. One of the reported rhinosinusitis cases was diagnosed as a mixed infection by both *C. lunata* and *C. hawaiiensis* (Taj-Aldeen et al. 2004). The causal agents were isolated from sinus mucus or sinus biopsies. Therapeutical interventions included surgery (Caldwell-Luc procedure, ethmoidectomy, turbinectomy, sinusotomy, polypectomy, sphenoidotomy, sinus trephination, endoscopic drainage), with or without the administration of single or combined antifungal therapy (local amphotericin B, oral fluconazole, oral itraconazole). Most of the patients could be cured successfully. Recurrence was documented in a case (Klapper et al. 1997), which was then successfully treated. The outcome was fatal for a patient where the allergic fungal sinusitis was complicated with central nervous system invasion (Viola and Sutton 2010).

Beside the numerous case descriptions, epidemiological data also underline the importance of *Curvularia* sinusitis. Bent and Kuhn (1994) studied 16 patients with AFS and found 7 of them having *Curvularia* infection. Schubert and Goetz (1998) found that with 67%, *C. spicifera* was the most prevalent fungus among 67 consecutive cases of allergic fungal sinusitis infection in the southwestern United States. McClay et al. (2002) compared the clinical presentation of allergic fungal sinusitis in children and adults and found that *Curvularia* species were the most frequently isolated causal agents in both groups with 28 of 32 (87.5%) and 52 of 70 (74.3%) cases, respectively. Further *C. hawaiiensis* and *C. spicifera* strains isolated from sinuses during unreported sinusitis infections were involved in the study of McGinnis et al. (1986).

Besides a large number of sinusitis cases caused by *Curvularia* species, cases of bronchopulmonary mycoses from Australia (Dolan et al. 1970; McAleer et al. 1981; Matthiesson 1981), USA (Hendrick et al. 1982; Halloran 1983; Halwig et al. 1985;

Adam et al. 1986; Travis et al. 1991; Mroueh and Spock 1992; Saenz et al. 2001; Hamilton et al. 2006; Dyer et al. 2008) and India (Chowdhary et al. 2011) were also reported in the literature (Table 4). Glancy et al. (1981) further reviewed 3 cases of bronchopulmonary mycoses caused by *C. hawaiiensis*, *C. lunata* as well as a mixed infection by these two species. Furthermore, Lake et al. (1991) performed serological testing between 1979 and 1986 in Western Australia, which revealed precipitins to *Bipolaris* and *Curvularia* species in 40 patients, 8 out of them were also diagnosed with allergic bronchopulmonary mycoses due to *Curvularia*. Symptoms reported in the literature included dyspnoea, shortness of breath, wheezing, coughing, sputum containing brown plugs and blood, haemoptysis, shoulder and back pain. Occasional bronchitis, allergic rhinitis and asthma were reported as the underlying conditions. In two of the reported cases the bronchopulmonary mycosis was suspected to be a secondary infection after AFS caused by *Curvularia* species (Travis et al. 1991). The infections could be characterized with raised direct eosinophil counts and sputum eosinophilia. The causal agents, *C. lunata* (McAleer et al. 1981; Halwig et al. 1985), *C. hawaiiensis* (McAleer et al. 1981; Saenz et al. 2001; Chowdhary et al. 2011), *C. spicifera* (Hamilton et al. 2006), *C. senegalensis* (Travis et al. 1991; Mroueh and Spock 1992) and *C. australiensis* (Dyer et al. 2008) were isolated from sputum, bronchoalveolar lavage and/or biopsy samples. One of the reported cases was diagnosed as a mixed infection by both *C. lunata* and *C. hawaiiensis* (McAleer et al. 1981). The applied therapeutic interventions were surgery, bronchoscopy with suctioning as well as antifungal treatment (oral miconazole, voriconazole, ketoconazole or itraconazole, voriconazole instillation and intravenous amphotericin B). Where reported, the therapy was successful. Further *C. hawaiiensis* and *C. spicifera* strains isolated from peritoneal dialysate in unreported peritonitis cases were involved in the study of McGinnis et al. (1986).

Alture-Werber and Edberg (1985) developed an animal model to study *C. geniculata* infection. A strain of the fungus deriving from the bronchial washings in a patient with pulmonary cavitary curvulariosis was intraperitoneally injected into mice, which resulted in the development of granulomas in the liver and spleen. The lesion in the patient's lungs and in the animal model were compared histopathologically and found to be similar. In a more recent study, Paredes et al. (2013) evaluated the virulence of *C. spicifera* isolates from sinuses as well as *C. hawaiiensis* isolates from sinuses and lung in an experimental model of disseminated infection in immunocompromised mice. Both species produced high mortality with the lung being the organ most affected.

Skin and Nail Infections

Case reports selected from the literature about *Curvularia* species causing skin and nail infections are summarized in Table 5. In the case of skin infections, both cutaneous (Estes et al. 1977; Agrawal and Singh 1995; Torda and Jones 1997; Tessari et al. 2003; Fan et al. 2009; Hiromoto et al. 2008; Vermeire et al. 2010; Moody et al. 2012) and subcutaneous (Costa et al. 1991; Subramanyam et al. 1993, Vásquez-del-Mercado et al. 2013) phaeohyphomycoses are known. A few cases of onychomycoses were

Table 5. Case reports selected from the literature about *Curvularia* species causing skin and nail infections.

Country	Age/Sex	Clinical diagnosis	Source of isolation	Etiology	Therapy	Outcome	Reference
Senegal	25/M	Black grain mycetoma	Foot abscess	*C. lunata*	ND	ND	Baylet et al. 1959
USA	5/M	Cutaneous phaeohyphomycosis	Lesion, excisional biopsy specimen	*C. spicifera*[a]	Excisional biopsy, systemic AMB	Healed	Estes et al. 1977
India	38/M	Onychomycosis	Nail plate	*C. lunata*	Antilepromatous therapy	ND	Barde and Singh 1983
India	46/F	Black verrucous lesions on the nasal conchae	Scraping of the nasal mucosa	*Curvularia* sp.[b]	Local excision of the crusted lesion, nystatin	Healed	Koshi et al. 1987
USA	28/M	Brown plaques on the scrotum resembling seborrheic keratosis	Plaques	*Curvularia* sp. (two strains)	ND	ND	Duvic et al. 1987
India	20/F	Skin infection	Skin lesions	*C. clavata*	Topical CLZ	Partial clearance, patient lost to follow up	Gugnani et al. 1990
Brazil	61/M	Subcutaneous phaeohyphomycosis in the first finger of the left foot	Biopsy material	*C. hawaiiensis*[c]	Electrocoagulation	Healed	Costa et al. 1991
India	51/F	Paronychia and black discoloration of a thumb nail	Thumb nail	*C. lunata*	Topical CLZ	Healed	Kamalam et al. 1992
USA	3/F	Invasive burn wound infection on the arms and legs	Full thickness biopsies	*Curvularia* sp.	AMB, surgeries	Healed	Still et al. 1993
India	21/M	Subcutaneous infection on the right leg and the dorsum of the right foot	Discharge and scrapings from the lesion	*Curvularia* sp.	KTZ, GRF	Healed	Subramanyam et al. 1993
India	50/M	Cutaneous phaeohyphomycosis on the ankle region of both feet and the medial aspect of the right thigh	Skin lesions	*C. pallescens*	None (patient failed to return for therapy)	Unknown	Agrawal and Singh 1995

Table 5. contd....

Table 5. contd.

Country	Age/Sex	Clinical diagnosis	Source of isolation	Etiology	Therapy	Outcome	Reference
India	35/M	Cutaneous phaeohyphomycosis on the posterior aspect of the left thumb	Skin biopsy specimen	*C. pallescens*	None (patient failed to return for therapy)	Unknown	Agrawal and Singh 1995
Canada	75/F	Ulcer on the proximal aspect of the right leg	Biopsy specimen	*C. pallescens*	Oral KTZ, excision and grafting	Healed	Berg et al. 1995
Australia	58/M	Necrotizing cellulitis involving both thighs	Surgical debridement specimens	*C. brachyspora*	Surgical debridements, hyperbaric oxygen therapy, intravenous AMB, grafting	Healed	Torda and Jones 1997
Brazil	50/F	Dermatomycosis of the toe web between the fourth and fifth toes of the right foot	Skin scrapings	*C. lunata*	Topical isoconazole	Healed	Lopes and Jobim 1998
India	27/M	Eumycetoma on the right foot	Discharged black granules, biopsy sample	*C. lunata*	Oral KTZ	ND (improving)	Janaki et al. 1999
Argentina	9/M	Ecthyma gangrenosum-like lesion on the right palm	Skin biopsy	*Curvularia* sp.	AMB, liposomal AMB, ITZ	Healed	Bonduel et al. 2001
Italy	69/M	Cutaneous infection with systemic dissemination	Skin biopsy	*C. lunata*	Intravenous AMB	Died	Tessari et al. 2003
USA	4/M	Scaly erythematous plaques of the ear	Plaque specimens	*Curvularia* sp.[b]	Topical KTZ	Healed	Robb et al. 2003
USA	38/F	Erythema and scaling on the scalp, central face and perioral areas	Facial scales	*Curvularia* sp.[b]	Oral ITZ	Healed	Robb et al. 2003
USA	43/F	Bilateral nail dystrophy and paronychia of 5 fingers	Nail culture	*Curvularia* sp.[b]	Oral ITZ	Healed	Robb et al. 2003
USA	41/M	Progressive verrucal distal onychomycosis of the left great toe	Biopsy sample	*Curvularia* sp.	Oral ITZ	Healed	Safdar 2003

USA	5/M	Erythematous patch with central punctate hemorrhage on the left cheek	Skin biopsy and tissue culture	*C. spicifera*[d]	Intravenous AMB, surgery, oral ITZ	Healed	Bilu et al. 2004
China	64/M	Cutaneous phaeohyphomycosis of foot	Exudate and biopsy specimen	*C. clavata*	Oral FLZ, topical ethacridine solution	Healed	Fan et al. 2009
India	65/M	Eumycetoma of the right foot	Black, irregular granules	*C. lunata*	Oral ITZ	ND (improving)	Garg et al. 2008
Japan	67/M	Cutaneous infection on the left forearm	Skin biopsy material	*Curvularia* sp.	Terbinafine, surgical resection, oral ITZ	Healed after recurrence	Hiromoto et al. 2008
Belgium	71/M	Cutaneous phaeohyphomycosis on the dorsum of the right hand	Punch biopsy specimen	*Curvularia* sp.	Oral ITZ, surgery	Healed	Vermeire et al. 2010
Argentina	72/M	Painful chronic varicose ulcer on the side of the left foot	Ulcer	*C. lunata*	Oral ITZ, topical sodium borate and KTZ	Healed	Russo et al. 2010
Japan	74/M	Keratotic brown lesions on the interdigital web between the toes	Scrapings from the tiny thick brown scales	*Curvularia* sp. KMU4944	Topical TRB and LNZ	Healed	Yanagihara et al. 2010
Mexico	25/M	Subcutaneous phaeohyphomycosis, ulcer with seropurulent discharge on the left leg	Secretion	*C. lunata*	Oral ITZ, excision of affected skin and subcutaneous tissue	Healed	Vásquez-del-Mercado et al. 2013
USA	73/M	Cutaneous infection of the right forearm	4-mm punch biopsies from the center of the lesion	*Curvularia* sp.	Oral ITZ	Healed	Moody et al. 2012
Sri Lanka	50/F	Black-grain eumycetoma on the left ankle	Incisional biopsy	*C. lunata*	Oral ITZ	Healed	Gunathilake et al. 2013

M: male, F: female, AMB: amphotericin B, CLZ: clotrimazole, FLZ: fluconazole, GRF:griseofulvin, ITZ: itraconazole, KTZ: ketoconazole, LNZ: lanoconazole, TRB: terbinafine; [a]reported as *Drechslera spicifera*; [b]reported as *Bipolaris* sp.; [c]reported as *Bipolaris hawaiiensis*; [d]reported as *Bipolaris spicifera*; ND: no data available also reported (Barde and Singh 1983; Kamalam et al. 1992; Robb et al. 2003; Safdar 2003), however, based on some epidemiological studies, 4.5% of the onychomycosis cases were caused by *Curvularia* species (Ramani et al. 1993; Gupta et al. 2007; Veer et al. 2007). Both immunocompetent (Harris and Downham 1978; Rohwedder et al. 1979; Fan et al. 2009) and immunocompromised (Barde and Singh 1983; Kiryu and Suenaga 1985; Berg et al. 1995; Tessari et al. 2003) patients were affected by *Curvularia* skin infections. Predisposing factors include leukemia, chemotherapy for non-Hodgkin's lymphoma, immunosuppression due to bone marrow transplantation for the treatment of severe aplastic anemia, transplantation of heart and kidney, acquired immunodeficiency syndrome, diabetes, chronic obstructive pulmonary disease, burn injuries, cryotherapy to pigmented lesions, viral vesicular dermatitis, traumatic injuries, application of topical or systemic corticosteroids and cocaine usage. Many of the cases occurred in patients who worked or played outdoors. Cases are known from many countries including the United States, Canada, Mexico, Brazil, Argentina, India, Senegal, Sri Lanka, China, Japan, Australia, Italy, Belgium and Sudan. The age of the patients was between 3 and 75 with an average of 42.6 (Table 5), however, extremely low birth weight neonates were also reported to be affected by invasive *Curvularia* dermatitis (Rowen et al. 1995). Based on the case reports processed in Table 5, males were more frequently affected (72%) than females (28%). Where the causal agent was identified to the species level, *C. spicifera* (Estes et al. 1977; Straka et al. 1989; Bilu et al. 2004; Dubois et al. 2008), *C. australiensis* (Chalet et al. 1986), *C. hawaiiensis* (Costa et al. 1991; Romano et al. 2004), *C. clavata* (Gugnani et al. 1990; Fan et al. 2009), *C. lunata* (Mahgoub 1973; Barde and Singh 1983; Kamalam et al. 1992; Grieshop et al. 1993; Lopes and Jobim 1998; Janaki et al. 1999; Fernandez et al. 1999; Tessari et al. 2003; Garg et al. 2008; Russo et al. 2010; Vásquez-del-Mercado et al. 2013; Gunathilake et al. 2013), *C. pallescens* (Agrawal and Singh 1995; Berg et al. 1995), *C. inaequalis* (Tanabe et al. 2010), *C. brachyspora* (Torda and Jones 1997), *C. trifolii* (Kyriu and Suenaga 1985) and the yet undescribed *Curvularia* sp. KMU4944 (Yanagihara et al. 2010) were reported, however, with the exception of the latter, these species level identifications were not confirmed with molecular techniques. Co-occurring infections with *Flavobacterium meningosepticum*, *Acinetobacter* sp. and *Staphylococcus aureus* were reported in the case of a burn wound infection (Still et al. 1993), with *P. aeruginosa*, *Klebsiella* sp. and *Staphylococcus* sp. in the case of a subcutaneous infection (Subramanyam et al. 1993) and with *Exserohilum rostratum* in a cutaneous infection of a cocaine user (Lavoie et al. 1993). The manifestations of *Curvularia* skin infection include different types of cutaneous (pigmented macules, scaling lesion, brown thick scales) and subcutaneous (abscesses, ulcers, papules, firm tender nodules and nodes, ecthyma gangrenosum-like lesion, macular lesion with central necrosis, hemorrhagic bullous lesion, elevated hyperpigmented scaly lesion) skin lesions involving the upper or lower extremities (Yanagihara et al. 2010). In some cases *Curvularia* species produced characteristic black granules that are discharged to the skin surface through sinus tracts, and are composed of a dense, interwoven mass of septate mycelium and thick-walled, chlamydospore-like cells embedded in a cement-like substance (Mahgoub 1973; Garg et al. 2008). The causal agents were isolated from scrapings or biopsy specimens of the skin lesions and from nail samples. Therapeutical interventions included surgery and the application of antifungal drugs (oral/topical itraconazole and ketoconazole, oral fluconazole and griseofulvin, topical clotrimazole, lanoconazole and terbinafine, intravenous amphotericin B). Antifungal chemotherapy combined with surgical excision seems to be the appropriate therapy (Hiromoto et al. 2008). The infection could be cured by therapy in most of the cases; a fatal outcome was reported only in a single case (Table 5) where the cutaneous infection was followed by systemic dissemination (Tessari et al. 2003). Further *C. hawaiiensis* strains isolated from lesions of the forehead or wounds on elbow and face, and *C. spicifera* strains deriving from femur or wounds (foot, arm) during unreported cutaneous infections were involved in the study of McGinnis et al. (1986).

Infections Complicating Peritoneal Dialysis

Peritonitis is a common complication of chronic ambulatory peritoneal dialysis (CAPD) and continuous cyclic peritoneal dialysis (CCPD) with bacteria as the principal pathogens, followed by *Candida* species and filamentous fungi (Ujhelyi et al. 1990). Risk factors for fungal peritonitis include recent bacterial peritonitis, immunosuppressive therapy, recent exposure to antibiotics and the presence of bowel perforation (Ujhelyi et al. 1990). Case reports in the literature about *Curvularia* species complicating peritoneal dialysis and/or causing peritonitis are summarized in Table 6. Most of these infections were related with CAPD (O'Sullivan et al. 1981; Guarner et al. 1989; Ujhelyi et al. 1990; Lopes et al. 1994; Bava et al. 2003; Pimentel et al. 2005; Kalawat et al. 2012) or CCPD (Gadallah et al. 1995; Canon et al. 2001; Vachharajani et al. 2005; Diskin et al. 2008) to treat chronic renal disease due to glomerulonephropathy, polycystic kidney disease, renovascular disease, posterior urethral valves or diabetes mellitus. In addition, cases of Tenckhoff catheter obstruction without peritonitis (De Vault et al. 1985; Unal et al. 2011) and a case of peritonitis unrelated with dialysis (Terada et al. 2014) were also reported. Most of the known cases are from warm and humid regions like the southern parts of the United States (O'Sullivan et al. 1981; De Vault et al. 1985; Brackett et al. 1988; Guarner et al. 1989; Ujhelyi et al. 1990; Gadallah et al. 1995; Canon et al. 2001; Vachharajani et al. 2005; Diskin et al. 2008; Charlton et al. 2010), as well as other countries including Brazil (Lopes et al. 1994), Argentina (Bava et al. 2003), Australia (Pimentel et al. 2005), Turkey (Unal et al. 2011), India (Varughese et al. 2011; Kalawat et al. 2012) and Japan (Terada et al. 2014). The age of the patients was between 3 (Bava et al. 2003) and 85 (Pimentel et al. 2005) with an average of 48, with both males and females equally affected. Where the causal agent was identified to the species level, *C. spicifera* (O'Sullivan et al. 1981; Bava et al. 2003), *C. hawaiiensis* (Gadallah et al. 1995), *C. lunata* (De Vault et al. 1985; Brackett et al. 1988; Lopes et al. 1994; Unal et al. 2011; Varughese et al. 2011; Kalawat et al. 2012), *C. geniculata* (Vachharajani et al. 2005; Terada et al. 2014), *C. pallescens* (Charlton et al. 2010) and *C. inaequalis* (Pimentel et al. 2005) were reported, however, a molecular, ITS sequence-based identification has been carried out only for a *C. inaequalis* isolate (Pimentel et al. 2005). A previous

Table 6. Case reports in the literature about *Curvularia* species complicating peritoneal dialysis and/or causing peritonitis.

Country	Age/Sex	Clinical diagnosis	Source of isolation	Etiology	Therapy	Outcome	Reference
USA	55/M	CAPD peritonitis	Peritoneal fluid	*C. spicifera*[a]	Intravenous and parenteral AMB	Infection resolved	O'Sullivan et al. 1981
USA	60/F	Tenckhoff catheter obstruction without peritonitis	Catheter, dialysate	*C. lunata*	Catheter removal	Infection resolved	De Vault et al. 1985
USA	46/M	Black material in effluent, fever	Bag, catheter, effluent	*C. lunata*	Catheter removal, intraperitoneal AMB changed to intravenous AMB	Infection resolved after relapse	Brackett et al. 1988
USA	53/M	CAPD peritonitis	Catheter, effluent	*Curvularia* sp.	Intravenous 5FC, catheter removal	Infection resolved	Guarner et al. 1989
USA	28/M	CAPD peritonitis	Catheter tip	*Curvularia* sp.	Intravenous AMB, oral 5FC, catheter removal	Infection resolved after relapse	Ujhelyi et al. 1990
Brazil	63/M	CAPD peritonitis	CAPD fluid, catheter extremity, exudate	*C. lunata* and *S. aureus*	Intravenous AMB	Infection resolved	Lopes et al. 1994
USA	73/F	CCPD peritonitis	Inner wall of the catheter	*C. hawaiiensis*[b]	Catheter removal, oral ITZ	Infection resolved	Gadallah et al. 1995
USA	11/M	CCPD peritonitis	Peritoneal fluid	*Curvularia* sp.	Catheter removal, intravenous AMB	Infection resolved	Canon et al. 2001
Argentina	3/F	CAPD peritonitis	Dialysate	*C. spicifera*[c]	Catheter removal, oral Fluconazole	Infection resolved	Bava et al. 2003
USA	45/M	CCPD peritonitis	Peritoneal fluid	*C. geniculata*	Intravenous AMB, oral ITZ, catheter removal	Infection resolved	Vachharajani et al. 2005
Australia	85/F	CAPD peritonitis	Peritoneal fluid	*C. inaequalis*	Intraperitoneal FLZ, intravenous AMB, catheter removal	Patient died	Pimentel et al. 2005
USA	53/F	CCPD peritonitis	Thick black material growing in the silicone tubing	*Curvularia* sp.	ND	ND	Diskin et al. 2008
Turkey	53/F	Tenckhoff catheter obstruction without peritonitis	Peritoneal effluent, wall of the Thenckhoff catheter	*C. lunata*	Catheter removal, intravenous AMB switched to oral Itraconazole	Infection resolved	Unal et al. 2011

USA	13/M	PD peritonitis	Peritoneal effluent	*C. pallescens*	Intravenous and intraperitoneal AMB, catheter removal, oral FLZ	Infection resolved	Charlton et al. 2010
India	61/M	PD peritonitis	Peritoneal lavage fluid, catheter tip	*C. lunata*	Intravenous liposomal AMB, catheter removal	Patient died	Varughese et al. 2011
India	55/F	CAPD peritonitis	CAPD fluid	*C. lunata*	Catheter removal, oral VRZ	Infection resolved	Kalawat et al. 2012
Japan	61/F	Non PD peritonitis	Specimens of ascitic fluid	*C. geniculata* and *Pithomyces* sp.	Oral VRZ	Infection resolved	Terada et al. 2014

M: male, F: female, CAPD: chronic ambulatory peritoneal dialysis, CCPD: continuous cycling peritoneal dialysis, 5FC: 5-fluorocytosine, AMB: amphotericin B, FLZ: fluconazole, ITZ: itraconazole, VRZ: voriconazole; [a]reported as *Drechslera spicifera;* [b]reported as *Bipolaris hawaiiensis*; [c]reported as *Bipolaris spicifera*; ND: no data available infection or coinfection by bacteria, e.g., *Staphylococcus aureus*, *Staphylococcus epidermidis*, *Enterobacter aerogenes*, *Enterococcus faecalis*, *Klebsiella pneumoniae*, *Proteus vulgaris* and *Pseudomonas* sp. was frequent. Reported symptoms included abdominal pain, fever, nausea, vomiting, diarrhea, anorexia, constipation, cloudy peritoneal dialysis fluid with black flakes/specks and Tenckhoff catheter malfunction. The possible source of infection is environmental contamination (e.g., soil contact) (Ujhelyi et al. 1990). Diskin et al. (2008) postulated the preference of *Curvularia* to colonize organic silicon as a possible predisposing condition. The causal agents were isolated from the peritoneal fluid as well as from the catheter and dialysis bag in most of the cases. Therapeutical interventions included catheter removal, as well as the administration of single or combined antifungal therapy (intravenous and/or parenteral amphotericin B, intravenous 5-fluorocytosin, intraperitoneal fluconazole, oral itraconazole, oral voriconazole). The early removal of the catheter and intravenous amphotericin B treatment are considered to be the first line of treatment in the case of *Curvularia* peritonitis, however, amphotericin B resistance has also been reported in one of the cases (Varughese et al. 2011), which suggests the necessity of routine susceptibility testing. Although the infection resolved upon therapy in most of the patients, relapse was documented in some cases (Brackett et al. 1988; Ujhelyi et al. 1990), and the outcome was fatal for two patients (Pimentel et al. 2005; Varughese et al. 2011). Further *C. hawaiiensis* and *C. spicifera* strains isolated from sinuses during unreported sinusitis infections were involved in the study of McGinnis et al. (1986).

Deep and Disseminated Infections

Case reports selected from the literature about *Curvularia* species causing deep and disseminated mycoses are summarized in Table 7.

A series of deep infections affected the central nervous system, these included cerebral abscesses (Fuste et al. 1973; Rohwedder et al. 1979; Filizzola et al. 2003; Gadgil et al. 2013; Gongidi et al. 2013; Skovrlj et al. 2013), cerebral mycetoma (Lampert et al. 1977; Friedman et al. 1981), encephalitis (Yoshimori et al. 1982), meningitis (Latham 2000; Singh et al. 2008) and meningoencephalitis (Fuste et al. 1973; Biggs et al. 1986; Adam et al. 1986). Some of these cases were disseminated infections also affecting other organs like lungs (Lampert et al. 1977; Rohwedder et al. 1979; Friedman et al. 1981; de la Monte and Hutchins 1985; Pierce et al. 1986), sinuses (McGinnis et al. 1992) and legs (Rohwedder et al. 1979). Both immunocompetent (Lampert et al. 1977; Friedman et al. 1981; Carter and Boudreaux 2004; Smith et al. 2007; Gadgil et al. 2013) and immunocompromised patients were affected (de la Monte and Hutchins 1985; Pierce et al. 1986; Singh et al. 2008). Reported symptoms included cough, low-grade fever, hemiparesis, weight loss, seizures, weakness, headaches, nausea, vomiting, bloody nasal drainage, dizziness, slurred speech, erectile dysfunction, hearing loss, memory loss, unsteady gait, anorexia, lethargy, night sweat, optic atrophy, progressive diplopia, eye ptosis and blurred or double vision. Most of the known cases are from the United States (Fuste et al. 1973; Lampert et al. 1977; Friedman et al. 1981; Biggs et al. 1986; Adam et al. 1986; Morton et al. 1986; McGinnis et al. 1992; Latham 2000; Filizzola et al. 2003; Carter and Boudreaux 2004; Singh et al. 2008;

Table 7. Case reports in the literature about *Curvularia/Bipolaris* species causing deep and disseminated infections.

Country	Age/Sex	Clinical diagnosis	Source of isolation	Etiology	Therapy	Outcome	Reference
USA	31/F	Meningoencephalitis	Autopsy specimen	*C. hawaiiensis*[a]	ND	Death	Fuste et al. 1973
USA	13/M	Pulmonary and cerebral mycetoma of the right frontoparietal lobe	Lung biopsy specimen, excised brain lesion, bronchoscopy, thoracotomy	*C. pallescens*	Surgery, intravenous AMB and MCZ	Recurrence after 3 years, death	Lampert et al. 1977; Friedman et al. 1981
USA	25/M	Progressive disseminated infection, chronic leg ulcers and cerebral abscess of the left parietal lobe	Cervical lymph node, pleural fluid, and leg ulcers	*C. lunata*	Surgery, AMB, MCZ, 5FC	ND	Rohwedder et al. 1979
Japan	21/F	Granulomatous encephalitis	Brain biopsy	*C. spicifera*[b]	AMB, 5FC	Recovery	Yoshimori et al. 1982
USA	41/M	Pulmonary infection and cerebral infection of the right parieto-occipital lobes	Lung tissue	*C. lunata*	Surgery, AMB, MCZ, KTZ	Recovery	de la Monte and Hutchins 1985; Pierce et al. 1986
USA	18/M	Granulomatous encephalitis	Brain biopsy specimen	*C. hawaiiensis*[c]	Surgery, systemic and intrathecal AMB	Recovery	Morton et al. 1986
USA	21/F	Meningo-encephalitis complicating compound skull fractures due to open head trauma	ND	*C. spicifera*[d]	ND	Death	Biggs et al. 1986
USA	49/F	Meningoencephalitis	ND	*C. spicifera*[b]	ND	Death	Adam et al. 1986
USA	26/F	Phaeohyphomycosis involving the brain and sinuses	Left temporal mass biopsy, aspiration specimen	*C. spicifera*[b]	AMB, KTZ	Recovery	McGinnis et al. 1992
USA	18/M	Meningitis complicating a neurosurgerical procedure	ND	*C. spicifera*[b]	AMB	Recovery	Latham 2000

Table 7. contd....

Table 7. contd.

Country	Age/Sex	Clinical diagnosis	Source of isolation	Etiology	Therapy	Outcome	Reference
USA	28/M	Brain abscess	Brain biopsy sample, sputum culture	*Curvularia* sp.[e]	ITZ, AMB, liposomal AMB, VRZ	Death	Filizzola et al. 2003
USA	21/M	Cerebral phaeohyphomycosis of the right basal ganglia	Brain biopsy specimen	*C. lunata*	AMB, liposomalAMB, vancomycin, ceftriaxone	Death	Carter and Boudreaux 2004
Australia	57/M	Pituitary fossa lesion resulting in optic nerve atrophy	Mucoid material	*C. lunata*	stereotactic transphenoidal resection, intravenous AMB, meropenem, liposomal AMB, oral VRZ	Recovery	Smith et al. 2007
USA	35/M	Cranial base meningioma	Biopsies	*C. geniculata*	limited transnasal endoscopic debulking, VRZ, intravenous AMB	Death	Singh et al. 2008
USA	55/M	Cerebral lesion in the left basal ganglia	Lesion biopsy	*C. spicifera*[b]	Intravenous liposomal AMB, oral VRZ	Recovery	Rosow et al. 2011
USA	50/F	Abscess of the skull base extending into the frontal lobe	Biopsy	*Curvularia* sp.	Surgery, VRZ, minocycline, levofloxacin	Recovery	Gadgil et al. 2013
USA	37/M	Multiple brain abscesses	Abscesses	*Curvularia* sp.	Surgical resections, intravenous antifungals	Recovery	Gongidi et al. 2013
USA	33/M	Abscess of the brainstem	Intraoperative biopsy of pus-filled, encapsulated lesion	*Curvularia* sp.	VRZ, flucytosine, liposomal AMB	Recovery	Skovrlj et al. 2013
USA	73/M	Fungal endarteritis	Blood, autopsy (suture material in the aortic arch)	*C. spicifera*[b]	ND	Death	Ogden et al. 1992
USA	44/M	Endocarditis	Removed valve	*C. lunata*	Surgery, AMB, KTZ, terbinafine	Recovery	Bryan et al. 1993
Canada	newborn/M	Sternal wound infection	Open sternotomy wound specimen	*C. lunata, Staphylococcus* sp.	Surgery, intravenous vancomycin, tobramycin, and ceftazidime	Death	Yau et al. 1994

Israel	43/F	Phaeohyphomycosis following cardiac surgery	Necrotic material in the sptum below the homograft	*C. spicifera*[b]	Surgery, AMB	Recovery	Pauzner et al. 1997
Brasil	43/F	Pericardium infection	Pericardial fluid	*Curvularia* sp.	ND	ND	Severo et al. 2012
USA	41/M	Disseminated infection (brain, heart valve, spleen, kidney)	Biopsies	*C. geniculata*	Surgery	Death	Kaufman 1971
USA	52/M	Disseminated infection (blood, heart, lung, kidney)	Biopsies	*C. spicifera*[b]	AMB, KTZ	Death	Adam et al. 1985
USA	76/F	Disseminated infection (skin)	Multiple skin biopsies	*C. spicifera*[b]	AMB	Death	Adam et al. 1985
Pakistan	24/M	Disseminated infection (bronchus, adrenal glands)	Bronchial and adrenal gland biopsy	*C. spicifera*[b]	Intravenous AMB	Recovery	Karim et al. 1993
USA	ND(child)/ ND	Fungemia, pneumonia	Blood, lumenal surfaces of Hickman catheter	*C. spicifera*[b]	AMB	ND	Walsh et al. 1995
Japan	10/F	Disseminated infection (liver, spleen)	ND	*C. geniculata*	AMB, MCZ	Partial response, death	Shigemori et al. 1996
UK	21/M	Disseminated infection (right lung, mediastinal lymphadenopathy, pericardial effusion, abdominal masses obstructing and invading the common bile duct and right ureter)	Biopsy specimen of the hilar mass	*C. australiensis*[f]	Intravenous AMB, oral ITZ	Recovery	Flanagan and Bryceson 1997
Pakistan	18/F	Disseminated infection (bronchus, lymph node)	Biopsies taken from the cervical lymph node and the right middle bronchus	*C. spicifera*[b]	ITZ	Recovery	Khan et al. 2000
USA	0/F	Disseminated infection (brain, lung, heart, liver, kidney)	Autopsy specimens	*Curvularia* sp.[e]	Lipid AMB, ITZ	Death	Bryan et al. 2000

Table 7. contd....

Table 7. contd.

Country	Age/Sex	Clinical diagnosis	Source of isolation	Etiology	Therapy	Outcome	Reference
USA	0/F	Disseminated infection (skin, brain, meninges, heart, lungs, and liver)	Shave biopsy of necrotic skin lesions, internal organ biopsies	*C. spicifera*[b], *Staphylococcus*	Liposomal AMB, ITZ	Death	Moore et al. 2001
USA	50/M	Disseminated infection (brain, lung, heart, liver, spleen, kidney)	ND	*C. spicifera*[b]	None	Death	Revankar et al. 2002
Japan	29/M	Disseminated infection (lymph node, lung, liver)	Lymph node biopsy specimen	*C. spicifera*[b]	Intravenous AMB, VRZ, oral ITZ	Recovery after recurrence	Kobayashi et al. 2008

M: male, F: female, 5FC: 5-fluorocytosine, AMB: amphotericin B, CSP: caspofungin, FLZ: fluconazole, ITZ: itraconazole, KTZ: ketoconazole, MCZ: miconazole, VRZ: voriconazole, ND: no data available; [a]reported as *Drechslera hawaiiensis;* [b]reported as *Bipolaris spicifera;* [c]reported as *Bipolaris hawaiiensis*; [d]reported as *Drechslera spicifera*; [e]reported as *Bipolaris* sp.; [f]reported as *Bipolaris australiensis*

Rosow et al. 2011; Gadgil et al. 2013; Gongidi et al. 2013; Skovrlj et al. 2013), but *Curvularia* CNS infections were also reported from Japan (Yoshimori et al. 1982) and Australia (Smith et al. 2007). The age of the patients was between 13 (Lampert et al. 1977; Friedman et al. 1981) and 57 (Smith et al. 2007) with an average of 32.2 (Table 7). Males were more frequently affected (66.7%) than females (33.3%). Where the causal agent was identified to the species level, *C. lunata* (Rohwedder et al. 1979; de la Monte and Hutchins 1985; Pierce et al. 1986; Carter and Boudreaux 2004; Smith et al. 2007), *C. spicifera* (Yoshimori et al. 1982; Adam et al. 1986; McGinnis et al. 1992; Latham 2000; Rosow et al. 2011), *C. hawaiiensis* (Fuste et al. 1973; Morton et al. 1986; Biggs et al. 1986), *C. geniculata* (Singh et al. 2008) and *C. pallescens* (Lampert et al. 1977; Friedman et al. 1981) were reported, but the identification of only a single isolate of *C. geniculata* was confirmed with molecular techniques (Singh et al. 2008). The causal agent was isolated from biopsy specimens. Applied therapies were surgery and the administration of antifungals (liposomal amphotericin B, systemic and intrathecal amphotericin B, miconazole, itraconazole, ketoconazole, voriconazole, 5-fluorocytosine). The outcome of *Curvularia* CNS infection was fatal in seven cases (Fuste et al. 1973; Lampert et al. 1977; Friedman et al. 1981; Biggs et al. 1986; Adam et al. 1986; Filizzola et al. 2003; Carter and Boudreaux 2004; Singh et al. 2008). A further *C. hawaiiensis* strain isolated from brain biopsy specimen and two *C. spicifera* strains from cerebrospinal fluid and brain biopsy specimen during unreported CNS infections were involved in the study of McGinnis et al. (1986).

Five cases of cardial infections due to *Curvularia* sp. are also known from the literature: a fatal endarteritis (Ogden et al. 1992) and a successfully treated endocarditis (Bryan et al. 1993) from adult males in the USA, a fatal sternal wound infection in a newborn from Canada (Yau et al. 1994), a successfully treated phaeohyphomycosis following cardiac surgery in a female from Israel (Pauzner et al. 1997) and a pericardial infection in a female in Brasil (Severo et al. 2012) (Table 7).

Regarding disseminated *Curvularia* infections, the available case descriptions report about the involvement of brain, heart valve, spleen, kidney, lung, blood, skin, liver, bronchus, adrenal glands, abdomen, bile duct, ureter and lymph nodes (Table 7). Causal agents reported from the genus were *C. spicifera* (Adam et al. 1986; Karim et al. 1993; Walsh et al. 1995; Khan et al. 2000; Moore et al. 2001; Revankar et al. 2002; Kobayashi et al. 2008), *C. australiensis* (Flanagan and Bryceson 1997) and *C. geniculata* (Kaufman 1971; Shigemori et al. 1996). Co-occurring infection with *Staphylococcus* was reported in one of the cases (Moore et al. 2001). Some cases were reported in immunocompetent hosts (Flanagan and Bryceson 1997; Kobayashi et al. 2008), while two of the patients were premature infants (Bryan et al. 2000; Moore et al. 2001). Despite surgical and antifungal therapies, disseminated *Curvularia* infections have a very high mortality rate, 7 patients (Kaufman 1971; Adam et al. 1986; Shigemori et al. 1996; Bryan et al. 2000; Moore et al. 2001; Revankar et al. 2002) including 6 from USA died out of the 12 cases processed.

In an experimental model, a *C. lunata* strain from a disseminated human infection was capable of infecting normal mice while three other strains of *C. lunata* and one each of *C. pallescens* and *C. spicifera* were not (Whitcomb et al. 1981). Hyphal filaments of the infective strain could be observed in liver and spleen abscesses. This study also reported that all examined strains of *Curvularia* infected mice were treated with 400 rads X-irradiation and 10.0 mg cortisone.

Antifungal Susceptibilities of *Curvularia* Strains

Antifungal susceptibility analysis is indispensible in any clinical microbiology laboratory as there is an increase in the number of antifungal agents being used and also for the generation of precise susceptibility data in order to help the clinicians to decide on the choice of antifungal therapy. Currently, the antifungal susceptibility data for *Curvularia* isolates are sparse, however, the obvious differences in the antifungal susceptibility patterns of *Curvularia* spp. highlight the importance of determining antifungal susceptibility isolate by isolate. Antifungal susceptibility data available for *Curvularia* isolates are summarized in Table 8. Among the studies reviewed, methods like E-test, broth macro- and microdilution have been applied to determine the minimal inhibitory concentration (MIC) values of the antifungal agents.

Table 8 reflects that most of the clinical *Curvularia* isolates are resistant to fluconazole and 5-fluorocytosine and susceptible or intermediate to amphotericin B, itraconazole, posaconazole, voriconazole, caspofungin, micafungin and anidulafungin; however, high MIC-levels of amphotericin B (Varughese et al. 2001; Espinel-Ingroff et al. 2002; Biancalana et al. 2011; Da Cunha et al. 2013), itraconazole (Radford et al. 1997; Guarro et al. 1999; Espinel-Ingroff 2001a; Espinel-Ingroff et al. 2002; Da Cunha et al. 2013), ketoconazole (Guarro et al. 1999), miconazole (Bryan et al. 1993; O'Sullivan et al. 1981), posaconazole (Espinel-Ingroff et al. 2002), voriconazole (Espinel-Ingroff et al. 2002; Odabasi et al. 2004), caspofungin (Lass-Flörl et al. 2008; Direkel et al. 2012) and anidulafungin (Da Cunha et al. 2013) were also reported in certain cases. Based on the available susceptibility data, micafungin and anidulafungin followed by itraconazole, caspofungin, voriconazole and amphotericin B might be important drugs in the treatment of *Curvularia* and *Bipolaris* infections (Ujhelyi et al. 1990; Del Poeta et al. 1997; Flanagan and Bryceson 1997; McGinnis and Pasarell 1998; Guarro et al. 1999; Wilhelmus and Jones 2001; Espinel-Ingroff 2001a; Espinel-Ingroff et al. 2002; Pfaller et al. 2002; Odabasi et al. 2004; Sabatelli et al. 2006; Lass-Flörl et al. 2008; Unal et al. 2011; Xu et al. 2010; Biancalana et al. 2011; Chowdhary et al. 2011; Da Cunha et al. 2012, 2013; Gajjar et al. 2013).

The MICs for micafungin and anidulafungin ranged between 0.015–>8 μg/ml (Da Cunha et al. 2012, 2013). However, from Table 8 it could be concluded that for *Curvularia* the lowest MIC of 0.06 μg/ml was obtained with amphotericin B (McGinnis and Pasarell 1998) and posaconazole (>0.007 μg/ml) (Espinel-Ingroff et al. 2002). Pfaller et al. (2003) compared the *in vitro* activity of posaconazole against *Curvularia* sp. by employing broth microdilution and E-test methods and the MICs obtained were 0.12–0.25 and 0.06–012 μg/ml, respectively. In a study conducted by Ben-Ami et al. (2009), 8 *Curvularia* isolates, but none of the strains previously known as *Bipolaris* (4), had minimum effective concentrations of caspofungin of >1.0 μg/ ml. The effects of caspofungin on the accumulation of whole-cell 1,3-β-D-glucan and the glucan synthase enzymatic activity of *C. geniculata* and *C. lunata* is well described by Kahn et al. (2006). Only very little data about natamycin susceptibility of *Curvularia* isolates

Table 8. Antifungal susceptibilities of clinical *Curvularia* isolates (minimum inhibitory concentration range in µg/ml).

Isolate	Clinical source	Method	AMB	NAT	5FC	FLZ	ITZ	KTZ	MCZ	PSZ	VRZ	CSP	5FLU	BMS	RVZ	MFG	PN	AFG	SC	MK 099	LY 0366	ECZ	BCZ	SPZ	CLZ	TRB	References
Curvularia sp. strain SC8156	Clinical	Broth macrodilution	0.5			8	0.5							2													Fung-Tomc et al. 1998
Curvularia sp. strain SC2475	Clinical	Broth macrodilution	0.13			32	0.25							1													Fung-Tomc et al. 1998
Curvularia sp.	Phaeohypomycosis in a tertiary care cancer center	Broth microdilution	≤ 1.0							≤ 1.0	≤ 1.0	>1															Ben-Ami et al. 2009
C. aeria (23)	Varying clinical sources	Broth microdilution	0.06–2		>64	4–64	0.25 to >16			0.125–2	0.5–16	0.5–1				0.03–0.125		0.03–0.25									Da Cunha et al. 2013
C. borreriae (3)			0.06–0.25		>64	8–16	0.5 to >16			0.125–0.5	1–4	0.5–1				<0.015 – 0.06		<0.015 0.06									
C. cf. *clavata* (1)			0.25		>64	8	0.5			0.25	0.5	1				0.06		0.06									
C. geniculata/ *C. senegalensis* (14)			0.06–0.5		64 to >64	2–16	0.06–1			<0.03–0.5	0.125–4	0.5–2				<0.015–0.06		<0.015 8									
C. cf. *inaequalis* (5)			0.125–0.25		>64	2–4	0.125–2			<0.03–1	0.5–2	1				0.03–0.125		0.06–0.125									
C. intermedia (2)			0.5–4		>64	2–8	0.125–0.25			0.06–0.125	0.125–0.5	0.5				0.06		0.03									
C. lunata (10)			0.125 to >16		>64	2 to 64	0.125 to >16			<0.03 to 0.5	0.25–1	0.5 to >8				0.015 to >8		<0.015 to >8									
C. protuberata (3)			0.25–0.5		>64	64 to >64	>16			0.5–1	8–16	0.5				<0.015–0.03		0.03–0.06									
C. pseudorobusta (1)			0.06		>64	64	>16			4	>16	0.5				0.03		0.06									
C. cf. *sorghina* (4)			0.125		>64	1–4	<0.03–0.5			0.06–0.025	0.25–2	0.5–1				0.015–0.06		0.03–0.5									
C. verruculosa (6)			0.125–0.25		>64	4–32	0.5–1			0.06–1	0.5–2	0.5–1				0.015–0.125		0.03–0.125									
Curvularia sp. I (15)			0.25–1		>64	2– 32	0.25–1			0.06–1	0.5–2	0.5–1				<0.015–0.125		0.03–									
Curvularia sp. II (9)			0.06–0.25		>64	2–16	0.125–2			<0.03–0.5	0.5–1	0.5–1				<0.015–0.03		0.03–0.125									
Curvularia sp. III (3)			0.125–0.25		>64	2–16	0.125–1			<0.03–0.25	0.5–1	1				0.06–0.125		0.06									
C. lunata (4)	Varying clinical sources	Broth macrodilution															≤0. 0.78										Del Poeta et al. 1997
C. lunata (4)	Keratitis	Broth microdilution	0.25*	2–32 *		≥128 *	≥128*																				Gajjar et al. 2013
C. brachyspora (4)	Keratitis	Broth microdilution	0.06–2		128–256	4–128	0.125–32	0.5–4	0.25–4																		Guarro et al. 1999
C. clavata (3)			0.25–0.5		256	16	0.125–0.5	0.5–1	1–2																		
C. geniculata (4)			0.06–0.25		128–256	4–32	0.125–0.5	0.5–2	0.5–4																		
C. lunata (3)			0.06–0.5		128–256	16–128	0.06–8	0.25–4	1–4																		
C. pallescens (4)			0.06–32		128–256	16–128	0.5–16	0.06–32	0.55–4																		
C. senegalensis (3)			0.25–2		128–256	16–128	0.25–1	1–2	1–2																		
C. verruculosa (4)			0.125–0.5		128–256	4–64	0.5–2	0.5–2	1–2																		

Table 8. contd....

Table 8. contd.

Isolate	Clinical source	Method	AMB	NAT	5FC	FLZ	ITZ	KTZ	MCZ	PSZ	VRZ	CSP	5FLU	BMS	RVZ	MFG	PN	AFG	SC	MK 099	LY 0366	ECZ	BCZ	SPZ	CLZ	TRB	References
Curvularia sp. (1)	Varying clinical sources	Broth microdilution	2				1				0.5																Espinel- Ingroff et al. 2008
Curvularia sp. (1)	Varying clinical sources	Broth microdilution								0.12–0.25																	Pfaller et al. 2003
		E- test								0.06–0.12																	
C. lunata (3)	Varying clinical sources	Agar dilution					≤0.03 to >64				0.12–0.5																Radford et al. 1997
Curvularia sp. (1)	Keratitis	-									0.06–0.25																Vemulakonda et al. 2008
Curvularia spp. (14)	Varying clinical sources	Broth microdilution	≤ 0.124–4	1–4					<0.25 –2																		Wilhelmus and Jones 2001
Curvularia spp. (12)					≤0.25–32																	0.12					
Curvularia spp. (11)																									<0.25 – 8		
Curvularia spp. (9)								<0.25 –4																			
Curvularia spp. (7)							≤ 0.25–1																				
Curvularia spp. (6)																							≤ 0.25–1				
Curvularia spp. (3)					>16																			≤ 0.06–0.5			
C. geniculata (1)	Varying clinical sources	Broth microdilution										16															Kahn et al. 2006
C. lunata (2)												16															
Curvularia spp. (2)												16															
C. lunata (2)	Varying clinical sources	Broth microdilution	0.5–1				0.03			0.25	>8	>8															Lass-Flörl et al. 2008
Curvularia spp. (1)	Keratitis	Broth microdilution		1–2																							Pradhan et al. 2011
Curvularia spp. (1)				2																							
C. lunata	Endophthalmitis	Broth microdilution	0.12			4	0.03	0.12			0.12		>128														Alex et al. 2013
C. lunata	Biopsy specimen of invasive phaeohypomycoses	Broth microdilution	0.25				0.5	2.0		0.125	2.0	<0.125			0.5											0.015	Safdar 2003

C. clavata (1)	Phaeohypomycoses	Broth microdilution	0.5				0.25				0.512															2.05	Biancalana et al. 2011
C. senegalensis (1)			2.0				0.5				0.256															4.10	
C. geniculata (1)			0.25				0.125				0.512															4.10	
C. lunata (4)			2.0–8.0				0.25–8				0.014–0.256															0.008 – 2.05	
Curvulatia sp. (1)	Endophthalmitis	-									0.06–0.25																Breit et al. 2005
C. lunata (3)	Ocular samples	Broth microdilution	0.05	2.0		4.0																					Xu et al. 2010
C. inaequalis	Human / animal origin	Broth macrodilution	0.125				0.5				0.25																McGinnis and Pasarell 1998
C. lunata			0.125–16				0.03–32				0.06–1																
C. senegalensis			0.03–0.25				0.06–2				0.06–0.25																
C. verruculosa			0.03–0.5				0.05–1				0.125																
C. lunata (1)	Peritonial lavage fluid	-	8								1.0																Varughese et al. 2001
C. lunata (1)	Darkly pigmented material from peritonial effluent and the wall of CAPD catheter	E test	0.064			>256	0.064				0.064																Unal et al. 2011
C. lunata (1)	Infected porcine heterograft valve	-	1.56				1.56	0.78	6.25																		Bryan et al 1993
Curvularia sp. (1)	Pritonitis	Broth microdilution	0.08			10	0.035						>100														Ujhelyi et al. 1990
Curvularia sp.[a] (2)	Clinical	E-Test	0.064			>256				0.50	0.123	>32															Direkel et al. 2012
C. australiensis[b] (1)	Biopsy specimen of hilarmass	Liquid double dilution	0.25				<0.25																				Flanagan and Bryceson 1997
C. australiensis[b] (2)	Varying clinical sources	Broth microdilution	0.06–0.125		>64	8–16	0.25–0.5			0.06	0.05–1	1				<0.015 –0.06		<0.015 0.06									Da Cunha et al. 2012
C. hawaiiensis[c] (14)			0.125–0.25		>64	2–32	<0.03–0.5			<0.03–0.5	0.25–2	0.5 –1				<0.015–0.06		<0.015 >8									
C. spicifera[d] (52)			<0.03–2		>64	4 to >64	<0.03–4			<0.03–1	0.25–4	0.2 5–2				<0.015–0.125		<0.015 >8									
C. spicifera[d] (1)	Varying clinical sources	Gordon *et al* Method for azole antifungal susceptibility testing				10			0.039																		Gordon et al. 1988
Curvularia sp.[a] (1)									0.156																		
C. hawaiiensis[c] (3)	Varying clinical sources	Broth Microdilution	0.12–0.25				0.12–0.25				0.5																Espinel-Ingroff 2001a
C. spicifera[d] (3)			0.25–2				0.5 to >8				2																

Table 8. contd....

Table 8. contd.

Isolate	Clinical source	Method	AMB	NAT	5FC	FLZ	ITZ	KTZ	MCZ	PSZ	VRZ	CSP	5FLU	BMS	RVZ	MFG	PN	AFG	SC	MK 099	LY 0366	ECZ	BCZ	SPZ	CLZ	TRB	References
Curvularia sp.[a] (6)	Clinical	Broth Microdilution with standard RPMI	0.06–4				0.03 to >8			>0.007 –8	0.12 to >8				0.2 to >8												Espinel-Ingroff 2002
Curvularia sp.[a] (6)	Clinical	Broth Microdilution with M3 medium	0.12 to >8				0.03 to >8			0.01 to >8	0.5 to >8				0.5 to >8												Espinel-Ingroff et al. 2002
Curvularia sp.[a] (6)	Clinical	Broth Microdilution																	0.0	1–2	1–4						Espinel- Ingroff 1998a
Curvularia sp.[a] (6)	Clinical	Broth Microdilution	0.5–1.0				<0.03–0.12				0.12–1																Espinel- Ingroff 1998b
C. spicifera[d] (4)	Clinical	Broth Microdilution	1–2								2-16							0.5–4									Odabasi et al. 2004
Curvularia sp. (1)	Clinical	Broth Microdilution	0.25				0.25			0.05	0.5				1												Pfaller et al. 2002
Curvularia sp.[a] (10)	Clinical	Broth Microdilution	0.25				0.063–0.25			0.063–0.125																	Sabatelli et al. 2006
Curvularia sp.[a] (1)	Keratomycosis	Broth Microdilution		2																							Pradhan et al. 2011
C.australiensis[b] (2)	Clinical	Broth Microdilution	1–2				0.5–2			0.125–0.5	>8	>8															Lass-Flörl et al. 2008
Curvularia sp.[a] *(1–3)*	Clinical	E test	0.06–4				0.01–0.12																				Espinel-Ingroff 2001b
		Broth Microdilution	0.12–4				0.06–2																				
Curvularia sp.[a] (2)	Phaeohypomycoses	Broth Microdilution	4.0				2.0				0.256															1.02	Biancalana et al. 2011
C. hawaiiensis[c]	Serial sputum and BAL specimens	Broth Microdilution	<0.03–0.125				<0.03–0.06				0.5–1	0.5				0.125		0.125									Chowdhary et al. 2011
C. hawaiiensis[c]	Granulomatus encephalitis	-	0.25						0.064				>100														Morton et al. 1986.
C.australiensis[b] (3)	Human / animal origin	Broth Macrodilution	0.125–0.25				0.03–0.125				0.125–0.25																McGinnis and Pasarell 1998
C. hawaiiensis[c] (17)			0.125–1				0.03–0.125				0.06–0.25																
C. spicifera[d] (24)			0.006–2				0.03–1				0.06–0.5																
C. spicifera[d] (1)	Blood/ peritonial fluid	-	0.3						>19.2				>323														O' Sullivan et al. 1981

AMB – Amphoterecin B; NAT – Natamycin; 5FC – 5 flucytosine; FLZ – Fulconazole; ITZ- Itraconazole; KTZ- Ketoconazole; MCZ – Miconazole; PSZ- Posoconazole; VRZ- Voriconazole; CSP- Caspofingin; 5FLU – 5 flurouracil; BMS – BMS-207147; RVZ – Ravuconazole; MFG –Micafungin; PNEU- Pneumocandin ; AFG – Anidulafungin; ECZ- Econazole; BCZ – Butaconazole; SPZ – Saperconazole; CLZ –Clotrimazole; TRB – Terbinafine

[a] reported as *Bipolaris* spp., [b] reported as *B. australiensis*, [c] reported as *B. hawaiiensis*, [d] reported as *B. soicifera*

* Geometric mean of MIC

is available (Wilhelmus and Jones 2001; Xu et al. 2010; Pradhan et al. 2011; Gajjar et al. 2013). Odabasi et al. (2004) reported that anidulafungin showed promising activity (0.5–4 µg/mL) against *C. spicifera*.

Biancalana et al. (2011) stated that a drastic reduction of amphotericin B MICs by the addition of terbinafine can be a sign that a combination of terbinafine with amphotericin B could be useful in the treatment of invasive infections caused by dematiaceous molds. But the antifungal susceptibility data of Gajjar et al. (2013) showed that *Curvularia* spp. isolated from keratitis were highly resistant to amphotericin B, natamycin, fluconazole and itraconazole, however, this could be due to the different methodology applied. In contrast to this study, Guarro et al. (1999) reported that amphotericin B, itraconazole, miconazole and ketoconazole are highly effective against almost all the species of *Curvularia* tested. Da Cunha et al. (2013) reported lower *in vitro* activities of itraconazole and voriconazole, which are the most commonly used drugs in *Curvularia* infections.

While analyzing the treatment aspects in the reported case studies, it was noted that ketoconazole (Koshy and Ebenezer 2002) and natamycin (Patil et al. 2011) were the drugs of choice in the treatment of *Curvularia* eye infections; itraconazole, amphotericin B or clotrimazole for cutaneous *Curvularia* infection (Sharma et al. 2014); itraconazole (Qureshi et al. 2006; Moody et al. 2012; Vasquez-del-Mercado et al. 2013; Chowdhary et al. 2011), amphotericin B (Morton et al. 1986; Ujhelyi 1990; Pimentel et al. 2005; Unal et al. 2011), caspofungin (Varughese et al. 2011), flucytosine (Pimentel et al. 2005) and itraconazole (Safdar 2003; Unal et al. 2011) for systemic infections with *Curvularia* spp. For ocular complications due to *Curvularia* spp., treatment with natamycin (Wilhelmus and Jones 2001; Pradhan et al. 2011) and voriconazole (Vemulakonda et al. 2008; Al-Badriyeh et al. 2010) have been recommended.

Conclusions and Future Perspectives

Curvularia strains are able to cause diseases varying from superficial infections to deep and disseminated diseases. There is a clearly emerging pattern of *Curvularia* infections: most of the cases are reported from patients with mycotic keratitis followed by respiratory tract infections. Despite surgical and antifungal therapies, disseminated *Curvularia* infections have a very high mortality rate.

C. lunata is the most frequently occurring etiologic agent within the genus causing eye, skin and nail infections. *C. spicifera* is the most isolated agent form respiratory tract infections, infections complicating peritoneal dialysis, peritonitis as well as deep and disseminated infections. Possible sources of infection include plant and animal materials, water, air, catheters and dialysis bags.

The exact potential virulence factors of *Curvularia* strains involved in opportunistic infections are not clearly understood. Sequencing the complete genome of *C. lunata* and *C. spicifera*—the most frequently occurring opportunistic human pathogens within the genus—would reveal the opportunity to identify further possible virulence factors by comparing their genome sequence with that of related species which are not known from clinical cases.

Antifungal susceptibility data available in the literature provide useful information for the planning of the therapy in cases of suspected or confirmed *Curvularia* infections. Currently, the antifungal susceptibility data for *Curvularia* isolates are sparse; however, the obvious differences in the antifungal susceptibility patterns of *Curvularia* spp. highlight the importance of determining antifungal susceptibilities isolate by isolate.

The identification of *Curvularia* species is problematic if only the morphological characters are considered; therefore, the identity of clinical isolates is suggested to be confirmed by the sequence analysis of ITS, *gpd*, LSU and EF1-α datasets. Molecular methods may also reveal diagnostic tools for rapid species identification in the clinical practice.

Acknowledgements

The authors of this study were supported by the Indian National Science Academy and the Hungarian Academy of Sciences within the framework of the Indian National Science Academy—Hungarian Academy of Sciences Indo-Hungarian bilateral exchange programme (SNK-49/2013) and by a grant of the Hungarian Scientific Research Fund (NN106394). The contribution of K. Krizsán was realized in the frames of TÁMOP 4.2.4. A/2-11-1-2012-0001 "National Excellence Program—Elaborating and operating an inland student and researcher personal support system". The project was subsidized by the European Union and co-financed by the European Social Fund. Csaba Vágvölgyi thanks the visiting professor program, Deanship of Scientific Research at King Saud University, Riyadh.

References

Adam, R.D., M.L. Paquin, E.A. Petersen, M.A. Saubolle, M.G. Rinaldi, J.G. Corcoran, J.N. Galgiani and R.E. Sobonya. 1986. Phaeohyphomycosis caused by the fungal genera *Bipolaris* and *Exserohilum*. A report of 9 cases and review of the literature. Medicine (Baltimore) 65: 203–217.

Agrawal, A. and S.M. Singh. 1995. Two cases of cutaneous phaeohyphomycosis caused by *Curvularia pallescens*. Mycoses 38: 301–303.

Agrawal, P.K., B. Lal, P.K. Shukla, Z.A. Khan and O.P. Srivastava. 1982. Clinical and experimental keratitis due to *Curvularia lunata* (Wakker) Boedijn var. *aeria* (Batista, Lima and Vasconcelos) Ellis. Sabouraudia. 20: 225–232.

Alex, D., D. Li, R. Calderone and S.M. Peters. 2013. Identification of *Curvularia lunata* by polymerase chain reaction in a case of fungal endophthalmitis. Med Mycol Case Rep. 2: 137–140.

Al-Badriyeh, D., C.F. Neoh, K. Stewart and D.C.M. Kong. 2010. Clinical utility of voriconazole eye drops in ophthalmic fungal keratitis. Clin Ophthalmol. 4: 391–405.

Alcorn, J.L. 1983. On the genera *Cochliobolus* and *Pseudocochliobolus*. Mycotaxon. 16: 353–379.

Alexander, B.D. and M.A. Pfaller. 2006. Contemporary tools for the diagnosis and management of invasive mycoses. Clin Infect. Dis. 43: S15–S27.

Alfonso, E.C., D. Miller, J. Cantu-Dibildox, T.P. O'brien and O.D. Schein. 2006. Fungal keratitis associated with non-therapeutic soft contact lenses. Am J Ophthalmol. 142(1): 154–5.

Alkatan, H., S. Athmanathan and C.C. Canites. 2012. Incidence and microbiological profile of mycotic keratitis in a tertiary care eye hospital: a retrospective analysis. Saudi J Ophthalmol. 26: 217–221.

Alture-Werber, E. and S.C. Edberg. 1985. An animal model of *Curvularia geniculata* and its relationship with human disease. Mycopathologia 89: 69–73.

Alvarez, V.C., L. Guelfand, J.C. Pidone, R. Soloaga, P. Ontivero, A. Margari and G. López Daneri. 2011. Allergic fungal rhinosinusitis caused by *Curvularia* sp. Rev Iberoam Micol. 28: 104–106 [Article in Spanish].

Amatya, R., S. Shrestha, B. Khanal, R. Gurung, N. Poudyal, S.K. Bhattacharya and B.P. Badu. 2012. Etiological agents of corneal ulcer: five years prospective study in eastern Nepal. Nepal Med Coll J. 14: 219–222.

Ambrosetti, D., V. Hofman, L. Castillo, M. Gari-Toussaint and P. Hofman. 2006. An expansive paranasal sinus tumour-like lesion caused by *Bipolaris spicifera* in an immunocompetent patient. Histopathology 49: 660–662.

Anandi, V., N.B. Suryawanshi, G. Koshi, A.A. Padhye and L. Ajello. 1988. Corneal ulcer caused by *Bipolaris hawaiiensis*. J Med Vet Mycol. 26: 301–306.

Anderson, B., S.S. Roberts, C. Gonzalez and E.W. Chick. 1959. Mycotic ulcerative keratitis. AMA Arch Ophthalmol. 62: 169–197.

Antoine, G.A. and M.H. Raterink. 1989. *Bipolaris*: a serious new fungal pathogen of the paranasal sinus. Otolaryngol. Head Neck Surg. 100: 158–162.

Anusuya, Devi D., R. Ambica and T. Nagarathnamma. 2013. The epidemiological features and laboratory diagnosis of keratomycosis. Int J Biol Med Res. 4: 2879–2883.

Aribandi, M. and C. Bazan. 2007. CT and MRI features in *Bipolaris* fungal sinusitis. Australas Radiol. 51: 127–132.

Ashraf, M.J., N. Azarpira, M. Pourjafar and B. Khademi. 2005. Allergic fungal sinusitis presenting as a paranasal sinus tumor. Iran J Allergy Asthma Immunol. 4: 193–195.

Balajee, S.A., A.M. Borman, M.E. Brandt, J. Cano, M. Cuenca-Estrella, E. Dannaoui, J. Guarro, G. Haase, C.C. Kibbler, W. Meyer, K. O'Donnell, C.A. Petti, J.L. Rodriguez-Tudela, D. Sutton, A. Velegraki and B.L. Wickes. 2009. Sequence-based identification of *Aspergillus*, *Fusarium*, and Mucorales species in the clinical mycology laboratory: where are we and where should we go from here? J Clin Microbiol. 47: 877–884.

Bandyopadhyay, S., D. Das, K.K. Mondal, A.K. Ghanta, S.K. Purkait and R. Bhaskar. 2012. Epidemiology and laboratory diagnosis of fungal corneal ulcer in the Sundarban Region of West Bengal, eastern India. Nepal J Ophthalmol. 4: 29–36.

Barde, A.K. and S.M. Singh. 1983. A case of onychomycosis caused by *Curvularia lunata* (Wakker) Boedijn. Mykosen. 26: 311–316.

Bartynski, J.M., T.V. McCaffrey and E. Frigas. 1990. Allergic fungal sinusitis secondary to dermatiaceous fungi—*Curvularia lunata* and *Alternaria*. Otolaryngol. Head Neck Surg. 103: 32–39.

Bashir, G., W. Hussain and A. Rizvi. 2009. *Bipolaris hawaiiensis* keratomycosis and endophthalmitis. Mycopathologia 167: 51–53.

Bava, A.J., A. Fayad, C. Céspedes and M. Sandoval. 2003. Fungal peritonitis caused by *Bipolaris spicifera*. Med Mycol Dec. 41(6): 529–31.

Baylet, J., R. Camain and G. Segretain. 1959. Identification des agents des maduromycoses du Senegal et de la Mauritaine. Description d'une espece nouvelle. Bull Soc Path Exot. 52: 448–477.

Beale, K.M. and D. Pinson. 1990. Phaeohyphomycosis caused by two different species of *Curvularia* in two animals from the same household. J Am Anim Hosp Assoc. 26: 67–70.

Begerow, D., H. Nilsson, M. Unterseher and W. Maier. 2010. Current state and perspectives of fungal DNA barcoding and rapid identification procedures. Appl Microbiol Biotechnol. 87: 99–108.

Ben-Ami, R., R.E. Lewis, I.I. Raad and D.P. Kontoyiannis. 2009. Phaeohyphomycosis in a tertiary care cancer center. Clin Infect Dis. 48: 1033–1041.

Ben-Shlomo, G., C. Plummer, K. Barrie and D. Brooks. 2010. *Curvularia* keratomycosis in a dog. Vet Ophthalmol. 13: 126–130.

Bent, J.P. and F.A. Kuhn. 1994. Diagnosis of allergic fungal sinusitis. Otolaryngol. Head Neck Surg. 111: 580–588.

Berbee, M.L., M. Pirseyedi and S. Hubbard. 1999. *Cochliobolus* phylogenetics and the origin of known, highly virulent pathogens inferred from ITS and glyceraldehydes-3-phosphate dehydrogenase gene sequences. Mycologia 91: 964–977.

Berbel, R.F., A.M. Casella, D. de Freitas and A.L. Höfling-Lima. 2011. *Curvularia lunata* endophthalmitis. J Ocul Pharmacol Ther. 27: 535–537.

Berg, D., J.A. Garcia, W.A. Schell, J.R. Perfect and J.C. Murray. 1995. Cutaneous infection caused by *Curvularia pallescens*: a case report and review of the spectrum of disease. J Am Acad Dermatol. 32: 375–378.

Berger, S.T., D.A. Katsev, B.J. Mondino and T.H. Pettit. 1991. Macroscopic pigmentation in a dematiaceous fungal keratitis. Cornea 10: 272–276.

Berry, A.J., T.M. Kerkering, A.M. Giordano and J. Chiancone. 1984. Phaeohyphomycotic sinusitis. Pediatr Infect Dis. 3: 150–152.
Bharathi, M.J., R. Ramakrishnan, S. Vasu, Meenakshi and R. Palaniappan. 2002. Aetiological diagnosis of microbial keratitis in South India—a study of 1618 cases. Indian J Med Microbiol. 20: 19–24.
Bharathi, M.J., R. Ramakrishnan, S. Vasu, R. Meenakshi and R. Palaniappan. 2003. Epidemiological characteristics and laboratory diagnosis of fungal keratitis. A three-year study. Indian J Ophthalmol. 51: 315–321.
Bharathi, M.J., R. Ramakrishnan, R. Meenakshi, S. Mittal, C. Shivakumar and M. Srinivasan. 2006. Microbiological diagnosis of infective keratitis: comparative evaluation of direct microscopy and culture results. Br J Ophthalmol. 90: 1271–1276.
Bharathi, M.J., R. Ramakrishnan, R. Meenakshi, C. Shivakumar and D.L. Raj. 2009. Analysis of the risk factors predisposing to fungal, bacterial & *Acanthamoeba* keratitis in south India. Indian J Med Res. 130: 749–57.
Biancalana, F.S.C., L. Lyra and A.Z. Schreiber. 2011. *In vitro* evaluation of the type of interaction obtained by the combination of terbinafine and itraconazole, voriconazole, or amphotericin B against dematiaceous molds. Antimicrob Agents Chemother. 55: 4485– 4487.
Biggs, P.J., R.L. Allen, J.M. Powers and H.P. Holley. 1986. Phaeohyphomycosis complicating compound skull fracture. Surg Neurol. 25: 393–396.
Bilu, D., S. Movahedi-Lankarani, R.A. Kazin, C. Shields and M. Moresi. 2004. Cutaneous *Bipolaris* infection in a neutropenic patient with acute lymphoblastic leukemia. J Cutan Med Surg. 8: 446–449.
Bonduel, M., P. Santos, C.F. Turienzo, G. Chantada and H. Paganini. 2001. Atypical skin lesions caused by *Curvularia* sp. and *Pseudallescheria boydii* in two patients after allogeneic bone marrow transplantation. Bone Marrow Transplant. 27: 1311–1313.
Boonpasart, S., N. Kasetsuwan, V. Puangsricharern, L. Pariyakanok and T. Jittpoonkusol. 2002. Infectious keratitis at King Chulalongkorn Memorial Hospital: a 12-year retrospective study of 391 cases. J Med Assoc Thai. 85: S217–S230.
Brackett, R.W., A.N. Shenouda, S.S. Hawkins and W.B. Brock. 1988. *Curvularia* infection complicating peritoneal dialysis. South Med J. 81: 943–944.
Breit, S.M., S.M. Hariprasad, W.F. Mieler, G.K. Shah, M.D. Mills and M.G. Grand. 2005. Management of endogenous fungal endophthalmitis with voriconazole and caspofungin. Am J Ophthalmol. 139: 135–40.
Bridges, C.H. 1957. Maduromycotic mycetomas in animals; *Curvularia geniculata* as an etiologic agent. Am J Pathol. 33: 411–427.
Brummund, W., V.P. Kurup, G.J. Harris, J.A. Duncavage and J.A. Arkins. 1986. Allergic sino-orbital mycosis. A clinical and immunologic study. JAMA 256: 3249–3253.
Bryan, C.S., C.W. Smith, D.E. Berg and R.B. Karp. 1993. *Curvularia lunata* endocarditis treated with terbinafine: case report. Clin Infect Dis. 16: 30–32.
Bryan, M.G., D.M. Elston, C. Hivnor and B.A. Honl. 2000. Phaeohyphomycosis in a premature infant. Cutis 65: 137–140.
Buzina, W., H. Braun, K. Schimpl and H. Stammberger. 2003. *Bipolaris spicifera* causes fungus balls of the sinuses and triggers polypoid chronic rhinosinusitis in an immunocompetent patient. J Clin Microbiol. 41: 4885–4887.
Canon, H.L., S.C. Buckingham, R.J. Wyatt and D.P. Jones. 2001. Fungal peritonitis caused by *Curvularia* species in a child undergoing peritoneal dialysis. Pediatr Nephrol. 16: 35–37.
Carter, E. and C. Boudreaux. 2004. Fatal cerebral phaeohyphomycosis due to *Curvularia lunata* in an immunocompetent patient. J Clin Microbiol. 42: 5419–5423.
Castelnuovo, P., F. De Bernardi, C. Cavanna, F. Pagella, P. Bossolesi, P. Marone and C. Farina. 2004. Invasive fungal sinusitis due to *Bipolaris hawaiiensis*. Mycoses 47: 76–81.
Cavanna, C., E. Seminari, A. Pusateri, F. Mangione, F. Lallitto, M.C. Esposto and F. Pagella. 2014. Allergic fungal rhinosinusitis due to *Curvularia lunata*. New Microbiol. 37: 241–245.
Chalet, M., D.H. Howard, M.R. McGinnis and I. Zapatero. 1986. Isolation of *Bipolaris australiensis* from a lesion of viral vesicular dermatitis on the scalp. J Med Vet Mycol. 24: 461–465.
Chander, J. and A. Sharma. 1994. Prevalence of fungal corneal ulcers in northern India. Infection 22: 207–209.
Chander, J., N. Singla, N. Agnihotri, S.K. Arya and A. Deep. 2008. Keratomycosis in and around Chandigarh: a five-year study from a north Indian tertiary care hospital. Indian J Pathol Microbiol. 51: 304–306.

Charlton, J.R., J.P. Barcia and V.F. Norwood. 2010. Black specks in dialysis fluid: an unusual case of peritonitis in a pediatric patient on peritoneal dialysis. Dial Transplant. 39: 445–448.
Chowdhary, A. and K. Singh. 2005. Spectrum of fungal keratitis in North India. Cornea 24: 8–15.
Chowdhary, A., H.S. Randhawa, V. Singh, Z.U. Khan, S. Ahmad, S. Kathuria, P. Roy, G. Khanna and J. Chandra. 2011. *Bipolaris hawaiiensis* as etiologic agent of allergic bronchopulmonary mycosis: first case in a paediatric patient. Med Mycol. 49: 760–765.
Clark, F.D., L.P. Jones and B. Panigrahy. 1986. Mycetoma in a grand Eclectus (*Eclectus roratus roratus*) parrot. Avian Dis. 30: 441–444.
Costa, A.R., E. Porto, A.H. Tabuti, S. Lacaz Cda, N.Y. Sakai-Valente, W.M. Maranhão and M.C. Rodrigues. 1991. Subcutaneous phaeohyphomycosis caused by *Bipolaris hawaiiensis*. A case report. Rev Inst Med. Trop. Sao Paulo. 33: 74–79.
Cruz, R., E. Barthel and J. Espinoza. 2013. Allergic rhinosinusitis by *Curvularia inaequalis* (Shear) Boedijn. Rev Chilena Infectol. 30: 319–322 [Article in Spanish].
Da Cunha, K.C., D.A. Sutton, A.W. Fothergill, J. Cano, J. Gené, H. Madrid, S. De Hoog, P.W. Crous and J. Guarro. 2012. Diversity of *Bipolaris* species in clinical samples in the USA and their antifungal susceptibility profiles. J Clin Microbiol. 50: 4061–4066.
Da Cunha, K.C., D.A. Sutton, A.W. Fothergill, J. Gené, J. Cano, H. Madrid, S. De Hoog, P.W. Crous and J. Guarro. 2013. *In vitro* antifungal susceptibility and molecular identity of 99 clinical isolates of the opportunistic fungal genus *Curvularia*. Diagn Microbiol Infect Dis. 76: 168–174.
de la Monte, S.M. and G.M. Hutchins. 1985. Disseminated *Curvularia* infection. Arch Pathol Lab Med. 109: 872–874.
Del Poeta, M., W.A. Schell and J.R. Perfect. 1997. *In vitro* antifungal activity of pneumocandin L-743, 872 against a variety of clinically important molds. Antimicrob Agents Chemother. 4: 1835–1836.
Deorukhkar, S., R. Katiyar and S. Saini. 2012. Epidemiological features and laboratory results of bacterial and fungal keratitis: a five-year study at a rural tertiary-care hospital in western Maharashtra, India. Singapore Med J. 53: 264–267.
Deshpande, S.D. and G.V. Koppikar. 1999. A study of mycotic keratitis in Mumbai. Indian J Pathol Microbiol. 42: 81–87.
DeVault, G.A., S.T. Brown, J.W. King, M. Fowler and A. Oberle. 1985. Tenckhoff catheter obstruction resulting from invasion by *Curvularia lunata* in the absence of peritonitis. Am J Kidney Dis. 6: 124–127.
Direkel, S., F. Otag, G. Aslan, M. Ulger and G. Emekdas. 2012. Identification of filamentous fungi isolated from clinical samples by two different methods and their susceptibility results. Mikrobiyol Bul. 46: 65–78.
Diskin, C.J., T.J. Stokes, L.M. Dansby, L. Radcliff and T.B. Carter. 2008. Case report and review: is the tendency for *Curvularia* tubular obstruction significant in pathogenesis? Perit Dial Int. 28: 678–679.
Dolan, C.T., L.A. Weed and D.E. Dines. 1970. Bronchopulmonary helminthosporiosis. Am J Clin Pathol. 53: 235–242.
del Palacio, A., M. Pérez-Simón, A. Arribi, A. Valle, S. Perea and A. Rodriguez-Noriega. 1997. *Bipolaris australiensis* in a Spanish patient with allergic chronic sinusitis. Rev Iberoam Micol. 14: 191–193 [Article in Spanish].
Dela Paz, M.A.G., P.H. Goodwin, A.K. Raymundo, E.Y. Ardales and C.M. Vera Cruz. 2006. Phylogenetic analysis based on ITS sequences and conditions affecting the type of conidial germination of *Bipolaris oryzae*. Plant Pathol. 55: 756–765.
Dorey, S.E., W.H. Ayliffe, C. Edrich, D. Barrie and P. Fison. 1997. Fungal keratitis caused by *Curvularia lunata*, with successful medical treatment. Eye (Lond) 11: 754–755.
Dubois, M., P. Brisou, B. Fournier, B. Guennoc and F. Carsuzaa. 2008. Cutaneous *Bipolaris spicifera* infection in a patient on systemic corticosteroids. Ann Dermatol Venereol. 135: 507–508 [Article in French].
Dunlop, A.A., E.D. Wright, S.A. Howlader, I. Nazrul, R. Husain, K. McClellan and F.A. Billson. 1994. Suppurative corneal ulceration in Bangladesh. A study of 142 cases examining the microbiological diagnosis, clinical and epidemiological features of bacterial and fungal keratitis. Aust N Z J Ophthalmol. 22: 105–110.
Durkin, S.R., T. Henderson, R. Raju and D. Ellis. 2008. Successful treatment of phaeohyphomycotic keratitis caused by *Bipolaris australiensis*. Clin Experiment. Ophthalmol. 36: 697–699.
Duvic, M., L. Lowe, A. Rios, E. MacDonald and P. Vance. 1987. Superficial phaeohyphomycosis of the scrotum in a patient with the acquired immunodeficiency syndrome. Arch Dermatol. 123: 1597–1599.

Dyer, Z.A., R.S. Wright, I.H. Rong and A. Jacobs. 2008. Back pain associated with endobronchial mucus impaction due to *Bipolaris australiensis* colonization representing atypical Allergic Bronchopulmonary Mycosis. Med Mycol. 46: 589–594.

Ebright, J.R., P.H. Chandrasekar, S. Marks, M.R. Fairfax, A. Aneziokoro and M.R. McGinnis. 1999. Invasive sinusitis and cerebritis due to *Curvularia clavata* in an immunocompetent adult. Clin Infect Dis. 28: 687–689.

Eghtedari, M. and K. Pakshir. 2006. Asyptomatic fungal cyst of conjunctiva caused by *Bipolaris spicifera.* Iran J Med Sci. 31: 56–58.

Ehlers, J.P., S.H. Chavala, J.A. Woodward and E.A. Postel. 2011. Delayed recalcitrant fungal endophthalmitis secondary to *Curvularia.* Can J Ophthalmol. 46: 199–200.

Elad, D., U. Orgad, B. Yakobson, S. Perl, P. Golomb, R. Trainin, I. Tsur, S. Shenkler and A. Bor. 1991. Eumycetoma caused by *Curvularia lunata* in a dog. Mycopathologia. 116: 113–118.

El-Zaatari, M.M., L. Pasarell and M.R. McGinnis. 1996. Recurrent allergic fungal sinusitis sequentially caused by *Exserohilum* and *Bipolaris*. Ann Saudi Med. 16: 564–567.

Espinel-Ingroff, A. 1998a. Comparison of *in vitro* activities of the new triazole SCH56592 and the echinocandins MK-0991 (L-743,872) and LY303366 against opportunistic filamentous and dimorphic fungi and yeasts. J Clin Microbiol. 36: 2950–2956.

Espinel-Ingroff, A. 1998b. *In vitro* activity of the new triazole voriconazole (UK-109, 496) against opportunistic filamentous and dimorphic fungi and common and emerging yeast pathogens. J Clin Microbiol. 36: 198–202.

Espinel-Ingroff, A. 2001a. *In vitro* fungicidal activities of voriconazole, itraconazole and amphotericin B against opportunistic moniliaceous and dematiaceous fungi. J Clin Microbiol. 39: 954–958.

Espinel-Ingroff, A. 2001b. Comparison of the E-test with the NCCLS M38-P method for antifungal susceptibility testing of common and emerging pathogenic filamentous fungi. J Clin Microbiol. 39: 1360–1367.

Espinel-Ingroff, A., V. Chatuevedi, A. Fothergill and M.G. Rinaldi. 2002. Optimal testing conditions for determining MICs and minimum fungicidal concentrations of new and established antifungal agents for uncommon molds: NCCLS collaborative study. J Clin Microbiol. 40: 3776–3781.

Espinel-Ingroff, A., E. Johnson, H. Hockey and P. Troke. 2008. Activities of voriconazole, itraconazole and amphotericin B *in vitro* against 590 moulds from 323 patients in the voriconazole Phase III clinical studies. J Antimicrob Chemother. 61: 616–620.

Estes, S.A., W.G. Merz and L.G. Maxwell. 1977. Primary cutaneous phaeohyphomycosis caused by *Drechslera spicifera.* Arch Dermatol. 113: 813–815.

Fan, Y.M., W.M. Huang, S.F. Li, G.F. Wu, W. Li and R.Y. Chen. 2009. Cutaneous phaeohyphomycosis of foot caused by *Curvularia clavata.* Mycoses 52: 544–546.

Fernandez, M., D.E. Noyola, S.N. Rossmann and M.S. Edwards. 1999. Cutaneous phaeohyphomycosis caused by *Curvularia lunata* and a review of *Curvularia* infections in pediatrics. Pediatr Infect Dis J. 18: 727–731.

Ferrer, C., F. Colom, S. Frasés, E. Mulet, J.L. Abad and J.L. Alió. 2001. Detection and identification of fungal pathogens by PCR and by ITS2 and 5.8S ribosomal DNA typing in ocular infections. J Clin Microbiol. 39: 2873–2879.

Filizzola, M.J., F. Martinez and S.J. Rauf. 2003. Phaeohyphomycosis of the central nervous system in immunocompetent hosts: report of a case and review of the literature. Int J Infect Dis. 7: 282–286.

Flanagan, K.L. and A.D. Bryceson. 1997. Disseminated infection due to *Bipolaris australiensis* in a young immunocompetent man: case report and review. Clin Infect Dis. 25: 311–313.

Frenkel, L., T.L. Kuhls, K. Nitta, M. Clancy, D.H. Howard, P. Ward and J.D. Cherry. 1987. Recurrent *Bipolaris* sinusitis following surgical and antifungal therapy. Pediatr Infect Dis J. 6: 1130–1132.

Friedman, A.D., J.M. Campos, L.B. Rorke, D.A. Bruce and A.M. Arbeter. 1981. Fatal recurrent *Curvularia* brain abscess. J Pediatr. 99: 413–415.

Fryen, A., P. Mayser, H. Glanz, R. Füssle, H. Breithaupt and G.S. de Hoog. 1999. Allergic fungal sinusitis caused by *Bipolaris (Drechslera) hawaiiensis.* Eur Arch Otorhinolaryngol. 256: 330–334.

Fu-Chin, H., H. Shen-Terng and T. Sung-Huei. 1998. Mycotic keratitis caused by *Bipolaris* species—report of two cases. Tzu Chi Med J. 10: 345–349.

Fung-Tomc, J.C., E. Huczko, B. Minassian and D.P. Bonner. 1998. *In vitro* activity of a new oral triazole, BMS-207147 (ER-30346). Antimicrob. Agents Chemother. 42: 313–318.

Fuste, F.J., L. Ajello, R. Threlkeld and J.E. Henry. 1973. *Drechslera hawaiiensis*: causative agent of a fatal fungal meningo-encephalitis. Sabouraudia. 11: 59–63.

Gadallah, M.F., R. White, M.A. el-Shahawy, F. Abreo, A. Oberle and J. Work. 1995. Peritoneal dialysis complicated by *Bipolaris hawaiiensis* peritonitis: successful therapy with catheter removal and oral itraconazole without the use of amphotericin-B. Am J Nephrol. 15: 348–352.

Gadgil, N., M. Kupferman, S. Smitherman, G.N. Fuller and G. Rao. 2013. *Curvularia* brain abscess. J Clin Neurosci. 20: 173–175.

Gajjar, D.U., A.K. Pal, B.K. Ghodadra and A.R. Vasavada. 2013. Microscopic evaluation, molecular identification, antifungal susceptibility, and clinical outcomes in *Fusarium*, *Aspergillus* and, dematiaceous keratitis. Biomed Res Int. 2013: 605308.

Garg, A., S. Sujatha, J. Garg, S.C. Parija and D.M. Thappa. 2008. Eumycetoma due to *Curvularia lunata*. Indian J Dermatol Venereol Leprol. 74: 515–516.

Georg, L.K. 1964. *Curvularia geniculata*, a cause of mycotic keratitis. J Med Assoc State Ala. 33: 234–236.

Giri, D.K., W.P. Sims, R. Sura, J.J. Cooper, B.K. Gavrilov and J. Mansell. 2011. Cerebral and renal phaeohyphomycosis in a dog infected with *Bipolaris* species. Vet Pathol. 48: 754–757.

Glancy, J.J., J.L. Elder and R. McAleer. 1981. Allergic bronchopulmonary fungal disease without clinical asthma. Thorax. 36: 345–349.

Goh, T.K., K.D. Hyde and D.K.L. Lee. 1998. Generic distinction in the *Helminthosporium*-complex based on restriction analysis of the nuclear ribosomal RNA gene. Fungal Divers. 1: 85–107.

Gongidi, P., D. Sarkar, E. Behling and J. Brody. 2013. Cerebral phaeohyphomycosis in a patient with neurosarcoidosis on chronic steroid therapy secondary to recreational marijuana usage. Case Rep Radiol. 2013: 191375.s

Gopalakrishnan, K., E. Daniel, R. Jacob, G. Ebenezer and M. Mathews. 2003. Bilateral *Bipolaris* keratomycosis in a borderline lepromatous patient. Int J Lepr Other Mycobact Dis. 71: 14–17.

Gopinathan, U., P. Garg, M. Fernandes, S. Sharma, S. Athmanathan and G.N. Rao. 2002. The epidemiological features and laboratory results of fungal keratitis: a 10-year review at a referral eye care center in South India. Cornea 21: 555–559.

Gopinathan, U., S. Sharma, P. Garg and G.N. Rao. 2009. Review of epidemiological features, microbiological diagnosis and treatment outcome of microbial keratitis: experience of over a decade. Indian J Ophthalmol. 57: 273–279.

Gordon, M.A., E.W. Lapa and P.G. Passero. 1988. Improved method for azole antifungal susceptibility testing. J Clin Microbiol. 26: 1874–1877.

Gourley, D.S., B.A. Whisman, N.L. Jorgensen, M.E. Martin and M.J. Reid. 1990. Allergic *Bipolaris* sinusitis: clinical and immunopathologic characteristics. J Allergy Clin Immunol. 85: 583–591.

Grieshop, T.J., D. Yarbrough and W.E. Farrar. 1993. Phaeohyphomycosis due to *Curvularia lunata* involving skin and subcutaneous tissue after an explosion at a chemical plant. Am J Med Sci. 305: 387–389.

Guarner, J., C. Del Rio, P. Williams and J.E. McGowan. 1989. Fungal peritonitis caused by *Curvularia lunata* in a patient undergoing peritoneal dialysis. Am J Med Sci. 298: 320–323.

Guarro, J., T. Akiti, R.A. Horta, L.A. Morizot Leite-Filho, J. Gené, S. Ferreira-Gomes, C. Aguilar and M. Ortoneda. 1999. Mycotic keratitis due to *Curvularia senegalensis* and *in vitro* antifungal susceptibilities of *Curvularia* spp. J Clin Microbiol. 37: 4170–4173.

Gugnani, H.C., C.N. Okeke and A. Sivanesan. 1990. *Curvularia clavata* as an etiologic agent of human skin infection. Lett Appl Microbiol. 10: 47–49.

Gunathilake, R., P. Perera and G. Sirimanna. 2013. *Curvularia lunata*: A rare cause of black-grain eumycetoma. J Mycol Med. 24: 158–160.

Gupta, M., N.L. Sharma, A.K. Kanga, V.K. Mahajan and G.R. Tegta. 2007. Onychomycosis: clinic-mycologic study of 130 patients from Himachal Pradesh, India. Indian J Dermatol Venereol Leprol. 73: 389–392.

Gupta, N., J.C. Samantaray, S. Duggal, V. Srivastava, C.S. Dhull and U. Chaudhary. 2010. *Acanthamoeba* keratitis with *Curvularia* co-infection. Indian J Med Microbiol. 28: 67–71.

Hagan, M., E. Wright, M. Newman, P. Dolin and G. Johnson. 1995. Causes of suppurative keratitis in Ghana. Br J Ophthalmol. 79: 1024–1028.

Halloran, T.J. 1983. Allergic bronchopulmonary helminthosporiosis. Am Rev Respir Dis. 128: 578.

Halwig, J.M., D.A. Brueske, P.A. Greenberger, R.B. Dreisin and H.M. Sommers. 1985. Allergic bronchopulmonary curvulariosis. Am Rev Respir Dis. 132: 186–188.

Hamilton, B.G., C.W. Humphreys, W.C. Conner and D.R. Hospenthal. 2006. Allergic bronchopulmonary disease secondary to *Bipolaris spicifera*. J Bronchol. 13: 77–79.

Harpster, W.H., C. Gonzalez and S.M. Opal. 1985. Pansinusitis caused by the fungus *Drechslera*. Otolaryngol. Head Neck Surg. 93: 683–685.

Harris, J.J. and T.F. Downham. 1978. Unusual fungal infections associated with immunologic hyporeactivity. Int J Dermatol. 17: 323–330.

He, D., J. Hao, B. Zhang, Y. Yang, W. Song, Y. Zhang, K. Yokoyama and L. Wang. 2011. Pathogenic spectrum of fungal keratitis and specific identification of *Fusarium solani*. Invest Ophthalmol Vis Sci. 52: 2804–2808.

Hemashettar, B.M., T.S. Veerappa, P.V. Verma, S. Hanchinamani, C.S. Patil and A. Thammayya. 1992. Mycotic keratitis caused by *Bipolaris spicifera*. Indian J Pathol Microbiol. 35: 274–277.

Hendrick, D.J., D.B. Ellithorpe, F. Lyon, P. Hattier and J.E. Salvaggio. 1982. Allergic bronchopulmonary helminthosporiosis. Am Rev Respir Dis. 126: 935–938.

Herraez, P., C. Rees and R. Dunstan. 2001. Invasive phaeohyphomycosis caused by *Curvularia* species in a dog. Vet Pathol. 38: 456–459.

Hiromoto, A., T. Nagano and C. Nishigori. 2008. Cutaneous infection caused by *Curvularia* species in an immunocompetent patient. Br J Dermatol. 158: 1374–1375.

Hosokawa, M., C. Tanaka and M. Tsuda. 2003. Conidium morphology of *Curvularia geniculata* and allied species. Mycoscience 44: 227–237.

Höfling-Lima, A.L., A. Forseto, J.P. Duprat, A. Andrade, L.B. Souza, P. Godoy and D. Freitas. 2005. Laboratory study of the mycotic infectious eye diseases and factors associated with keratitis. Arq Bras Oftalmol. 68: 21–27 [Article in Portuguese].

Isa-Isa, R., C. García, M. Isa and R. Arenas. 2012. Subcutaneous phaeohyphomycosis (mycotic cyst). Clin Dermatol. 30: 425–431.

Ismail, Y., R.H. Johnson, M.V. Wells, J. Pusavat, K. Douglas and E.L. Arsura. 1993. Invasive sinusitis with intracranial extension caused by *Curvularia lunata*. Arch Intern Med. 153: 1604–1606.

Iyer, S.A., S.S. Tuli and R.C. Wagoner. 2006. Fungal keratitis: emerging trends and treatment outcomes. Eye Contact Lens 32: 267–271.

James, W.D., T. Berger and D.M. Elston. 2006. Andrews' Diseases of the Skin: Clinical Dermatology. Saunders Elsevier, USA.

Janaki, C., G. Sentamilselvi, V.R. Janaki, S. Devesh and K. Ajithados. 1999. Case report. Eumycetoma due to *Curvularia lunata*. Mycoses 42: 345–346.

Jaramillo, S. and C. Varon. 2013. *Curvularia lunata* endophthalmitis after penetrating ocular trauma. Retin. Cases Brief Rep. 7: 315–318.

Jay, W.M., R.W. Bradsher, B. LeMay, N. Snyderman and E.J. Angtuaco. 1988. Ocular involvement in mycotic sinusitis caused by *Bipolaris*. Am J Ophthalmol. 105: 366–370.

Kahn, J.N., M. Hsu, F. Racine, R. Giacobbe and M. Motyl. 2006. Caspofungin susceptibility in *Aspergillus* and non-*Aspergillus* molds: Inhibition of glucan synthase and reduction of β-D-1,3 glucan levels in culture. Antimicrob Agents Chemother. 50: 2214–2216.

Kalawat, U., G.S. Reddy, Y. Sandeep, P.R. Naveen, Y. Manjusha, A. Chaudhury and V.S. Kumar. 2012. Successfully treated *Curvularia lunata* peritonitis in a peritoneal dialysis patient. Indian J Nephrol. 22: 318–319.

Kamalam, A., K. Ajithadass, G. Sentamilselvi and A.S. Thambiah. 1992. Paronychia and black discoloration of a thumb nail caused by *Curvularia lunata*. Mycopathologia 118: 83–84.

Kaplan, W., F.W. Chandler, L. Ajello, R. Gauthier, R. Higgins and P. Cayouette. 1975. Equine phaeohyphomycosis caused by *Drechslera spicifera*. Can Vet J. 16: 205–208.

Karim, M., H. Sheikh, M. Alam and Y. Sheikh. 1993. Disseminated *Bipolaris* infection in an asthmatic patient: case report. Clin Infect Dis. 17: 248–253.

Kaufman, S.M. 1971. *Curvularia* endocarditis following cardiac surgery. Am J Clin Pathol. 56: 466–470.

Kaushik, S., J. Ram, A. Chakrabarty, M.R. Dogra, G.S. Brar and A. Gupta. 2001. *Curvularia lunata* endophthalmitis with secondary keratitis. Am J Ophthalmol. 131: 140–142.

Khan, J.A., S.T. Hussain, S. Hasan, P. McEvoy and A. Sarwari. 2000. Disseminated *Bipolaris* infection in an immunocompetent host: an atypical presentation. J Pak Med Assoc. 50: 68–71.

Killingsworth, S.M. and S.J. Wetmore. 1990. *Curvularia/Drechslera* sinusitis. Laryngoscope 100: 932–937.

Kim, K.S., S.K. Chung, Y.W. Myung and N.H. Baek. 1999. A case of fungal keratitis caused by *Curvularia* species. J Korean Ophthalmol Soc. 40: 3224–3228.

Kindo, A.J., K. Suresh, Premamalini, S. Anita and J. Kalyani. 2009. Fungus as an etiology in keratitis- our experience in SRMC. Sri Ramachandra J Med. 2: 14–17.

Kiryu, H. and Y. Suenaga. 1985. A case of cutaneous *Curvularia* infection caused by *Curvularia trifolii*. pp. 63–66. *In*: Fukushiro, R., S. Kagawa and K. Kitamura (eds.). Typical Cutaneous Fungus Disease—

Rare Case (in Japanese). Proceedings of the 37th Meeting of Japanese Society of Medical Mycology, Yokohama.
Klapper, S.R., A.G. Lee, J.R. Patrinely, M. Stewart and E.L. Alford. 1997. Orbital involvement in allergic fungal sinusitis. Ophthalmology 104: 2094–2100.
Kobayashi, H., A. Sano, N. Aragane, M. Fukuoka, M. Tanaka, F. Kawaura, Y. Fukuno, E. Matsuishi and S. Hayashi. 2008. Disseminated infection by *Bipolaris spicifera* in an immunocompetent subject. Med Mycol. 46: 361–365.
Koshi, G., V. Anandi, M. Kurien, M.G. Kirubakaran, A.A. Padhye and L. Ajello. 1987. Nasal phaeohyphomycosis caused by *Bipolaris hawaiiensis*. J Med Vet Mycol. 25: 397–402.
Koshy, S. and D. Ebenezer. 2002. *Biporalis* keratomycosis in a leprosy patient: a case report. Lepr Rev. 73: 76–78.
Kotigadde, S., M. Ballal, Jyothirlatha, A. Kumar, R. Srinivasa and P.G. Shivananda. 1992. Mycotic keratitis: a study in coastal Karnataka. Indian J Ophthalmol. 40: 31–33.
Kumar, A., S. Pandya, G. Kavathia, S. Antala, M. Madan and T. Javdekar. 2011. Microbial keratitis in Gujarat, Western India: findings from 200 cases. Pan Afr Med. J. 10: 48.
Kumar, M. and P.K. Shukla. 2005. Use of PCR targeting of internal transcribed spacer regions and single-stranded conformation polymorphism analysis of sequence variation in different regions of rRNA genes in fungi for rapid diagnosis of mycotic keratitis. J Clin Microbiol. 43: 662–668.
Kumar, M., N.K. Mishra and P.K. Shukla. 2005. Sensitive and rapid polymerase chain reaction based diagnosis of mycotic keratitis through single stranded conformation polymorphism. Am J Ophthalmol. 140: 851–957.
Kunimoto, D.Y., S. Sharma, P. Garg, U. Gopinathan, D. Miller and G.N. Rao. 2000. Corneal ulceration in the elderly in Hyderabad, south India. Br J Ophthalmol. 84: 54–59.
Krishna, S., S. Shafiyabi, S. Liba, R. Ramesha and D. Pavitra. 2013. Microbial keratitis in Bellary district, Karnataka, India: influence of geographic, climatic, agricultural and occupational risk factors. Int J Pharm Biomed Res. 4: 189–193.
Kwochka, K.W., M.B. Calderwood Mays, L. Ajello and A.A. Padhye. 1984. Canine phaeohyphomycosis caused by *Drechslera spicifera*: A case report and literature review. J Am Anim Hosp Assoc. 20: 625–633.
Lake, F.R., J.H. Froudist, R. McAleer, R.L. Gillon, A.E. Tribe and P.J. Thompson. 1991. Allergic bronchopulmonary fungal disease caused by *Bipolaris* and *Curvularia*. Aust N Z J Med. 21: 871–874.
Lampert, R.P., J.H. Hutto, W.H. Donnelly and S.T. Shulman. 1977. Pulmonary and cerebral mycetoma caused by *Curvularia pallescens*. J Pediatr. 91: 603–605.
Laspina, F., M. Samudio, D. Cibils, C.N. Ta, N. Farina, R. Sanabria, V. Klauss and H. Mino de Kaspar. 2004. Epidemiological characteristics of microbiological results on patients with infectious corneal ulcers: a 13-year survey in Paraguay. Graefe's Archive for Clinical and Experimental Ophthalmology 242: 204–209.
Lass-Flörl, C., A. Mayr, S. Perkhofer, G. Hinterberger, J. Hausdorfer, C. Speth and M. Fille. 2008. Activities of antifungal agents against yeast and filamentous fungi: assessment according to the methodology of the European Committee on antimicrobial susceptibility testing. Antimicrob Agents Chemother. 52: 3637–3641.
Latham, R.H. 2000. *Bipolaris spicifera* meningitis complicating a neurosurgerical procedure. Scand J Infect Dis. 32: 102–103.
Lavoie, S.R., A. Espinel-Ingroff and T. Kerkering. 1993. Mixed cutaneous phaeohyphomycosis in a cocaine user. Clin Infect Dis. 17: 114–116.
Leck, A.K., P.A. Thomas, M. Hagan, J. Kaliamurthy, E. Ackuaku, M. John, M.J. Newman, F.S. Codjoe, J.A. Opintan, C.M. Kalavathy, V. Essuman, C.A. Jesudasan and G.J. Johnson. 2002. Aetiology of suppurative corneal ulcers in Ghana and south India, and epidemiology of fungal keratitis. Br J Ophthalmol. 86: 1211–1215.
Liu, D. and J. Gray. 2011. *Bipolaris* and *Drechslera*. pp. 49–55. *In*: Liu, D. (ed.). Molecular Detection of Human Fungal Pathogens. CRC Press, Taylor and Francis Group, Boca Raton, FL.
Lopes, J.O. and N.M. Jobim. 1998. Dermatomycosis of the toe web caused by *Curvularia lunata*. Rev Inst Med Trop Sao Paulo 40: 327–328.
Lopes, J.O., S.H. Alves, J.P. Benevenga, F.B. Brauner, M.S. Castro and E. Melchiors. 1994. *Curvularia lunata* peritonitis complicating peritoneal dialysis. Mycopathologia 127: 65–67.
Loveless, M.O., R.E. Winn, M. Campbell and S.R. Jones. 1981. Mixed invasive infection with *Alternaria* species and *Curvularia* species. Am J Clin Pathol. 76: 491–493.

Luque, A.G., R. Nanni and B.J. de Bracalenti. 1986. Mycotic keratitis caused by *Curvularia lunata* var. *aeria*. Mycopathologia 93: 9–12.
MacMillan, R.H., P.H. Cooper, B.A. Body and A.S. Mills. 1987. Allergic fungal sinusitis due to *Curvularia lunata*. Hum Pathol. 18: 960–964.
Mahgoub, E.S. 1973. Mycetomas caused by *Curvularia lunata, Madurella grisea, Aspergillus nidulans*, and *Nocardia brasiliensis* in Sudan. Sabouraudia 11: 179–182.
Manamgoda, D.S., L. Cai, E.H.C. McKenzie, P.W. Crous, H. Madrid, E. Chukeatirote, R.G. Shivas, Y.P. Tan and K.D. Hyde. 2012. A phylogenetic and taxonomic re-evaluation of the *Bipolaris-Cochliobolus-Curvularia* Complex. Fungal Divers. 56: 131–144.
Marangon, F.B., D. Miller, J.A. Giaconi and E.C. Alfonso. 2004. *In vitro* investigation of voriconazole susceptibility for keratitis and endophthalmitis fungal pathogens. Am J Ophthalmol. 137: 820–825.
Marcus, L., H.F. Vismer, H.J. van der Hoven, E. Gove and P. Meewes. 1992. Mycotic keratitis caused by *Curvularia brachyspora* (Boedjin). A report of the first case. Mycopathologia. 119: 29–33.
Maskin, S.L., R.J. Fetchick, C.R. Leone, Jr., P.K. Sharkey and M.G. Rinaldi. 1989. *Bipolaris hawaiiensis*-caused phaeohyphomycotic orbitopathy. A devastating fungal sinusitis in an apparently immunocompetent host. Ophthalmology 96: 175–179.
Matthiesson, A.M. 1981. Allergic bronchopulmonary disease caused by fungi other than *Aspergillus*. Thorax 36: 719.
McAleer, R., D.B. Kroenert, J.L. Elder and J.H. Froudist. 1981. Allergic bronchopulmonary disease caused by *Curvularia lunata* and *Drechslera hawaiiensis*. Thorax 36: 338–344.
McClay, J.E., B. Marple, L. Kapadia, M.J. Biavati, B. Nussenbaum, M. Newcomer, S. Manning, T. Booth and N. Schwade. 2002. Clinical presentation of allergic fungal sinusitis in children. Laryngoscope 112: 565–569.
McGinnis, M.R. and L. Pasarell. 1998. *In vitro* testing of susceptibilities of filamentous ascomycetes to voriconazole, itraconazole, and amphotericin B, with consideration of phylogenetic implications. J Clin Microbiol. 36: 2353–2355.
McGinnis, M.R., M.G. Rinaldi and R.E. Winn. 1986. Emerging agents of phaeohyphomycosis: pathogenic species of *Bipolaris* and *Exserohilum*. J Clin Microbiol. 24: 250–259.
McGinnis, M.R., G. Campbell, W.K. Gourley and H.L. Lucia. 1992. Phaeohyphomycosis caused by *Bipolaris spicifera*: an informative case. Eur J Epidemiol. 8: 383–386.
Minckler, D., K.W. Small and T.J. Walsh. 2014. Clinical and pathologic features of *Bipolaris* endophthalmitis after intravitreal triamcinolone. JAMA Ophthalmol. in press. doi: 10.1001/jamaophthalmol.2014.257.
Miño de Kaspar, H., G. Zoulek, M.E. Paredes, R. Alborno, D. Medina, M. Centurion de Morinigo, M. Ortiz de Fresco and F. Aguero. 1991. Mycotic keratitis in Paraguay. Mycoses 34: 251–254.
Moody, M.N., J. Tschen and M. Mesko. 2012. Cutaneous *Curvularia* infection of the forearm. Cutis. 89: 65–68.
Moore, M.L., G.R. Collins, B.J. Hawk and T.S. Russell. 2001. Disseminated *Bipolaris spicifera* in a neonate. J Perinatol. 21: 399–401.
Morton, S.J., K. Midthun and W.G. Merz. 1986. Granulomatous encephalitis caused by *Bipolaris hawaiiensis*. Arch Pathol Lab Med. 110: 1183–1185.
Mroueh, S. and A. Spock. 1992. Allergic bronchopulmonary disease caused by *Curvularia* in a child. Pediatr Pulmonol. 12: 123–126.
Muller, G.H., W. Kaplan, L. Ajello and A.A. Padhye. 1975. Phaeohyphomycosis caused by *Drechslera spicifera* in a cat. J Am Vet Med Assoc. 166: 150–154.
Myers, D.A., R. Isaza, G. Ben-Shlomo, J. Abbott and C.E. Plummer. 2009. Fungal keratitis in a gopher tortoise (*Gopherus polyphemus*). J Zoo Wildl Med. 40: 579–582.
Naggie, S. and J.R. Perfect. 2009. Molds: hyalohyphomycosis, phaeohyphomycosis, and zygomycosis. Clin Chest Med. 30: 337–353.
Nakada, M., C. Tanaka, K. Tsunewaki and M. Tsuda. 1994. RFLP analysis for species separation in the genera *Bipolaris* and *Curvularia*. Mycoscience 35: 271–278.
Nath, R., S. Baruah, L. Saikia, B. Devi, A.K. Borthakur and J. Mahanta. 2011. Mycotic corneal ulcers in upper Assam. Indian J Ophthalmol. 59: 367–371.
Nhung, P.H., T.A. Thu, L.H. Ngoc, K. Ohkusu and T. Ezaki. 2012. Epidemiology of fungal keratitis in North Vietnam. J Clin Exp Ophthalmol. 3: 238.
Nishioka, G., J.G. Schwartz, M.G. Rinaldi, T.B. Aufdemorte and E. Mackie. 1987. Fungal maxillary sinusitis caused by *Curvularia lunata*. Arch Otolaryngol Head Neck Surg. 113: 665–666.

Nityananda, K., P. Sivasubramaniam and L. Ajello. 1962. Mycotic keratitis caused by *Curvularia lunata*. Sabouraudia 2: 35.
Nityananda, K., P. Sivasubramaniam and L. Ajello. 1964. A case of mycotic keratitis caused by *Curvularia geniculata*. Arch Ophthalmol. 71: 456–458.
Odabasi, Z., V.L. Paetznick, J.R. Rodriguez, E. Chen and L. Ostrosky-Zeichner. 2004. *In vitro* activity of anidulafungin against selected clinically important mold isolates. Antimicrob. Agents Chemother. 48: 1912–1915.
Ogden, P.E., D.L. Hurley and P.T. Cain. 1992. Fatal fungal endarteritis caused by *Bipolaris spicifera* following replacement of the aortic valve. Clin Infect Dis. 14: 596–598.
O'Sullivan, F.X., B.R. Stuewe, J.M. Lynch, J.W. Brandsberg, T.B. Wiegmann, R.V. Patak, W.G. Barnes and G.R. Hodges. 1981. Peritonitis due to *Drechslera spicifera* complicating continuous ambulatory peritoneal dialysis. Ann Intern Med. 94: 213–214.
Paredes, K., J. Capilla, D.A. Sutton, E. Mayayo, A.W. Fothergill and J. Guarro. 2013. Virulence of *Curvularia* in a murine model. Mycoses 56: 512–515.
Parmjeet, K.G. and D. Pushpa. 2011. Keratomycosis-A retrospective study from a North Indian tertiary care institute. JIACM 12: 271–273.
Parva, P., R. Rojas and E. Palacios. 2005. Unusual rhinosinusitis caused by *Curvularia* fungi. Ear Nose Throat J. 84: 270–275.
Pasarell, L., M.R. Mcginnis and P.G. Standard. 1990. Differentiation of medically important isolates of *Bipolaris* and *Exserohilum* with exoantigens. J Clin Microbiol. 28: 1655–1657.
Pathengay, A., G.Y. Shah, T. Das and S. Sharma. 2006. *Curvularia lunata* endophthalmitis presenting with a posterior capsular plaque. Indian J Ophthalmol. 54: 65–66.
Pathengay, A., D.M. Miller, H.W. Flynn, Jr. and S.R. Dubovy. 2012. *Curvularia* endophthalmitis following open globe injuries. Arch Ophthalmol. 130: 652–654.
Patil, S., S. Kulkarni, S. Gadgil and A. Joshi. 2011. Corneal abscess caused by *Bipolaris spicifera*. Indian J Pathol Microbiol. 54: 408–410.
Pauzner, R., A. Goldschmied-Reouven, I. Hay, Z. Vered, Z. Ziskind, N. Hassin and Z. Farfel. 1997. Phaeohyphomycosis following cardiac surgery: case report and review of serious infection due to *Bipolaris* and *Exserohilum* species. Clin Infect Dis. 25: 921–923.
Pavan, P.R. and C.E. Margo. 1993. Endogenous endophthalmitis caused by *Bipolaris hawaiiensis* in a patient with acquired immunodeficiency syndrome. Am J Ophthalmol. 116: 644–645.
Peden, M.C., A. Neelakantan, C. Orlando, S.A. Khan, A. Lessner and M.T. Bhatti. 2008. Breaking the mold of orbital cellulitis. Surv Ophthalmol. 53: 631–635.
Penrith, M., J.J. Van der Lugt, M.M. Henton, J.A. Botha and J.C. Stroebel. 1994. A review of mycotic nasal granuloma in cattle, with a report on three cases. J S Afr Vet Assoc. 65: 179–183.
Pfaller, M.A., S.A. Messer, R.J. Hollis and R.N. Jones. 2002. Antifungal activities of posaconazole, ravuconazole and voriconazole compared to those of itraconazole and amphotericin B against 239 clinical isolates of *Aspergillus* spp. and other filamentous fungi: Report from SENTRY antimicrobial surveillance program, 2000. Antimicrob Agents Chemother. 46: 1032–1037.
Pfaller, M.A., S.A. Messer, L. Boyken, R.J. Hollis and D.J. Diekema. 2003. *In vitro* susceptibility testing of filamentous fungi: comparison of Etest and reference M38-A microdilution methods for determining posaconazole MICs. Diagn Microbiol Infect Dis. 45: 241–244.
Pfaller, M.A., L.N. Woosley, S.A. Messer, R.N. Jones and M. Castanheira. 2012. Significance of molecular identification and antifungal susceptibility of clinically significant yeasts and moulds in a global antifungal surveillance programme. Mycopathologia 174: 259–271.
Pierce, N.F., J.C. Millan, B.S. Bender and J.L. Curtis. 1986. Disseminated *Curvularia* infection. Additional therapeutic and clinical considerations with evidence of medical cure. Arch Pathol Lab Med. 110: 959–961.
Pimentel, J.D., K. Mahadevan, A. Woodgyer, L. Sigler, C. Gibas, O.C. Harris, M. Lupino and E. Athan. 2005. Peritonitis due to *Curvularia inaequalis* in an elderly patient undergoing peritoneal dialysis and a review of six cases of peritonitis associated with other *Curvularia* spp. J Clin Microbiol. 43: 4288–4292.
Pingree, T.F., G.R. Holt, R.A. Otto and M.G. Rinaldi. 1992. *Bipolaris*-caused fungal sinusitis. Otolaryngol. Head Neck Surg. 106: 302–305.
Poria, V.C., V.R. Bharad, D.S. Dongre and M.V. Kulkarni. 1985. Study of mycotic keratitis. Indian J Ophthalmol. 33: 229–231.

Posteraro, B., E. Scarano, M. La Sorda, R. Torelli, E. De Corso, A. Mulé, G. Paludetti, G. Fadda and M. Sanguinetti. 2010. Eosinophilic fungal rhinosinusitis due to the unusual pathogen *Curvularia inaequalis*. Mycoses 53: 84–88.

Pradhan, L., S. Sharma, S. Nalamada, S.K. Sahu, S. Das and P. Garg. 2011. Natamycin in the treatment of keratomycosis: Correlation of treatment outcome and *in vitro* susceptibility of fungal isolates. Indian J Ophthalmol. 59: 512–514.

Prajna, N.V., R.A. Rao, M.M. Mathen, L. Prajna, C. George and M. Srinivasan. 2002. Simultaneous bilateral fungal keratitis caused by different fungi. Indian J Ophthalmol. 50: 213–214.

Prajna, N.V., P.K. Nirmalan, R. Mahalakshmi, P. Lalitha and M. Srinivasan. 2004. Concurrent use of 5% natamycin and 2% econazole for the management of fungal keratitis. Cornea 23: 793–796.

Pratt, M.F. and J.R. Burnett. 1988. Fulminant *Drechslera* sinusitis in an immunocompetent host. Laryngoscope 98: 1343–1347.

Qualls, C.W., F.W. Chandler, W. Kaplan, E.B. Breitschwerdt and D.Y. Cho. 1985. Mycotic keratitis in a dog: concurrent *Aspergillus* sp and *Curvularia* sp. infections. J Am Vet Med Assoc. 186: 975–976.

Quraishi, H.A. and H.H. Ramadan. 1997. Endoscopic treatment of allergic fungal sinusitis. Otolaryngol. Head Neck Surg. 117: 29–34.

Qureshi, S., S.A. Wani and S. Beg. 2006. *Curvularia* dermatomycosis in a Jersey heifer: a case report. 2006. Pakistan Vet J. 26(3): 149–150.

Radford, S.A., E.M. Johnson and D.W. Warnock. 1997. *In vitro* studies of activity of voriconazole (UK-109,496), a new triazole antifungal agent, against emerging and less-common mold pathogens. Antimicrob Agents Chemother. 41: 841–843.

Rachitskaya, A.V., A.K. Reddy, D. Miller, J. Davis, H.W. Flynn, W. Smiddy, W. Lara, S. Lin, S. Dubovy and T.A. Albini. 2014. Prolonged *Curvularia* endophthalmitis due to organism sequestration. JAMA Ophthalmol. In press. doi: 10.1001/jamaophthalmol.2014.1069.

Rahman, M.R., G.J. Johnson, R. Husain, S.A. Howlader and D.C. Minassian. 1998. Randomised trial of 0.2% chlorhexidine gluconate and 2.5% natamycin for fungal keratitis in Bangladesh. Br J Ophthalmol. 82: 919–925.

Ragini, T., S. Abhisek, S.M. Om Prakash, C. Abhishek, T. Vijai and G. Anil Kumar. 2010. Mycotic keratitis in India: a five-year retrospective study. J Infect Dev Ctries. 4: 171–174.

Ramani, R., C.R. Srinivas, A. Ramani, T.G. Kumari and P.G. Shivananda. 1993. Molds in onychomycosis. Int J Dermatol. 32: 877–878.

Rao, A., R. Forgan-Smith, S. Miller and H. Haswell. 1989. Phaeohyphomycosis of the nasal sinuses caused by *Bipolaris* species. Pathology 21: 280–281.

Rautaraya, B., S. Sharma, S. Kar, S. Das and S.K. Sahu. 2011. Diagnosis and treatment outcome of mycotic keratitis at a tertiary eye care center in eastern India. BMC Ophthalmol. 11: 39.

Revankar, S.G. and D.A. Sutton. 2010. Melanized fungi in human disease. Clin Microbiol Rev. 23: 884–928.

Revankar, S.G., J.E. Patterson, D.A. Sutton, R. Pullen and M.G. Rinaldi. 2002. Disseminated phaeohyphomycosis: review of an emerging mycosis. Clin Infect Dis. 34: 467–476.

Rinaldi, M.G., P. Phillips, J.G. Schwartz, R.E. Winn, G.R. Holt, F.W. Shagets, J. Elrod, G. Nishioka and T.B. Aufdemorte. 1987. Human *Curvularia* infections. Report of five cases and review of the literature. Diagn Microbiol Infect Dis. 6: 27–39.

Robb, C.W., P.J. Malouf and R.P. Rapini. 2003. Four cases of dermatomycosis: superficial cutaneous infection by *Alternaria* or *Bipolaris*. Cutis 72: 313–316.

Robson, A.M. and R.D. Craver. 1994. *Curvularia* urinary tract infection: a case report. Pediatr Nephrol. 8: 83–84.

Robson, J.M., P.G. Hogan, R.A. Benn and P.A. Gatenby. 1989. Allergic fungal sinusitis presenting as a paranasal sinus tumour. Arch Dermatol. 125: 1383–1386.

Rohwedder, J.J., J.L. Simmons, H. Colfer and B. Gatmaitan. 1979. Disseminated *Curvularia lunata* infection in a football player. Arch Intern Med. 139: 940–941.

Rolston, K.V., R.L. Hopfer and D.L. Larson. 1985. Infections caused by *Drechslera* species: case report and review of the literature. Rev Infect Dis. 7: 525–529.

Romano, C., A. Ghilardi and L. Massai. 2004. Subungual hyperkeratosis of the big toe due to *Bipolaris hawaiiensis*. Acta Derm Venereol. 84: 476–477.

Rosow, L., J.X. Jiang, T. Deuel, M. Lechpammer, A.A. Zamani, D.A. Milner, R. Folkerth, F.M. Marty and S. Kesari. 2011. Cerebral phaeohyphomycosis caused by *Bipolaris spicifera* after heart transplantation. Transpl Infect Dis. 13: 419–423.

Rowen, J.L., J.T. Atkins, M.L. Levy, S.C. Baer and C.J. Baker. 1995. Invasive fungal dermatitis in the < or = 1000-gram neonate. Pediatrics 95: 682–687.

Ruan, B., J. Ma and J. Sui. 2005 [One case of allergic sinusitis due to *Bipolaris spicifera*]. Zhonghua Er Bi Yan Hou Tou Jing Wai Ke Za Zhi. 40: 951–952 [Article in Chinese].

Russo, J.P., R. Raffaeli, S.M. Ingratta, P. Rafti and S. Mestroni. 2010. Cutaneous and subcutaneous phaeohyphomycosis. Skinmed 8: 366–369.

Sabatelli, F., R. Patel, P.A. Mann, C.A. Mendrick, C.C. Norris, R. Hare, D. Loebenberg, T.A. Black and P.M. McNicholas. 2006. *In vitro* activities of posaconazol, fluconazole, itraconazole, voriconazole and amphotericin B against a large collection of clinically important molds and yeasts. Antimicrob Agents Chemother. 50: 2009–2015.

Saenz, R.E., W.D. Brown and C.V. Sanders. 2001. Allergic bronchopulmonary disease caused by *Bipolaris hawaiiensis* presenting as a necrotizing pneumonia: case report and review of literature. Am J Med Sci. 321: 209–212.

Safdar, A. 2003. *Curvularia*—favorable response to oral itraconazole therapy in two patients with locally invasive phaeohyphomycosis. Clin Microbiol Infect. 9: 1219–1223.

Saha, R. and S. Das. 2005. *Bipolaris* keratomycosis. Mycoses 48: 453–455.

Saha, R. and S. Das. 2006. Mycological profile of infectious Keratitis from Delhi. Indian J Med Res. 123: 159–164.

Santos, D.W.C.L., A.C.B. Padovan, A.S.A. Melo, S.S. Goncalves, V.R. Azevedo, M.M. Ogawa, T.V.S. Freitas and A.L. Colombo. 2013. Molecular identification of melanised non-sporulating moulds: a useful tool for studying the epidemiology of phaeohyphomycosis. Mycopathologia 175: 445–454.

Schell, W.A. 2002. Dematiaceous hyphomycetes. pp. 565–637. *In*: Howard, D.H. (ed.). Pathogenic Fungi in Humans and Animals, Marcel Dekker, Inc., USA.

Schoch, C.L., K.A. Seifert, S. Huhndorf, V. Robert, J.L. Spouge, C.A. Levesque and W. Chen. 2012. Nuclear ribosomal internal transcribed spacer (ITS) region as a universal DNA barcode marker for fungi. Proc Natl Acad Sci USA 109: 6241–6246.

Schroeder, W.A., Jr., D.G. Yingling, P.C. Horn and W.D. Stahr. 2002. Frontal sinus destruction from allergic eosinophilic fungal rhinosinusitis. Mo Med. 99: 197–199.

Schubert, M.S. and D.W. Goetz. 1998. Evaluation and treatment of allergic fungal sinusitis. I. Demographics and diagnosis. Allergy Clin Immunol. 102: 387–394.

Schuetz, A.N. 2011. *Curvularia*. pp. 71–82. *In*: Liu, D. (ed.). Molecular Detection of Human Fungal Pathogens. CRC Press, Taylor and Francis Group, Boca Raton, FL.

Sengupta, S., S. Rajan, P.R. Reddy, K. Thiruvengadakrishnan, R.D. Ravindran, P. Lalitha and C.M. Vaitilingam. 2011. Comparative study on the incidence and outcomes of pigmented versus non pigmented keratomycosis. Indian J Ophthalmol. 59: 291–296.

Severo, C.B., M. Oliveira Fde, E.F. Pilar and L.C. Severo. 2012. Phaeohyphomycosis: a clinical-epidemiological and diagnostic study of eighteen cases in Rio Grande do Sul, Brazil. Mem Inst Oswaldo Cruz. 107: 854–858.

Sharma, P., P. Kaur and A. Aggarwal. 2014. *Bipolaris spicifera*: An unusual case of non-healing cutaneous ulcers in a patient with diabetes and alcohol abuse. J Microbiol Infect Dis. 4: 33–35.

Sharma, V., M. Purohit and S. Vaidya. 2008. Epidemiological study of mycotic keratitis. Internet J Ophthalmol Vis Sci. 6: 2.

Sherwal, B.L. and A.K. Verma. 2008. Epidemiology of ocular infection due to bacteria and fungus—a prospective study. JK Sci. 10: 127-131.

Shetty, K., P. Giannini and R. Achong. 2006. Chronic craniofacial dematiaceous fungal infection: a case report. Spec. Care Dentist. 26: 155–158.

Sheyman, A.T., B.Z. Cohen, A.H. Friedman and J.M. Ackert. 2013. An outbreak of fungal endophthalmitis after intravitreal injection of compounded combined bevacizumab and triamcinolone. JAMA Ophthalmol. 131: 864–869.

Shigemori, M., K. Kawakami, T. Kitahara, O. Ijichi, M. Mizota, N. Ikarimoto and K. Miyata. 1996. Hepatosplenic abscess caused by *Curvularia* boedijn in a patient with acute monocytic leukemia. Pediatr Infect Dis J. 15: 1128–1129.

Shimizu, K., C. Tanaka, Y.-L. Peng and M. Tsuda. 1998. Phylogeny of *Bipolaris* inferred from nucleotide sequences of *Brn1*, a reductase gene involved in melanin biosynthesis. J Gen Appl Microbiol. 44: 251–258.

Shin, J.H., J.H. Sung, S.J. Park, J.A. Kim, J.H. Lee, D.Y. Lee, E.S. Lee and J.M. Yang. 2003. Species identification and strain differentiation of dermatophyte fungi using polymerase chain reaction amplification and restriction enzyme analysis. J Am Acad Dermatol. 48: 857–865.
Shukla, P.K., M. Kumar and G.B.S. Keshava. 2008. Mycotic keratitis: an overview of diagnosis and therapy. Mycoses 51: 183–199.
Singh, G., M. Palanisamy, B. Madhavan, R. Rajaraman, K. Narendran, A. Kour and N. Venkatapathy. 2006. Multivariate analysis of childhood microbial keratitis in South India. Ann Acad Med Singapore 35: 185–189.
Singh, H., S. Irwin, S. Falowski, M. Rosen, L. Kenyon, D. Jungkind and J. Evans. 2008. *Curvularia* fungi presenting as a large cranial base meningioma: case report. Neurosurgery 63: E177.
Sirikul, T., T. Prabriputaloong, A. Smathivat, R.S. Chuck and A. Vongthongsri. 2008. Predisposing factors and etiologic diagnosis of ulcerative keratitis. Cornea 27: 283–287.
Sivanesan, A. 1987. Graminicolous Species of *Bipolaris*, *Curvularia*, *Drechslera*, *Exserohilum* and their Teleomorphs. Mycol Pap. 158: 1–261.
Skovrlj, B., M. Haghighi, M.E. Smethurst, J. Caridi and J.B. Bederson. 2013. *Curvularia* abscess of the brainstem. World Neurosurg. in press. doi: 10.1016/j.wneu.2013.07.014.
Small, K.W., C.K. Chan, R. Silva-Garcia and T.J. Walsh. 2014. Onset of an outbreak of *Bipolaris hawaiiensis* fungal endophthalmitis after intravitreal injections of triamcinolone. Ophthalmology 121: 952–958.
Smith, T., T. Goldschlager, N. Mott, T. Robertson and S. Campbell. 2007. Optic atrophy due to *Curvularia lunata* mucocoele. Pituitary 10: 295–297.
Sobol, S.M., R.G. Love, H.R. Stutman and T.J. Pysher. 1984. Phaeohyphomycosis of the maxilloethmoid sinus caused by *Drechslera spicifera*: a new fungal pathogen. Laryngoscope 94: 620–627.
Sodhi, P.K. and R. Kaur. 2000. *Curvularia* dacryocystitis: report of two cases. Orbit. 19: 45–50.
Sodhi, P.K. and D.K. Mehta. 2003. Fluconazole in management of ocular infections due to *Curvularia*. Ann Ophthalmol. 35: 68–72.
Somabhai, Katara R., Patel N. Dhanjibhai and M. Sinha. 2013. A clinical microbiological study of corneal ulcer patients at western Gujarat, India. Acta Med Iran. 51: 399–403.
Srinivasan, M. 2004. Fungal keratitis. Curr Opin Ophthalmol. 15: 321–327.
Srinivasan, M., C.A. Gonzales, C. George, V. Cevallos, J.M. Mascarenhas, B. Asokan, J. Wilkins, G. Smolin and J.P. Whitcher. 1997. Epidemiology and aetiological diagnosis of corneal ulceration in Madurai, south India. Br J Ophthalmol. 81: 965–971.
Still, J.M., Jr., E.J. Law, G.I. Pereira and E. Singletary. 1993. Invasive burn wound infection due to *Curvularia* species. Burns 19: 77–79.
Straka, B.F., P.H. Cooper and B.A. Body. 1989. Cutaneous *Bipolaris spicifera* infection. Arch Dermatol. 125: 1383–1386.
Subramanian, C.V. and B.L. Jain. 1966. A revision of some graminicolous *Helminthosporia*. Curr Sci. 14: 352–325.
Subramanyam, V.R., C.C. Rath, M. Mishra and G.P. Chhotrai. 1993. Subcutaneous infection due to *Curvularia* species. Mycoses 36: 449–450.
Sun, G., S. Oide, E. Tanaka, K. Shimizu, C. Tanaka and M. Tsuda. 2003. Species separation in *Curvularia* "*geniculata*" group inferred from *Brn1* gene sequences. Mycoscience 44: 239–244.
Sunay, S.W., N.N. Neena, R.T. Vilas and M. Hema. 2014. *Curvularia* a most common missed occulomycosis in ocular trauma. JMSCR 2: 1344–1348.
Sundaram, B.M., S. Badrinath and S. Subramanian. 1989. Studies on mycotic keratitis. Mycoses 32: 568–572.
Swift, I.M., A. Griffin and M.A. Shipstone. 2006. Successful treatment of disseminated cutaneous phaeohyphomycosis in a dog. Aust Vet J. 84: 431–435.
Taguchi, K., T. Kawabata, M. Wakayama, T. Oharaseki, Y. Yokouchi, K. Takahashi, S. Naoe, T. Ogoshi, S. Iwabuchi, K. Shibuya and K. Nishimura. 2004. A case of allergic fungal sinusitis caused by *Bipolaris spicifera*. Nihon Ishinkin Gakkai Zasshi 45: 239–245 [Article in Japanese].
Taj-Aldeen, S.J., A.A. Hilal and W.A. Schell. 2004. Allergic fungal rhinosinusitis: a report of 8 cases. Am J Otolaryngol. 25: 213–218.
Tanabe, K., M. Seino and S. Senda. 2010. Superficial mycosis of the breast caused by *Curvularia inaequalis*. Eur J Dermatol. 20: 658–659.
Tessari, G., A. Forni, R. Ferretto, M. Solbiati, G. Faggian, A. Mazzucco and A. Barba. 2003. Lethal systemic dissemination from a cutaneous infection due to *Curvularia lunata* in a heart transplant recipient. J Eur Acad Dermatol Venereol. 17: 440–442.

Terada, M., E. Ohki, Y. Yamagishi, Y. Nishiyama, K. Satoh, K. Uchida, H. Yamaguchi and H. Mikamo. 2014. Fungal peritonitis associated with *Curvularia geniculata* and *Pithomyces* species in a patient with vulvar cancer who was successfully treated with oral voriconazole. J Antibiot. (Tokyo) 67: 191–193.
Tewari, A., N. Sood, M.M. Vegad and D.C. Mehta. 2012. Epidemiological and microbiological profile of infective keratitis in Ahmedabad. Indian J Ophthalmol. 60: 267–272.
Thomas, P.A. 2003. Current perspectives on ophthalmic mycoses. Clin Microbiol Rev. 16: 730–797.
Torda, A.J. and P.D. Jones. 1997. Necrotizing cutaneous infection caused by *Curvularia brachyspora* in an immunocompetent host. Australas. J Dermatol. 38: 85–87.
Toul, P., L. Castillo, V. Hofman, J.P. Bouchara, S. Chanalet and M. Gari-Toussaint. 2006. A pseudo tumoral sinusitis caused by *Bipolaris* sp. J Infect. 53: e235–e237.
Travis, W.D., K.J. Kwon-Chung, D.E. Kleiner, A. Geber, W. Lawson, H.I. Pass and D. Henderson. 1991. Unusual aspects of allergic bronchopulmonary fungal disease: report of two cases due to *Curvularia* organisms associated with allergic fungal sinusitis. Hum Pathol. 22: 1240–1248.
Tsuda, M., A. Ueyama and N. Nishihara. 1977. *Pseudocochliobolus nisikadoi*, the perfect state of *Helminthosporium coicis*. Mycologia 69: 1109–1120.
Tuli, S.S. and S.H. Yoo. 2003. *Curvularia* keratitis after laser *in situ* keratomileusis from a feline source. J Cataract Refract Surg. 29: 1019–1021.
Ujhelyi, M.R., R.H. Raasch, C.M. van der Horst and W.D. Mattern. 1990. Treatment of peritonitis due to *Curvularia* and *Trichosporon* with amphotericin B. Rev Infect Dis. 12: 621–627.
Unal, A., M.H. Sipahioğlu, M.A. Atalay, F. Kavuncuoglu, B. Tokgoz, A.N. Koc, O. Oymak and C. Utas. 2011. Tenckhoff catheter obstruction without peritonitis caused by *Curvularia* species. Mycoses 54: 363–364.
Usha, A., K.G. Parmjeet and C. Sandeep. 2009. Fungal profile of keratomycosis. Bombay Hosp J. 51: 325–327.
Vachharajani, T.J.1., F. Zaman, S. Latif, R. Penn and K.D. Abreo. 2005. *Curvularia geniculata* fungal peritonitis: a case report with review of literature. Int Urol Nephrol. 37: 781–784.
Vajpayee, R.B., T. Dada, R. Saxena, M. Vajpayee, H.R. Taylor, P. Venkatesh and N. Sharma. 2000. Study of the first contact management profile of cases of infectious keratitis: a hospital-based study. Cornea 19: 52–56.
Varughese, S., V.G. David, M.S. Mathews and V. Tamilarasi. 2011. A patient with amphotericin-resistant *Curvularia lunata* peritonitis. Perit Dial Int. 31: 108–109.
Vásquez-del-Mercado, E., L. Lammoglia and R. Arenas. 2013. Subcutaneous phaeohyphomycosis due to *Curvularia lunata* in a renal transplant patient. Rev Iberoam Micol. 30: 116–118.
Veer, P., N.S. Patwardhan and A.S. Damie. 2007. Study of onychomycoses: prevailing fungi and pattern of infection. Indian J Med Microbiol. 25: 53–56.
Vemulakonda, G.A., S.M. Hariprasad, W.F. Mieler, R.A. Prince, G.K. Shah and R.N. Van Gelder. 2008. Aqueous and vitreous concentrations following topical administration of 1% voriconazole in humans. Arch Ophthalmol. 126: 18–22.
Venugopal, P.L., T.L. Venugopal, A. Gomathi, E.S. Ramakrishna and S. Ilavarasi. 1989. Mycotic keratitis in Madras. Indian J Pathol Microbiol. 32: 190–197.
Vermeire, S.E., H. de Jonge, K. Lagrou and D.R. Kuypers. 2010. Cutaneous phaeohyphomycosis in renal allograft recipients: report of 2 cases and review of the literature. Diagn Microbiol Infect Dis. 68: 177–180.
Viola, G.M. and R. Sutton. 2010. Allergic fungal sinusitis complicated by fungal brain mass. Int J Infect Dis. 14: e299–e301.
Walsh, T.J., C. Gonzalez, E. Roilides, B.U. Mueller, N. Ali, L.L. Lewis, T.O. Whitcomb, D.J. Marshall and P.A. Pizzo. 1995. Fungemia in children infected with the human immunodeficiency virus: new epidemiologic patterns, emerging pathogens, and improved outcome with antifungal therapy. Clin Infect Dis. 20: 900–906.
Wan, Z., J. Yü and X.H. Wang. 2006. One case of allergic sinusitis due to *Bipolaris spicifera*. Zhonghua Er Bi Yan Hou Tou Jing Wai Ke Za Zhi. 41: 226–227 [Article in Chinese].
Wang, H., M. Xiao, F. Kong, S. Chen, H.-T. Dou, T. Sorrell, R.-Y. Li and Y.-C. Xu. 2011. Accurate and practical identification of 20 *Fusarium* species by seven-locus sequence analysis and reverse line blot hybridization, and an *in vitro* antifungal susceptibility study. J Clin Microbiol. 49: 1890–1898.
Wang, L., Y. Zhang, Y. Wang, G. Wang, J. Lu and J. Deng. 2000. Spectrum of mycotic keratitis in China. Zhonghua Yan Ke Za Zhi. 36: 138–140 [Article in Chinese].
Wanmali, S.S., N.N. Nagdeo, V.R. Thombare and H. Mathurkar. 2014. *Curvularia* a most common missed occulomycosis in ocular trauma. J Med Sci Clin Res. 2: 1344–1348.

Waurzyniak, B.J., J.P. Hoover, K.D. Clinkenbeard and R.D. Welsh. 1992. Dual systemic mycosis caused by *Bipolaris spicifera* and *Torulopsis glabrata* in a dog. Vet Pathol. 29: 566–569.
Whitcomb, M.P., C.D. Jeffries and R.W. Weise. 1981. *Curvularia lunata* in experimental phaeohyphomycosis. Mycopathologia 75: 81–88.
Wilhelmus, K.R. and D.B. Jones. 2001. *Curvularia* keratitis. Trans Am Ophthalmol Soc. 99: 111–132.
Willard, C.C., V.D. Eusterman and P.L. Massengil. 2003. Allergic fungal sinusitis: report of 3 cases and review of the literature. Oral Surg. Oral Med. Oral Pathol. Oral Radiol Endod. 96: 550–560.
Williams, G., F. Billson, R. Husain, S.A. Howlader, N. Islam and K. McClellan. 1987. Microbiological diagnosis of suppurative keratitis in Bangladesh. Br J Ophthalmol. 71: 315–321.
Wind, C.A. and F.M. Polack. 1970. Keratomycosis due to *Curvularia lunata*. Arch Ophthalmol. 84: 694–696.
Wong, T.Y., K.S. Fong and D.T.H. Tan. 1997. Clinical and microbial spectrum of fungal keratitis in Singapore: a 5-year retrospective study. Int Opthalmol. 21: 127–130.
Xu, Y., G.R. Pang, D.Q. Zhao, C.W. Gao, L.T. Zhou, S.T. Sun, B.L. Wang and Z.J. Chen. 2010. Activity of butenafine against ocular pathogenic filamentous fungi *in vitro*. Zhonghua Yan Ke Za Zhi. 46: 38–42.
Yanagihara, M., M. Kawasaki, H. Ishizaki, K. Anzawa, S. Udagawa, T. Mochizuki, Y. Sato, N. Tachikawa and H. Hanakawa. 2010. Tiny keratotic brown lesions on the interdigital web between the toes of a healthy man caused by *Curvularia* species infection and a review of cutaneous *Curvularia* infections. Mycoscience 51: 224–233.
Yau, Y.C., J. de Nanassy, R.C. Summerbell, A.G. Matlow and S.E. Richardson. 1994. Fungal sternal wound infection due to *Curvularia lunata* in a neonate with congenital heart disease: case report and review. Clin Infect Dis. 19: 735–740.
Young, C.N., J.G. Swart, D. Ackermann and K. Davidge-Pitts. 1978. Nasal obstruction and bone erosion caused by *Drechslera hawaiiensis*. J Laryngol Otol. 152: 137–143.
Yoshimori, R.N., R.A. Moore, H.H. Itabashi and D.G. Fujikawa. 1982. Phaeohyphomycosis of brain: granulomatous encephalitis caused by *Drechslera spicifera*. Am J Clin Pathol. 77: 363–370.
Zhang, G. and M.L. Berbee. 2001. *Pyrenophora* phylogenetics inferred from ITS and glyceraldehydes-3-phosphate dehydrogenase gene sequences. Mycologia 93: 1048–1063.
Zhang, Y., C.L. Schoch, J. Fournier, P.W. Crous, J. De Gruyter, J.H.C. Woudenberg, K. Hirayama, K. Tanaka, S.B. Pointing, J.W. Spatafora and K.D. Hyde. 2009. Multi-locus phylogeny of Pleosporales: a taxonomic, ecological and evolutionary re-evaluation. Stud Mycol. 64: 85–102.

SECTION III

Classic Mycoses Caused by Dimorphic Fungi

CHAPTER 8

Classic Histoplasmosis

Ricardo Negroni

Introduction

Classic histoplasmosis or histoplasmosis capsulati is a systemic endemic mycosis, caused by the thermally dimorphic fungus *Histoplasma capsulatum* var. *capsulatum.* This microorganism lives in the environment, especially in the soil, where it exists as mould. In blood-agar medium at 37°C and in tissues it grows as a budding yeast. In the infected organs these yeasts are inside the cells of the reticuloendothelial system (Arenas 2011; Bonifaz 2012).

Histoplasmosis has been registered in more than 60 countries, but it is more frequent in the middle east area of U.S.A. and in Latin America (Borelli 1970). The infection is produced by inhalation of microconidia and the lungs are its portal of entry. The majority of the infections in immunocompetent individuals are asymptomatic or mild and self-limited (Larsh 1970). The severity of the respiratory manifestations is related to the amount of conidia inhaled (Goodwin et al. 1981). Chronic progressive pulmonary histoplasmosis is detected in males above 50 years of age with chronic obstructive pulmonary disease (Goodwin et al. 1981; George and Penn 1993). Acute or chronic disseminated histoplasmosis occurs in patients with cell-mediated immunity failures and it is a life-threatening disease (Goodwin et al. 1980; Alsip and Dismukes 1986). Amphotericin B and itraconazole have been successfully applied in the treatment of this mycosis (Wheat 2002).

History

In 1904, when Samuel Darling was working at the Ancon Canal zones Hospital, he observed the first fatal case of disseminated histoplasmosis. He performed the autopsy

Consultant MD, Mycology Unit, Hospital de Infecciosas Francisco J. Muñiz, Buenos Aires, Argentina. Director de la Maestría en Micología Médica de la Facultad de Medicina de la Universidad Nacional del Nordeste.
E-mail: ricnegroni@hotmail.com

of a black man from Martinique who had died of an infection resembling a severe tuberculosis, but microscopically he observed an intracellular parasite very similar to *Leishmania.* Upon closer examination he noted that this microorganism lacked kinetoplasts. Darling thought that this new infective agent was a protozoa and named it *Histoplasma capsulatum* (Negroni 1965). In 1906 Darling reported two new cases of disseminated histoplasmosis, one in a black man from Martinique and the other in a Chinese who had lived in Panama for 15 years (Kwon-Chung and Bennett 1992).

In 1912 the eminent Brazilian parasitologist, Henrique Da Rocha Lima, who was working in Hamburg, suggested that the microorganism described by Darling was a budding yeast and not a protozoa (Negroni 1965).

In 1934 Dodd and Tompkins reported a new case of histoplasmosis and studied the growth of *H. capsulatum* in blood-agar at 37°C. In these cultures they obtained the growth of the yeast form of this dimorphic fungus and confirmed Da Rocha Lima's hypothesis. They also reproduced the disease in *Macacus rhesus* (Kwon-Chung and Bennett 1992). In the same year at Vanderbilt University De Mombreun discovered the dimorphism of *Histoplasma capsulatum* and thought that the mycelial form of this fungus probably existed in nature (Rippon 1988).

Christie and Peterson in 1945 reported many patients with pulmonary calcifications who reacted negatively to tuberculin and positively to histoplasmin. Palmer (1945–1946) carried out a nation-wide project of histoplasmin skin testing and found a particular geographic distribution of histoplasmin hypersensitivity in U.S.A. (George and Penn 1993).

Furcolow (1945) made very important contributions to the knowledge of the epidemiology and ecology of histoplasmosis. Emmons (1948) isolated *H. capsulatum* from soil in a sample collected near a rat burrow under the edge of a poultry house (Negroni 1965).

The first outbreak of histoplasmosis was detected in soldiers at Camp Grubber (Oklahoma, U.S.A.) (Rippon 1988).

Kwon-Chung (1972) discovered the sexual reproduction of *H. capsulatum* and identified the two mating types (+) and (–). She named the teleomorphic form *Emmonsiella capsulata.* In 1979, McGinnis and Katz transferred *E. capsulata* to the genus *Ajellomyces,* now named *Ajellomyces capsulatus* (Kwon-Chung and Bennett 1992; Deepe 2012).

Pablo Negroni (1940–1941) published a mycological study of the first Argentinean case of histoplasmosis. He performed a very careful study of the isolated fungus and was able to obtain the yeast form *in vitro* and employed a sterile extract of the mycelial form of *H. capsulatum* (histoplasmin) for the first time (Negroni 1965).

Etiology

Histoplasma capsulatum grows at 28°C after a 15 day incubation in several culture media such as Sabouraud dextrose-agar, dextrose-potato-agar and Borelli lactrimel. It presents a cottony aerial mycelia, white to tan in color. The reverse is uncolored or brown. Two types of colonies are found: white and brown. White colonies grow faster and lose the capacity of producing spores after several subcultures. The brown type is more virulent for mice and produces a great amount of conidia (Kwon-Chung and Bennett 1992; Kauffman 2011; Deepe 2012).

Microscopically vegetative mycelia consist of septated, branched, hyaline hyphae of 2 to 5 µm in diameter. Three types of asexual conidia are observed: (1) large, spherical or pear shape spores, 10 to 25 µm in diameter with a thick cell wall covered by tubercles, of which some are like a digital protuberance in shape, 1 to 3 µm in length. This cell wall has a thin inner layer and a thick verrucous part. These spores are named macroconidia, they are found in the aerial mycelia and are born on short sporophores (Fig. 2); (2) conidia similar to the previous one, 5 to 20 µm in diameter, spherical to oval with thin walls, usually born on short hyphae, and found in the submerged mycelia; (3) microconidia which are spherical or pyriform with thin walls, 2 to 5 µm in diameter, sessile or on short sporophores (Negroni 1965).

H. capsulatum yeast form develops in rich culture media such as brain heart infusion-agar with 5% of rabbit blood incubated at 37°C. Blood cisteine-agar is also an excellent culture medium for this purpose. After 4 to 5 days of incubation colonies are visible as whitish, wrinkled or cerebriform, 2 or 3 mm in diameter, moist, glossy and of cream consistency growth. Microscopically, small single budding yeasts, 3 to 5 µm in diameter, are observed. They multiply by polar budding and the connection between the mother and daughter cells is narrow. These yeast cells are uninucleated. The transformation from the mycelial growth to yeast form is not easy and may require several attempts (Rippon 1988; Kwon-Chung and Bennett 1992).

Mycelial form is also named saprophytic phase because it is found in the soil as well as in bat and bird droppings. The yeast form is called tissues phase. In animal tissues it is found in the form of small yeast-like elements of spherical to oval shape, single budding; they are 3–5 µm in diameter and have a thin cell wall which does not take aniline stains. Due to this characteristic it was initially mistaken for a capsule. The majority of these budding yeasts are found inside macrophages or giant cells in the granulomas and they are of the same shape and size (Salfelder et al. 1990; Negroni 2004). In smears stained by Giemsa or Wright techniques the cell wall does not take up the stain and appears as a clear halo, the cytoplasm has a single distinct mass, half moon shaped, placed at the opposite side of the bud, which is darker than the rest of the cytoplasm (Fig. 1) (Negroni 1965). *H. capsulatum* is Gram positive, stains red with periodic acid Schiff (P.A.S.) and dark brown or black with Grocott methenamine-silver technique (G.M.S.) (Salfelder et al. 1990).

H. capsulatum sexual reproduction follows the heterothallic conjunction of compatible mating types (+) and (–), paired in poor culture media as yeast extract-agar or soil extract-agar at 28°C for several weeks. The young cleistothecia are globose, 100 to 150 µm in diameter; they become irregularly stellate with age because of the radiated spinal peridial hyphae. Asci are club to pear shape, 3–5 x 10–15 µm in diameter and contain 8 oval ascospores. This sexual (teleomorphic) form is called *Ajellomyces capsulatus* (Kwon-Chung and Bennett 1992; Kauffman 2011; Deepe 2012).

H. capsulatum has 4 to 7 chromosomes. According to the number and characteristics of these chromosomes the strains of this fungus were initially divided into two chemo types, but new molecular biology techniques have allowed the

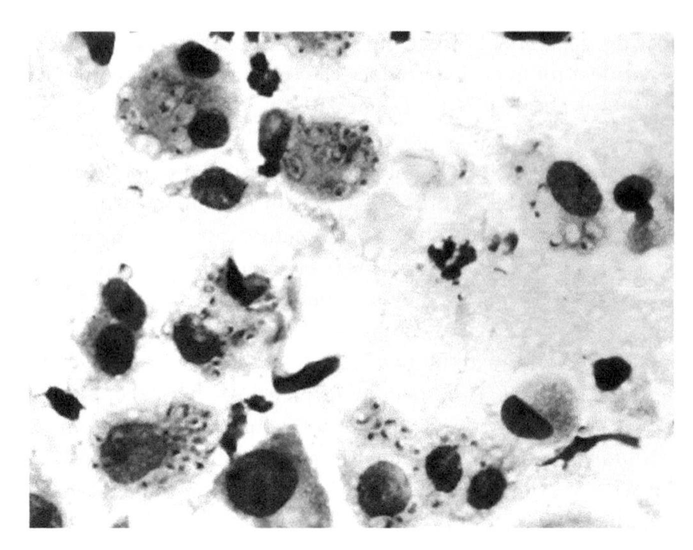

Figure 1. Giemsa stained smear of a cutaneous lesion showing yeast-like elements of *H. capsulatum* inside macrophages, X 1,000.

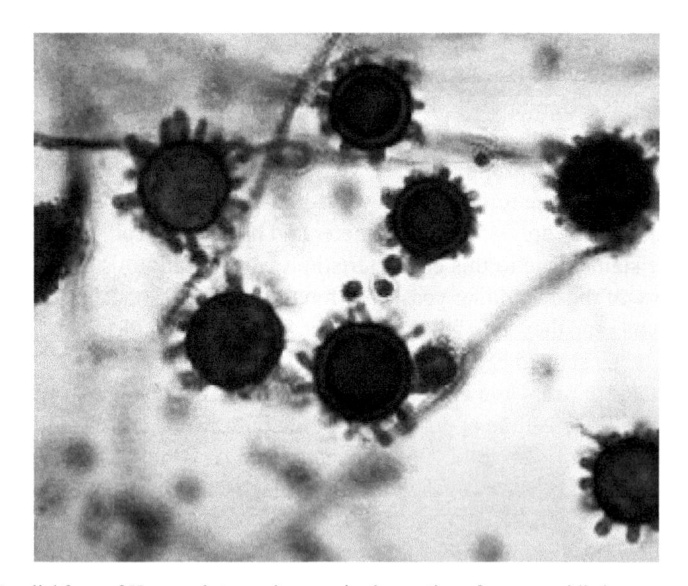

Figure 2. Mycelial form of *H. capsulatum*, microscopic observation of macroconidia in a preparation with lacto phenol cotton blue, X 400.

identification of 8 clades: 2 from North America, 2 from Latin America and 1 from each of these regions: Australia, Indonesia and Eurasia (Muniz et al. 2001; Kauffman 2011; Deepe 2012).

The genetic differences between *H. capsulatum* strains are associated with different clinical manifestations; strains from South America produce more mucocutaneous lesions than those from North America. This wide genetic variety of *H. capsulatum* seems to have originated through a sexual recombination of the strains (Zancopé-Oliveira et al. 1994; Negroni et al. 2010a).

The mating type (–) is isolated 2 to 5 times more frequently in patients than the (+) mating type (Kauffman 2008).

Some *Chrysosporium* and *Sepedonium* species present mycelial growth with micromorphological findings very similar to the mycelial form of *H. capsulatum*. However, *H. capsulatum* produces microconidia in addition to tubercular macroconidia; it is dimorphic and pathogenic for laboratory animals (Negroni 1965). Atypical *H. capsulatum* strains can be identified by nucleic acid hybridization test, Gen-probe (San Diego, CA), kit or exoantigen testing; both have a good sensitivity and specificity (>90 %) (Wheat and Kauffman 2003; Negroni et al. 2010).

H. capsulatum belongs to Ascomycotina sub-division, to Onygenaceae family and *Ajellomyces capsulatus* specie. It is genetically related to *Blastomyces dermatitidis (Ajellomyces dermatitis)* and *Paracoccidiodes brasiliensis* (Deepe 2012).

There are three varieties of *Histoplasma capsulatum: H. capsulatum* var. *capsulatum* (the one we have described), *H. capsulatum* var. *duboisii,* the etiologic agent of African histoplasmosis, which has a larger yeast form, and *H. capsulatum* var. *farcinimosum*, which produces epizootic lymphangitis in horses and mules in southern and central Europe, North of Africa, the Middle East and Southern Asia (Arenas 2011; Bonifaz 2012).

Some minerals and vitamins are necessary for *H. capsulatum* growth, such as thiamine, biotin and iron. Amino acids containing sulfhydrilic groups are required for the growth and survival of the yeast form. The mycelial to yeast transformation is related to the hyper expression of mRNA of calcium binding protein. This transition is stimulated by a temperature of 37°C, which gives way to an increase in cell membrane fluidity. This process is very complex and produces several biochemical, physical and genetic changes triggered by a transcription factor called Ryp_1 (Deepe 2012).

The *H. capsulatum* cell wall polysaccharides are chitin and glucans; they are very important virulent factors, especially α-glucan, which is the most abundant polysaccharide of the yeast form; ß–glucan is predominant in mycelial form (Negroni 2000; Kauffman 2011).

Epidemiology

Histoplasmosis is an endemic mycosis with a wide geographical distribution. It has been reported in more than 60 countries in the temperate and tropical zones throughout the world. It is most common in the American Continent from Canada to Argentina. The most important endemic areas are located along the great river valleys in the eastern and central parts of the United States as well as in La Plata River and Serra do Mar in South America. In the highly endemic zones 80% of adults react positive to histoplasmina skin test (Borelli 1970; Negroni 2004; Negroni 2010a). It should be noted that the incidence of histoplasmin reactivity at any time underestimates the true prevalence of the infection since the skin test may give negative results 2 to 4 years after the infection if the person has not been exposed to new infections (Rippon 1988; George and Penn 1993). Autochthonous cases of histoplasmosis have also been reported in Africa, Australia, India and the Far East (Kauffman 2011).

Histoplasma capsulatum has been isolated from the soil in several endemic regions. This microorganism grows well in rich soils with high nitrogen concentration and acid pH, particularly if they are contaminated with bird or bat excreta. The majority of the endemic areas are near great rivers or like shores where the mean annual temperature varies between 15 and 20ºC and the annual rainfall oscillates between 800–1,200 mm. Some "epidemic spots" have been recorded inside or outside the endemic zones (Larsh 1970; Rubinstein and Negroni 1981; Deepe 2012). These are characterized by heavy soil contamination with *H. capsulatum* and may produce small outbreaks. These outbreaks have appeared after the cleaning of poultry or pigeon houses, after entering caves or mines where bats have nested or after any action which leads to the disturbance of the soil where black birds (stornins) or pigeons have roosted (George and Penn 1993; Wheat and Kauffman 2003; Negroni 2010b). *H. capsulatum* may infect bats, which are able to spread the fungus to new locations, both inside or outside the endemic zones (Wheat and Kauffman 2003).

The asymptomatic or mild self-limited infections are very common in urban or rural areas of the endemic zones. In these areas 20 to 80% of the adult population has positive histoplasmin skin tests. The prevalence of this sensitization increases from childhood to 30 years of age (Negroni 1965; Rubinstein and Negroni 1981; Kauffman 2011).

Histoplasmosis is not usually transmitted from man to man or from animal to man, but a few cases of histoplasmosis have been recorded in liver transplant recipients who received the liver from an infected person (Silveira and Husain 2007; Deepe 2012).

Natural infection with *H. capsulatum* has been found in several animal species, most frequently in dogs and rodents (Rippon 1988).

Chronic cavitary pulmonary histoplasmosis is more often observed in adult males above 50 years of age with chronic obstructive pulmonary disease. This clinical form is more frequent in Caucasians (Goodwin et al. 1981; Rubinstein and Negroni 1981; Negroni 2000).

The progressive disseminated forms are related to predisposing factors; they are observed in children under 6 years of age and in adults above 50 years of age, especially in Caucasian males. Persons considered at risk of suffering disseminated histoplasmosis are those who have various alterations of cell-mediated immunity such as chronic alcoholism, diabetes mellitus, long-term therapy with corticosteroids, leukemia, lymphomas, treatment with TNF-α inhibitors and those infected with HIV, with less than 150 CD_4 + cells/μL (Goodwin et al. 1980; Rubinstein and Negroni 1981; Alsip and Dismukes 1986; Negroni et al. 1987; Negroni et al. 2010a).

Those who handle the mycelial form of *H. capsulatum* at the laboratory are at a risk of acquiring a massive primary infection. This job should be done in laminar flow chambers (BSL 3), under very strict biosafety conditions (Kauffman 2009).

Pathogenesis

The infective elements of *H. capsulatum* are the microconidia of the mycelial form which live in the environment, especially in soils with organic decay. These spores are inhaled by humans and other animal species, as they float in the air (Negroni 1965; Negroni 1989a; Kwon-Chung and Bennett 1992).

The inhaled spores penetrate the airways until they reach the pulmonary alveoli, where they are phagocytosed, but not lysed, by the alveolar macrophages. The phagocytosis occurs after the binding of these spores to the CD_{18} and CD_{11} families, adhesion promoting glicoproteins of neutrophils and macrophages (Deepe 2012). Inside the alveolar macrophages the conidia transform into yeast-like elements which multiply by budding. In this phase of the infection the yeast form of *H. capsulatum* is able to survive in the phagolysosomes of macrophages by using several mechanisms, including its ability to resist being killed by oxygen radicals and to modulate the intraphagosome pH. Iron and calcium acquisition by yeasts is also an important survival tool that allows the growth of this microorganism inside the macrophages (Negroni 2000; Kauffman 2009; Kauffman 2011; Deepe 2012).

The regulatory genes controlling production of 60 KDa heat-shock proteins by the yeast cell wall play a fundamental role in the transformation of the mycelial form into the yeast form. The enzymes produced during this process are involved in the sulfhydrilic link amino acids, like cisteine and biotin. These, in addition to the temperature of 37°C, are the two most important factors in the dimorphic process (Negroni et al. 2010a; Kauffman 2011; Deepe 2012).

The yeast form of *H. capsulatum* is initially found inside alveolar macrophages and polymorphonuclear neutrophils, as we have already mentioned. The initial stage of the inflammatory response involves polymorphonuclear neutrophils, these cells are able to decrease the growth of *H. capsulatum* by producing a respiratory burst and by releasing azurophil granules, but they are unable to control the progress of the infection. The fungistatic activity of the azurophil granules of the neutrophils is not related to nitric oxide or to the action of toxic radicals of oxygen. In this early part of the infection macrophages permit a rapid reproduction of the *H. capsulatum* yeast form inside the phagolysosomes of macrophages, as we have previously said. When the number of yeasts inside these cells is very high, they are liberated from them and rapidly captured by other macrophages to which they adhere by $ß_2$ integrins (Negroni 2000; Negroni 2010; Deepe 2012). These yeast cells also adhere to dendritic cells by binding fibronectin. The dendritic cells are in charge of the antigens presentation to the TCD_4+ lymphocytes, giving way to the specific cell-mediated immunity (Negroni 2000; Deepe 2012). Lung infection during this initial phase progresses by contiguity, continuing through the lymphatic system to mediastinal lymph nodes and finally, to the blood stream. This hematogenous dissemination is asymptomatic in the majority of the infections and *H. capsulatum* yeasts colonize the reticuloendothelial system (Kwon-Chung and Bennett 1992; George and Penn 1993).

The activation of specific cell-mediated immunity in immunocompetent hosts is evident within two to three weeks after infection. The immunological response involves the production of Th_1 type of cytokines, which effectively dominate the infectious process. In this stage of the infection, macrophages become activated, especially by the action of the INF-γ , IL_3, IL_{12} and TNF-α produced by T CD_4+ cells. Activated macrophages are able to kill yeast cells through nitric oxide activity. The maturation of cell-mediated immunity becomes evident by the production of compact epithelioid granulomas in affected tissues, by the delayed hypersensitivity skin test with histoplasmin turning positive, and by the blastogenic response of the lymphocytes against specific antigens becoming evident. Other cytokines such as IL_1 and GM-CSF also aid to contain the infection (Negroni 2000; Kauffman 2009; Deepe 2012).

The role of T CD_4 + cells in the defensive mechanisms against histoplasmosis is seen in athymic mice and AIDS patients, both of whom have serious forms of the disease. The role of T CD_8 + cells does not appear to be significant in host survival, but, apparently, these cells are necessary for optimal defense. The importance of NK cells, which kill extra cellular yeast, is not absolutely clear nor is the mechanism of cytokines in the transformation of macrophages into activated macrophages (Wheat and Kauffman 2003; Negroni et al. 2010a; Deepe 2012). Nevertheless, γ-INF plays a protective role in experimental animal models of histoplasmosis.

When cell-mediated immunity mechanisms are normal the infection progresses to a latent stage which probably persists for a lifetime. In this latent stage epithelioid granulomas with a caseous center, which present viable yeast-like elements inside, can be detected in different organs (Salfelder et al. 1990). These granulomas are surrounded by a fibrous capsule which calcifies with time. The reactivation of this latent infection may occur if the host becomes immunocompromised (Negroni 2000).

According to this pathogenesis model, several clinical forms of histoplasmosis have been accepted. They are presented in Table 1 (Goodwin et al. 1980; Goodwin et al. 1981; Alsip and Dismukes 1986; Negroni 2000).

Table 1. Histoplasmosis Clinical Forms.

1. Histoplasmosis in immunocompetent host.
   1.1 Asymptomatic or mild self-limited respiratory primary infection.
   1.2 Symptomatic pulmonary primary infection.
   1.3 Primary infection complications and sequelae.
   1.4 Re-infection.
   1.5 Progressive chronic pulmonary disease.
2. Histoplasmosis in immunocompromised host.
   2.1 Acute disseminated histoplasmosis.
   2.2 Subacute disseminated histoplasmosis.
   2.3 Chronic disseminated histoplasmosis.
3. Immunologically-mediated disease.

Clinical Manifestations

Asymptomatic or mild respiratory infection

More than 95% of primary infections belong to this group. Sometimes, they present mild respiratory alterations like those of influenza, which are self-limited. The course of the primary infection depends upon the number of inhaled macroconidia as well as on age and previous clinical and immune status of the host. Occasionally, they may produce pneumonitis and enlargement of the hiliar lymph nodes (Negroni 1965; Goodwin et al. 1981; Negroni 2000).

The incubation period varies from 3 to 21 days; it is shorter in re-infections and in massive infections.

These mild cases have retrospectively been identified during the epidemiological research by the positive histoplasmin skin tests and calcified lesions in the lungs, lymph nodes or spleen. These calcifications appear in approximately a quarter of those infected, one or two years after the infection. Only 20% of such cases present positive serologic reaction to histoplasmin. As we have already said, healing is spontaneous (Kauffman 2006; Deepe 2012).

Acute pulmonary infection

These are symptomatic primary infections; their severity is related to the quantity of inhaled macroconidia. Their clinical characteristics are similar to the pneumonia produced by *Legionella, Mycoplasma, Chlamydia* or viral infections. The symptoms most often observed are fever, asthenia, myalgia, night sweats, dry cough, dyspnea and pleuritic or non-pleuritic chest pain (Rubinstein and Negroni 1981; Negroni 1989a). Chest radiographs frequently show bilateral patches of pulmonary infiltration and hiliar or mediastinal adenopathies (Deepe 2012).

Severe acute pulmonary infections are observed in people who are exposed to heavy inoculum of *H. capsulatum.* In these cases, dyspnea, cough and fever are more severe, hepatosplenomegaly is detected and the chest radiology exam shows a micronodular intersticiopathy with a diffuse, bilateral, reticulonodular pattern and hiliar adenomegalies (Figs. 3 and 4). Most of the serious cases are reported during histoplasmosis outbreaks and some of these patients develop an adult respiratory distress, which requires mechanical respiratory assistance. This extremely severe infection may rarely be fatal (Kauffman 2009; Negroni 2010b).

Approximately 6%–10% of the infected persons may present clinical manifestations associated with hypersensitivity, such as erythema nodosum or multiform, polyarthritis, pleural and pericardial effusion. Joint involvement is usually symmetric and the fluids recovered from arthritis, pleural and pericardial effusions are xantochromic or serofibrinous containing lymphocytes and polymorphonuclear

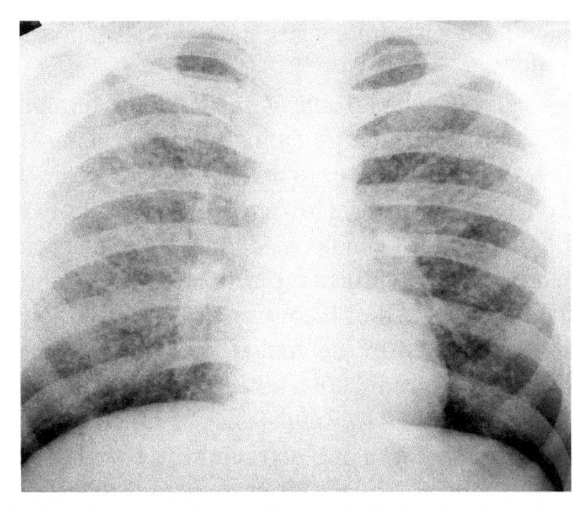

Figure 3. X-ray examination of a severe primary pulmonary infection, showing micronodularinterstitiopathy.

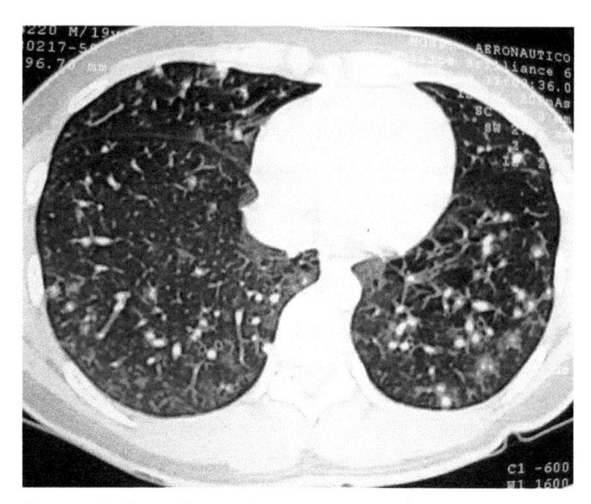

Figure 4. CT scan of a case similar to the previous one.

leukocytes (Goodwin et al. 1981; Kauffman 2011). Some cases of primary pulmonary infection may present arthralgias, erythema nodosum and hiliar adenopathies without pulmonary infiltrates, mimicking sarcoidosis (Deep 2012).

Independent of the severity of the disease, these acute respiratory infections tend to remit spontaneously in four to six weeks. The sequelae of the primary infections include fibrotic lung nodules which calcify over time, and contain a caseous center, with yeast-like elements of *H. capsulatum* which are either dead or alive. Approximately one-third of the infected patients present calcified nodules in the lungs and hiliar or mediastinal lymph nodes; less frequently, these nodules are also seen in liver and spleen. The pulmonary calcified nodules can be multiple and uniformly distributed in both lungs (Negroni 1965; Salfelder et al. 1990).

The histoplasmin skin test becomes positive three or four weeks after the infective contact, usually at the onset of clinical symptoms. This specific delayed hypersensitivity reaction is maintained for two or more years and vanishes if no new infection occurs (George and Penn 1993).

Specific serological tests with *H. capsulatum* antigens turn positive two or three weeks after the infection. The tube precipitation test recognizes IgM antibodies, reaches its higher titers between three or four weeks after exposition to *H. capsulatum* and becomes negative 3 months later. This serology test is rarely used now due to its difficult reading. ELISA for IgM can be used, but it is less specific. IgG antibodies can be demonstrated by immunodiffusion reaction, counter immunoelectrophoresis and complement fixation tests. All of them turn positive only in moderate to severe infections, the titers are related to the fungal burden and diminish after clinical remission of the infection (Negroni 2000; Wheat and Kauffman 2003).

In severe primary infections *H. capsulatum* can be isolated from sputum, bronchial secretions, blood cultures and urine. For blood cultures a lysis-centrifugation technique should be used due to its higher efficacy (Bianchi et al. 2000).

Re-infection

Cases of acute respiratory re-infection have been reported in patients who suffered a primary infection some years earlier. In these cases the incubation period is shorter, only 4 to 5 days. The clinical manifestations are more serious, especially respiratory symptomatology; chest X- ray and CT scan studies show military-type micronodules and hiliar adenopathies. Clinical regression is more rapid; it usually takes place within seven to fourteen days.

Complication of primary Infections

Complications of primary infections are rare. Granulomatous mediastinitis is produced by the invasion of mediastinum lymph nodes, which gives way to the compression of the esophagus, bronchi, trachea and large blood vessels, especially the superior vena cava. When spontaneous remission occurs, fibrosis replaces the granulomas. Fibrosis of the peribronchial region results in stenosis, bronchiectasis, pneumonia and bronchopleural fistulae. Calcified granulomas may be eliminated via the bronchi and generate broncholithiasis, which may produce cough, hemoptysis and atelectasis, but some cases are asymptomatic. Periesophageal fibrosis may cause lumen stenosis, diverticuli and bronchoesophageal fistulae (Goodwin et al. 1981; Alsip and Dismukes 1986; Negroni 2000).

Serofibrinous pericarditis is usually the consequence of granulomas in the carina lymph nodes. The inflammatory response is caused by hypersensitivity to *H. capsulatum* antigens; the cultures of pericardial fluid are negative. The clinical outcome is usually benign, pericarditis remits in few weeks and only rarely does it cause cardiac tamponade or constrictive pericarditis. Nevertheless, this complication incapacitates patients for several weeks (Wheat and Kauffman 2003).

In patients with very high hypersensitivity to *H. capsulatum* antigens a massive mediastinal fibrosis and extrinsic compression of important structures in the area may be observed, especially in the superior vena cava.

Histoplasmomas are residual lesions from the pneumonitis which occur during the primary infection. These lesions are stable or slow-growing. They are usually asymptomatic and an X-ray or a CT scan of the chest show a solitary, subpleural, spherical nodule, 1–4 cm in diameter. These nodules may be confused with lung neoplasms, particularly when they are not calcified. Calcification is often centric or in target configuration. As in the cases of mediastinal fibrosis, histoplasmoma is produced by the release of *H. capsulatum* antigens from fibrous or caseous nodules in sensitized patients (Goodwin et al. 1981; Negroni 1989a).

Chronic Pulmonary Histoplasmosis

In approximately 10% of the patients with symptomatic primary infection the spontaneous remission does not take place and the respiratory disease adopts a chronic and progressive course. This clinical form of histoplasmosis is clinically and radiologically identical to advanced tuberculosis in adult patients (Rubinstein and Negroni 1981; Negroni 1989a). Nearly all cases are of Caucasian males over 50 years of age, heavy smokers who suffer from chronic obstructive pulmonary disease. Defective lung architecture is considered the most important risk factor for this clinical form of histoplasmosis. These defects impede the complete resolution of the mycosis, even in immunologically normal hosts. This chronic pulmonary form may result from exogenous re-infection or reactivation of endogenous foci (Alsip and Dismukes 1986; Negroni 2000).

Histopathologically inflammatory infiltrates, consisting of macrophages and lymphocytes that subsequently give rise to the formation of epithelioid granulomas are observed. With chronic evolution, caseous material in the central part of the granulomas, fibrosis in their periphery and pulmonary emphysematous bullae around the granulomas appear. When granulomas evolve, lung parenchyma is destroyed and areas of fibrosis develop. This is a continuous process in which the same cycle repeats in adjacent zones and gives way to extensive areas of compromise in both lungs. Involvement is usually symmetrical, affecting the apices and producing pleural thickening. With time this inflammatory process causes cavitations whose walls progressively thicken (Goodwin et al. 1981; Salfelder et al. 1990; George and Penn 1993).

Although the clinical manifestations are similar to those observed in pulmonary tuberculosis, histoplasmosis is less severe and presents a chronic evolution over several years with periods of progression and remission. More than 50% of the cases with lung infiltrates without cavitations remit spontaneously. Similar findings are seen in patients with cavitations with thin walls measuring 1–2 mm. On the other hand, the disease is chronic and progressive in cases that exhibit cavitations with walls measuring 3–4 mm thick (Goodwin et al. 1981; Negroni 2000).

The most frequent symptoms are evening fever, cough, mucopurulent or bloody expectoration, thoracic pain, dyspnea on excerption, asthenia, anorexia and weight loss.

Radiologically, heterogeneous, diffuse or nodular infiltrates are seen, mainly in the upper lobes, accompanied by pleural thickness. Cavitations in one or both pulmonary apices are observed, fibrosis and emphysema develop over time. Calcified lung nodes are detected in one-third of the cases.

The functional capacity of the lungs is significantly reduced as is demonstrated by the functionally respiratory capacity tests.

The complementary laboratory tests tend to reveal a marked acceleration of the sedimentation rate, mild normocytic anemia, neutrophilia in one-third of the cases and elevation of alkaline phosphatase levels.

Diagnosis is ascertained by the microscopic observation of *H. capsulatum* yeast form in mycological or histopathological studies or when positive cultures of sputum, bronchoalveolar lavage and surgical specimens are obtained. The scarcity of yeasts in the affected tissues and the rapid development of contaminant fungi in the upper airways make both the microscopic observation and the isolation in cultures of *H. capsulatum* difficult (Negroni 1989a).

Serological studies, especially immunodiffusion and complement fixation tests, constitute a valuable aid in the diagnosis of this clinical form. Immunodiffusion is less sensitive but more specific than complement fixation tests. The result is defined as positive when it shows M and H bands against histoplasmosis antigen. Complement fixation tests may produce 5% of false positive results in endemic regions as well as

25% of cross-reactions with other mycotic antigens. All positive reactions with titer equal or greater than 1/32, or reactions with progressive elevations in the titers, are strong indicators of progressive diseases (Ajello et al. 1962; George and Penn 1993).

Chronic pulmonary histoplasmosis is an invalidating disease leading to functional respiratory insufficiency, fatal hemoptysis, secondary bacterial infection, pulmonary hypertension and "cor pulmonale". The spontaneous evolution of the disease is often fatal, but exhibits little or no tendency to disseminate beyond the lungs and contiguous lymph nodes (Negroni 1989a; Kauffman 2009).

Disseminated Histoplasmosis

The progressive disseminated forms are seen in 1: 2000 infected persons in USA; smaller proportions have been detected in other parts of the world, according to the endemic zones. The majority of the patients suffering from this clinical form are immunocompromised. Sex and age are important predisposing factors; the majority of the patients with progressive disseminated manifestations are under one or over 53 years of age; the latter being predominantly males in a proportion of 3 : 1 or 10 : 1 (Goodwin et al. 1980; Kwon-Chung and Bennett 1992; Kauffman 2009). The most important conditioning factor is a deficit of cell-mediated immunity. It may be mild, such as that produced by advanced age, type 2 diabetes, the use of non-steroid anti-inflammatory agents or of low doses of corticosteroids, alcoholism and chronic smoking. These predisposing factors usually give way to the chronic disseminated forms (Negroni 1989a; Negroni 2000). More serious defects of cell-mediated immunity are observed in AIDS patients with low CD_4+ counts (<150/μL), in organ transplant recipients, in those undergoing chemotherapy for onco-hematological diseases, in patients receiving high doses of corticosteroids or treatment with TNF-α antagonists. The latter risk factors are responsible for acute or subacute disseminated histoplasmosis (Wheat and Kauffman 2003; Kauffman 2008; Deepe 2012). Progressive disseminated histoplasmosis may result from exogenous re-infection or from the reactivation of latent foci after a prolonged period of asymptomatic infection and the evolution of the disease is conditioned by the degree of immunity alterations (Negroni 1965; Negroni et al. 2010a).

Acute Disseminated Form

This clinical form is often found in early childhood and in patients with onco-hematological diseases or advanced HIV infection. These acute cases represent approximately 10% of the patients with disseminated histoplasmosis. Non-focal clinical manifestations of a severe infectious disease predominate upon the focal signs. Clinical signs include high fever, weight loss, a rapid deterioration of the general conditions, purpuric skin lesions, pancytopenia, diarrhea, cough, dyspnea, acute respiratory insufficiency and shock. The presentation of these cases is similar to that of an acute septic syndrome with multiorgan failure, shock and intravascular disseminated coagulopathy or to an adult acute respiratory distress (Negroni 1989a; Kauffman 2009; Negroni et al. 2010a).

Chest radiographs show diffuse interstitial or reticulonodular infiltrates, but rapidly progress to the findings associated with an acute respiratory distress.

The evolution is often fatal in less than a month (Negroni 2008a).

Subacute Disseminated Form

AIDS constitutes the most important risk factor for this clinical form. In South America more than 90% of the cases are HIV-positive patients with CD_4 cell counts below 150/μL. Most of them live in the endemic area. Histoplasmosis is associated with an estimated tenfold increase in frequency when HIV-positive patients of endemic and non-endemic areas are compared. At the beginning of the AIDS pandemia, 5% of the AIDS cases that required assistance for infectious diseases complications in Buenos Aires, exhibited subacute disseminated histoplasmosis. After the introduction of High Active Antiretroviral Therapy (HAART) this percentage decreased to 2.5% (Corti et al. 2004; Negroni et al. 2004; Negroni 2008a). In other endemic regions a higher proportion of AIDS-related histoplasmosis can be observed: in Indianapolis, U.S.A., 27% of the HIV-positive patients requiring hospitalization suffer from this mycosis (Wheat et al. 1985; Wheat and Kauffman 2003). The clinical manifestations are similar to those of other serious infectious processes: prolonged fever, weight loss, asthenia, anorexia, diarrhea, vomiting, hepatosplenomegaly, multiple adenomegalies, cough, expectoration, dyspnea, skin and mucus membranes lesions and pancitopenia (Kauffman 2006; Kauffman 2008; Negroni et al. 2008a). In Latin America skin or mucus membrane lesions appear in 80% of the patients in this clinical condition. In U.S.A. only 6% of these cases present skin alterations. Skin lesions are usually multiple and exhibit a wide spectrum of clinical aspects. Very frequently, they manifest as small papules, 3–4 mm in diameter, on various parts of the body; the vertex are usually ulcerated and covered with scabs (Figs. 6 and 7). Others are large ulcers with granulomatous bases and sharp edges, vegetated ulcers, nodules or diffuse hypodermitis, moluscoid papules or lupoid lesions. Mucosal lesions are less frequently observed and appear like ulcers covered by white secretions, localized on the oropharynx, on the larynx or on the penis (Fig. 8) (Negroni 1978; Corti et al. 2004; Huber et al. 2008; Negroni 2008b; Arenas 2011; Bonifaz 2012).

Chest radiological studies show interstitial micronodular infiltrates or diffuse shadows in both lungs. Pleural involvement is rare.

Central nervous system compromise is seen in less than 20% of the patients with acute or subacute disseminated histoplasmosis. Clinically, it is meningoencephalitis which compromises the basal nuclei of the brain, the most frequent symptoms and signs being headaches, convulsions, alterations of the state of consciousness, behavior changes, nucal rigidity, intracranial hypertension and cranial nerves paralysis. Encephalic magnetic resonance shows focal lesions in the brain's basal nuclei. Cerebrospinal fluid presents an increase in the level of proteins, positive globulin reaction and discrete lymphocytic pleocytosis (50–100 cells/μL). *H. capsulatum* can rarely be isolated from CSF, but cultures are more frequently positive than in the chronic meningoencephalitis (Negroni et al. 1997; Corti et al. 2004).

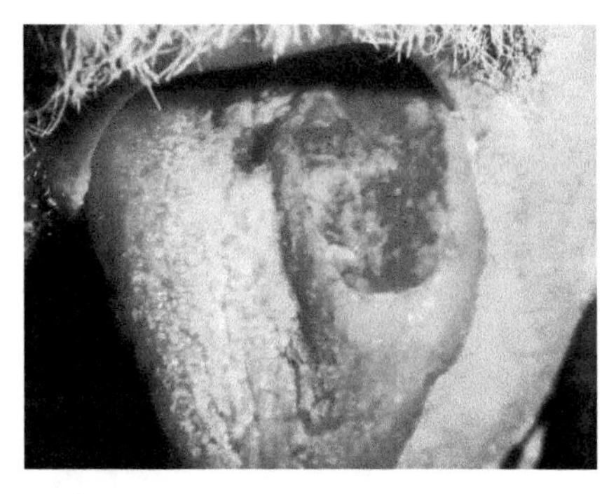

Figure 5. Tongue ulcer in a chronic disseminated histoplasmosis.

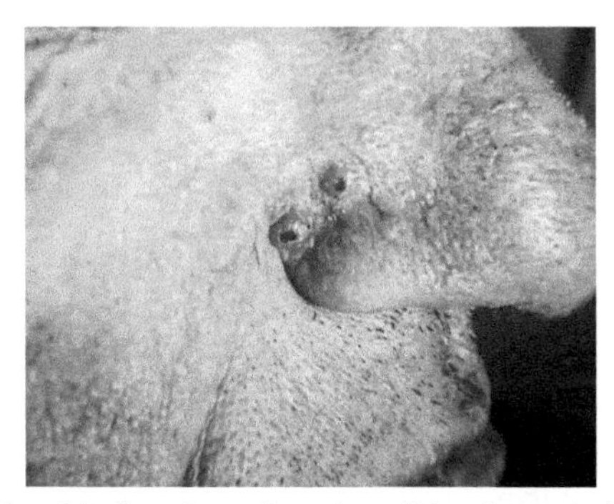

Figure 6. Ulcerated papules of the face, due to disseminated histoplasmosis, in a 74-year-old man suffering a thymoma.

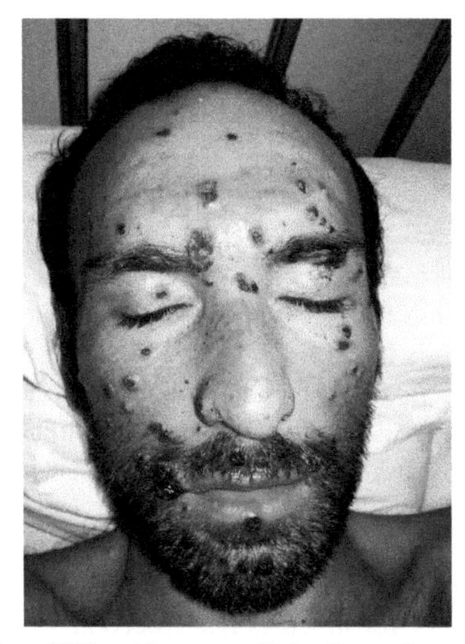

Figure 7. Ulcerated papules in an HIV-positive man suffering from subacute disseminated histoplasmosis.

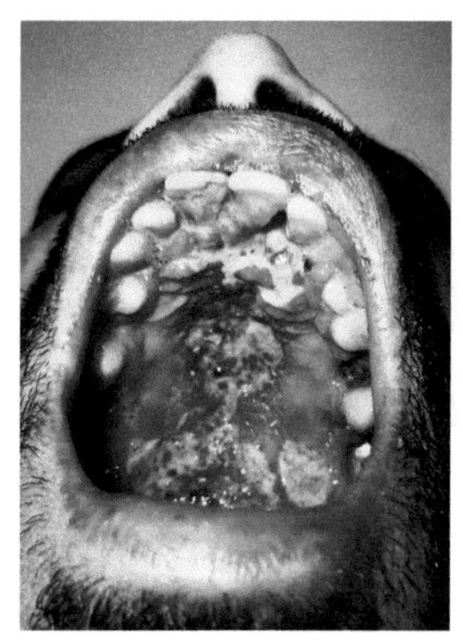

Figure 8. Oral ulcerated lesion in the same case.

Gastrointestinal attack is observed in this clinical form, the most common clinical manifestations being diarrhea, abdominal or stomach aches, hematemesis, melena, ulcers and intestinal perforations with very serious peritonitis. Ulcerations of the mucous membranes of the stomach and gut are detected by endoscopic studies (Negroni 2000).

Bone lesions are rare; they are more often located in the long bones and can be seen in radiology examinations. Clinically, they produce pain, functional impotence and swelling of the soft tissues adjacent to the bone lesions.

Ultrasonography and CT scan of the abdominal cavity frequently show heterogeneous hepatomegaly, homogeneous splenomegaly and abdominal and retroperitoneal adenopathies (Negroni 2008a).

Complementary laboratory studies reveal an accentuated acceleration of the erythrocyte sedimentation rate, thrombocytopenia, anemia and elevations of hepatic enzymes (especially alkaline phosphates).

In HIV-positive patients, the association of subacute histoplasmosis with other diseases is frequent. In 93 cases observed in the Infectious Diseases Muñiz Hospital of Buenos Aires City, Argentina, the following co-morbidities were detected: tuberculosis (32 patients), B viral hepatitis 19, C viral hepatitis 16, herpes simplex 12, oropharyngeal candidiasis 10, *Pneumocystis jiroveci* pneumonia 9, cerebral toxoplasmosis 8, herpes zoster 5, gastric and esophageal candidiasis 6, neurosyphilis 6; pulmonary nocardiosis, meningeal cryptococcosis, *Streptococcus pneumoniae* septicemia and Kaposi sarcoma with 2 cases each (Negroni et al. 1997).

In solid organ transplants, recipients fungal infections are usually detected in 5.3% of the cases, of which 22% are histoplasmosis. Previous CMV infection is a very important risk factor for mycosis (La Rocco and Burgert 1997; Paddi et al. 1996; Marques et al. 2008). Subacute disseminated histoplasmosis observed in these patients shows clinical symptoms and signs similar to those observed in AIDS-related histoplasmosis, but skin nodules, gummas and diffuse hypodermitis are more often detected. These types of skin lesions are also seen in patients submmited with high doses of corticosteroids for other reasons. Cutaneous lesions are observed in 57% of the cases (Negroni et al. 2010a).

Histoplasmosis is a late complication of solid organ transplantations; it may occur approximately 130 days after the graft. In the last 10 years a decrease in the incidence of histoplasmosis in these patients has been detected, probably due to less aggressive immunosuppression therapy (Kauffman 2009; Negroni 2010a).

Due to the high incidence of skin and mucous membranes lesions and the high fungal burden of this clinical form, the diagnosis of subacute disseminated histoplasmosis is easy. Approximately 80% of the cases of AIDS-related histoplasmosis are diagnosed by Tzanck's cytodiagnosis method. The clinical sample is obtained by scraping the base of the ulcers and preparing smears on slides, which are finally stained by Giemsa technique (Arechavala et al. 1993; Negroni 2008a; Negroni 2008b). Skin and mucus membrane biopsies are also very useful for mycological and histopathological studies. Biopsies for mycological examination should be sent to the laboratory in a sterile receptacle with isotonic saline solution. Blood cultures by lysis-centrifugation technique have proved to be highly efficient, yielding positive results in 75% of the patients; and blood cultures were the first diagnostic element in 20% of them (Negroni et al. 1997; Bianchi et al. 2000). Other very useful diagnosis procedures, which are not routinely carried out, are: bone marrow and lymph node aspiration, microscopic observation of the leukocyte layer of the hematocrit (buffy coat) and bronchoalveolar lavage. All these methods allow the microscopic observation of *H. capsulatum* or its isolation in cultures (Negroni 2008a).

Classic serology studies searching for antibodies frequently yield false-negative results in this clinical form. Immunodiffusion, counter immunoelectrophoresis and complement fixation tests give positive results in less than 30% of the patients (Arechavala et al. 1993). The ELISA technique, using an exo-antigen of the yeast phase of *H. capsulatum,* increases the positive results to approximately 75% in patients with AIDS-related histoplasmosis, but cross-reactions with other fungal antigens have often been observed with this method (Arechavala et al. 1997).

Radioimmunoassay and ELISA techniques searching for antigens of *H. capsulatum* in urine and serum are very important diagnostic tools in this clinical form. Urine samples are better than serum samples and sensitivity increases when several concentrated urine samples from the same patient are studied (Wheat and Kauffman 2003; Wheat et al. 2007; Kauffman 2008).

Histoplasmin skin test is usually negative in these patients.

Chronic Disseminated Form

This clinical form predominates in adult males over 53 years of age. The main predisposing factors include mild immunological deficiency caused by advanced age, chronic alcoholism, type II diabetes, prolonged used of low doses of corticosteroids, solid neoplasms and chronic lymphomas (Negroni et al. 1987; Negroni et al. 1994).

The most important clinical features include asthenia, weight loss and usually, muco-cutaneous lesions. These are monomorphic, ulcerative or vegetating and are localized in the buccal mucosa, the tongue, the pharynx or the nasal septum. These alterations have been detected in 40% of the cases. The ulcers are sharp-edged with a smooth base or with peaked red granulomas, partially covered with yellow-white secretions (Fig. 5). Occasionally, the base of the ulcer is vegetated. White lesions resembling lichen planus or leukoplasia are also observed. These alterations are produced by a superficial necrobiosis of the mucus membrane (Negroni 2000; Negroni 2008b).

In one-third of the patients lesions are polymorphic, showing nodules, erosions with granulomatous bases and hemorrhagic dotting similar to the mulberry-like stomatitis of paracoccidiodomycosis. Chancroid and aphtae lesions are rarely present (Negroni 1978).

Tongue involvement occurs in 10% of the patients. Fissured ulcers situated on the central and posterior parts of the tongue are the most characteristic lesions. Sublingual and lateral tongue ulcers are also common. Oropharyngeal ulcers are accompanied by pain, odynophagia, sialorrhea, macroglosia and poor dental conditions (Negroni 2000).

The destruction of sub-nasal septum by granulomatous, ulcerated, scabbed lesions, which mimics lesions produced by muco-cutaneous leishmaniasis, occurs in less than 10% of the patients (Negroni 1978).

Laryngeal compromise is detected in half of the cases. The symptoms are dysphonia, odynophagia, obstructive dyspnea, cough and mucopurulent expectoration. Laryngoscopy reveals a predominance of supraglottic lesions with red-violet infiltrates of the epiglottis and ventricular bands. Erythematous nodules and ulcers with granulomatous bases partially covered by yellow secretions are also often seen. Infraglottic lesions are infiltrative or granulomatous in nature and may produce laryngeal stenosis. Sometimes, tracheotomy is necessary. Signs and symptoms are very similar to those caused by laryngeal cancer (Negroni 1989a).

Skin lesions are less frequent, appearing in only 10% of the cases. They may present ulcers with clean edges, several centimeters in diameter, with a red granulomatous base partially covered by a brown-yellow scab. Ulcerated papules and nodules, diffuse infiltration of the lips and chancroid lesions of the external genitalia have been observed in some patients. Those persons under prolonged corticosteroid treatment may present a nodular necrotizing cellulitis (Negroni 2008b).

Pulmonary lesions are detected in less than 20% of the patients. They consist of diffuse interstitial infiltrates, localized in the middle and lower fields of the lung.

Hepatosplenomegaly is not prominent and can be detected by ultrasonography or CT scan.

Adrenal insufficiency is also observed in this clinical form. It may develop into Addison's syndrome and enlargement or destruction of the adrenal glands can be found using abdominal CT scan. Although adrenal insufficiency is detected in 10% of the cases, autopsy studies reveal that adrenal involvement is more frequent (Negroni 1989a).

Chronic meningoencephalitis due to *H. capsulatum* is rarely observed. Headache and mental confusion are the first symptoms; later on, seizures and nuchal rigidity may appear. Cerebrospinal fluid shows hyperproteinrrhachia and lymphocytic pleocytosis, but the most important diagnosis finding is the presence of specific antibodies detected by complementary fixation tests or immunodiffusion. *H. capsulatum* is rarely isolated from the CSF in this clinical form (Negroni et al. 1995).

Immunologically-Mediated Diseases

This includes histoplasmoma, mediastinal fibrosis and the ocular syndrome presumably associated with histoplasmosis (Goodwin et al. 1981; Alsip and Dismukes 1986).

Histoplasmomas are asymptomatic nodular lesions of the lungs, which appear on chest X-rays as coin lesions of variable sizes. They may grow slowly; central and peripheral calcifications are often found. Mediastinal fibrosis has already been described. Histoplasmomas and mediastinal fibrosis are sequelae of primary infections and are considered to be mediated by exaggerated hypersensitivity to *H. capsulatum* antigens which are released from the latent foci of infection (Alsip and Dismukes 1986; Negroni 1989a).

The presumed ocular histoplasmosis is a type of choriorretinitis often observed in some endemic areas; its incidence oscillates between 1% and 10%, predominantly affecting Caucasian women between 30 and 40 years of age who have the HLA-B7 histocompatibility antigens. The histoplasmin skin test as well as the *in vitro* lymphoblast transformation with histoplasmin are often strongly positive, but *H. capsulatum* has never been found in histopathological sections nor has it been cultivated from the enucleated eyes of these patients. This choriorretinitis might result from the deposits on the choroids of antigens liberated from pulmonary or ganglionic foci. However, an identical syndrome may also occur outside the endemic zones, suggesting that there may be other causes of this clinical picture (Negroni 2000). Clinically, a non-specific inflammatory reaction producing local hemorrhages and retinal detachment is the most frequent finding. Later on, yellow scars with sharp edges surrounded by inflammatory choroiditis can be observed. Patients exhibit loss of visual acuity and permanent scotomas. Blindness occurs in 50% of the non-treated cases.

Histopathological studies show mononuclear cellular infiltrates in which B lymphocytes and T CD_8-positive cells predominate (Deepe 2012).

This clinical form of the disease does not cause a generalized compromise of the patient. Corticosteroid treatment and laser photocoagulation are effective in the control of these ocular signs; however, these treatments are not advisable when lesions occur near the fovea (Negroni 2000).

Laboratory Diagnosis

The laboratory diagnosis of histoplasmosis is made by direct or indirect methods. The first ones include microscopic examination, cultures and, rarely, animal inoculation in mice or hamsters of specimens obtained from human lesions. The indirect methods are those related to specific immune responses, including serologic and skin tests with histoplasmin as well as the detection of *H. capsulatum* antigens in organic fluids (Ajello et al. 1962; Arechavala et al. 1993; Wheat and Kauffman 2003; Negroni et al. 2004).

The clinical samples used for diagnosis are obtained from biopsies of mucocutaneous, lymph node, hepatic or pulmonary lesions as well as from sputum, bronchoalveolar lavage, bone marrow aspiration, etc.

These specimens should be placed in sterile containers with an isotonic saline solution for mycological study and in 10% formaldehyde for histopathology.

The microscopic observation in the mycological study is performed by examining Giemsa stained smears with X 1,000 magnification. The histopathological examination must be carried out on histological sections stained by P.A.S. and Grocott, since hematoxylin-eosin renders poor results. The microscopic morphology of the etiologic agent has already been described.

The clinical samples should be cultured in different media such as Sabouraud dextrose-agar, potato-dextrose agar or Borelli lactrimel, to all of which antibacterial antibiotics and cycloheximide should be added; the incubation is at 28°C for 3 to 4 weeks. In addition, a part of the sample must be seeded in brain-heart infusion agar with 5% of rabbit blood and antibacterial antibiotics and incubated at 37°C.

Animal inoculation of the clinical specimens is usually done by intraperitoneal route in mice or hamsters with a suspension of the sample in isotonic saline solution with antibacterial antibiotics. Three weeks later the animals are killed, and the liver and spleen pieces are cultured in Sabouraud-dextrose agar and brain-heart infusion agar for *H. capsulatum* isolation (Negroni 1965; Kown-Chung and Bennett 1992).

Blood cultures are an important diagnostic tool in immunocompromised hosts, especially HIV-positive patients. In Muñiz Hospital of Buenos Aires City, 72.8% of the cases suffering AIDS-related histoplasmosis present positive blood cultures. A lysis-centrifugation technique should be used because it is seventimes more sensitive than classical blood cultures. Patients with positive blood cultures usually present a worse evolution (Bianchi et al. 2000; Negroni 2008a).

Bone marrow aspiration allows the microscopic observation of *H. capsulatum* in Giemsa stained smears and its isolation in cultures. This diagnosis method is advised in patients with pancytopenia and/or big splenomegaly (Negroni 2010a).

During the last decade several molecular biology methods for *H. capsulatum* DNA detection in clinical samples have been used in various laboratories; nested PCR and real time PCR were the techniques more often employed, but commercial kits are not yet available (Muniz et al. 2001; Bracca et al. 2003; Kauffman 2009; Deepe 2012).

Histoplasmin skin test is used in epidemiologic surveys since it is useful for diagnosing both recent and past infections. As we have already said, a high percentage of healthy individuals in endemic areas have positive reactions. The skin test is not useful in the diagnosis of the active disease, but it has an important prognostic value since it presents negative results in severe cases of disseminated histoplasmosis (Ajello et al. 1962; Negroni et al. 1987).

The usefulness of classical serology tests has been previously explained; modern methods, such as ELISA for antibody detection and Western blot, are used in some centers. ELISA is more sensitive than immunodiffusion but less specific, showing cross reactions with other fungal antigens (George and Penn 1993; Zancopé-Oliveira et al. 1994; Arechavala et al. 1997; Kauffman 2008). Western blot presents 90% sensitivity and 100% specificity and was effective in the diagnosis of acute pulmonary forms in an outbreak of histoplasmosis (Pizzini et al. 1999).

Three ELISA commercial kits for *H. capsulatum* antigens detection in serum and urine are now available, all of them presenting cross reactions with *Blastomyces dermatitidis* and *Paracoccidiodes brasiliensis.* The sensitivity of ELISA techniques for antigens may be improved using a previous treatment of the sample with EDTA and heat with the purpose of breaking antibody-antigen complexes. The use of monoclonal antibodies against species-specific epitopes of *H. capsulatum* in an inhibition ELISA technique for antigen detection has shown 71% sensitivity and 98% specificity (Gomez et al. 1999; Wheat et al. 2007). This kit is also useful for antigen detection in CSF and bronchoalveolar lavage. In Muñiz Hospital we are using *Histoplasma* galactomannan antigen kit, IMMY, Immuno Mycology. Norman. OK, U.S.A. with good results; urine is better than serum sample for antigen detection.

Chemical techniques for α-1.3 D glucan dosage in serum, which are frequently employed in the diagnosis of some invasive fungal infections in compromised patients, may render positive results in patients with disseminated histoplasmosis, but it lacks specificity (Egan et al. 2007).

Histopathology

Histopathological studies have differentiated three primary aspects in this mycosis: inflammation, necrosis and reparation (Negroni 1965).

In the inflammatory response three different patterns have been recognized: a) suppurative, with production of abscesses; b) macrophagic hyperplasia and c) epithelioid granuloma with giant cells (Chandler et al. 1980).

Necrosis is characterized by the presence of karyolysis, karyorrhesis and pygnosis of the cells. Caseation is observed in lymph nodes, adrenal glands and especially in the lungs.

In the severe primary infections a particular reaction occurs in the lungs: the alveolar spaces of the entire lungs are filled with proteinaseous substances, which may be differentiated from common edematous fluid. In these cases the pneumonocytes producing the so-called alveolar surfactant proteins are stimulated by fungal antigens. After stimulation they proliferate and produce an excess of these substances, which may not be reabsorbed at the time. The etiologic agent grows rapidly in the alveolar spaces (Salfelder et al. 1990).

The reparation consists of a proliferation of fibroblasts, which give way to collagenous fibrosis, calcification and rarely ossification.

There is an important correlation between the histopathological reactions and the degree of immunity in the hosts. In severely immunocompromised patients a great proliferation of histiocytes with foamy appearance due to the presence of numerous phagolysosomes is detected (Negroni 2000). In HIV-positive patients the fungal burden is so high that the yeast-like cells are liberated when the macrophages burst (Negroni 2008a). In cases with a minor deficit of cell-mediated immunity the histopathological response consists of the production of epithelioid granulomas with giant cells and a lymphocyte and plasmocyte layer in the periphery. In chronic cases a caseous necrobiosis is observed in the central part of the granulomas and collagenous fibrosis can be seen surrounding the whole inflammatory reaction.

H. capsulatum is difficult to recognize in H & E stains, except when the fungal burden is very high. In these cases several corpuscular structures are observed inside the phagosomes in macrophagic cells. As we have already pointed out, P.A.S. and Grocott stains are useful for *H. capsulatum* observation; Gram stains may also be employed, as all fungi in this microorganism are Gram-positive. *H. capsulatum* should be differentiated from *Candida glabrata* and *Penicillium marneffei,* both present in the same size and shape, but *C. glabrata* is usually outside the cells and *P. marneffei* presents a septum instead of a bud. *Leishmania* and *Toxoplasma gondii* may be confounded with *H. capsulatum,* but they do not take Grocott stains. Some non-living corpuscular structure particles and inclusion bodies, which may be found in tissues, may look like small yeast cells because they are P.A.S.-positive, especially in myosphesulosis, a disease produced by the intramuscular infection of oily suspension of antibiotics and other drugs. These structures are Grocott-negative (Salfelder et al. 1990).

Differential Diagnosis

Symptomatic primary infections resemble viral, *Mycoplasma* or bacterial pneumonia. The chronic pulmonary form is identical to that of fibrocaseous tuberculosis (Rubinstein and Negroni 1981). The mucocutaneous manifestations of chronic disseminated forms may be confused with several diseases, including carcinoma, primary or tertiary syphilis, paracoccidiodomycosis, tuberculosis, leukoplasia and lichen planus (Negroni 2008b). The acute and subacute disseminated histoplasmosis resemble leukemia, lymphomas, visceral leishmaniasis, pneumocytosis and bacterial sepsis (Negroni 1989a). The pulmonary and lymph node calcifications are similar to those produced by tuberculosis, coccidiodomycosis, brucellosis and silicosis. Histoplasmomas are difficult to differentiate from benign or malignant tumors of the lung; mediastinal and pericardial fibrosis resemble those caused by tuberculosis (Rubinstein and Negroni 1981; Deepe 2012).

Prognosis

The prognosis of histoplasmosis is greatly variable according to the different clinical forms. It is good in symptomatic primary infections and poor in acute disseminated forms. The central nervous system and adrenal compromise turns the prognosis worse.

In general, the antifungal treatment is very effective in this mycosis. In Argentina, more than 70% of the cases suffering from AIDS-related histoplasmosis present a good clinical response to specific treatments. The same is observed in patients with other types of immunodeficiency (Negroni 2008a; Negroni et al. 2010a).

Treatment

Azolic compounds such as itraconazole, ketoconazole, fluconazole, voriconazole and posaconazole and the polyenic antibiotic amphotericin B are active *in vitro* and *in vivo* against *H. capsulatum* (Wheat 2002). The indications of antifungal drugs vary according to the clinical form of histoplasmosis and the individual characteristics of the patient. The therapeutic schemes most frequently used are summarized in Table 2.

Symptomatic primary infections do not usually require antifungal treatment. Only those cases which do not present spontaneous remission within four to six weeks after the infection or patients with immunological compromise are treated with antifungal drugs, usually with itraconazole (Negroni 2000; Wheat and Kauffman 2003).

Severe cases with marked respiratory insufficiency need mechanical respiratory assistance as well as corticosteroids in doses equivalent to 60 or 80 mg/day of prednisone. During corticosteroid treatment the patients should be protected with itraconazole in doses of 200–400 mg/day (Negroni 2000).

Hypersensitivity reactions such as erythema nodosum, arthritis, pericarditis or pleurisy are treated with non-steroidal anti-inflammatory drugs and, in serious cases, with corticosteroids. Antifungal protection is mandatory in these cases (Wheat 2002).

Mediastinal granulomas are very difficult to treat. Initially, itraconazole by oral route at a daily dose of 200–400 mg is the treatment of choice; it should be maintained for 18 months. If this medication fails, surgical intervention should be evaluated very carefully because it is a high-risk procedure and it frequently fails (Kauffman 2009). No medical treatment is useful for mediastinal fibrosis. Surgery is considered to be risky and it is not advisable in these cases. The placement of intravascular stents into obstructed arteries or veins may be helpful in very serious cases (Deepe 2012).

Histoplasmomas are often surgically resected by lobectomy because of the high risk of being confused with lung neoplasms. Those with calcified centers are not usually considered for surgical resection because the risk of being confused with lung cancer is much less (Negroni 1989a).

Chronic cavitary histoplasmosis of the lung is treated successfully with antifungal drugs in 90% of the cases; recurrences are reported in 20% of these cases. Itraconazole is the drug more often indicated, but it requires attention to drug interactions. Surgical resection of the cavitations is rarely indicated due to the high risk of developing chronic respiratory insufficiency and bronchopleural sinus tracts (Negroni 2000; Kauffman 2011).

Table 2. Therapeutic schemes for histoplasmosis.

Clinical form	Drug	Administration	Daily Dose	Duration
Symptomatic Primary infection	Itraconazole	Oral	400 mg	4 weeks
Pulmonary Chronic Cavitary	Itraconazole Amphotericin B	Oral 0.7 mg/Kg	200–400 mg 0.7 mg/Kg	12 months 35 mg/Kg
Chronic Disseminated	Itraconazole Amphotericin B	Oral I.V.	100–200 mg 0.7 mg/Kg	6 months 35 mg/Kg
Subacute Disseminated	Itraconazole Amphotericin B	Oral I.V.	400 mg 0.7 mg/Kg	12 months 35 mg/Kg
Acute disseminated	L-amphotericin B	I.V.	3–5 mg/Kg	2 months
Mediastinal Granulomas	Itraconazole	Oral	400 mg	18 months

In Argentina the treatment of choice for chronic disseminated histoplasmosis is itraconazole at a daily dose of 200 mg for 6 to 12 months. It is effective and well tolerated, but it should not be used in patients receiving other drugs which interact with itraconazole, such as rifampin, phenytoin, antacids, H_2 receptor blockers, cyclosporine, terfenadine, etc., or in those patients suffering from cardiac insufficiency or liver alterations (Negroni et al. 1987; Negroni et al. 1989b). Amphotericin B is indicated in these cases. Although ketoconazole has shown to be active in this clinical form of the disease, it is less effective and worse tolerated than itraconazole; in some countries it is no longer available for oral use.

Patients with adrenal insufficiency can usually be stabilized with 30 mg/day of oral hydrocortisone.

Histoplasma meningitis responds poorly to medical treatment. Liposomal amphotericin B at a daily dose of 5 mg/kg during 4–6 weeks is the treatment of choice. After this initial part of the treatment, itraconazole by oral route at a dose of 400–600 mg/day for 12 months is indicated in order to avoid relapses. Although fluconazole reaches higher levels in CSF it does not seem to be more effective than itraconazole in this clinical form (Negroni 2000). Due to the risk of hydrocephalus, ventricular-atrial or peritoneal shunts are often required. Intratecal or intraventricular administration of amphotericin B is rarely indicated because of the high risk of arachnoiditis or meningeal hemorrhage. This drug should be administered in doses of 0.1 to 0.5 mg dissolved in 5 ml of a 10% dextrose solution with 20 mg of hydrocortisone twice a week. After the injection the patient should be put in the Trendelenburg position for 30 minutes (Negroni 1965; Negroni 2010a).

Endocarditis is a rare clinical event associated with chronic disseminated histoplasmosis. The prognosis is very poor; the mortality rate in treated patients is 50%. Treatment consists of amphotericin B-deoxycholate in a total dose of 35 mg/kg. Surgical valve replacement is often required. Secondary prophylaxis for one year with itraconazole in doses of 200 mg/day is advisable (Wheat 2002; Wheat and Kauffman 2003).

In patients with subacute disseminated histoplasmosis, amphotericin B by intravenous route is the treatment of choice when the patient presents diarrhea and emesis, receives drugs which interact with itraconazole or suffers serious clinical manifestations or meningoencephalitis. In South America, the remaining cases respond well to itraconazole. During the first three or four days of treatment with itraconazole, doses of 600 mg/day are required to achieve a rapidly effective tissue concentration of this azolic compound (Negroni 2010a).

Liposomal amphotericin B, as well as amphotericin B bound to other lipid formulations, is not routinely indicated for the treatment of this mycosis in Latin America due to its elevated cost. They should be used when amphotericin B-deoxycholate fails, or in patients with renal failure and creatinine blood levels equal or higher than 3 mg/dL or with severe anemia (Negroni 2000).

In AIDS-related histoplasmosis a secondary prophylaxis with itraconazole in doses of 200 mg/day is usually indicated. This treatment should be administered until the patient presents CD_4 + cell counts higher than 150/µL in two controls as a consequence of HAART. Protease inhibitors interact negatively with itraconazole.

If itraconazole is contraindicated 50 mg of amphotericin B should be administered twice a week (Negroni 2008a).

Conclusions

It is not possible to avoid *H. capsulatum* infections completely. It is recommended to avoid being unnecessarily exposed to caves with bats, poultry houses or to neighborhood constructions where the soil has been turned over, especially in the case of persons suffering immune deficiencies (Negroni 1989a).

In highly contaminated areas the following actions are recommended:

1. Areas containing black birds' roosts or bat excreta should not be unnecessarily disturbed.
2. If such sites have to be disturbed a 3% formaldehyde solution may be applied for soil decontamination.
3. The workers have to wear masks and protective clothing (Negroni et al. 2010a). No effective vaccine for histoplasmosis is available, but the 62 and 80 kDa thermal shock glucoproteins are considered important candidates for this vaccine (Deepe 2012).

References

Ajello, L., L. George, L. Kaufman and W. Kaplan. 1962. Laboratory Manual for Medical Mycology. Communicable Diseases Center, Atlanta, Georgia, USA.

Alsip, S. and W. Dismukes. 1986. Approach to the patients with suspected histoplasmosis. pp. 254–296. *In*: Remington, J.S. and M.N. Swartz (eds.). Current Topics in Infectious Diseases. Mc Graw-Hill, New York.

Arechavala, Robles A.M., R. Negroni, Bianchi and A. Taborda. 1993. Valor de los métodos directos e indirectos de diagnóstico en las micosis sistémicas asociadas al sida. Rev Med Trop Sao Paulo 35: 163–169.

Arechavala, A., K. Euguchi, C. Iovannitti and R. Negroni. 1997. Utilidad del enzimo inmunoensayo para el serodiagnóstico de la histoplasmosis asociada al sida. Rev Argent Micol. 20: 24–28.

Arenas, R. 2011. Histoplasmosis. In: Arenas R. Micología Médica Ilustrada. 4° Edición. Mc Graw-Hill Interamericana. México pp. 192–202.

Baddley, J.M., I.R. Sankara, J.M. Rodriguez, P.G. Pappas and W. Many, Jr. 2008. Histoplasmosis in HIV-infected patients in a Southern regional medical center: poor prognosis in the era of highly active antiretroviral therapy. Diag Microbiol Infect Dis. 62: 151–156.

Bianchi, M.H., A.M. Robles, R. Vitale, S. Helou, A. Arechavala and R. Negroni. 2000. Usefulness of blood culture in the diagnosing HIV-related systemic mycoses: evaluation of a manual lysis-centrifugation method. Medical Mycology 32: 77–80.

Bonifaz, A. 2012. Histoplasmosis. pp. 279–296. *In*: A. Bonifaz (ed.). Micología Médica Básica. 4° Edición. Mc Graw-Hill Interamericana, México.

Borelli, D. 1970. Prevalence of systemic mycoses in Latin America. pp. 28–38. *In*: Pan American Health Organization. Proceeding of the International Symposium on the Mycoses; Publication n° 205.

Bracca, A., M.E. Tosello, J.E. Girardini, S.L. Amigot, C. Gomez and E. Serra. 2003. Molecular detection of *Histoplasma capsulatum* var. *capsulatum* in human clinical samples. J Clin Microbiol. 41: 1753–1755.

Chandler, F.W., W. Kaplan and L. Ajello. 1980. A Colour Atlas and Text book of the Histopathology of Mycotic Diseases. Wolfe Medical, Lochem, Netherlands. pp. 64–69.

Corti, M., R. Negroni, P. Esquivel and M.F. Villafañe. 2004. Histoplasmosis diseminada en pacientes con sida: Análisis epidemiológico, clínico, microbiológico e inmunológico de 26 pacientes. Enfermedades Emergentes 6: 8–15.

Deepe, G.E. 2012. *Histoplasma capsulatum*. pp. 3299–3313. *In*: Mandell, G.L., J.E. Bennett and R. Dolin (eds.). Mandell, Douglas, Bennett. Enfermedades Infecciosas. Principios y práctica. Séptima Edición. Elsevier España, Barcelona.

Egan, L., P. Connolly, L.J. Wheat, D. Fuller, T.E. David, K. Knox and C.A. Hage. 2007. Histoplasmosis as a cause for positive Fungitell ™ (1-3)-D-glucan test. Med Mycology 45: 1.

George, R.B. and R.L. Penn. 1993. Histoplasmosis. pp. 39–50. *In*: Sarosi, G.A. and S.F. Davies (eds.). Fungal Diseases of the Lung, Second edition. Raven Press Ltd., New York.

Gomez, B.L., J.L. Figueroa, A.J. Hamilton, S. Diez, M. Rojas, A.M. Tobon, A. Restrepo and R.J. Hay. 1999. Detection of the 70-kilodalton *Histoplasma capsulatum* antigen in serum of histoplasmosis patients: correlation between antigenemia and therapy during follow-up. J Clin Microbiol. 37: 675–680.

Goodwin, J.E., J.L. Shapiro, G.H. Thurman, S.S. Thurman and R. Des Prez. 1980. Disseminated histoplasmosis. Clinical and pathologic correlations. Medicine 59: 1–33.

Goodwin, J.E., R. Lloyd and R. Des Prez. 1981. Histoplasmosis in normal host. Medicine 60: 231–266.

Huber, F., M. Nacher, C. Aznar, M. Pierre-Demar, M. El Guedi, T. Vaz et al. 2008. AIDS- related *Histoplasma capsulatum* var. *capsulatum* infection: 26 years experience of French Guiana. AIDS 22: 1047–1053.

Kauffman, CA. 2006. Endemic Mycoses: blastomycosis, histoplasmosis and sporotrichosis. Dis Clin North Amer. 3: 645–662.

Kauffman, C.A. 2008. Diagnosis of histoplasmosis in immunosupressed patients. Curr Opin Infect Dis. 21: 421–425.

Kauffman, C.A. 2009. Histoplasmosis. Clin Chest Med. 30: 17–225.

Kauffman, C.A. 2011. Histoplasmosis. pp. 321–336. *In*: Kauffman, C.A., P.G. Pappas, J.D. Sobel and W.E. Dismukes (eds.). Essential in Clinical Mycology, Second edition. Springer, New York, Dordrecht, Heidelberg, London.

Kwon-Chung, K.J. and J.E. Bennett. 1992. Histoplasmosis. pp. 464–513. *In*: Kwon-Chung, K.J. and J.E. Bennett (eds.). Medical Mycology. Lea & Febiger, Philadelphia.

La Rocco, M.T. and M. Burgert. 1997. Fungal infections in the transplant recipient and laboratory methods for diagnosis. Rev Iberoamer Micol. 14: 143–146.

Larsh, H.W. 1970. Ecology and epidemiology of histoplasmosis. pp. 59–63. *In*: Pan American Health Organization. Proceedings of the International Symposium on Mycoses. Publication nº 205.

Lindsley, M.D., H.L. Holland, S.L. Bragg, S.F. Hurst, K.A. Wannemuchler and C.J. Morison. 2007. Production and evaluation of reagents for detection of *Histoplasma capsulatum* antigenuria by enzyme immunoassay. Clin Vaccine Immunol. 14: 700.

Marques, S.A., S. Hozumi, R.M.P. Camargo, M.F.C. Carvalho and M.E. Marques. 2008. Histoplasmosis presenting as cellulitis 18 years after renal transplantation. Medical Mycology 46: 725–728.

Muniz, M.M., C.V. Pizzini, J.M. Peralta, E. Reiss and R.M. Zancopé-Oliveira. 2001. Genetic diversity of *Histoplasma capsulatum* strains isolated from soil, animals and clinical especiments in Rio de Janeiro state, Brazil, by a PCR-Based random amplified polymorphic DNA assay. J Clin Microbiol. 39: 4487–4494.

Negroni, P. 1965. Histoplasmosis Diagnosis and treatment. Charles Thomas Publisher Springfield, Illinois, USA.

Negroni, R. 1978. Manifestaciones cutáneas de la histoplasmosis. Revista Argentina de Micología. 1: 5–16.

Negroni, R. 1989a. Histoplasmosis. *In*: Hay, R.J. (ed.). Tropical Fungal Infections Bailliere's Clinical Tropical Medicine and Communicable Diseases. Bailliere Tindall, London. 4: 169–183.

Negroni, R. 2000. Clinical Spectrum and treatment of classic histoplasmosis. Rev Iberoamer Micol. 17: 159–167.

Negroni, R. 2004. Histoplasmose. Em: Cimerman S, Cimerman B. Condutas em Infectología. Atheneu. Sao Paulo, Brazil pp. 376–383.

Negroni, R. 2008a. Manifestaciones cutáneo-mucosas de la histoplasmosis diseminada (histoplasmosis clásica o histoplasmosis capsulati). Dermatol Argent. 14: 104–110.

Negroni, R. 2008b. Micosis asociadas al sida. pp. 325–351. *In*: Benetucci J. SIDA y enfermedades asociadas. 3º Edición. FUNDAI, Buenos Aires, Argentina.

Negroni, R., A. Arechavala and A.M. Robles. 1987a. Histoplasmosis diseminada crónica como afección oportunista. Medicina Cutánea Ibero-latino-Americana. 15: 377–383.

Negroni, R., O. Palmieri and F. Koren. 1987b. Oral treatment of paracoccidiodomycosis and histoplasmosis with itraconazole in humans. Rev Infect Dis. 9(Suppl. 1): 47–50.

Negroni, R., A.M. Robles, A. Arechavala and A. Taborda. 1989b. Itraconazole in human histoplasmosis Mycoses. 32: 123–130.

Negroni, R., A.M. Robles and A. Arechavala. 1994. Histoplasmosis progresiva. Estudio de 10 años. Revista Argentina de Micología. 17: 14–21.

Negroni, R., A.M. Robles, A. Arechavala, C. Iovannitti, S. Helou and L. Kaufman. 1995. Chronic meningoencephalitis due to *Histoplasma capsulatum*. Usefulness of serodiagnostic procedures in diagnosis. Serodiagn & Immunotherap 7: 84–88.

Negroni, R., A.M. Robles, A. Arechavala, H.M. Bianchi and S. Helou. 1997. Histoplasmosis relacionada al SIDA, su estado actual en la Argentina. Prensa Médica Argentina 84: 696–700.

Negroni, R., A.I. Arechavala and E.I. Maiolo. 2010a. Histoplasmosis clásica en pacientes inmunocomprometidos. Medina Cutánea. Ibero- Latino-Amicana 38: 59–69.

Negroni, R., R. Dure, A. Ortiz-Naredo, E.I. Maiolo, A. Arechavala, G. Santiso, C. Iovannitti, B. Ibarra-Camou and C.E. Canteros. 2010b. Brote de histoplasmosis en la Escuela de Cadetes de la Base Aérea de Morón, Provincia de Buenos Aires, Republica Argentina. Rev Argent Microbiol. 42: 254–260.

Paddi, V.R., S. Hariharan and M.R. First. 1996. Disseminated histoplasmosis in renal allograft recipients. Clin Transpl. 10: 160–165.

Pizzini, I., R.M. Zancopé-Oliveira, E. Reiss, R. Haijeeh, L. Kaufman and J.M. Peralta. 1999. Evaluation of a Western blot tests in an outbreak of acute pulmonary histoplasmosis. Clin Diagn Lab Immunol. 6: 20–23.

Rippon, J.W. 1988. Histoplasmosis (*Histoplasmosis capsulati*). *In*: J.W. Rippon (ed.). Tratado de Micología Médica. Interamericana McGraw Hill, México, pp. 411–456.

Rubinstein, P. and R. Negroni. 1981. Histoplasmosis. pp. 249–290. *In*: Rubinstein, P. and R. Negroni (eds.). Micosis Broncopulmonares de adulto y del niño. Segunda edición. Editorial Beta, Buenos Aires, Argentina.

Salfelder, K., T.R. de Liscano and E. Sauerteig. 1990. Histoplasmosis capsulati. *In*: K. Salfeldre, T.R. de Liscano and E. Sauerteig. Atlas of Fungal Pathology, Current Histopathology series. Kluwer Academic Publisher, Dordrecht, Boston, London. 17: 73–97.

Silveira, F.P. and S. Husain. 2007. Fungal infections in solid organ transplantation. Medical Mycology 46: 725–728.

Wheat, L.J., T.G. Slama and M.L. Zeckel. 1985. Histoplasmosis in the acquired immunodeficiency syndrome. Amer J Med. 78: 203–210.

Wheat, L.J. 2002. *Histoplasma capsulatum*. pp. 1069–1079. *In*: Yu, V., R. Weber and D. Raoult (eds.). Antimicrobial Therapy and Vaccine, 2nd edition. Apple Tree Productions, LLC. New York.

Wheat, L.J. and C.A. Kauffman. 2003. Histoplasmosis. *In*: Walsh, T. and J. Rex (eds.). Fungal Infections. Part II. Recent Advances in Diagnosis, Treatment and Prevention of Endemic and Cutaneous Mycoses. Infect Dis Clin N Amer. 17: 1–9.

Wheat, L.J., J. Witt, M. Dirkin and P. Connely. 2007. Reduction of false antigenemia in the second generation Histoplasma antigen assay. Med Mycology 45: 169–171.

Zancopé-Oliveira, R.M., S.L. Bragg, E. Reiss, B. Wanke and J.M. Peralta. 1994. Effects of histoplasmin M antigen chemical and enzymatic deglycosylation on cross-reactivity in enzyme-link immunoelectrotransfer blot method. Clin Diagn Lab Immunol. 1: 390–393.

CHAPTER 9

Sporotrichosis: The-State-of-The-Art

Alexandro Bonifaz,[1,2,*] *Rubí Rojas-Padilla,*[3]
Andrés Tirado-Sánchez[2] and *Rosa M. Ponce*[2]

Introduction

Sporotrichosis is an implantation mycosis caused by direct inoculation of the causative agents into the skin (by implantation); it is a subacute or chronic disease caused due to *Sporothrix* spp. complex. The most important etiologic agent is *Sporothrix schenckii*. It is the most widespread implantation mycoses in the world; many cases have been reported in virtually every continent, especially in the tropical and subtropical areas. Schenk described the first report of sporotrichosis in 1898 in the United States of America (USA); it corresponded to a classic lymphangitic sporotrichosis. This case was classified later as *Sporotrichum*. Hektoen and Perkins (Hektoen and Perkins 1900), who included the fungus in the genus *Sporothrix* and described the species *schenckii*, also described the next cases in USA. In France, de Beurmann (de Beurmann and Ramnond 1903) isolated a pigmented strain from a case of sporotrichosis, which was macroscopically different from that obtained from the American cases; he considered it as *Sporotrichum beurmanni*. Carmichael (Carmichael 1962), in 1962, made the unification of the different names as *Sporothrix schenckii*. It is important to mention that, in France, de Beurmann and Gougerot (de Berurmann and Gougerot 1912) reported more than 200 cases in 1912; nowadays, this mycosis is extremely rare in this French region due to an ecologic change or modification in the work conditions, it also draws attention to the epidemic in Transvaal, Africa, reported by Simson

[1] Department of Mycology, Dermatology Service.Hospital General de México. Mexico City.
[2] Dermatology Service.Hospital General de México. Mexico City.
[3] Dermatology Service, Hospital Infantil de México. Mexico City.
* Corresponding author: a_bonifaz@yahoo.com.mx

(Simson 1947); it occurred in gold mines and more than 3,000 cases were reported (pulmonary and cutaneous). It was later demonstrated that the fungus had infested the wood inside the mines.

Small epidemics have been reported in recent years, with familial cases that shared a common source. It is important to pay particular attention to the epidemic of southern Brazil, which was related to cat disease (zoonosis). It affected more than 3,000 cats and 1,000 humans.

A proposal on the classification of *Sporothrix*, based on various studies of molecular biology has been made (Marimon et al. 2007; Marimon et al. 2008). They proposed a cryptic complex denominated *Sporothrix schenckii*, which includes six species well differentiated in phylogenetic terms.

Etiology and Mycological Characteristics

Sporothrix complex is integrated by dimorphic fungi identified by genetic and molecular studies, based on specific gene sequences (chitin synthase, β tubulin, and calmodulin), including the following species that are classified into five different clades: *Sporothrix brasiliensis* (Clade I); *Sporothrix schenckii* (*sensu stricto*) (Clade II); *Sporothrix globosa* (Clade III); *Sporothrix mexicana* (Clade IV), and *Sporothrix pallida* (formerly *S. albicans*) (Clade V). The molecular identification of the species is the most useful tool for their classification.

In addition to the small morphological differences, different behavior has been reported from the species of the *Sporothrix schenckii* complex with respect to their growth patterns in different temperatures or media, as well as in their assimilation of carbohydrates but, undoubtedly, the most important difference between them is that the five species demonstrate distinct sensitivities to the majority of systemic antifungal agents.

Epidemiology

Gender and age

There are two peaks in the age distribution curve for sporotrichosis: the first occurs in school-aged children (30% of all cases), and the second is found in young adults, between 16 to 35 years of age (50% of all cases). In fact, there are cases reported in all age groups; for example, the youngest case ever reported in the literature is that of a newborn (2 days old) who was bitten by a rat and 8 days later developed the disease. Recently a large number of pediatric cases in Brazil have been reported, as part of the Brazilian outbreak (Barros et al. 2008). In this report, of the 81 children who were affected, 54% related to the coexistence with an infected cat, 30% due to feline scratch, and 6% due to bite.

There is no gender difference regarding this disease; the majority of adults have reported a 1:1 ratio between males and females.

Habitat and ecological conditions

The species belonging to the *Sporothrix schenckii* complex live on soil, in decaying matter, wood, leaves, and branches. In 1988, the largest documented USA-outbreak of cutaneous sporotrichosis occurred, with 84 culture-confirmed cases among persons from 15 states who were exposed to Wisconsin-growing sphagnum-moss used in packing evergreen tree seedlings. After 1941 there have been 11 reported epidemics in the USA literature; 8 were associated with exposure to sphagnum moss and 3 were related to hay exposure. Also, thorny-yard plants possessing rigid needles such as roses, bayberry, *Berberisthunbergui*, hawthorns, acacia, and shrubbery are a risk factor for contracting sporotrichosis.

The principal entrance route for the pathogen is the skin, typically following localized trauma or excoriations involving contaminated material, therefore sporotrichosis is considered a fungal infection by implantation, which enables the causative agent to penetrate the skin. However, it is now proven that in individuals who live in highly endemic areas, the fungal conidia can also enter into the body through respiratory tract and are able to reach the alveoli, where they produce primary pulmonary and upper respiratory airways infections. Animal vectors are also a significant source of infection; they can act either indirectly, with the fungus being isolated from the animals' paws, teeth, etc., or directly through animal bites, especially in the case of rodents such as rats, mice, and squirrels. There have also been cases reported that were associated with insect and reptile bites, ants, spiders, wasps, flies, horses, dogs, and birds. The role of felines in the transmission of this mycosis has gained attention since 1980, when (Read and Sterling 1982) reported an outbreak that involved five persons who had contact with a cat who had the disease. Since then a new group at risk of acquiring the disease has been well defined: owners of cats and veterinarians. The first epidemic of zoonotic sporotrichosis was detected in Rio de Janeiro, Brazil. Since 1998 and until 2009, more than 2,000 cases in humans and more than 3,000 cases in cats have been documented in Brazil (Freitas et al. 2010).

Although the disease can present at any time during the year, the majority of cases occur at the end of fall and beginning of winter, which is the end of the rainy season; this produces the optimal temperature and humidity for the development of the fungus. The strains of the *Sporothrix schenckii* complex are generally found in temperate and humid climates, with an average temperature between 20–25°C, and a relative humidity above 90%. Also, during this time of the year, the probability of humans coming into contact with decaying vegetable matter increases, since farming and hunting activities increase in many countries.

Endemic zoonotic sporotrichosis

Different animal species can be attacked by *S. schenckii* usually as isolated cases, and are rarely transmitted to man by accident. Animal vectors have been described in the previous section.

In some countries, sporotrichosis is an emerging disease. From 1998 to 2003, 497 positive cultures in humans and 1,056 in cats were recorded at the Infectious Dermatology Service of the Research Center, Evandro Chagas Hospital in Rio de

Janeiro, Brazil (Schubach et al. 2005). A total of 421 (67.4%) human patients had a history of a cat's scratch or bite, or reported contact with infected cats. The patients had an age range from 5 to 89 years; 68% were women. Housewives (30%) and students (18%) were the most attacked groups. Two mycological studies of hundreds of cats were done to determine the sources of the zoonotic transmission: *S. schenckii* was recovered from the skin swabs, aspirates or biopsies in 96–100%, nasal swab specimens in 66–70%, oral specimens in 41–49%, and nail fragments in 39% of the samples. Also, the fungus was cultured from the oral and nasal cavities of 10 cats (9.9%) from 101 apparently healthy cats that lived with sporotrichosis-infected felines.

The isolation of the fungus from different clinical specimens obtained from cats during both the preclinical and clinical phase of sporotrichosis provides support for and strong evidence of the zoonotic potential and reinforces the necessity of individual's protection specially those handling cats in endemic areas.

Entrance Route and Incubation Period

The usual mode of infection is by cutaneous inoculation of the organism. The most common form of trauma to the skin involves punctures from sharp thorns, splinters, cuts or handling of reeds, grasses, and corn stalks. Pulmonary and disseminated forms of infection, although uncommon, can occur when *S. schenckii* conidia are inhaled. Infections are most often sporadic and are usually associated with trauma during the course of outdoor work. Infections can also be related to zoonotic spread from infected cats or scratches from digging animals, such as armadillos. Outbreaks have been well described and often are traced back to activities that involved contaminated sphagnum moss, hay, or wood. For the cases involving cutaneous or mucosal infection, the incubation period is variable, ranging from 7 to 30 days.

Geographical distribution

Sporotrichosis is the most widespread implantation mycosis in the world, with cases reported in every continent although there are some very specific endemic areas. In America, the highest numbers of cases are found in Peru, Mexico, Colombia, Uruguay, Guatemala, Brazil, Venezuela, Costa Rica, and USA. In Brazil, the number of cases has increased in recent times, with significant epidemics, especially related to the direct transmission by animal vectors in the State of Rio de Janeiro. In Asia, in India, several cases have been reported; most are located in the sub-Himalayan region in the north, the north-east states and certain cases have been reported in south Karnataka. A few cases have been reported in other Asiatic countries such as Thailand and Vietnam; China also has a large number of reports, majority of them are located in the north (Jill in region) and to a lesser extent in the Yangtze River (Sichuan, Jiangsu) and in southeast (Guangxi and Guangdong); Japan represents the highest number of cases in this continent, where all the clinical variants have been reported. In Africa, the most significant epidemic (3000 cases) occurred in the old province of Transvaal (South Africa) among workers of the Witwatersrand mines between 1941 and 1944. In Europe, sporadic cases have been reported in the Mediterranean regions of Spain and France (Barile et al. 1983; Ventin et al. 1987; Magand et al. 2009).

Occupation

Sporotrichosis is an occupational disease most commonly seen in manual workers such as farmers and rural corn cultivators; greenhouse and nursery workers; floral workers; masonry and construction workers; horticulturists, orchid growers, coffee-garden workers; outdoor laborers, tree planters, and forestry workers; and people involved in activities that expose them to contaminated soil and vegetation such as sphagnum moss, salt marsh hay, prairie hay, and thorny plants.

Predisposing Factors

These include the ones related with occupation although malnutrition, chronic alcoholism and other debilitating diseases may exacerbate the disease. Sporotrichosis affects predominantly farm workers, housewives, school-age children, people who grow or sell flowers, hunters, fishermen, miners, and those who package glass, among others.

Pathogenesis

Among the physiopathogenic mechanisms and virulence factors of *S. schenckii*, most occur in concert, the most important are:

***Dimorphism*:** First of all is its very own phenomenon, which means that the fungus presents two antigens on its cell wall: one from its mycelial phase and the other from its yeast phase. Both antigens are composed of a glycopeptide made up of a polysaccharide fraction that contains rhamnomannan polysaccharide, which is responsible for its antigenicity. This fraction sets off the primary cellular immune response and plays a role in the phenomenon of adhesion of the fungus to the host cells. The peptide fraction is composed primarily of threonine, serine, aspartic acid, and glutamic acid. Also, ergosterol peroxide has been identified in *S. schenckii* yeast cells. This compound can be converted to ergosterol when in contact with an enzyme extract from the fungus. The ergosterol peroxide is formed as a protective mechanism to evade reactive oxygen species during phagocytosis and may also represent a virulence factor.

***Melanin*:** Another important virulence factor found in the majority of the species of the *S. schenckii* complex is the production of melanin on their cell wall, which confers protection to the fungus by retaining and neutralizing free radicals. Melanin production in *S. schenckii* dematiaceous conidia occurs through the 1,8-dihydroxynaphtalene (DHN) pentaketide pathway. Macroscopically, only the mycelial phase of the fungus is melanized. Recently, it has been demonstrated that *S. schenckii* can also produce melanin using phelonic compounds such as 3,4-dihydroxy-L-phenylalanine (L-DOPA) as a substrate both in filamentous and yeast forms. *In vitro* studies indicate that melanization in *S. schenckii* is controlled by several factors, such as temperature, pH, and nutrient conditions. It has been shown that conidial melanization enhances *S. schenckii* resistance to macrophage phagocytosis, allowing the first steps of infection. Melanization also has a role in the pathogenesis of cutaneous sporotrichosis, since pigmented isolates had a greater invasive ability than the albino mutant strain in an experimental rat model of sporotrichosis.

Thermotolerance: is one of the putative *S. schenckii* virulence factors. In fact, isolates able to grow at 35°C but not at 37°C are incapable of causing lymphatic sporotrichosis and produce fixed cutaneous lesions instead. The fungi isolated from lymphatic, disseminated, and extracutaneous lesions show tolerance and growth at 37°C.

Following the processing and the presenting of the antigens, a Th1-type lymphocytic cellular response is set off and, clinically, a lesion called sporotrichotic chancre is formed after some kind of trauma involving contaminated materials, which inoculates the fungus in the skin. Primary adhesion to endothelial and epithelial cells as well as on extracellular matrix components is essential to an effective invasion of host tissues by pathogens. Both conidia and yeast cells from *S. schenckii* are able to recognize three important glycoproteins from the extracellular matrix: fibronectin, laminin, and type II collagen.

Primary pulmonary sporotrichosis starts and follows a similar course of pulmonary tuberculosis. The initial contact comes at the lungs, and then, 98% of cases remain asymptomatic; only the remaining 2% manifest as pneumonia with the possibility of systemic dissemination.

Clinical Features

One of the most used classifications of sporotrichosis describes the clinical aspects of the disease in relation to the immunologic status of the host. Based on this system, it can be divided into the cutaneous-lymphatic, fixed-cutaneous, disseminated-cutaneous, mucosal, and cutaneous-hematogenous forms; and the extra-cutaneous forms: conjunctival, pulmonary, and osteo-articular.

Cutaneous-Lymphatic or lymphangitic sporotrichosis

In cutaneous forms, the infection usually appears after minor trauma with disruption of epidermis integrity. After penetrating through the skin, the fungus converts into the yeast form and may remain localized in the subcutaneous tissue or extend along the adjacent lymphatic vessels, constituting the fixed or the lymphocutaneous form, respectively.

This is the most classic and frequent form of the disease, accounting for up to 75% of all cases of sporotrichosis in some countries such as Mexico and Peru (Bonifaz et al. 2007; Bustamante et al. 2001). One or two weeks after the initial inoculation of the fungus on the skin, the sporotrichotic chancre forms at the site of inoculation, and consists of a discrete increase in volume, erythema, nodular gumma lesions, and ulcers. The chancre is not painful and rarely does it produce pruritus. Over the course of the next two weeks, similar lesions appear in a linear and step-like fashion, following the course of the lymphatic vessels toward the main regional ganglion. The nodules may ulcerate due to trauma or due to superimposed bacterial infections until they form vegetative, verrucous, tuberous or infiltrated plaques. The primary lesion is usually located on the extremities, especially hands and forearms, corresponding to the sites most exposed to trauma. On the lower extremities it is possible to observe a mycetoma-like variety, which originates from multiple inoculations.

In children, it is common for cutaneous-lymphatic sporotrichosis to affect the face (up to 40% of all cases), and it can manifest unilaterally or bilaterally.

With the most chronic cases of cutaneous-lymphatic sporotrichosis, it is possible to develop lymphostasis, which may lead to significant fibrosis and to a large increase in volume (elephantiasis) (Fig. 1).

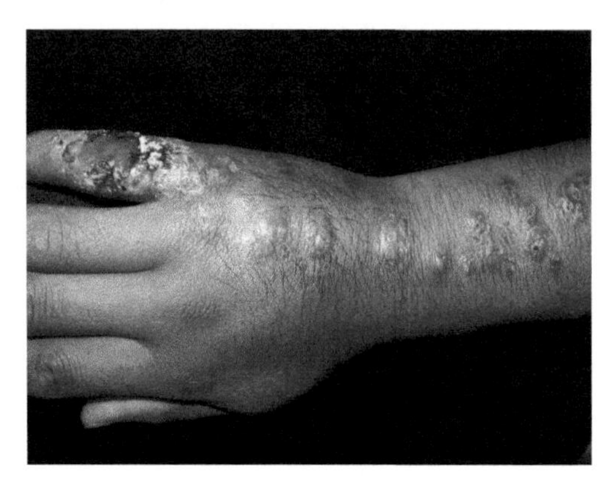

Figure 1. Cutaneous-lymphatic case of the arm. Chancre and nodular lesions (gummas).

Fixed-cutaneous sporotrichosis

The limited nature of this clinical variant is due to the fact that the patient has a good immune response, even hyperergic, and thus has a good probability of experiencing a spontaneous cure. This is a chronic and localized form of sporotrichosis; most authors report it as the main presentation in children. It is observed in approximately the remaining 25%, but in some countries such as Japan or Costa Rica, it may account for up to 60% of all cases.

This clinical form is represented by a single lesion or a few lesions at the inoculation site, which is often ulcerated, with well-defined borders surrounded by an erythematous-violaceous halo, covered with scales and bloody. It is generally asymptomatic. The morphology can also be vegetative, verrucous, plaque infiltrated, or tuberous, without lymphatic involvement. It is a clinical variant which does not tend to disseminate. Some cases may spontaneously regress (Fig. 2).

Cutaneous-disseminated sporotrichosis

Most of the affected patients are hypoergic or anergic so it never spontaneously remits without treatment. It is characterized by multiple skin lesions at noncontiguous sites without extracutaneous involvement. It consists of erythematous, scaly, violaceous and pruritic plaques that typically affect the face; this plaque does not remain fixed but rather advances slowly without affecting the lymphatic vessels. This is the rarest of the cutaneous variants, and some authors consider it a variant of the fixed-cutaneous type. It also receives the name of superficial dermoepidermic, or scrofulous, sporotrichosis (Fig. 3).

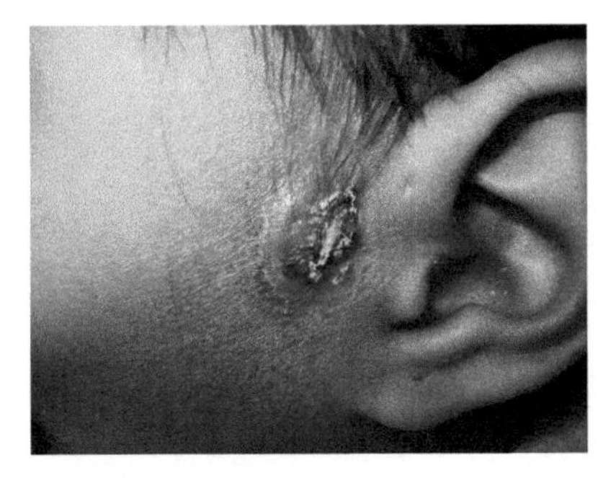

Figure 2. Chronic fixed-cutaneous verrucose case in child.

Mucosal sporotrichosis

Some authors consider it a variant of the cutaneous form. It can be a consequence of self-inoculation through hands contaminated with the fungus, hematogenous dissemination, and inhalation of conidia. In the nasal mucosa, the lesions often involve the septum, with drainage of bloody secretions and detachment of crusts. In the conjunctiva, the granulomatous lesion is accompanied by a serous-purulent discharge, redness, and presence or absence of lid edema. An interesting fact is that in these cases there has not been any report of prior trauma to the ocular mucosa which could explain the direct inoculation of the fungus in the conjunctiva, in contrast, for example, with cases of mycotic keratitis. Mucosal forms are frequently accompanied by preauricular and submandibular lymph node enlargement. Although rare, this form has been described even in pediatric cases.

Cutaneous-hematogenic sporotrichosis

In these cases, the causative agent acts as an opportunist, and the host immune response is practically anergic. The characteristic cutaneous lesions such as nodules, gummas, ulcers, and verrucous plaques disseminate throughout the cutaneous surface, affecting mucosal surfaces as well. This clinical variant has a great tendency to disseminate into the bones and joints (particularly in the elbows and knees) as well as into other organs. Some extraordinary cases present with central nervous system manifestations, or even a fungemia, which is usually lethal. This variety is not commonly cured and generally is associated with a poor prognosis.

This is a rare clinical variant (1–2% of all cases), and it is well correlated with significant states of immunosuppression. The most commonly associated causes of immune suppression related to this infection are those that affect the cellular immune response (diabetes, HIV/AIDS, hematological neoplasias, and some other states of

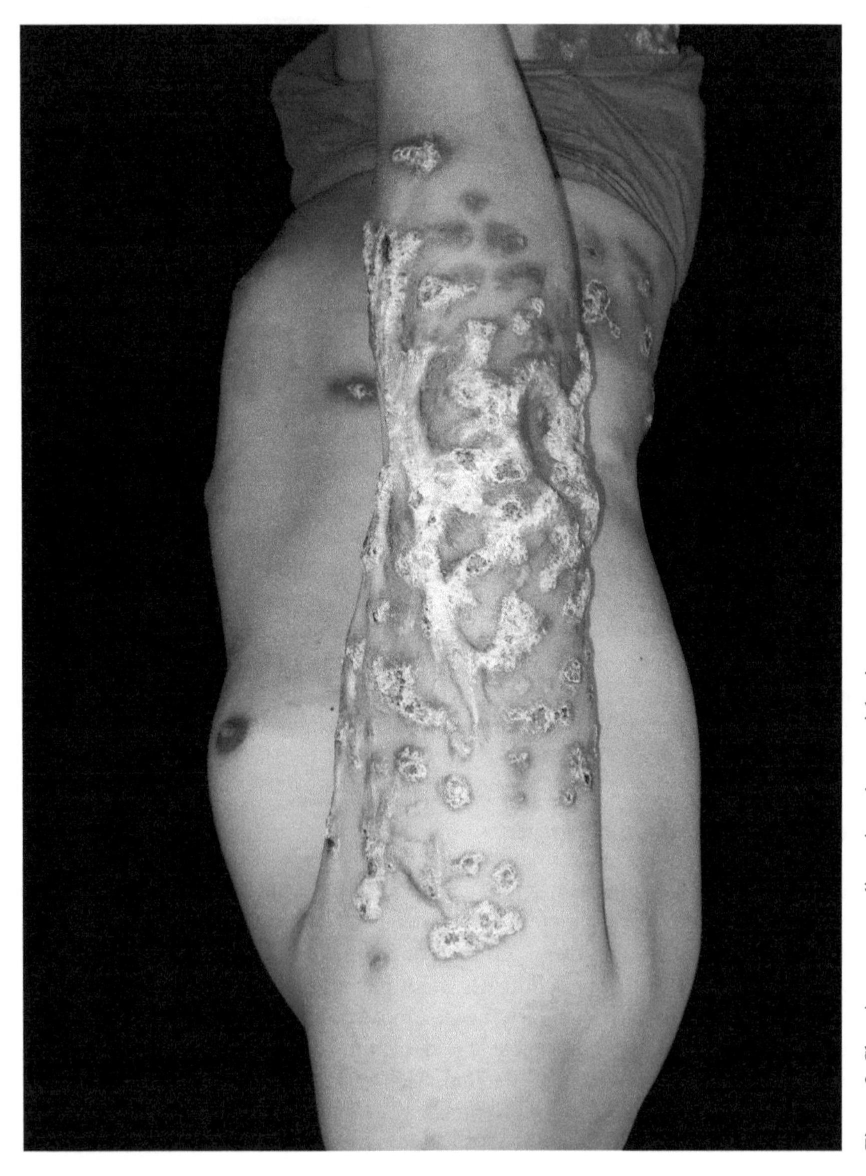

Figure 3. Chronic cutaneous-disseminated sporotrichosis.

partial immunosuppression, such as pregnancy, treatment with systemic corticosteroids, malnutrition, and chronic alcoholism).

Extracutaneous sporotrichosis

The extracutaneous forms are rare and difficult to diagnose, although they are more frequent after the onset of HIV/AIDS. Besides AIDS, other conditions such as diabetes, alcoholism, granulomatous diseases, cirrhosis, renal transplantation, malignancies, corticosteroids, and use of immunosuppressive agents are commonly reported in patients with extracutaneous sporotrichosis.

The most commonly reported clinical variants are the pulmonary cases, of which approximately 100 cases have been reported worldwide. Primary pulmonary sporotrichosis, resulting from inhalation of the fungus, is usually associated with chronic obstructive pulmonary disease, alcoholism, chronic use of corticosteroids, and immunosuppressive diseases. The cases of primary pulmonary sporotrichosis can be divided into two types: the most common is the chronic type (98% of cases), being asymptomatic, self-limited, and with cavitary areas, similar to tuberculosis. The symptomatic cases have a clinical course like pneumonia, with a discrete cough and scant expectoration. Radiological patterns include cavitary disease, tracheobronchial lymph nodes enlargement, nodular lesions or even miliary infiltrates. In contrast, the second clinical variant is acute and progressive; it involves the hiliar lymphatic ganglia, especially the tracheobronchial ones, and may even cause bronchial obstructions. The symptoms vary widely but significant weight loss, cough with expectoration, dyspnea, and fatigue are constant. In this clinical variant, chest radiographs will reveal hiliaradenopathies and, on rare occasion, mediastinal widening.

The osteoarticular form may occur by contiguity or hematogenous spread. The lesions may vary from small granulomas to large lytic lesions identical to osteomyelitis. One or several joints and bones can be involved, as well as tenosynovitis or bursitis. In immunocompetent patients, monoarthritis is more frequent than multiple articular involvements.

Reports on meningitis associated with *Sporothrix* infections are not frequent. Diagnosis of this form of chronic meningitis is challenging because of the rarity of demonstration of *Sporothrix schenckii* in smears of cerebrospinal fluid and the difficulty in isolating the yeast on culture.

Equally uncommon is a case of primary laryngeal sporotrichosis in a pediatric patients whose only manifestation was laryngeal stridor, making the diagnosis in cases like this one quite a challenge. There are also some rare clinical forms which constitute atypical presentations of the disease; for example, disseminated ulcers and disseminated nodular lesions which do not follow any lymphatic tracts.

Laboratory Diagnosis

Direct examination and staining

These are not very useful for establishing the diagnosis because the yeasts are not seen, and the conventional stains (Gram, Giemsa, PAS, Grocott) do not make the fungal structures visible. Only in about 1–2% of the cases the structures are observed in the form of "cigars".

Culture

Definitive diagnosis is based on the isolation and identification of the etiological agent in culture and is consider the gold standard test. The samples are exudates from the lesions, scales, tissue fragments or sputum; they are spread on Sabouraud dextrose agar and Sabouraud dextrose agar with antibiotics. The cultures are incubated at a temperature of 25–28°C in order to obtain the filamentous phase, which is the most useful in terms of allowing for the micromorphologic identification of the fungus. After 5 to 8 days of incubation, we can observe colonies which are initially limited, membranous, radiated, and of a whitish or beige color. Afterwards, they develop aerial mycelium, and the colony becomes acuminated with a dark brown color due to the gathering of the proliferative conidia. This depends on the media that is used and on the strain itself, as is the case with some strains of *Sporothrix albicans* that do not produce melanin pigment (Fig. 4).

Microscopically, it consists of very thin hyphae (1–3 μm), branched, hyaline, and with septae; their asexual reproduction is by way of ovoid, round, elongated, and piriformmicroconidia which are formed in one of the two ways: based on a conidiophore of approximately 10–30 μm in length (sympoduloconidia), arranging around it in such a way that it resembles a "peach flower" or a "daisy flower", or being born directly from the hyphae (aleurioconidia or raduloconidia) (Fig. 5).

The yeast phase is obtained at 37°C when using nutrient-rich media as blood agar, chocolate agar, and BHI agar, being able to stimulate growth by adding 5% CO_2. Within 3 to 5 days, one can observe creamy, yellowish-white, slightly acuminated colonies which are quite similar to bacterial colonies.

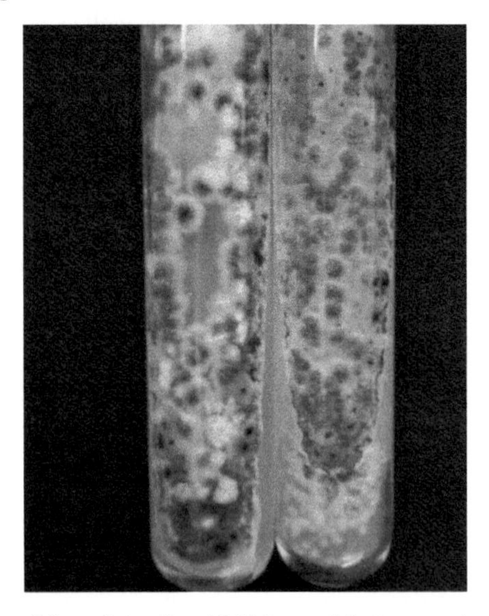

Figure 4. Mycelial culture of *Sporothrix schenckii* (Sabouraud dextrose agar).

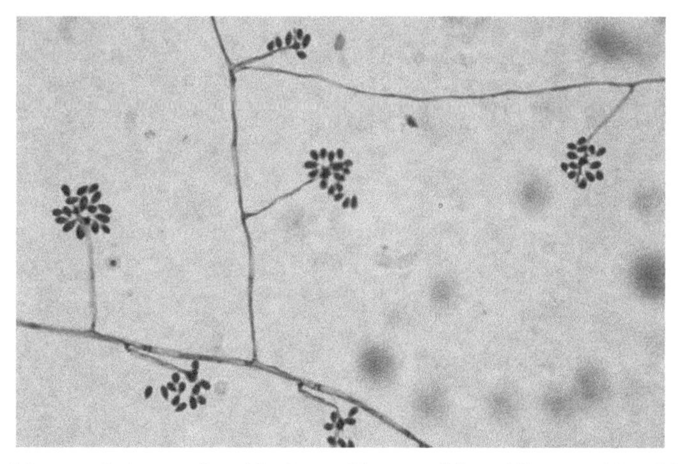

Figure 5. Microscopic image of conidiophore with sympoduloconidia, resemble a "daisy flowers" (Erythrosine, magnification: 60X).

Histopathology

Tissue reaction must also be evaluated in histopathological examinations from patients with sporotrichosis. *S. schenckii* usually causes a mixed suppurative and granulomatous inflammatory reaction in the dermis and subcutaneous tissue, frequently accompanied by microabscess and fibrosis. Besides intact polymorphonuclear cells, granulomas usually contain cellular debris, caseous material, giant and epithelioid cell lymphocytes, plasmocytes, and fibroblasts as well as *S. schenckii* yeast cells within phagocytic cells or in the extracellular medium.

The pyogenous reaction is divided into three different zones: the central or chronic zone containing microabscesses of neutrophils, histiocytes, and lymphocytes, and it is in this area where sometimes it is possible to observe the so called asteroid bodies (budding cells in the center with a radiating halo composed of eosinophilic material); the second zone surrounds the central one and presents a tuberculoid image formed by epithelioid cells, multinucleated giant cells (strange body and Langhans type); the third zone is composed of lymphocytes, plasmocytes, and fibroblasts.

Although *S. schenckii* may be seen in tissue with the routinely used hematoxylin and eosin (H&E) stain, other special stains such as Gomorimethenamine silver (GMS) or periodic acid-Schiff (PAS) stain can be employed to enhance fungal detection (Fig. 6). It is unknown why the cutaneous lymphatic and fixed cutaneous form is very difficult to observe and recognize the yeasts, perhaps because most are phagocytosed because these clinical forms are of immunocompetent patients.

Molecular detection

It is useful for a rapid diagnosis of sporotrichosis and is also valuable in cases of negative cultures due to low fungal burden or secondary infections. Sandhu and

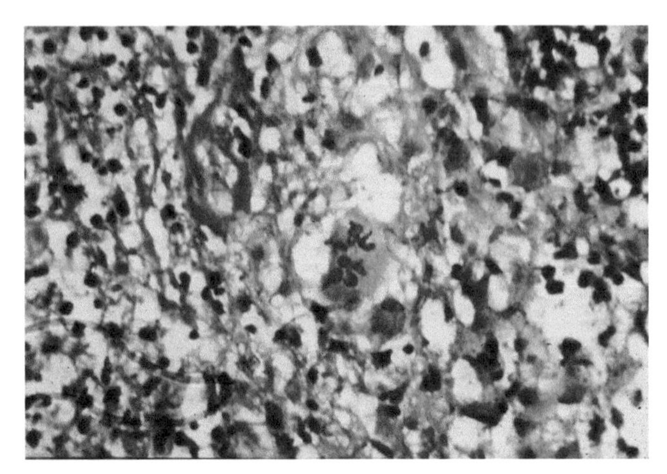

Figure 6. Histopathology, with multiple elongated yeasts (H&E, magnification: 100X).

collaborators reported the development of 21 specific nucleotide probes targeting the large subunit rRNA genes from several fungi, including *S. schenckii* (Sandhu et al. 1955). There is also a protocol for DNA extraction from clinical specimens that consists of boiling the specimens in an alkaline guanidine-phenol-Tris reagent, followed by amplification of a variable region of the 28S rRNA gene with universal primers and amplicon identification using the specific probes.

Immunological testing

Epidemiological studies usually involve the sporotrichin skin testing of individuals living or working in a determinate area together with attempts to isolate the fungus from the soil in that area. The intradermal reaction with sporotrichin M (mycelial) detects delayed hypersensitivity, and can be a useful diagnostic tool. It is performed with the polysaccharide metabolic fraction obtained from *S. schenckii.* A positive result involves an indurated, erythematous, and painful area of more than 5 mm in diameter at the site of the injection 48 hours after inoculation. This test is fairly specific; however, there are rare false negative results which may occur in anergic or immunosuppressed patients. False positives may present in currently healthy individuals who have had the disease previously, since the response remains positive nearly for life (immunologic memory). It is important to emphasize that positive sporotrichin skin-tests are also observed in people living in endemic areas without a history of previous sporotrichosis, due to asymptomatic infections.

Antibody detection

Precipitation and agglutination techniques were first adopted. Immunoelectrophoresis, tube agglutination, and latex agglutination have also been utilized. Immunoenzymatic assays are currently being used more frequently for serodiagnosis purposes. Importantly, the results from all antibody detection tests provide a presumptive diagnosis of sporotrichosis and require clinical and epidemiological correlation for an accurate evaluation and determination of the final diagnosis.

The main differential diagnosis of sporotrichosis types are presented in Table 1.

Table 1. Main differential diagnosis of sporotrichosis (Bonifaz et al. 2010).

Sporotrichosis types	**Differential diagnosis**
Cutaneous-lymphatic or lymphangitic sporotrichosis:	Cutaneous tuberculosis, tuberculoid leprosy, mycetoma, tertiary syphilis, chromoblastomycosis, tularemia, infections caused by atypical mycobacteria, leishmaniasis, keloid scars, acne, cutaneous cryptococcosis.
Fixed-cutaneous sporotrichosis:	Verrucous tuberculosis, chromoblastomycosis, leishmaniasis, squamous cell carcinoma, non-tuberculosis mycobacteriosis, Orf disease.
Cutaneous-disseminated sporotrichosis:	Psoriasis, dermatophytosis, non-tuberculosis mycobacteriosis.
Cutaneous-hematogenic sporotrichosis:	Tuberculosis, syphilitic gummas, coccidioidomycosis.
Pulmonary sporotrichosis:	Bacterial pneumonia, tuberculosis.

Treatment

In general, the treatment requires chemotherapy once the diagnosis is confirmed. The majority of cases respond in a favorable way, and only a few cases recur and require adjustment of the dose, time of therapy or a change in medication.

Chemotherapy

Potassium iodide (KI): it is the treatment of choice for sporotrichosis in many developing countries, especially for lymphocutaneous and fixed cases, because it has high efficacy, minimal side effects, is easy to administer, and has a low cost. Some authors attribute to KI antifungal and immune-stimulant effects, increasing the number of monocytes and neutrophils. However, the exact mechanism of action remains unknown.

Although there is no agreement about precise dosage, the majority of reports mention for adults 3 to 6 g per day, and for children 1 to 3 g per day. The drug is administered orally in two ways: a) saturated solution KI (SSKI), in which each drop of the solution is equal to 50 mg; the recommended starting dose for adults is 5 drops 3 times a day increasing to 40–50 drops 3 times a day and, for children, 3 drops 3 times a day until a maximum of 25 drops 3 times a day is achieved; and b) non-saturated solution of KI. In the latter, 20 grams of KI are dissolved in 300 ml of water in a dark flask; thus, 15 ml of the solution (approximately 1 tablespoon) provides a dose of 1 g of KI. In case of gastritis and nausea it is recommendable to administer it with milk or juice fruit.

KI is generally well tolerated by patients; however, the main side effects are gastritis, nausea, diarrhea, abdominal pain, metallic taste, urticarial, angioedema, rhinitis, bronchitis, and erythema nodosum. For many authors it cannot be considered the therapy of choice because there is not a precise dosage and it is not a patented medication.

The duration of treatment is until clinical and mycological cure is achieved, with an average of 2 to 3 months, although it is desirable to continue with therapy for an additional 1–2 months in order to avoid relapses. Some series report cure rates of 89.6% and 94.7% after 3 to 4 months of treatment. It is important to emphasize that therapy with KI is not recommended for disseminated or extra-cutaneous cases, and in immunosuppressed patients. In pregnant patients treatment with KI may produce hyperthyroidism.

Amphotericin B

It is the treatment of choice in the cases of systemic or anergic sporotrichosis, particularly when there is bone, visceral, and pulmonary involvement. In pregnant women, amphotericin B may be used after 12 weeks of pregnancy, but this medication is to be reserved for pulmonary and disseminated forms for which treatment cannot be delayed. It should be administered in a hospital setting and with all the measures indicated for this drug. The dose of amphotericin B deoxycholate is 0.25 to 0.75 mg/kg/day, and in some cases it may be as high as 1 mg/kg/day. For the rest of the amphotericins B currently available (lipidic, liposomal, and colloidal dispersion), the average dose is between 3 and 6 mg/kg/day.

Sulfamethoxazole-trimethoprim

It has been reported to be quite useful at doses of 400 mg/80 mg, respectively, during a period of 3 to 4 months. It has been used in association with KI in cases of cutaneous-osteoarticular sporotrichosis, obtaining good results.

Itraconazole

The “Clinical practice guidelines for the management of sporotrichosis” (CPGMS) and a majority of authors consider itraconazole, an oral use triazole, the drug of choice for the therapy of sporotrichosis due to its cost-benefit relation, easy way of administration and fewer side effects. The MICs for *S. schenckii* strains are 1–32 μg/ml, which indicates some strains are very sensible and some resistant. The dosage is 100 to 200 mg per day, but in relapsing cases or with low response the dosage can be duplicated to 300–400 mg/day. For children up to 20 kg weight, the dosage is 5–10 mg/kg/day. It is important to emphasize that itraconazole is metabolized by the cytochrome p450 3A4; thus, it presents multiple drug interactions and is forbidden during pregnancy. The main side effects with this treatment are nausea, epigastric pain, dizziness, diarrhea, and headache.

The use of itraconazole in intermittent or pulse therapy can be achieved using dosages of 400 mg/day per one week and 3 weeks of rest, obtaining satisfactory results with an average of 4 pulses.

Fluconazole

it is also a triazole of oral and intravenous use. Its MIC is 128 μg/ml for most of *S. schenckii* strains. Recommended dosage is 400 mg/day, although some reports mention dosages from 150 until 800 mg/day. It is considered a second line of therapy, less effective than itraconazole, and will be used when this one cannot be tolerated. The time period of therapy is variable in diverse reports, having an average of 6 months.

Ketoconazole

This imidazole, it is no longer used because of its low efficacy and frequent and elevated side effects.

Regarding other triazole derivatives, voriconazole has high MICs and there are no reports in literature of its clinical use. Posaconazole has a similar spectrum of action than itraconazole (and similar MICs), and there are a few reports of its good activity in experimental sporotrichosis (murine). Sensible testing with ravuconazole and other new antifungal agents is still incipient, and so far there is no indication for use of these drugs for the treatment of sporotrichosis.

Terbinafine

It is an oral alilamine considered as an alternative option for treatment of sporotrichosis. It is a fungistatic and fungicidal antimycotic that interacts with the synthesis of ergosterol; one of the most important characteristics of this drug is that it is metabolized via cytochrome CYP 2D6, reducing the number of drug interactions, especially if compared with itraconazole and fluconazole. This drug presents the minor MICs for *Sporothrix* sp. strains, being for *S. schenckii* and *S. brasiliensis* of 0.06–0.5 and 0.06–0.25 μg/ml, respectively. This indicates that, *in vitro*, those strains are much more sensible to terbinafine than to itraconazole and amphotericin B; however, in practice, it is less effective than itraconazole and there are only a few reports of isolated cases and one case series.

Recommended dosage is 250 to 500 mg/day, during a time period of 4 to 6 months of treatment as the CPGMS refer. For adults suggested dosage is 500 mg/day, but some studies have used a maximum of 1,000 mg/day. For children, it is also a good therapeutic alternative at doses between 125 and 250 mg/day. The three studies using terbinafine in human patients demonstrated a good efficacy for doses ranging from 250 to 1,000 mg/day. A study comparing 250 mg/day of terbinafine to 100 mg/day of itraconazole resulted in healing in 92.7% and 92% of patients, respectively, indicating terbinafine to be an effective and well-tolerated option for the treatment of cutaneous sporotrichosis. Nevertheless, according to the little experience obtained from case series, terbinafine is considered a second line drug, even more effective than fluconazole and is highly recommended when itraconazole cannot be used.

Physical method

Thermotherapy

This therapy is recommendable for very limited cases (fixed cutaneous), as a concomitant treatment (lymphocutaneous form) or in pregnant women, in whom the drugs mentioned previously cannot be used. These patients can use thermotherapy with daily application of local heat (42 to 43°C) through a hot water bag, a source of infrared, or a similar method until healing of the lesions. The mechanism of action of local heat has been demonstrated in the laboratory. When cells of *S. schenckii* in serum are incubated with neutrophils at 40°C and 37°C, there is no difference related to phagocytosis in the two groups. However, once the cells are phagocytosed, the death rate of the fungus is higher at 40°C than at 37°C.

Cryotherapy

It is an invasive and painful method with less effective results than thermotherapy. It has been used to treat lymphocutaenous and fixed sporotrichosis, alone or in combination with diverse proved therapies (KI, itraconazole, terbinafine, etc.).

Combined therapy

Overall lymphocutaneous sporotrichosis responds well to monotherapy with the medications described above; nevertheless, the combination of itraconazole, KI, and cryotherapy reduces the time of treatment. It has been demonstrated that itraconazole plus terbinafine synergize their effects, so they can be used together in selected cases with poor responses to monotherapy, especially for disseminated or extracutaneous cases.

In Table 2, the immunological classification of sporotrichosis and its behavior is presented.

Table 2. Immunological classification of sporotrichosis and its behavior (Saul 1990; Bonifaz et al. 2010).

Types/Variants	Hyperergic and normoergic cases	Anergic and Hypoergic cases
Clinical features	Fixed-cutaneous Lymphangitic-cutaneous (classical)	Cutaneous disseminated Osteo-articular Pulmonary Visceral
Parasite forms on direct exam and in tissues	Rare	Frequent
Histopathology	Suppurative granuloma	Non-specific granuloma
Sporotrichine (intradermal reaction)	Always positive	Frequently negative
Resistance	High	Low
Prognosis	Very good	Bad
Spontaneous regression	Probably	Never
Response to treatment	Very good	Bad

Prophylaxis

In highly endemic areas, it is necessary to promote preventive measures to avoid direct contact with the causal agents of sporotrichosis, such as the use of gloves, masks, and special footwear for the rural workers, as well as the control of the wild fauna that acts as indirect vectors, and of the epidemics in the domestic animals that are direct transmitters of the disease.

Sporotrichosis in cats requires preventive measures to avoid transmission within the species and from animals to humans. Cats with sporotrichosis should be correctly treated and kept isolated in a proper place. Any physical contact with the animal should be avoided until complete healing of the lesions. When handling the sick cat, during either injury or treatment of medication administration, protocols must be adopted to reduce exposure to the fungus, such as using latex gloves.

Conclusions and Future Perspectives

We can assume that this disease will continue to be present in poor and underdeveloped countries, places where the fungus lives (due to ecologic conditions) in which the people dedicated to agriculture and rural work do not follow protective measures in order to avoid contact with it. Focusing on the zoonotic epidemic of the southern region of Brazil (Rio de Janeiro and Rio Grande do Sul), it is necessary to have an adequate control over dogs and cats to avoid slow dissemination of the disease to adjacent places and the rest of countries of South America.

Speaking of the *Sporothrix* complex, it is possible to find one more cryptic species; nevertheless, two species will continue to predominate: *S. schenckii* (ss) and *S. brasiliensis*. Recently, Rodrigues et al. (2013) studied the chromosomal polymorphism, which will help to explain the genetic diversity of the complex. When the complete genome of *S. schenckii* is described, we will have a more precise knowledge of the pathogenicity and adaptation factors of the fungus.

Although the actual treatments for this fungus are effective, there always exists the possibility of adaptation of the strains with subsequent resistance; because of this, the investigation of new medications is necessary to achieve the control of sporotrichosis.

References

Alba-Fierro, C.A., A. Pérez-Torres, E. López-Romero, M. Cuéllar-Cruz and E. Ruiz-Baca. 2014. Cell wall proteins of *Sporothrix schenckii* as immunoprotective agents. Rev Iberoam Micol. 31(1): 86–89.

Alencar-Marques, S., R.M. Pires de Camargo and V. Haddad-Junior. 1998. Human sporotrichosis: transmitted by feline. An Bras Dermatol. 73: 559–562.

Almeida, H.L., Jr., C.B. Lettnin, J.L. Barbosa and M.C. Dias. 2009. Spontaneous resolution of zoonotic sporotrichosis during pregnancy. Rev Inst Med Trop Sao Paulo. 51: 237–238.

Arenas, R. 2005. Sporotrichosis. pp. 367–384. *In*: Merz, W.G. and R. Hay (eds.). Topley & Wilson's. Microbiology and Microbial Infections. 10th ed. Hodder-Arnold, London.

Aung, A.K., B.M. Teh, C. McGrath and P.J. Thompson. 2013. Pulmonary sporotrichosis: case series and systematic analysis of literature on clinico-radiological patterns and management outcomes. Med Mycol. 51: 534–544.

Barile, F., M. Mastrolonardo, F. Loconsole and F. Rantuccio. 1993. Cutaneous sporotrichosis in the period 1978–1992 in the province of Bari, Apulia, Southern Italy. Mycoses. 36: 181–185.

Barros, M.B., D.L. Costa and T.M. Schubach. 2008. Endemic of zoonotic sporotrichosis: profile of cases in children. Pediatr Infect Dis J. 2: 246–250.
Bonifaz, A. and D. Vázquez-González. 2010. Sporotrichosis: an update. G Ital Dermatol Venereol. 145: 659–673.
Bonifaz, A. and D. Vázquez-González. 2013. Diagnosis and treatment of sporotrichosis lymphocutaneous: what are the options. Curr Fungal Infect Rev. 7: 252–259.
Bonifaz, A., L. Fierro, A. Saúl and R.M. Ponce. 2007a. Cutaneous sporotrichosis. Intermittent treatment (pulses) with itraconazole. Eur J Dermatol. 18: 61–64.
Bonifaz, A., A. Saúl, V. Paredes-Solis, L. Fierro, A. Rosales, C. Palacios and J. Araiza. 2007b. Sporotrichosis in childhood. Clinical and therapeutic experience in 25 cases. Pediatr Dermatol. 24: 369–372.
Bonifaz, A., D. Vázquez-González and A.M. Perusquía-Ortiz. 2010. Subcutaneous mycoses: chromoblastomycosis, sporotrichosis and mycetoma. J Dtsch Dermatol Ges. 8: 619–627.
Bonifaz, A., J. Araiza, A. Pérez-Mejía, L.A. Ochoa and C. Toriello. 2013. Prueba intradérmica con esporotricina en una comunidad de la Sierra Norte de Puebla. Dermatol Rev Mex. 57: 428–432.
Borges, T.S., C.N. Rossi, J.D. Fedullo, C.P. Taborda and C.E. Larsson. 2013. Isolation of *Sporothrix schenckii* from the claws of domestic cats (indoor and outdoor) and in captivity in São Paulo (Brazil). Mycopathologia 176: 129–137.
Bustamante, B. and P.E. Campos. 2001. Endemic sporotrichosis. Curr Opin Infect Dis. 14: 145–149.
Bustamante, B. and P.E. Campos. 2004. Sporotrichosis: a forgotten disease in the drug research agenda. Expert Rev Anti Infect Ther. 2: 85–94.
Campos, P., R. Arenas and H. Coronado. 1994. Epidemic cutaneous sporotrichosis. Int J Dermatol. 33: 38–41.
Carmichael, J.W. 1962. *Chrysosporium* and some other aleurosporic hyphomycetes. Can J Bot. 40: 1137–1173.
Castro, R.A., P.H. Kubitschek-Barreira, P.A. Teixeira, G.F. Sanches, M.M. Teixeira, L.P. Quintella et al. 2013. Differences in cell morphometry, cell wall topography and gp70 expression correlate with the virulence of *Sporothrix brasiliensis* clinical isolates. PLoS One 8(10): e75656.
Chaves, A.R., M.P. de Campos, M.B. Barros, C.N. do Carmo, I.D. Gremião, S.A. Pereira and T.M. Schubach. 2013. Treatment abandonment in feline sporotrichosis—study of 147 cases. Zoonoses Public Health 60: 149–153.
De Araujo, T., A.C. Marques and F. Kerdel. 2001. Sporotrichosis. Int J Dermatol. 40: 737–742.
de Beurmann, L. and H. Gougerot. 1912. Les sporotrichoses. Paris, Felix Alcan.
de Beurmann, L. and L. Ramond. 1903. Abcés sous-cutanes multiples d'origine mycosique. Ann Dermatol Syphilogr. 4: 678–685.
De Lima-Barros, M.B., A.O. Schubach, C. de Vasconcellos, R. de Oliveira, E.B. Martins, J.L. Teixeira and B. Wanke. 2011. Treatment of cutaneous sporotrichosis with itraconazole: study of 645 patients. Clin Infect Dis. 52: 200–206.
Fernandes, G.F., P.O. dos Santos, A.M. Rodrigues, A.A. Sasaki, E. Burger and Z.P. de Camargo. 2013. Characterization of virulence profile, protein secretion and immunogenicity of different *Sporothrix schenckii* sensu stricto isolates compared with *S. globosa* and S. *brasiliensis* species. Virulence 4: 241–249.
Francesconi, G., A.C. Valle, S. Passos, R. Reis and M.C. Galhardo. 2009. Terbinafine (250 mg/day): an effective and safe treatment of cutaneous sporotrichosis. J Eur Acad Dermatol Venereol. 23: 1273–1276.
Freitas, D.F., A.C. do Valle, R. de Almeida Paes, F.I. Bastos and M.C. Galhardo. 2010. Zoonotic Sporotrichosis in Rio de Janeiro, Brazil: a protracted epidemic yet to be curbed. Clin Infect Dis. 50: 453.
Hay, R.J. and R. Morris-Jones. 2008. Outbreaks of sporotrichosis. Curr Opin Infect Dis. 21: 119–121.
Hektoen, L. and C.F. Perkins. 1900. Refractory subcutaneous abscesses caused by Sporothrix schenckii. J Exp Med. 5: 77–79.
Hu, S., W.H. Chung and S.I. Hung. 2003. Detection of *Sporothrix schenckii* in clinical samples by a nested PCR assay. J Clin Microbiol. 41: 1414–1418.
Kauffman, CA. 1999. Sporotrichosis. Clin Infect Dis. 29: 231–237.
Kauffman, C.A., R. Hajjeh and S.W. Chapman. 2000. Practice guidelines for the management of patients with sporotrichosis. Clin Infect Dis. 30: 684–687.
Lavalle, P. and F. Mariat. 1983. Sporotrichosis. Bull Inst Pasteur. 81: 295–322.
Liu, X., Z. Zhang, B. Hou, D. Wang, T. Sun, F. Li, H. Wang and S. Han. 2013. Rapid identification of *Sporothrix schenckii* in biopsy tissue by PCR. J Eur Acad Dermatol Venereol. 27: 1491–1497.

López-Romero, E., M.R. Reyes-Montes, A. Pérez-Torres, E. Ruiz-Baca, J.C. Villagómez-Castro et al. 2011. *Sporothrix schenckii* complex and sporotrichosis, an emerging health problem. Future Microbiol. 6: 85–102.

Madrid, H., J. Cano, J. Gené, A. Bonifaz, C. Toriello and J. Guarro. 2009. *Sporothrix globosa*, a pathogenic fungus with widespread geographical distribution. Rev Iberoam Micol. 26: 218–222.

Magand, F., J.L. Perrot, F. Cambazard, M.H. Raberin and B. Labeille. 2009. Autochthonous cutaneous sporotrichosis in France. Ann Dermatol Venereol. 136: 273–275.

Marimon, R., J. Cano, J. Gené, D.A. Sutton, M. Kawasaki and J. Guarro. 2007. *Sporothrix brasiliensis, S. globosa,* and *S. mexicana*, three new *Sporothrix* species of clinical interest. J Clin Microbiol. 45: 3198–3206.

Marimon, R., J. Gene, J. Cano and J. Guarro. 2008a. *Sporothrix luriei*: a rare fungus from clinical origin. Med Mycol. 10: 1–5.

Marimon, R., C. Serena, J. Gené, J. Cano and J. Guarro. 2008b. *In vitro* antifungal susceptibilities of five species of *Sporothrix*. Antimicrob Agents Chemother. 52: 732–734.

Mayorga, R., A. Cáceres and C. Toriello. 1978. Etude d'une zone d'endemie sporotrichosi que au Guatemala. Sabouraudia. 16: 185–198.

Mesa-Arango, A.C., R. Reyes-Montes and A. Pérez-Mejia. 2002. Phenotyping and genotyping of *Sporothrix schenckii* isolates according to geographic origin and clinical form of Sporotrichosis. J Clin Microbiol. 40: 3004–3011.

Oliveira, M.M., S.B. Maifrede, M.A. Ribeiro and R.M. Zancope-Oliveira. 2013. Molecular identification of *Sporothrix* species involved in the first familial outbreak of sporotrichosis in the state of Espírito Santo, southeastern Brazil. Mem Inst Oswaldo Cruz. 108: 936–938.

Pappas, P.G., I. Tellez, A.E. Deep, D. Nolasco, W. Holgado and B. Bustamante. 2000. Sporotrichosis in Peru: description of an area of hyperendemicity. Clin Infect Dis. 30: 65–70.

Quintella, L.P., S.R. Passos, A.C. do Vale, M.C. Galhardo, M.B. Barros, T. Cuzzi, S. Reis Rdos et al. 2011. Histopathology of cutaneous sporotrichosis in Rio de Janeiro: a series of 119 consecutive cases. J Cutan Pathol. 38: 25–32.

Read, S.I. and L.C. Sperling. 1982. Feline sporotrichosis. Transmission to man. Arch Dermatol. 118: 429–431.

Rodrigues, A.M., S. de Hoog and Z.P. de Camargo. 2013a. Emergence of pathogenicity in the *Sporothrix schenckii* complex. Med Mycol. 51: 405–12.

Rodrigues, A.M., M. de Melo Teixeira, G.S. de Hoog, T.M. Schubach, S.A. Pereira, G.F. Fernandes et al. 2013b. Phylogenetic analysis reveals a high prevalence of *Sporothrix brasiliensis* in feline sporotrichosis outbreaks. PLoS Negl Trop Dis. 20; 7(6): e2281.

Sandhu, G.S., B.C. Kline, L. Stockman and G.D. Roberts. 1995. Molecular probes for diagnosis of fungal infections. J Clin Microbiol. 33: 2913–2919.

Saúl, A. 1990. Sporotrichosis. pp. 53–59. *In*: P.H. Jacobs and L. Nall (eds.). Antifungal Drug Therapy. Marcel Dekker, Inc., New York.

Schechtman, R.C. 2010a. Sporotrichosis: part I. Skinmed. 8: 216–220.

Schechtman, R.C. 2010b. Sporotrichosis: part II. Skinmed. 8: 275–280.

Schubach, A., T.M. Schubach, M.B. Barros and B. Wanke. 2005. Cat-transmitted sporotrichosis, Rio de Janeiro, Brazil. Emerg Infect Dis. 11: 1952–1954.

Schubach, A., M.B. Barros and B. Wanke. 2008. Epidemic sporotrichosis. Curr Opin Infect Dis. 21: 129–133.

Simson, S.W. 1947. Sporotrichosis infection in mines of the Witwatersrand. A symposium. Proc Transv. Mine Med. Officers Assoc.

Song, Y., S.X. Zhong, L. Yao, Q. Cai, J.F. Zhou, Y.Y. Liu, S.S. Huo and S.S. Li. 2011. Efficacy and safety of itraconazole pulses vs. continuous regimen in cutaneous sporotrichosis. J Eur Acad Dermatol Venereol. 25: 302–305.

Tlougan, B.E., J.O. Podjasek, S.P. Patel, X.H. Nguyen and R.C. Hansen. 2009. Neonatal sporotrichosis. Pediatr Dermatol. 26: 563–565.

Ventin, M., C. Ramírez, M. Ribera, C. Ferrandiz, R. Savall and J. Peyri. 1987. A significant geographical area for the study of the epidemiological and ecological aspect of Mediterranean sporotrichosis. Mycopathologia. 99: 41–43.

Xu, T.H., J.P. Lin, X.H. Gao, H. Wei, W. Liao and H.D. Chen. 2010. Identification of *Sporothix schenckii* of various mtDNA types by nested PCR assay. Med Mycol. 48: 161–165.

CHAPTER 10

Paracoccidioidomycosis: An Endemic Mycosis in the Americas

Carlos Pelleschi Taborda,[1,2,*] *Martha Eugenia Uran J.*[2,3] and *Luiz R. Travassos*[4]

Introduction

Paracoccidioidomycosis (PCM), is a systemic mycosis with clinical manifestations of a granulomatous disease, caused by thermally dimorphic *Paracoccidioides* spp. Adolpho Lutz first described it in 1908, while examining oral lesions in two patients. The first patient, a Spanish man, 40 years old, reported oral lesions for six months and the second patient, a 30 year old man, reported a four-year history of multiple oral lesions. Five months later, Lutz reported that the patient returned quite emaciated, with hoarseness and diarrhea (Reviewed by Marques 2008). Primarily named as "pseudococcidial hyphoblastomycosis" (pseudococcidial to differentiate from coccidioidomycosis and hifoblastomycosis), the mycosis was then called South American Blastomycosis or Lutz and Splendore-Almeida disease (Lacaz et al. 1991). The actual term Paracoccidioidomycosis was established in 1971 in Medellín, Colombia, during a meeting of the American Continent mycologists and has been widely accepted since then (Lacaz 1982).

[1] Institute of Biomedical Sciences, Department of Microbiology – University of São Paulo, SãoPaulo, Brazil.
[2] Laboratory of Medical Mycology IMTSP/LIM53/HCFMUSP University of São Paulo, São Paulo, Brazil.
[3] Medical and Experimental Mycology Group, Corporaciónpara Investigaciones Biológicas, Medellín, Colombia.
[4] Department of Microbiology, Immunology and Parasitology, Federal University of São Paulo, São Paulo, Brazil.
* Corresponding author: taborda@usp.br

After the description of *Paracoccidioides brasiliensis* by Adolpho Lutz, studies on the main characteristics of the infectious agent took place, involving morphological phases and growth temperature. Splendore in 1912, suggested the classification of the agent in the genus *Zymonema*, thus creating the name *Zymonema brasiliense*. The disease began to be called "Brazilian blastomycosis" and, soon after, "South American blastomycosis" since isolated cases had been reported in other countries in South America. After systematic studies, a new gender was created within the kingdom Fungi—*Paracoccidioides*, revalidating the species name created by Splendore in 1912. In 1930, Floriano Paulo de Almeida officially named the fungal agent as *Paracoccidioides brasiliensis* (Almeida 1930). Taxonomically, the fungus *P. brasiliensis* was initially classified by Ajello (1977) as follows: Kingdom Fungi, Phylum Eumycota, Subdivision Deuteromycotina, class Hyphomycetes, Order Moniales, Moniliaceae Family, Genus and Species *Paracoccidioides brasiliensis*. However, another classification was proposed because phylogenetic studies using molecular tools positioned the etiologic agent of PCM along with other dimorphic fungi (*Coccidioides posadasii*, *Coccidioides immitis*, *Blastomyces dermatitidis* and *Histoplasmacapsulatum*) as belonging to the following taxonomic category: Kingdom Fungi, Phylum Ascomycota, Pleomycetes class, Order Onigenales, Onygenaceae Family, Genus and Species *Ajellomyces brasiliensis*, with the fungus again being called *Paracoccidioides brasiliensis* (San-Blas et al. 2002).

In recent studies, based on the analysis of the genetic variability, the existence of three distinct phylogenetic groups was proposed, namely: S1—paraphyletic group with 38 isolates from Argentina, Brazil, Peru and Venezuela and one isolate from a penguin from Antarctica; PS2—monophyletic group with six isolates, five from Brazil and one in Venezuela; PS3—monophyletic group with 21 isolates from Colombia (Matute et al. 2006). Subsequently, seven other isolates that were not included in previous studies were analyzed and grouped into clades S1 and PS3. Isolate Pb01 was separated from the other groups, based on the cladogram (Desjardins et al. 2011). It is estimated that the Pb01-like monophyletic group was separated from the other groups S1, PS2 and PS3 approximately 30 million years ago. Based on this data, a new classification for Pb01-like isolates was proposed, placing them as a new species within the genus, named *Paracoccidioides lutzii*. The Pb01-like group is endemic in North and Central-West regions of Brazil (States of Rondônia, MatoGrosso and Goiás) and shares some geographical areas with group S1 (Teixeira et al. 2009 and 2013) (Fig. 1).

No one knows for sure the natural reservoir of the fungus. Over the years, researchers have succeeded in isolating the fungus from different soils in Venezuela (Albornoz 1971), Argentina (Negroni 1968) and Brazilian States such as Minas Gerais (Vergara et al. 1998) and Pará (Naiff et al. 1986). There are reports on isolations of *P. brasiliensis* from animals such as armadillos (Bagagli et al. 1998), bats (Grose and Tamsitt 1965) and even dog food (Ferreira et al. 1990). These samples were collected near rivers and other water bodies. Growth of the fungus, however, appears to occur at temperatures ranging from 18°C to 24°C (Restrepo-Moreno 1994). The main form of infection by *Paracoccidioides* spp. is through the respiratory tract by inhalation of fungal propagules (conidios), which reach the lung parenchyma were transformed into yeast-like forms (Franco 1987). Activities related to agriculture, gardening and

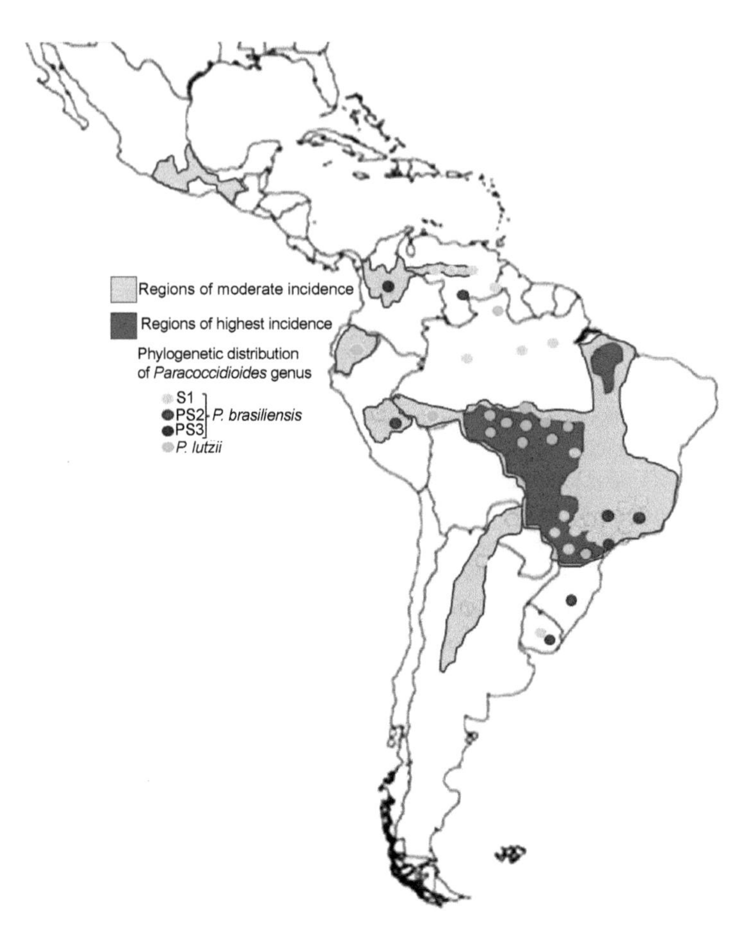

Figure 1. Map depicts the high and moderate incidence regions of paracoccidioidomycosis in the Americas. Dots indicate phylogenetic distribution of *Paracoccidioides* genus.

developmental work in rural or near-forest environments are associated with cases of the disease. The majority of patients report having undertaken these activities during certain periods of their lives.

The disease represents a major public health problem, since the resultant high degree of incapacitation and sometimes even death, causes the affected individuals to interrupt their activities, having both social and economic impact (Shikanai-Yasuda et al. 2006).

The aim of this chapter is to update the novelties of the new species-*P. lutzii*, description of different genotypes and mechanisms of pathogenicity in the genus *Paracoccidioides*. Advances in clinical and laboratory diagnosis, antifungal drugs used in the treatment and vaccine development have been included to give a complete overview of the complexity of this endemic systemic mycosis in Latin America.

Morphology

The *Paracoccidioides* spp., show thermal-dimorphism, the fungus develops as mold at room temperature or between 18–25°C (saprophytic phase), and at temperatures 35–37°C, has the shape of unicellular yeast (parasitic phase) (Lacaz et al. 1991; Franco et al. 1994).

Colonies of different morphological types can be visualized depending on the culture medium and especially, on the temperature of incubation. *Paracoccidioides* spp., at room temperature shows white colonies adhering to the medium (Fig. 2). These colonies grow slowly. When examined under the microscope, the colonies show fine, septated mycelial filaments with arthroconidia, globose club-shaped aleurioconidia and arthroaleurioconidia (Fig. 3).

The first report on conidia was that of Conant and Howell in 1942. They described the fungus of South American Blastomycosis producing spores (Conant and Howell 1942). Later, in 1951, Neves and Bogliolo in Brazil described the production of "aleurias" (Neves and Bogliolo 1951) and some years later Borrelli (1955), noted abundant production of "aleurias" in microcultures from differents substrates (Borelli 1955) and described their role in *P. brasiliensis* reproduction. But it was only in 1971 that Pollak obtained conidia from poor-medium supplemented cultures (Pollak 1971).

Restrepo has studied these structures since 1970, comparing them to the lateral and intercalary arthroconidia produced by molds and other pathogens like *Coccidioides immitis* (Negroni 1966; Restrepo 1970; Cole and Sun 1985). Conidia are small, less than 5 µm, have various shapes and different locations on the parental mycelium (Bustamante-Simon et al. 1985). The conidia-to-yeast transition occurs in 96 h at 37°C

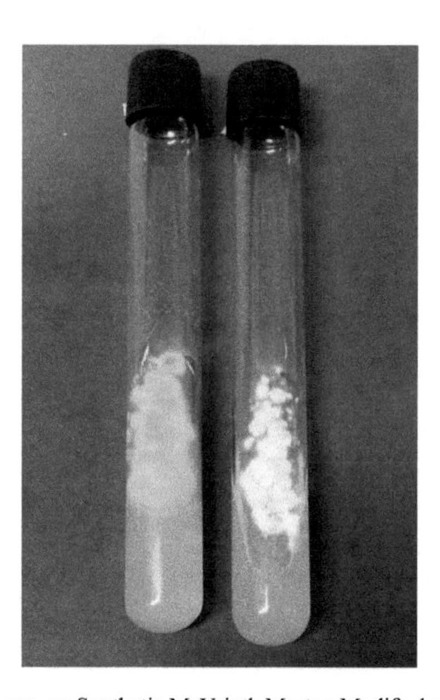

Figure 2. *Paracoccidioides* spp. on Synthetic McVeigth Morton Modified agar at 18°C.

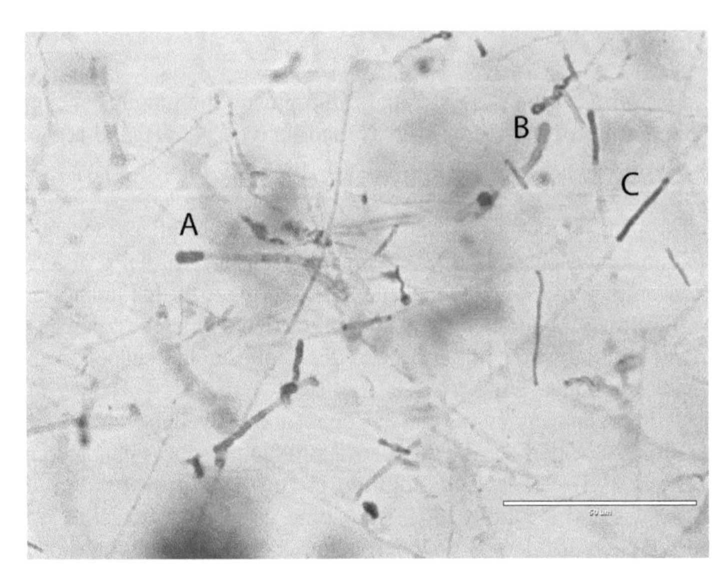

Figure 3. Mycelial form of *Paracoccidioides* spp. Mycelial filaments with globose club-shaped aleurioconidia (A), arthroaleurioconidia (B) and arthroconidia (C).

but cream colonies called cerebriform or yeast-like appear after 10–20 days (Fig. 4). Round cells, some with germinal tubes and cell forms similar to those detected in tissue are observed. The cells are spherical (3–30 μm) with thin and thick walls. Along

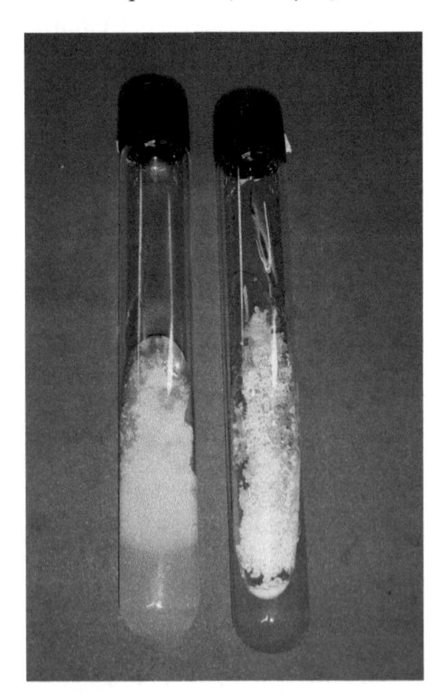

Figure 4. *Paracoccidioides* spp. on Sabouraud's Dextrose agar at 37°C.

the surface, spherical buds (2–10 μm), with a narrow connection with the mother cell give the characteristic "pilot wheel" appearance (Fig. 5) (Brummer et al. 1993).

P. lutzii, in addition to the typical conidia, are also produced by *P. brasiliensis*, which frequently produces elongated, rod-shaped conidia. Yeast cells from both species show no significant variation in size and shape, with the exception of *P. lutzii* Pb01 isolate, which exhibits large yeast cells and *P. brasiliensis* PS2, which commonly presents elongated yeast cells, similar to pseudohyphae (Teixeira et al. 2009; Theodoro et al. 2012; Teixeira et al. 2013).

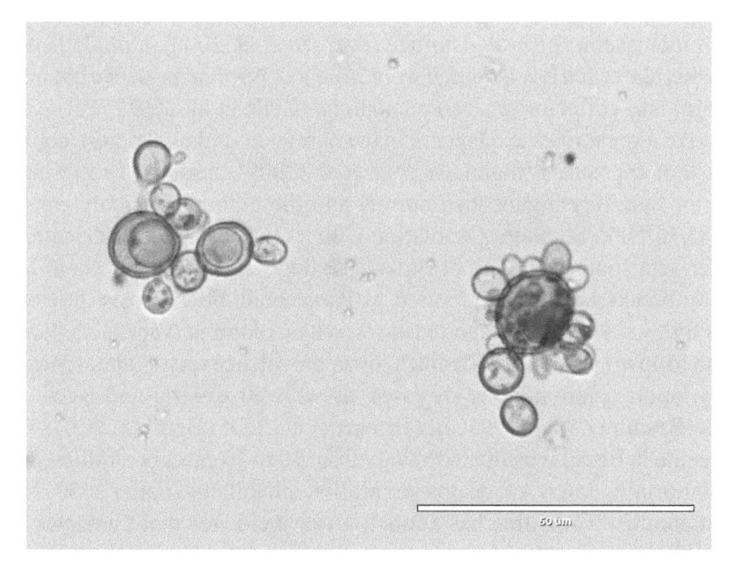

Figure 5. Yeast cells of *Paracoccidioides* spp. Spherical cells with thin and thick walls. Along the surface, spherical buds, with a narrow connection with mother cell give the characteristic "pilot wheel" appearance.

Epidemiology

Paracoccidioidomycosis is endemic and considered the most prevalent systemic mycosis in Latin America. It is reported from Mexico to Argentina and in Brazil, accounts for 80% of cases, followed by Colombia and Venezuela (Casquero et al. 1998; Lazo 1998; Calle et al. 2001; Shikanai-Yasuda et al. 2006; Theodoro et al. 2012; Martínez et al. 2013; López-Martínez et al. 2014). Atypical or non-indigenous cases were reported in Chile, Suriname, Guyana, Nicaragua and Belize (Restrepo et al. 2001; Restrepo et al. 2008). An interesting aspect in the distribution of the disease is that countries where PCM is endemic, there is no homogeneous distribution of the disease, which rather occurs in certain regions that offer favorable conditions for the fungus (Wanke and Londero 1994; Restrepo et al. 2001) (Fig. 1).

The accidental or permanent presence of fungi in animals, plants, soils and watercourses should not be taken too lightly because they constitute the source where potential pathogens will be contracted (Restrepo et al. 2001).

Indeed, the presence of the fungus in animals, plants, soils or waterways, has been reported, and suggests an important link between nature localization and place of infection. Analysis of *P. brasiliensis* mycelial form reveals that this form has the required capabilities to be the natural infectious form. Requirements of high relative humidity for cultures *in vitro* or for growth in the environment have been reported, which correlates with the endemic areas, called "reservareas" by Borrelli (Restrepo 1985).

Such specific conditions defined in different studies, correlate with enhanced fungal growth in localities with abundance of vegetation and watercourses and after long term increase in soil water storage (Barrozo et al. 2009), altitude from 1,000 to 1,499 metres above sea level, rainfall from 2000 to 2999 mm, presence of humid forests (Holdridge) and coffee or tobacco plantations (Calle et al. 2001). Those conditions could lead to a great release of spores in the humid air and under these conditions the largest risk of exposure is found. Barrozo et al. (2009), associated soil water storage, absolute air humidity higher than normal and the climatic anomaly caused by the 1982/1983 El Niño Southern Oscillation with a cluster of acute/subacute cases 1–2 years later, in a southern region of Brazil (Restrepo 2000a; Barrozo et al. 2009).

No outbreaks have been reported to Paracoccidioidomycosis, however Brazil accounts for over 80% of all reported cases, with Colombia, Venezuela, Ecuador, and Argentina following suit but with much lower prevalence rates. Other Latin American countries report a low number of cases, as with all Central and North American countries (Brummer et al. 1993; Bittencourt et al. 2005; Bonifaz 2010). The annual incidence rate in Brazil is estimated to vary from 10 to 30 cases per million inhabitants; the mean mortality rateis 1.4 deaths per million inhabitants (Santo 2008; Nucci et al. 2009). In contrast, Colombia has a much lower incidence that fluctuates from year to year, with the highest being 2.4 cases per million inhabitants (Torrado et al. 2000; Restrepo et al. 2011).

Studies on the specific mortality due to systemic mycoses began to be published in 2002 by Coutinho et al. and revealed that paracoccidioidomycosis had been the eighth cause of death by predominantly chronic or recurrent diseases, infectious and parasitic diseases and the leading cause of death among the systemic mycoses in the 1980–1995 period (Coutinho et al. 2002).

In Brazil, the high mortality rates of this mycosis, were found in the Southeast, especially in the States of São Paulo, Minas Gerais and Rio de Janeiro followed by Paraná and Rio Grande do Sul (Prado et al. 2009). It is considered a serious public health problem due to the existence of extensive endemic areas, with a number of premature deaths especially among rural workers (Blotta et al. 1999; Shikanai-Yasuda et al. 2006). The annual incidence of new cases vary within endemic areas ranging from 1 to 3 new cases per 10^5 inhabitants (Shikanai-Yasuda et al. 2006).

Independent of the geographical area, high incidence is observed in male patients, aged 30 and 50, the ratio of male to female is 13:1. The gender difference is not observed in children or adolescents, where the mycosis is evenly distributed between genders, with a slight predominance in young adult males (Brummer et al. 1993; Restrepo and Tobon 2005; Restrepo et al. 2008).

Women are more resistant to the disease due to the presence of endogenous estrogens which bind to proteins in the cytosol of the fungus and inhibit the transformation of hyphae into yeast forms (Stover et al. 1986). Without hormone interference, infection may appear before puberty or after menopause, since *P. brasiliensis* have receptors for 17-β-estradiol (E2) in the cytoplasm (Shankar et al. 2011a,b).

Epidemiological studies show a striking predominance of paracoccidioidomycosis in adult men compared to premenopausal women. *In vitro* and *in vivo* studies suggest that the female hormone (17β-estradiol, E(2)) regulates or inhibits M-or-C-to-Y transition. The effect of E(2) at the molecular level on the inhibition of the M-to-Y transition involves signaling genes that regulate dimorphism such as palmitoyltransferase (erf2), small GTPase RhoA, phosphatidylinositol-4-kinase, and protein kinase (serine/threonine), which have low expression in the presence of E(2) (Shankar et al. 2011a,b). The increased susceptibility of individuals to paracoccidioidomycosis may also have a genetic basis, affecting the cellular immune response in susceptible patients. Resistance is associated with cytokine-stimulated granuloma formation and nitric oxide production (Aristizabal et al. 2002).

The disease in humans may develop shortly after the fungus comes into contact with the host, or after a latency period which can last from months to several years. This has been demonstrated in patients who have moved away from the recognized endemic areas (Brummer et al. 1993; Restrepo 1994; Restrepo 2000b).

It should be noted that PCM has a higher incidence in populations with low economic resources, who do not have access to health services, especially those who perform activities in agriculture (Marques et al. 1983). Reports of patients with PCM living in large urban centers or in non-endemic countries such as USA, Africa, Europe and Asia reflect disease importation by patients who have lived in endemic areas prior to the clinical manifestation of the mycosis (Forjaz et al. 1999).

Pathogenesis

Fungal virulence factors should favor the adhesion, colonization, dissemination and the ability of the fungus to survive in hostile environments and escape the host immune response. Interactions with the host defense mechanisms are complex and define the pathogenesis of systemic mycoses (Kurokawa et al. 1998; Casadevall and Pirofski 2001). Pathogenicity of dimorphic fungi depends on their ability to grow at 37°C (thermotolerance), the expression of adhesins, production of toxins and lytic enzymes and melanization capacity, among other characteristics (San-Blas et al. 2000; Mendes-Giannini et al. 2008).

Little is known, however, about the virulence mechanisms that allow the fungus to adapt to the host tissue, start colonization and proceed to invasion (San-Blas et al. 2002; Restrepo et al. 2008). Nevertheless, one can assume that dimorphic transition is an important virulence factor, since the conversion of conidia to yeast in the lung initiates the infection. Dimorphic conversion is the rule in experimental models, so its significance in the pathogenesis of PCM is clear (Mendes-Giannini et al. 2000; Restrepo 2003; Brigido et al. 2005).

The cell wall components of yeast and filamentous forms of *P. brasiliensis* comprise of glucans, chitin, proteins and lipids (Kanetsuna et al. 1969; Carbonell et al. 1970; Kanetsuna et al. 1972). The cell wall polysaccharides may have a role in virulence. α-1,3 Glucan is the major polysaccharide in the cell wall of yeast forms, which have only traces of β-glucans (de Moraes and Schäffer 2002). In the mycelium cell wall, β-glucan is the only glucan present. This has raised the hypothesis that dimorphic transformation of *P. brasiliensis* requires strict control over the synthesis of glucans (Brummer et al. 1993).

Studies conducted *in vitro* with *P. brasiliensis* isolates suggested that α-1,3-glucan protects the fungus against digestive enzymes of the host as well as leukocytes and macrophages (San-Blas and San-Blas 1977). Brummer et al. (1990) also observed that the virulence of isolates of *P. brasiliensis* was attenuated or lost when it was maintained by *in vitro* sub-culturing for a long time.

The fungal cell wall represents a protection barrier against physical and chemical processes, and also a contact structure for adhesion and subsequent invasion of the host tissue (Reiss et al. 1992). *P. brasiliensis* synthesizes many proteolytic enzymes capable of hydrolyzing ECM and even membrane components of the host cells, which may facilitate the invasion of their tissues (Puccia et al. 1998; Mendes-Giannini et al. 2004).

Salem-Izacc et al. (1997) found that the mycelium exhibited a range of protein patterns suggesting that the heterogeneous expression may reflect the adaptability of the saprophytic phase in different environments.

The genetic polymorphism of the fungus was related to its ability to invade tissues by "virulent" strains, which differ in the experimental infection by their ability to produce granulomas and invade tissues (Franco et al. 1993; Molinari-Madlum et al. 1999).

P. brasiliensis is a fungus able to synthesize melanin (Gomez et al. 2001). Uran et al. (2011) demonstrated that *P. brasiliensis* conidia synthesize melanin without chemical induction and that both conidia and yeast are able to synthesize melaninin a culture medium containing L-DOPA or L-epinephrine and use these or similar metabolites from infected tissues. *Paracoccidioides* melanin induces monoclonal antibody production, of IgG and Ig Misotypes, that react with most of the pathogenic fungi melanins (Uran et al. 2011).

The *P. brasiliensis* melanin appears to contribute to the virulence of the fungus by reducing phagocytosis of yeast cells by peritoneal, alveolar, and primary macrophages (J774.16 and MH-S). The cells of melanized fungi are also less susceptible to antifungal drugs, particularly amphotericin B (da Silva et al. 2006).

Fungal adhesion is mediated by a 32-kDa protein, a member of the haloaciddehalogenase (HAD) superfamily of hydrolases, which binds to the extracellular matrix (ECM) and modulates the initial immune response for evasion of the host defense. Its role as a virulence factor was demonstrated using an antisense RNA (aRNA) which consistently reduced PbHAD32 gene expression (Hernandez et al. 2010).

In addition to being essential for the diagnosis of PCM, gp43, the major diagnostic antigen, has been suggested to be a virulence factor, based on its specific binding to laminin. The binding to laminin induces an increase in adhesion to epithelial cells of the fungus (Vicentini et al. 1994). Furthermore, it was observed that peripheral blood mononuclear cells (PBMC) from patients with PCM when stimulated with gp43 increase the secretion of IL-10, an important immunosuppressive cytokine that can inhibit the expression of proinflammatory cytokines.

Clinical Manifestations

The clinical forms of PCM were classified in the International Colloquium on PCM, held in 1986 in Medellín, Colombia, where the relationship between clinical aspects and natural history of the disease were established (Franco et al. 1987). Definitions for the course of the disease, established that:

1. *Paracoccidioidomycosis-infection* corresponds to the patient without signals and symptoms of the disease but with positive paracoccidioidin skin test reaction.
2. *Paracoccidioidomycosis-disease* is divided in two main forms: soon after primary infection, acute/subacute type, or after a long period of latency, chronic type, with localization in different organs and the degree of severity based on general and nutritional status and organ dysfunction. The acute/subacute type (also known as juvenile) can be classified as moderate or severe.

   Acute and subacute forms have high levels of specific antibodies but mostly depend on the cellular immune system to combat the infection. Lung granulomas, associated with a high number of viable yeast cells, are formed to contain the fungus (Franco et al. 1987). The acute form, which represents 15–20% of PCM cases, has rapid evolution, with frequent lymphadenopathy, hepatosplenomegaly, osteo-articular involvement, mediastinal, intra-abdominal and skin lesions (Shikanai-Yasuda et al. 2006). It affects young persons of both genders and immunocompromised persons, and immunologically the disease is characterized by antigenemia, polyclonal B cell activation, anergy and fungal dissemination (Montenegro and Franco 1994; de Camargo and Franco 2000).

   The chronic form (adult type) is characterized by unifocal and multifocal lesions and sequels. Patients with the chronic form of PCM, who represent approximately 90% of cases, have a prolonged disease with slow and gradual installation, tending to localize in organs and tissues in a focal way (unifocal or multifocal), affecting mainly the lungs and the lining of the airways with maintenance of cellular immune response and low titers of specific antibodies. This is the hyperergic or localized form (Mendes 1994; Montenegro and Franco 1994). The chronic form progresses slowly, silently, affecting the lungs (90% of patients), mucosal and skin sites (Shikanai-Yasuda et al. 2006).
3. *Sequelae*, particularly involve pulmonary chronic obstructive disease, stenosis and obstruction of the superior airways and adrenal insufficiency (Tobon et al. 2003). Although the fungal infection can be controlled using conventional chemotherapy, sequels such as microstomia, laryngeal or tracheal stenosis and intestinal obstruction can be observed in some patients (Bocca et al. 2013). Pulmonary fibrosis is usually associated with a previous smoking habit (Restrepo et al. 2008).

Fibrosis was present in 30% of patients with severe infiltration and in 12.5% of patients with minor infiltration. Among patients with severe infiltration, fibrosis could increase (to 75%) when bullae were concomitantly present at diagnosis. Prompt initiation of treatment is necessary to avoid the development of fibrosis (Tobon et al. 2003).

Co-infection with tuberculosis is observed in 10 to 15% of pulmonar PCM cases. HIV co-infection is not a problem as observed with other systemic mycoses such as cryptococcosis and histoplasmosis. Cases of PCM in HIV patients have been reported and some cases of immune reconstitution inflammatory syndrome (IRIS) were reported (Gryschek et al. 2010).

Laboratorial Diagnosis

According to Guidelines in paracoccidioidomycosis (Shikanai-Yasuda et al. 2006) laboratorial tests and image detection such as X-ray, abdominal ultrasound, blood counts, erythrocyte sedimentation rate, biochemical liver tests (transaminases, alkaline phosphatase), protein electrophoresis and kidney metabolism (creatinine and Na and K) are needed for evaluation of patients. Patients with central nervous system (headache, motor deficit, convulsive syndrome, behavior modification and / or level of consciousness), gastrointestinal disorders symptoms, adrenal dysfunction, respiratory failure and lesions of bone or muscle require more complex exams.

The definitive diagnosis of PCM can be made by direct examination of sputum, biopsy specimens or crusts/pus from suppurated lymph nodes, which typically contain the yeast forms of *P. brasiliensis/P. lutzii.* This is combined with culturing the fungus from any clinical specimen. Characteristically, the morphology of *P. brasiliensis/ P. lutzii* fungal elements in its parasitical form shows large globose cells with narrow-necked multiple budding yeasts or mother cells with only two buddings. Routine methods for wet preparations use KOH treatment or calcofluor fluorescent stain. Histopathological preparations are usually stained with Grocott-Gomori stain or PAS for better identification of the fungal elements (Travassos et al. 2008).

Culturing of *Paracoccidioides* spp. from clinical samples is achieved on Sabouraud's agar or yeast extract agar containing chloramphenicol and cycloheximide. Recovery rates of the pathogen may be limited by the overgrowth of bacteria present in potentially contaminated organic fluids (ex: sputum and skin lesions).

Serological tests are important for the diagnosis and follow-up of PCM including the double-immunodiffusion test, immunoenzymatic assays and counter-immunoelectrophoresis. In clinical practice, detection of specific antibodies is used in the screening of patients suspected to be infected by *P. brasiliensis* as well as for monitoring the clinical response to therapy (Travassos et al. 2008; de Camargo 2008). The serological diagnosis of PCM by double-immunodiffusion test and counter-immunoelectrophoresis was accomplished using only the standard exoantigen from *P. brasiliensis* B339 isolate, which has a high concentration of the 43 kDa glycoprotein known as gp43, the immunodominant and specific antigen for PCM diagnosis. However, patients infected with *P. lutzii*, may give false-negative serology (de Camargo 2008; Gegembauer et al. 2014).

The follow-up of patients under treatment for PCM has shown that many times the antibody titers obtained by the ID test do not correlate with the clinical status of the patient, such discrepancy is probably related to the fact that the cellular response and not the humoral response is the main immunologic mechanism able to contain *P. brasiliensis* in the infected organism (de Camargo 2008). In the medical practice, clinical, radiological, mycological and serological aspects must be evaluated over a long period of observation in order to assess the result of treatment (Tobon et al. 2003; Restrepo and Tobon 2005).

Skin tests using both crude antigen preparations and purified antigen (gp43) from *P. brasiliensis* are used in epidemiological studies and do not have diagnostic value (Kalmar et al. 2004).

PCR methods for DNA amplification have also been used to identify patients with PCM. *P. brasiliensis* DNA sequences of potential diagnostic use have been described by different authors, including the 5.8S rRNA gene and ITS regions (Motoyama et al. 2000), as well as the gp43 gene (Gomes et al. 2000). Specific primers designed on a 0.72-kb DNA fragment of *P. brasiliensis* that were useful for identification of this pathogen in sputum and cerebrospinal fluid of PCM patients were also described (San-Blas et al. 2005). However, the real application of PCR for PCM diagnosis involves high costs and diagnostic laboratories that are not available in all endemic regions. Others strategies such as Matrix-Assisted Laser Desorption Ionization–Time of Flight Mass Spectrometry or MALDI-TOF MS for diagnosis of PCM is currently in development.

Conventional Treatment with Antifungal Drugs

According to Guidelines in paracoccidioidomycosis (Shikanai-Yasuda et al. 2006), recently reviewed (Bocca et al. 2013) there are many therapeutic options available, which include sulfone derivatives (sulfadiazine, sulfadoxine, sulfamethoxypyridazine, cotrimazine and trimethoprim–sulfamethoxazole), amphotericin B, azoles (ketoconazole, itraconazole, fluconazole, voriconazole and posaconazole) and terbinafine.

The treatment option must consider the severity of disease:

a. **For mild-to-moderate clinical forms:** the standard treatment is itraconazole. A randomized trial comparing itraconazole, ketoconazole and sulfadiazine showed that all drugs are efficient at promoting the clinical cure of severe PCM. Voriconazole is also effective for treating patients with mild infection, and can be used in patients with neuroPCM, since it penetrates better into the CNS than other drugs. Patients infected with *P. lutzii* isolates demonstrated better responses to trimethoprim–sulfamethoxazole than patients infected with *P. brasiliensis* isolates.
b. **Severe and disseminated forms:** amphotericin B in conventional or lipid formulations is a treatment option, despite its known adverse effects.
c. **Maintenance treatment:** after having the disease controlled, it is recommended that the treatment should consist of less toxic drugs to be administered (once or twice daily), like sulfamethoxazole/trimethoprim which is slowly excreted, thus favoring the deployment scheme for long-term treatment.

Vaccine and Antibody Therapy

Immunotherapy against PCM has been actively investigated aiming at the reduction of the treatment period and protection against relapses and sequels. Different methodologies have been used to induce immuneprotection in experimental models of *P. brasiliensis* infection.

A major line of investigation focused on purified antigens in the attempt to develop a peptide vaccine. The 43 kDa glycoprotein (gp43), isolated from *P. brasiliensis* culture supernatant fluid, is recognized by antibodies from most patients with PCM, except those exposed to the isolate recently designated as *Paracoccidioides lutzii* that shows an irregular reactivity with this antigen (de Camargo 2008; Gegem bauer et al. 2014). Epitopes in gp43 that elicited strong antibody response are peptidic in nature (reviewed by Travassos and Taborda 2012) and different isoforms of gp43 varied in their reactivity with patients' sera. The gp43 gene sequence, expression and polymorphism have been reviewed (Travassos et al. 2008).

The first evidence that gp43 carries an immunodominant epitope able to elicit DTH reactions was shown in guinea pigs and later in patients with PCM using the purified antigen. The T cell epitope responsible for DTH reactions, and $CD4^+$ T-cell proliferation, has been mapped and a peptide called P10 with the sequence: QTLIAIHTLAIRYAN was identified. By using the TEPITOPE algorithm additional peptides in the gp43 were also shown to bind HLA-DR antigens (Taborda et al. 1998; Travassos et al. 2008).

The immunoprotective efficacy of P10 was shown by immunization with the peptide associated with chemotherapy in intratracheally infected BALB/c mice. Animals were challenged with yeast cells of a highly virulent *P. brasiliensis* (Pb18) isolate and were treated with P10 and/or chemotherapeutic drug. The treatment was done for 30 days, during which groups of mice received intraperitoneal doses of itraconazole, fluconazole, ketoconazole, sulfamethoxazole or TMP/SMZ every 24 hr. Amphotericin B was administered every 48 hr. Immunization with P10 was carried out weekly for 4 weeks, once in complete Freund's adjuvant (CFA) and three times in incomplete Freund's adjuvant. Animals were sacrificed at different times of infection and a significantly reduced fungal load was observed, with additive protective effect obtained with the combination of P10 and the antifungal drug (Travassos et al. 2008).

In an attempt to reproduce acute/subacute forms of PCM, or the anergic state, BALB/c mice were treated with dexamethasone-21 phosphate added to drinking water. After 30 days animals showed negative DTH to fungal antigens. Mice were then infected with virulent *P. brasiliensis* (Pb18) and after fifteen days underwent chemotherapy and/or P10 immunization. The association of drugs and P10 immunization conferred additive protection. A significant increase in IL-12 and IFN-γ and decrease of IL-4 and IL-10 was observed in mice immunized with P10 alone or in association with antifungal drugs (Marques et al. 2006).

The therapeutic and prophylactic effects of P10 were compared with the peptide alone or mixed with different adjuvants: flagellin, aluminum hydroxide, cationic lipid or CFA. A vaccine formulation based on the intranasal administration of gp43 or P10 in combination with the *Salmonella enterica* FliC flagellin was evaluated in BALB/c mice. Immunization with synthetic P10 admixed with purified FliC followed by intratracheal infection with *P. brasiliensis* (Pb18) showed reduced lung burden and elicited a predominant Th1-type immune response (Braga et al. 2009). The combination of P10 entrapped within Poly (lactic acid-glycolic acid) nanoparticles (PLGA) with TMP/SMZ was tested in an experimental therapeutic protocol of PCM. The incorporation of P10 into PLGA (1 μg/50 μL) reduced the amount of this peptide necessary to decrease the fungal load in the infected animals and avoid disease relapse when compared with P10 emulsified in Freund's adjuvant (20 μg/50 μL) (Amaral et al. 2010). The P10 was also used in the subcutaneous immunization of BALB/c in the presence of different adjuvants followed by intratracheal infection with *P. brasiliensis* (Pb18). A reduction in the fungal burden was observed with all adjuvants used with P10. Particularly, the cationic lipid proved to be very efficient with encouraging results (Mayorga et al. 2012).

The use of the mammalian expression vector (VR-gp43) carrying the full gene of gp43 with CMV promoter induced B and T cell-mediated immune responses protective against the intratracheal challenge by virulent *P. brasiliensis* yeast forms (Pinto et al. 2000). In order to develop a more specific DNA vaccine predominantly based on T cell-mediated immune response, a plasmid encoding P10 in pcDNA3 expression vector was tested in intratracheally infected BALB/c and B10.A mice. The vaccination with plasmid encoding P10 induced a significant reduction in the fungal burden in the lung. Co-vaccination with a plasmid encoding mouse IL-12 proved to be even more effective in the elimination of the fungus with virtual sterilization in the long term (5 months) infection and treatment assay. The immunization elicited significant production of IL-12 and IFN-γ and reduction of IL-4 levels in lung homogenates (Rittner et al. 2012). The immunization with plasmid encoding P10 was also able to induce memory cells (de Amorim et al. 2013).

The use of dendritic cells pulsed with P10 to protect mice infected with *P. brasiliensis* was also investigated. BALB/c mice were infected with virulent *P. brasiliensis* Pb18 and were treated with dendritic cells pulsed with P10. The adoptive transference of pulsed dendritic cells in mice previously infected with the fungus rendered significant protection with reduced CFU in the lungs of the animals (Magalhães et al. 2012).

Other antigens apart from gp43 and P10 have also been investigated as alternative vaccines against *P. brasiliensis.* cDNA encoding the 27 kDa protein present at the surface and cytosol of *P. brasiliensis* was cloned into pGEX 4T-2 plasmid and expressed in *Escherichia coli* (rPb27). BALB/c mice immunized subcutaneously with purified rPb27 in the presence of *Propionibacterium acnes* and aluminum hydroxide as adjuvant prior to intravenous infection in the orbital plexus with the virulent isolate of *P. brasiliensis* (Pb18) developed high levels of IgG2b, moderate levels of IgG1 and low levels of IgG2a. Reduced fungal load in the lung, liver and spleen was observed (Santos et al. 2012). Radio-attenuated yeast cells have also been under investigation with promising experimental results (do Nascimento-Martins et al. 2007).

Although the humoral immune response in PCM has been regarded as non protective in the acute and subacute forms of the disease, it has been reported that monoclonal antibodies against gp70 were protective in the experimental PCM (de Mattos Grosso et al. 2003). The protective effects of a panel of anti-gp43 MAbs were also examined in BALB/c mice intratracheally infected with *P. brasiliensis*

(Buissa-Filho et al. 2008). Protective and non-protective mAbs that recognized gp43 in ELISA were identified. MAb 3E, was the most efficient in reducing the fungal burden *in vivo* and promoting fungal phagocytosis *in vitro*. The recognized epitope sequence was identified as NHVRIPIGYWAV (Buissa-Filho et al. 2008) and this peptide is also a candidate for a peptide vaccine against PCM. The association of P10 preimmunization and MAb 3E administered 24 hr before intratracheal challenge with virulent *P. brasiliensis* yeasts resulted in additive protection using a short-term protocol in comparison with a non protective mAb (Reviewed by Travassos et al. 2007; Travassos and Taborda 2012).

Recently, Thomaz et al. (2014) demonstrated that mAbs generated against the heat shock protein 60 from *Histoplasma capsulatum* interact with *P. lutzii* yeast cells and enhance phagocytosis by macrophages cells. The passive transference of Hsp60 7B6 and 4E12 mAbs were protective and reduced the lung fungal burden in BALB/c mice intratracheally infected with *P. lutzii*.

Conclusion

In 2008 the first description of patients with paracoccidioidomycosis completed 100 years. Despite this, paracoccidioidomycosis remains a neglected fungal disease in most countries where it is endemic. One of common factors is that the most patients are poor and live in rural areas. Although, the available treatments are efficient, they take a long time and adhesion becomes a serious concern since the lesions heal much before the fungus is eradicated, remaining in unapparent foci. The laboratory diagnosis and follow up of patients has evolved to account for the new species and to achieve better sensitivity and specificity standards. A prophylactic vaccine appears to be an in achievable task now; however, a therapeutic vaccine using peptides or transference of monoclonal antibodies would be an adjunct to the treatment of patients with this mycosis in the near future.

References

Ajello, L. 1977. Medically infectious fungi. Control Microbiology and Immunology. 3: 7–19.

Albornoz, M.B. 1971. Isolation of *Paracoccidioides brasiliensis* from rural soil in Venezuela. Sabouradia 9: 248–53.

Almeida, F. 1930. Estudos comparativos do granuloma coccidióidico nos Estados Unidos e no Brasil. Novo gênero para o parasito brasileiro. An Fac Med São Paulo 5: 125–141.

Amaral, A.C., A.F. Marques, J.E. Munoz, A.L. Bocca, A.R. Simioni, A.C. Tedesco, P.C. Morais, L.R. Travassos, C.P. Taborda and M.S. Felipe. 2010. Poly(lactic acid-glycolic acid) nanoparticles markedly improve immunological protection provided by peptide P10 against murine paracoccidioidomycosis. Br J Pharmacol. 159(5): 1126–32.

Aristizabal, B.H., K.V. Clemons, A.M. Cock, A. Restrepo and D.A. Stevens. 2002. Experimental paracoccidioides brasiliensis infection in mice: influence of the hormonal status of the host on tissue responses. Med Mycol. 40(2): 169–78.

Bagagli, E., A. Sano, K.I. Coelho, S. Alquati, M. Miyaji, Z.P. de Camargo, G.M. Gomes, M. Franco and M.R. Montenegro. 1998. Isolation of *Paracoccidioides brasiliensis* from armadillos (*Dasypus noveminctus*) captured in an endemic area of paracoccidioidomycosis. Am J Trop Med Hyg. 58(4): 505–12.

Barrozo, L.V., R.P. Mendes, S.A. Marques, G. Benard, M.E. Silva and E. Bagagli. 2009. Climate and acute/subacute paracoccidioidomycosis in a hyper-endemic area in Brazil. Int J Epidemiol. 38(6): 1642–1649.

Bittencourt, J.I., R.M. de Oliveira and Z.F. Coutinho. 2005. Paracoccidioidomycosis mortality in the State of Parana, Brazil, 1980/1998. Cad Saude Publica. 21(6): 1856–1864.
Blotta, M.H., R.L. Mamoni, S.J. Oliveira, S.A. Nouer, P.M. Papaiordanou, A. Goveia and Z.P. Camargo. 1999. Endemic regions of paracoccidioidomycosis in Brazil: a clinical and epidemiologic study of 584 cases in the southeast region. Am J Trop Med Hyg. 61(3): 390–394.
Bocca, A.L., A.C. Amaral, M.M. Teixeira, P.K. Sato, M.A. Shikanai-Yasuda and M.S.S. Felipe. 2013. Paracoccidioidomycosis: eco-epidemiology, taxonomy and clinical and therapeutic issues. Future Microbiol. 8(9): 1177–1191.
Bonifaz, A. 2010. Micología Médica Básica. McGraw Hill Interamericana Editores, Mexico City, Mexico.
Borelli, D. 1955. Las aleurias de *Paracoccidioides brasiliensis*. VI Cong Venezolano Cienc Med; Venezuela.
Braga, C.J., G.M. Rittner, J.E.H. Muñoz, A.F. Teixeira, L.M. Massis, M.E. Sbrogio-Almeida, C.P. Taborda, L.R. Travassos and L.C. Ferreira. 2009. *Paracoccidioides brasiliensis* vaccine formulations based on the gp43-derived P10 sequence and the *Salmonella enterica* FliC flagellin. Infect Immun. 77(4): 1700–1707.
Brigido, M.M., M.E. Walter, A.G. Oliveira, M.K. Inoue, D.S. Anjos, E.F. Sandes, J.J. Gondim, M.J. Carvalho, N.F. Almeida, Jr. and M.S. Felipe. 2005. Bioinformatics of the *Paracoccidioides brasiliensis* EST Project. Genet Mol Res. 4(2): 203–125.
Brummer, E., A. Restrepo, L.H. Hanson and D.A. Stevens. 1990. Virulence of *Paracoccidioides brasiliensis*: the influence of *in vitro* passage and storage. Mycopathologia 109(1): 13–17.
Brummer, E., E. Castaneda and A. Restrepo. 1993. Paracoccidioidomycosis: an update. Clin Microbiol Rev. 6(2): 89–117.
Buissa-Filho, R., R. Puccia, A.F. Marques, F.A. Pinto, J.E. Muñoz, J.D. Nosanchuk, L.R. Travassos and C.P. Taborda. 2008. The monoclonal antibody against the major diagnostic antigen of *Paracoccidioides brasiliensis* mediates immune protection in infected BALB/c mice challenged intratracheally with the fungus. Infect Immun. 76(7): 3321–3328.
Bustamante-Simon, B., J.G. McEwen, A.M. Tabares, M. Arango and A. Restrepo-Moreno. 1985. Characteristics of the conidia produced by the mycelial form of *Paracoccidioides brasiliensis*. Sabouraudia 23(6): 407–414.
Calle, D., D.S. Rosero, L.C. Orozco, D. Camargo, E. Castaneda and A. Restrepo. 2001. Paracoccidioidomycosis in Colombia: an ecological study. Epidemiol Infect. 126(2): 309–315.
Carbonell, L.M., F. Kanetsuna and F. Gil. 1970. Chemical morphology of glucan and chitin in the cell wall of the yeast phase of *Paracoccidioides brasiliensis*. J Bacteriol. 101(2): 636–642.
Casadevall, A. and L. Pirofski. 2001. Host-pathogen interactions: the attributes of virulence. J Infect Dis. 184(3): 337–344.
Casquero, J., J. Demarini, M. Castillo, J. Candella and S. Zurita. 1998. Nuevos Casos de Paracoccidioidomicosis. Rev Med Exp. INS, XV (1/2).
Cole, G.T. and S.H. Sun. 1985. Arthroconidium-spherule-endospore transformation in *Coccidioides immitis*. pp. 282–333. *In*: Szaniszlo, P.J. (ed.). Fungal Dimorphism. Plenum Press, New York.
Conant, N.F. and A. Howell. 1942. The similarity of the fungi causing South American Blastomycosis (paracoccidioidomycosis) and North American blastomycosis (Gilchrist disease). J Investigat Dermatol. 5: 353–370.
Coutinho, Z.F., D. Silva, M. Lazera, V. Petri, R.M. Oliveira, P.C. Sabroza and B. Wanke. 2002. Paracoccidioidomycosis mortality in Brazil (1980–1995). Cad Saude Publica. 18(5): 1441–1454.
da Silva, M.B., A.F. Marques, J.D. Nosanchuk, A. Casadevall, L.R. Travassos and C.P. Taborda. 2006. Melanin in the dimorphic fungal pathogen *Paracoccidioides brasiliensis*: effects on phagocytosis, intracellular resistance and drug susceptibility. Microbes Infect. 8(1): 197–205.
de Amorim, J., A. Magalhaes, J.E. Muñoz, G.M. Rittner, J.D. Nosanchuk, L.R. Travassos and C.P. Taborda. 2013. DNA vaccine encoding peptide P10 against experimental paracoccidioidomycosis induces long-term protection in presence of regulatory T cells. Microbes Infect. 15(3): 181–191.
de Camargo, Z.P. 2008. Serology of paracoccidioidomycosis. Mycopathologia 165(4-5): 289–302.
de Camargo, Z.P. and M.F. de Franco. 2000. Current knowledge on pathogenesis and immunodiagnosis of paracoccidioidomycosis. Rev Iberoam Micol. 17(2): 41–48.
de Mattos Grosso, D., S.R. de Almeida, M. Mariano and J.D. Lopes. 2003. Characterization of gp70 and anti-gp70 monoclonal antibodies in *Paracoccidioides brasiliensis* pathogenesis. Infect Immun. 71(11): 6534–6542.
de Moraes, B.C. and G.M. Schäffer. 2002. *Paracoccidioides brasiliensis*: virulence and an attempt to induce the dimorphic process with fetal calf serum. Mycoses 45(5-6): 174–179.

Desjardins, C.A., M.D. Champion, J.W. Holder, A. Muszewska, J. Goldberg, A.M. Bailao, M.M. Brigido, M.E. Ferreira, A.M. Garcia, M. Grynberg, S. Gujja, D.I. Heiman, M.R. Henn, C.D. Kodira, H. León-Narváez, L.V. Longo, L.J. Ma, I. Malavazi, A.L. Matsuo, F.V. Morais, M. Pereira, S. Rodríguez-Brito, S. Sakthikumar, S.M. Salem-Izacc, S.M. Sykes, M.M. Teixeira, M.C. Vallejo, M.E. Walter, C. Yandava, S. Young, Q. Zeng, J. Zucker, M.S. Felipe, G.H. Goldman, B.J. Haas, J.G. McEwen, G. Nino-Vega, R. Puccia, G. San-Blas, C.M. Soares, B.W. Birren and C.A. Cuomo. 2011. Comparative genomic analysis of human fungal pathogens causing paracoccidioidomycosis. PLoS Genet. 7(10): e1002345.

do Nascimento-Martins, E.M., B.S. Reis, V.C. Fernandes, M.M. Costa, A.M. Goes and A.S. de Andrade. 2007. Immunization with radioattenuated yeast cells of *Paracoccidioides brasiliensis* induces a long lasting protection in BALB/c mice. Vaccine 25(46): 7893–7899.

Ferreira, M.S., L.H. Freitas, C.S. Lacaz, G.M. del Negro, N.T. de Melo, N.M. Garcia, C.M. de Assis, A. Salebian and E.M. Heins-Vaccari. 1990. Isolation and characterization of a *Paracoccidioides brasiliensis* strain from a dogfood probably contaminated with soil in Uberlandia, Brazil. J Med Vet Mycol. 28(3): 253–6.

Forjaz, M.H., O. Fischman, Z.P. de Camargo, J.P. Vieira-Filho and A.L. Colombo. 1999. Paracoccidioidomycosis in Brazilian Indians of the Surui tribe: clinical-laboratory study of 2 cases. Rev Soc Bras Med Trop. 32(5): 571–575.

Franco, M. 1987. Host-parasite relationships in paracoccidioidomycosis. J Med Vet Mycol. 25(1): 5–18.

Franco, M., M.R. Montenegro and R.P. Mendes. 1987. Paracoccidioidomycosis: a recently proposed classification of its clinical forms. Rev Soc Bras Med Trop. 20: 129–132.

Franco, M., M.T. Peracoli, A. Soares, R. Montenegro, R.P. Mendes and D.A. Meira. 1993. Host-parasite relationship in paracoccidioidomycosis. Curr Top Med Mycol. 5: 115–149.

Franco, M., C. Lacaz, A. Restrepo and G. del Negro. 1994. Paracoccidioidomycosis. CRC Press, Boca Raton—Florida.

Gegembauer, G., L.M. Araujo, E.F. Pereira, A.M. Rodrigues, A.M. Paniago, R.C. Hahn and Z.P. de Camargo. 2014. Serology of paracoccidioidomycosis due to *Paracoccidioides lutzii*. PLoS Negl Trop Dis. 8(7): e2986.

Gomes, G.M., P.S. Cisalpino, C.P. Taborda and Z.P. de Camargo. 2000. PCR for diagnosis of paracoccidioidomycosis. J Clin Microbiol. 38(9): 3478–3480.

Gomez, B.L., J.D. Nosanchuk, S. Diez, S. Youngchim, P. Aisen, L.E. Cano, A. Restrepo, A. Casadevall and A.J. Hamilton. 2001. Detection of melanin-like pigments in the dimorphic fungal pathogen *Paracoccidioides brasiliensis in vitro* and during infection. Infect Immun. 69(9): 5760–5767.

Grose, E. and J.R. Tamsitt. 1965. *Paracoccidioides brasiliensis* recovered from the intestinal tract of three bats (*Artibeus lituratus*) in Colombia, S.A. Sabouraudia 4(2): 124–125.

Gryschek, R.C., R.M. Pereira, A. Kono, R.A. Patzina, A.T. Tresoldi, M.A. Shikanai-Yasuda and G. Benard. 2010. Paradoxical reaction to treatment in 2 patients with severe acute paracoccidioidomycosis: a previously unreported complication and its management with corticosteroids. Clin Infect Dis. 50(10): e56-8.

Hernandez, O., A.J. Almeida, A. Gonzalez, A.M. Garcia, D. Tamayo, L.E. Cano, A. Restrepo and J.G. McEwen. 2010. A 32-kilodalton hydrolase plays an important role in *Paracoccidioides brasiliensis* adherence to host cells and influences pathogenicity. Infect Immun. 78(12): 5280–5386.

Kalmar, E.M., F.E. Alencar, F.P. Alves, L.W. Pang, G.M. Del Negro, Z.P. Camargo and M.A. Shikanai-Yasuda. 2004. Paracoccidioidomycosis: an epidemiologic survey in a pediatric population from the Brazilian Amazon using skin tests. Am J Trop Med Hyg. 71(1): 82–86.

Kanetsuna, F., L.M. Carbonell, R.E. Moreno and J. Rodriguez. 1969. Cell wall composition of the yeast and mycelial forms of *Paracoccidioides brasiliensis*. J Bacteriol. 97(3): 1036–1041.

Kanetsuna, F., L.M. Carbonell, I. Azuma and Y. Yamamura. 1972. Biochemical studies on the thermal dimorphism of *Paracoccidioides brasiliensis*. J Bacteriol. 110(1): 208–218.

Kurokawa, C.S., M.F. Sugizaki and M.T. Peracoli. 1998. Virulence factors in fungi of systemic mycoses. Rev Inst Med Trop São Paulo 40(3): 125–135.

Lacaz, C.S. 1982. Evolução dos conhecimentos sobre a paracoccidioidomicose. Um pouco da história. pp. 1–9. *In*: Del Negro, G., C.S. Lacaz and A.M. Fiorillo (eds.). Paracoccidioidomicose Blastomicose Sul-Americana. Sarvier, São Paulo.

Lacaz, C.S., E. Porto and J.E.C. Martins. 1991. Paracoccidioidomicose, 8th ed. Sarvier, Sao Paulo—Brazil.

Lazo, S.R.F. 1998. Historia de la micología ecuatoriana. Rev Iberoam Micol. 15: 248–252.

López-Martínez, R., F. Hernández-Hernández, L.J. Méndez-Tovar, P. Manzano-Gayosso, A. Bonifaz, R. Arenas, M.C. Padilla-Desgarennes, R. Estrada and G. Chávez. 2014. Paracoccidioidomycosis in Mexico: clinical and epidemiological data from 93 new cases (1972–2012). Mycoses 57(9): 525–530.
Lutz, A. 1908. Uma mycose pseudococcidica localisada na bocca e observada no Brasil. Contribuição ao conhecimento das hyphoblastomycoses americanas. Bras Med. 22: 121–124.
Magalhães, A., K.S. Ferreira, S.R. Almeida, J.D. Nosanchuk, L.R. Travassos and C.P. Taborda. 2012. Prophylactic and therapeutic vaccination using dendritic cells primed with peptide 10 derived from the 43-kilodalton glycoprotein of *Paracoccidioides brasiliensis*. Clin Vaccine Immunol. 19(1): 23–29.
Marques, A.F., M.B. da Silva, M.A. Juliano, L.R. Travassos and C.P. Taborda. 2006. Peptide immunization as an adjuvant to chemotherapy in mice challenged intratracheally with virulent yeast cells of *Paracoccidioides brasiliensis*. Antimicrob Agents Chemother. 50(8): 2814–2819.
Marques, S.A. 2008. Paracoccidioidomycosis: a century from the first case report. An Bras Dermatol. 83: 271–273.
Marques, S.A., M.F. Franco, R.P. Mendes, N.C. Silva, C. Baccili, E.D. Curcelli, A.C. Feracin, C.S. Oliveira, J.V. Tagliarini and N.L. Dillon. 1983. Epidemiologic aspects of paracoccidioidomycosis in the endemic area of Botucatu (São Paulo—Brazil). Rev Inst Med Trop São Paulo 25(2): 87–92.
Martínez, D.M., R.V. Hernández, P. Alvarado and M. Mendoza. 2013. Mycoses in Venezuela: Working Groups in Mycology reported cases (1984–2010). Rev Iberoam Micol. 30(1): 39–46.
Matute, D.R., J.G. McEwen, R. Puccia, B.A. Montes, G. San-Blas, E. Bagagli, J.T. Rauscher, A. Restrepo, F. Morais, G. Nino-Vega and J.W. Taylor. 2006. Cryptic speciation and recombination in the fungus Paracoccidioides brasiliensis as revealed by gene genealogies. Mol Biol Evol. 23: 65–73.
Mayorga, O., J.E. Muñoz, N. Lincopan, A.F. Teixeira, L.C. Ferreira, L.R. Travassos and C.P. Taborda. 2012. The role of adjuvants in therapeutic protection against paracoccidioidomycosis after immunization with the P10 peptide. Front Microbiol. 3: 154.
Mendes, R.P. 1994. The gamut of clinical manifestations. pp. 233–252. *In*: Franco, M.F., C.S. Lacaz, A. Restrepo and G. Del Negro (eds.). Paracoccidioidomycosis. CRC Press, Boca Ratón, Florida.
Mendes-Giannini, M.J., M.L. Taylor, J.B. Bouchara, E. Burger, V.L. Calich, E.D. Escalante, S.A. Hanna, H.L. Lenzi, M.P. Machado, M. Miyaji, J.L. Monteiro Da Silva, E.M. Mota, A. Restrepo, S. Restrepo, G. Tronchin, L.R. Vincenzi, C.F. Xidieh and E. Zenteno. 2000. Pathogenesis II: fungal responses to host responses: interaction of host cells with fungi. Med Mycol. 38(1): 113–123.
Mendes-Giannini, M.J., S.A. Hanna, J.L. da Silva, P.F. Andreotti, L.R. Vincenzi and R. Benard. 2004. Invasion of epithelial mammalian cells by *Paracoccidioides brasiliensis* leads to cytoskeletal rearrangement and apoptosis of the host cell. Microbes Infect. 6(10): 882–891.
Mendes-Giannini, M.J., J.L. Monteiro-da-Silva, J. de Fatima-da-Silva, F.C. Donofrio, E.T. Miranda, P.F. Andreotti and C.P. Soares. 2008. Interactions of *Paracoccidioides brasiliensis* with host cells: recent advances. Mycopathologia 165(4-5): 237–248.
Molinari-Madlum, E.E., M.S. Felipe and C.M. Soares. 1999. Virulence of *Paracoccidioides brasiliensis* isolates can be correlated to groups defined by random amplified polymorphic DNA analysis. Med Mycol. 37(4): 269–276.
Montenegro, M.R. and F. Franco. 1994. Pathology. pp. 121–130. *In*: Franco, M.F., C.S. Lacaz, A. Restrepo and G. Del Negro (eds.). Paracoccidioidomycosis. CRC Press, Boca Ratón, Florida.
Motoyama, A.B., E.J. Venancio, G.O. Brandao, S. Petrofeza-Silva, I.S. Pereira, C.M. Soares and M.S. Felipe. 2000. Molecular identification of *Paracoccidioides brasiliensis* by PCR amplification of ribosomal DNA. J Clin Microbiol. 38(8): 3106–3109.
Naiff, R.D., L.C. Ferreira, T.V. Barrett, M.F. Naiff and J.R. Arias. 1986. Enzootic paracoccidioidomycosis in armadillos (*Dasypus novemcinctus*) in the State of Pará. Rev Inst Med Trop São Paulo 28(1): 19–27.
Negroni, P. 1966. Patología y Micología de la micosis de Lutz. *In*: Científica CdI, editor. Micosis Profundas Las Blastomicosis y Coccidioidomicosis. Provincia de Buenos Aires, Argentina.
Negroni, P. 1968. Studies on the ecology of *Paracoccidioides brasiliensis* in Argentina. Torax. 17(1): 60–63.
Neves, J.S. and L. Bogliolo. 1951. Researches on the etiological agents of the American blastomycosis. I Morphology and systematics of the Lutz′ disease agent. Mycopathol Mycol Appl. 5: 133–142.
Nucci, M., A.L. Colombo and F. Queiroz-Telles. 2009. Paracoccidioidomycosis. Curr Fungal Infect Rep. 3: 15–20.
Pinto, A.R., R. Puccia, S.N. Diniz, M.F. Franco and L.R. Travassos. 2000. DNA-based vaccination against murine paracoccidioidomycosis using the gp43 gene from *Paracoccidioides brasiliensis*. Vaccine 18(26): 3050–3058.
Pollak, L. 1971. Aleuriospores of *Paracoccidioides brasiliensis*. Mycopathol Mycol Appl. 45(3): 217–219.

Prado, M., M.B. da Silva, R. Laurenti, L.R. Travassos and C.P. Taborda. 2009. Mortality due to systemic mycoses as a primary cause of death or in association with AIDS in Brazil: a review from 1996 to 2006. Mem Inst Oswaldo Cruz. 104(3): 513–521.

Puccia, R., A.K. Carmona, J.L. Gesztesi, L. Juliano and L.R. Travassos. 1998. Exocellular proteolytic activity of *Paracoccidioides brasiliensis*: cleavage of components associated with the basement membrane. Med Mycol. 36(5): 345–348.

Reiss, E., V.M. Hearn, D. Poulain and M.G. Shepherd. 1992. Structure and function of the fungal cell wall. J Med Vet Mycol. 30(1): 143–156.

Restrepo, A. 1970. A reappraisal of the microscopical appearance of the mycelial phase of *Paracoccidioides brasiliensis*. Sabouraudia 8(2): 141–144.

Restrepo, A. 1985. The ecology of *Paracoccidioides brasiliensis*: a puzzle still unsolved. Sabouraudia 23(5): 323–334.

Restrepo, A. 1994. Treatment of tropical mycoses. J Am Acad Dermatol. 31(3 Pt 2): S91–102.

Restrepo, A. 2000b. Morphological aspects of *Paracoccidioides brasiliensis* in lymph nodes: implications for the prolonged latency of paracoccidioidomycosis. Med Mycol. 38(4): 317–322.

Restrepo, A. 2000a. *Paracoccidioides brasiliensis*. pp. 2768–2772. *In*: Mandell, G.L., J.E. Bennett and R. Dolin (eds.). Principles and Practice of Infections Diseases. Churchill Livingstone, New York.

Restrepo, A. 2003. Paracoccidioidomycosis. pp. 328–345. *In*: Dismukes, W.E. (ed.). Clinical Mycology. Oxford University Press, New York.

Restrepo, A. and A.M. Tobon. 2005. *Paracoccidioides brasiliensis*. pp. 3062–3068. *In*: Mandell, G.L., J.E. Bennett and R. Dollin (eds.). Principles and Practice of Infectious Diseases, 6 ed. Elsevier, Philadelphia.

Restrepo, A., J.G. McEwen and E. Castaneda. 2001. The habitat of *Paracoccidioides brasiliensis*: how far from solving the riddle? Med Mycol. 39(3): 233–241.

Restrepo, A., G. Benard, C.C. de Castro, C.A. Agudelo and A.M. Tobon. 2008. Pulmonary paracoccidioidomycosis. Semin Respir Crit Care Med. 29(2): 182–197.

Restrepo, A., A. Gonzalez and C.A. Agudelo. 2011. Paracoccidioidomycosis. pp. 376–386. *In*: Kauffman, C.A., P.G. Pappas, J.D. Sobel and W.E. Dismukes (eds.). Essentials of Clinical Mycology, 2nd Edition. Springer, New York.

Restrepo-Moreno, A. 1994. Ecology of Paracoccidioides brasiliensis. pp. 121–130. *In*: M. Franco, C.S. Lacaz, A. Restrepo-Moreno and G. Del Negro (eds.). Paracoccidioidomycosis, CRC Press, Boca Raton.

Rittner, G.M., J.E. Muñoz, A.F. Marques, J.D. Nosanchuk, C.P. Taborda and L.R. Travassos. 2012. Therapeutic DNA vaccine encoding peptide P10 against experimental paracoccidioidomycosis. PLoS Negl Trop Dis. 6(2): e1519.

Salem-Izacc, S.M., R.S. Jesuino, W.A. Brito, M. Pereira, M.S. Felipe and C.M. Soares. 1997. Protein synthesis patterns of *Paracoccidiodes brasiliensis* isolates in stage-specific forms and during cellular differentiation. J Med Vet Mycol. 35(3): 205–211.

San-Blas, G. and F. San-Blas. 1977. *Paracoccidioides brasiliensis*: cell wall structure and virulence. A review. Mycopathologia 62(2): 77–86.

San-Blas, G., L.R. Travassos, B.C. Fries, D.L. Goldman, A. Casadevall, A.K. Carmona, T.F. Barros, R. Puccia, M.K. Hostetter, S.G. Shanks, V.M. Copping, Y. Knox and N.A. Gow. 2000. Fungal morphogenesis and virulence. Med Mycol. 38 (1): 79–86.

San-Blas, G., G. Niño-Vega and T. Iturriaga. 2002. *Paracoccidioides brasiliensis* and paracoccidioidomycosis: molecular approaches to morphogenesis, diagnosis, epidemiology, taxonomy and genetics. Med Mycol. 40(3): 225–242.

San-Blas, G., G. Niño-Vega, L. Barreto, F. Hebeler-Barbosa, E. Bagagli, R. Olivero de Briceno and R.P. Mendes. 2005. Primers for clinical detection of *Paracoccidioides brasiliensis*. J Clin Microbiol. 43(8): 4255–4257.

Santo, A.H. 2008. Paracoccidioidomycosis-related mortality trend, state of Sao Paulo, Brazil: a study using multiple causes of death. Rev Panam Salud Publica. 23: 313–324.

Santos, L.S., V.C. Fernandes, S.G. Cruz, W.C. Siqueira, A.M. Goes and E.R. Pedroso. 2012. Profile of total IgG, IgG1, IgG2, IgG3 and IgG4 levels in sera of patients with paracoccidioidomycosis: treatment follow-up using Mexo and rPb27 as antigens in an ELISA. Mem Inst Oswaldo Cruz. 107(1): 1–10.

Shankar, J., A. Restrepo, K.V. Clemons and D.A. Stevens. 2011a. Hormones and the resistance of women to paracoccidioidomycosis. Clin Microbiol Rev. 24(2): 296–313.

Shankar, J., T.D. Wu, K.V. Clemons, J.P. Monteiro, L.F. Mirels and D.A. Stevens. 2011b. Influence of 17beta-estradiol on gene expression of *Paracoccidioides* during mycelia-to-yeast transition. PLoS One 6(12): e28402.

Shikanai-Yasuda, M.A., F.Q. Telles-Filho, R.P. Mendes, A.L. Colombo and M.L. Moretti. 2006. Guidelines in paracoccidioidomycosis. Rev Soc Bras Med Trop. 39(3): 297–310.

Stover, E.P., G. Schar, K.V. Clemons, D.A. Stevens and D. Feldman. 1986. Estradiol-binding proteins from mycelial and yeast-form cultures of *Paracoccidioides brasiliensis*. Infect Immun. 51(1): 199–203.

Taborda, C.P., M.A. Juliano, R. Puccia, M. Franco and L.R. Travassos. 1998. Mapping of the T-cell epitope in the major 43 Kda glycoprotein of *Paracoccidioidesbrasiliensis* which induces a Th-1 response protective against fungal infection in BALB/c mice. Infect Immun. 66: 786–793.

Teixeira, M.M., R.C. Theodoro, M.J. de Carvalho, L. Fernandes, H.C. Paes, R.C. Hahn, L. Mendoza, E. Bagagli, G. San-Blas and M.S. Felipe. 2009. Phylogenetic analysis reveals a high level of speciation in the Paracoccidioides genus. Mol Phylogenet Evol. 52(2): 273–283.

Teixeira, M.M., R.C. Theodoro, L.S. Derengowski, A.M. Nicola, E. Bagagli and M.S. Felipe. 2013. Molecular and morphological data support the existence of a sexual cycle in species of the genus *Paracoccidioides*. Eukaryot Cell. 12(3): 380–389.

Theodoro, R.C., M.M. Teixeira, M.S. Felipe, K.S. Paduan, P.M. Ribolla, G. San-Blas and E. Bagagli. 2012. Genus *Paracoccidioides*: Species recognition and biogeographic aspects. PLoS One 7(5): e37694.

Thomaz, L., J.D. Nosanchuk, D.C.P. Rossi, L.R. Travassos and C.P. Taborda. 2014. Monoclonal antibodies to heat shock protein 60 induce a protective immune response against experimental *Paracoccidioides lutzii*. Microbes Infect pii: S1286-4579(14)00104-X. doi: 10.1016/j.micinf.2014.08.004 [Epub ahead of print].

Tobon, A.M., C.A. Agudelo, M.L. Osorio, D.L. Alvarez, M. Arango, L.E. Cano and A. Restrepo. 2003. Residual pulmonary abnormalities in adult patients with chronic paracoccidioidomycosis: prolonged follow-up after itraconazole therapy. Clin Infect Dis. 37(7): 898–904.

Torrado, E., E. Castañeda, F. de-la-Hoz and A. Restrepo. 2000. Paracoccidioidomicosis: definición de las áreas endémicas de Colombia. Biomédica 20: 327–334.

Travassos, L.R. and G. Goldman. 2007. Insights in *Paracoccidioides brasiliensis* pathogenicity. pp. 241–265. *In*: Kavanagh, K. (ed.). New Insights in Medical Mycology. Springer, Dordrecht, The Netherlands.

Travassos, L.R. and C.P. Taborda. 2012. New advances in the development of a vaccine against paracoccidioidomycosis. Front Microbiol. 3: 212.

Travassos, L.R., C.P. Taborda and A.L. Colombo. 2008. Treatment options for paracoccidioidomycosis and new strategies investigated. Expert Rev Anti Infect Ther. 6(2): 251–262.

Uran, M.E., J.D. Nosanchuk, A. Restrepo, A.J. Hamilton, B.L. Gomez and L.E. Cano. 2011. Detection of antibodies against *Paracoccidioides brasiliensis* melanin *in vitro* and *in vivo* studies during infection. Clin Vaccine Immunol. 18(10): 1680–1688.

Vergara, M.L. and R. Martinez. 1998. Role of the armadillo Dasypus novemcinctus in the epidemiology of paracoccidioidomycosis. Mycopathologia 144(3): 131–133.

Vicentini, A.P., J.L. Gesztesi, M.F. Franco, W. de Souza, J.Z. de Moraes, L.R. Travassos and J.D. Lopes. 1994. Binding of *Paracoccidioides brasiliensis* to laminin through surface glycoprotein gp43 leads to enhancement of fungal pathogenesis. Infect Immun. 62(4): 1465–1469.

Wanke, B. and A.T. Londero. 1994. Epidemiology and paracoccidioidomycosis infection. pp. 109–120. *In*: Franco, M., C.S. Lacaz, A. Restrepo-Moreno and G. Del-Negro (eds.). Paracoccidiodomycosis. CRC Press, Flórida.

CHAPTER 11

Increased Cases of Valley Fever Disease in Central California: An Update

Tara Dubey

Introduction

Mold in indoor air has been an important health issue since a long time. Many mold species can cause serious respiratory and other allergic diseases to the sensitive persons due to toxins produced by them (Rodrick et al. 1977; Hoog et al. 1995; German and Summerbell 1996; Scott 2001).

Millions of fungal species occur in natural outdoor environments because they have favorable nutritional, physical and chemical conditions in these surroundings but only a small amount of these species are identified. They produce spores or conidia to multiply on new substratum either in the same environment or search additional sources to survive. They can be carried to distant places by air, soil, water or by contacts. Part of these spores can enter inside the buildings through various sources and stay to multiply if moisture, temperature and nutritional conditions are suitable. They can use various substrates rich in carbohydrates, cellulose, chitin or proteins as source of food and start multiplying to produce more spores. Major factors that support inside mold growth and the corresponding potential for human exposure to mold are:

1. Building materials as source of nutrition for mold.
2. Moisture from leaking roofs, pipes, or from condensation on or water intrusion through walls or basements.

Vice President, TBON LAB LLC, 3526-214, Investment Boulevard, Hayward, CA 94545.
E-mail: tdubey@hotmail.com

3. Inadequate or poorly maintained ventilation system that may not provide enough air for dilution or dehumidification that can harbor sources of mold spores to disperse into the occupants' breathing zones (Eileen et al. 2004).

There are several diseases related to mold exposure among animals and human beings but unlike bacterial and viral diseases, there are fewer evidences to prove a direct co-relation between a specific human disease symptom and a specific fungal species. In spite of a large amount of research, there are no specific standards for concentration of mold spores responsible for causing a disease. This lack of standard data is the major factor for not having an environmental regulation against the exposure of mold spores for outdoors or indoors.

The purpose of this chapter is to increase the awareness of this fungal species and related health issues generated due to inhalation of its spores by human body. This chapter also provides an update on various aspects of pathogenic *Coccidioides immitis* including preventive actions to control the disease outbreak among the sensitive populations.

Valley Fever Issue

Valley Fever Fungus *Coccidioides immitis* is a unique fungal species known to produce spores in outdoor environment and has created serious health issues among the residents of the Central Valley of California. Presence of Valley fever fungus, *C. immitis* is one of the strongest evidence which proves that outdoor mold spores can be dangerous to human health if allowed to accumulate or ingested by human body. There are several reports and publications on the potential risk of disease among the local employees, farmers, construction workers, football players, lab technicians and other sensitive persons, who are exposed to this fungus.

Major publications on various aspects of Valley fever and its prevention include those from Mayo Clinic (Steckelberg 2012), UC Davis Health System Publication (2014), Valley fever Center of Excellence in Arizona (2009–2014), and California Department of Public Health Publications (2014).

History of Disease Incidence

Coccidiomycosis was first noted in 1890s in Argentina and it was considered that the disease was caused by coccidian (Protozoa). Further research during 1896–1900 proved that it was caused by fungus, not Protozoa and used the term Coccidiomycosis (Medicinenet.com 2013).

More cases of the disease were observed among prisoners of war, from 1940–1944, the Army established the Western Flying Training Command, a program with bases in Arizona and the San Joaquin Valley. The region's climate was reputedly healthful but a fledgling awareness of Valley fever made the military cautious. Major population was affected in 1944 when Cocci (arthrospores of *C. immitis*) exposure to prisoners of war in Arizona was complained by Germany where German POW were kept. This issue was discussed at the Geneva Convention. The Secretary of War immediately asked Charles E. Smith to monitor the soldiers for signs of the disease. Smith found the conditions alarmingly conducive to the spread of Cocci. Reporting in 1958 on his wartime work in the San Joaquin Valley, Smith wrote, “There were vast earth scars where Minter and Gardner Fields were being built. As there was no dust control in operation, the locally generated dust billowed in clouds over the areas.”

The first major steps taken by California Department of Public Health were in December 1977 when San Joaquin valley that runs through Central California was designated as disaster area. In Bakersfield (city of southern valley) area, a very strong wind of 192 miles per hour speed, followed by rain and mud caused a sudden increase in incidence of Cocci in Sacramento area. CDC (Center of Disease Control) statistics also indicated an increase of disease incidence up to 42.6% (about tenfold to 22000 diagnosed individuals per year) during 1998 to 2011 in South West regions (CDCP 2012).

Recent Action in California

New Yorker, a famous news agency called Valley Fever a Silent Epidemic in its featured article (2014). Based on these reports to California Court, a federal official ordered the transfer of more than 3,000 highly vulnerable inmates from two San Joaquin Valley prisons where several dozen have died of Valley Fever in recent years.

NIOSH Evaluation

NIOSH-HHE (National Institute for Occupational Safety and Health-Health Hazard Evaluation): During their investigation in two prisons (A and B), NIOSH-HHE also reported 65 confirmed cases of Coccidiomycosis among prison “A” employees and 38 among prison “B” employees. During 4.5 years of evaluation, they confirmed that prison employees may be potentially exposed to *Coccidioides immitis* in outdoor and indoor work environment as well as outside of work. In January 2014, a report on employee exposure to the fungus *Coccidioides* at two state prisons in California (Report H.2013-0113-3198-Jan2014) was also released.

Progress of Disease

Symptoms

The fungus *C. immitis* causes either no symptoms or mild symptoms and those infected never seek medical care; when symptoms are more pronounced, they usually present as lung problems (cough, shortness of breath, sputum production, fever, and chest pains). The disease can progress to chronic or progressive lung disease and may even become disseminated to the skin, lining tissue of the brain (meninges), skeleton, and other body areas. The disease can also infect many animal types (for example, dogs, cattle, otters, and monkeys).

History and Nomenclature of Valley Fever

Valley fever or Coccidioidomycosis?

Most microbiologists and infectious disease physicians prefer the name Coccidioidomycos because the word describes the disease as a specific fungal disease. This disease has several commonly used names (Valley fever, San Joaquin valley fever, California valley fever, acute valley fever, and desert fever).

Coccidioidomycosis was first noted as a disease caused by coccidia (protozoa) because tissue biopsies of people with the disease showed pathogens that resembled coccidia (protozoa). During 1896–1900, investigators learned a fungus caused the disease, not protozoa, so the term "mycosis" was eventually added to "coccidia." The disease is often noted to occur in outbreaks, usually when soil is disturbed and dust arises, and when groups of people visit an endemic region (such as San Joaquin Valley or Bakersfield [in Kern County], Calif., and Tucson, Arizona, or parts of southern New Mexico or West Texas) during late summer and early fall. The disease is not transmitted from person to person; it is acquired from the environment via contaminated soil and dust. About 150,000 individuals are estimated to become infected each year in the U.S.

Figure 1. Dust storms in the West stir up microscopic spores of soil-dwelling fungus *Coccidioides immitis*. The Centers for Disease Control reports a tenfold increase in infections, some of them fatal.

Favorable Conditions for Disease Development

Inhaling fungus spores in airborne dust can cause the infection. Experts say a hotter, dryer climate has increased the dust carrying the spores of a fungus called *Coccidioides*

(Maddy 1957). There was a sharp increase in the incidence of Valley Fever in California's agricultural heartland in 2010 and 2011.

According to Dr. Gil Chavez, deputy director of the Center for Infectious Diseases at the California Department of Public Health; when soil is dry and it is windy, more spores are likely to become airborne in endemic areas. The incidence of fungal species, *Coccidioides*, which lives in the soil of relatively arid regions (southwest U.S.) is increasing. People are infected by inhaling dust contaminated with *Coccidioides*; the fungus is not transmitted from person to person.

Symptoms may or may not be visible; they usually occur in the lung and initially resemble the flu or pneumonia (cough, fever, malaise, sputum production, and shortness of breath). Immunocompressed people or those suffering from HIV or cancer, or pregnant women are more susceptible to infection and and may develop this disease. Diagnosis is usually easy to accomplish, and the disease can be treated by several antifungal medications. The U.S. Center for Disease Control and Prevention found that in Arizona, California, Nevada, New Mexico and Utah, the number of cases climbed from less than 2,300 in 1998 to more than 22,000 in 2011. During that time, Arizona and California had the largest average increases in Valley Fever incidence, at 66% and 31% per year, respectively. Valley Fever (Coccidioidomycosis) disease is caused by inhaling a fungus called *Coccidioides*, which lives in the soil in southwestern states. Not everyone who is exposed to the fungus gets sick, but those who do become ill typically have flu-like symptoms that can last for weeks or months. More than 40 percent of patients who get sick may require hospitalization, with an average cost of nearly $50,000 per visit.

Between 1998 and 2011, nearly 112,000 cases of Valley Fever were reported in 28 states and Washington, D.C., but 66% of the cases were in Arizona, 31% were in California, 1% in Nevada, New Mexico and Utah, and about 1% in all other states combined.

Nature of the Pathogen

Coccidioides immitis is a dimorphic pathogen with two different stages as arthroconidia and as spherule (Volk 2002).

Taxonomy: Based on 18S rDNA sequence, this species has shown closeness to *Ajellomyces dermatitidis*, and *A. capsulatus* (Greene et al. 2000) and has been placed under following classification:
Group: Ascomycota
Class: Arthrodermataceae and
Order: Onygenales

Colony characteristics: On SGA culture medium, colonies are expanding, glabrous to felty, whitish to greyish, turning to tan at maturity. On reverse, they start as cream colored and turn brown with age (Hoog et al. 1996).

Microscopy: Fertile hyphae grow at 24°C originating at right angles. Arthroconidia are short, cylindrical to barrel-shaped. Smooth-walled arthroconidia have wall thickness of 3–8 x 3.5–4.5 μm, and alternating with empty disjuncture cells. On liberation, conidia show a frill-like structure on both ends as remains of the adjacent cell (Fig. 1 A, B and C)

Physiologically this species is intolerant to Benomyl.

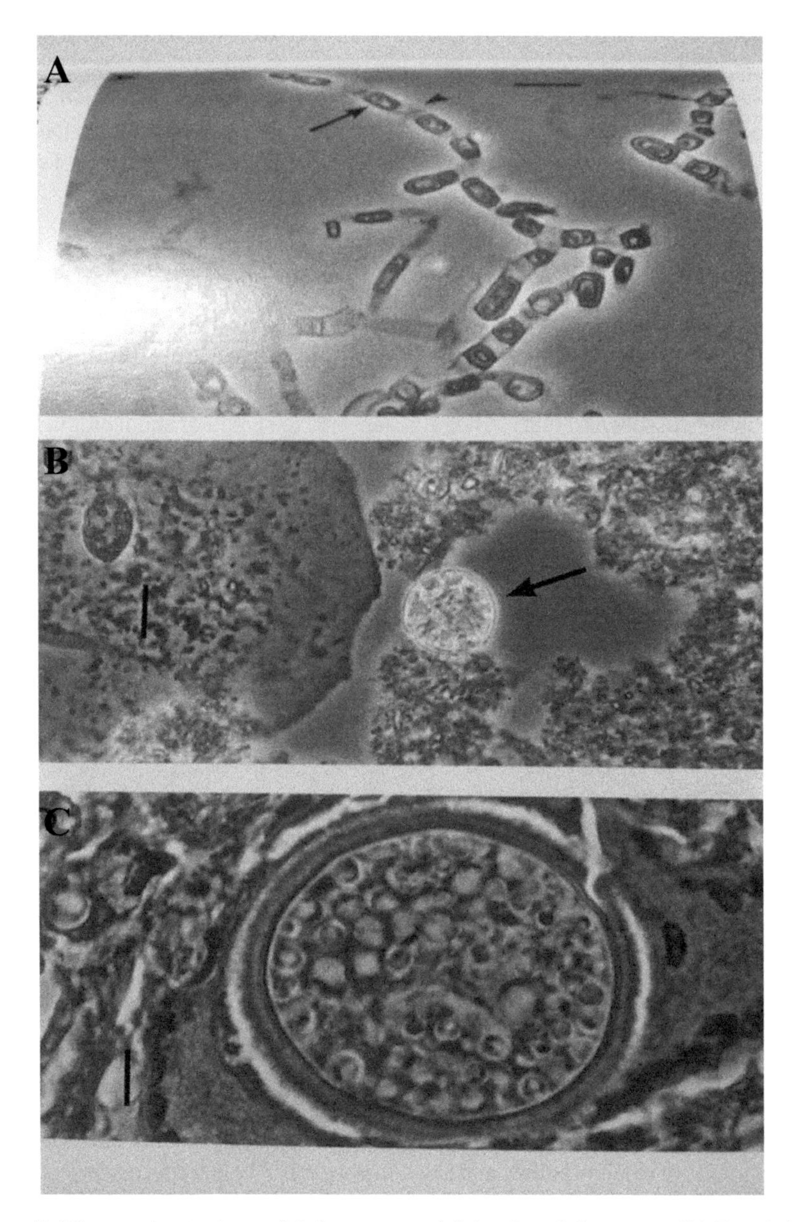

Figure 2. Microscopic structures of Arthrospores and Spherules of *C. immitis:* (A) Barrel shaped Arthrospores, (B) Spherules in human sputum, (C) Spherules in human body (modified from Germaine and Summerbell 1996).

Diagnosis

Fungus shows hyphal stage at room temperature at 24°C *in vitro* to form extremely small barrel-shaped arthroconidia (Fig. 1A) as airborne source of infection. After ingestion of arthroconidia, endogenous reactivation spherules are produced. Spherules are sporangia-like stage similar to Zygomycota, but Woronin bodies near septa are similar to Ascomycota.

Spherules are inhaled by animal/human body to produce endospore at body temperature (Fig. 1B and C.). Endospores are disseminated haematogenously and may cause fever and relevant symptoms. Endospores are characterized by not showing yeast like budding at 37°C.

Pathogenicity

Primary infection occurs through inhalation of airborne spores of fungus. This may lead to non-specific fever with bronchopneumonia followed by conjunctivitis and erythema. Generally, the disease resolves spontaneously and disseminates to skin, bones, joints, sub-cutaneous tissues and other visceral organs due to spread of endoconidia through blood circulation (Ampel 2010).

The Prognosis Outcome of Valley Fever

Majority of infected patients have a good prognosis, as the disease infection is self-limiting. This stage may cause a few small calcified areas in the lung, without any serious problem to the patient. Chronic disease may produce more nodules and cavities in the lung and take one or two years to dissolve. On the other hand, patients with disseminated Coccidioidomycosis can have rapid development of disease symptoms and may die if disease is not treated rapidly in an appropriate manner.

Ecology and Distribution of Fungus

Cocci are endemic to the deserts of Southwest California, Arizona, New Mexico, Nevada, Texas—and to the semi-arid parts of Central and South America (Hoog et al. 1995). Digging-building, drilling, tilling, clearing—stirs it up, and dry, hot, windy conditions, a regional feature intensified by climate change, disperse it. In recent years, infections have risen dramatically. This fungus can survive 20 cm deep under soil but is absent on the surface during hot dry weather. The fungus is absent from aerosols during dry wet winter and spring but grows again in summer. September–November is peak endemic period.

Other factors like alkaline soil, with high carbonated organic matter, and high concentrations of calcium sulfate and calcium borate also help for recurrence of fungus after rainy season. In addition, rodent burrows and rodent movements also support the spread of fungal spores (Standaert et al. 1995; Ampel et al. 1998).

Antje Lauer et al. (2012, 2014), studied the distribution pattern of this fungus. She observed that little is known about where the fungus thrives and why because a spot that tests positive once may subsequently come up negative; a positive site can be separated from a negative one by a matter of yards. She found that the cocci season peaks in the fall. By continuous sampling on Coles Levee Road (a desolate strip owned by Los Angeles County), which uses part of it as a sewage dump, she found the presence of fungus nearly every time she looked. She also collected samples from West side near the oil drilling area with fences. In this area small oil drills operate in the middle of neighborhoods and cocci infection has long been considered an occupational hazard for oil workers. The most sensitive method that was successfully used to detect *C. immitis* in soil samples was based on a nested PCR approach. With this method, followed by sequencing of the obtained PCR products, 6% of 150 soil samples collected from the Antelope Valley Area (Edwards Air force Base, Rosamond, Lancaster and Palmdale) indicated the presence of the pathogen.

Prevention

Pre-Screening of high risk employees: James Scott, a professor at the University of Toronto's Dalla Lana School of Public health, pointed out that some populations like Filipinos and Blacks are more sensitive than others. He recommended to Pre-Screen and identify workers with high risk and exclude them from risk-type work from endemic zones (Synergist 2014). He also suggested that pregnant women must be excluded from potential risk in research labs handling this mold species. NIOSH-HHE (Health Hazard Evaluation) team found 103 confirmed cases of Valley fever in prison employees.

Major Actions to Prevent the Disease

I. Control the spread of disease from one to other neighboring areas by

1. Restricted movement during peak season for dusty wind
2. Reducing the soil contamination in the sensitive area
3. Outdoor workers at prison must wet the soil before disturbing it (NIOSH-HHE 2014 report)
4. Must keep doors and windows shut to minimize the spore deposition inside the buildings
5. Reducing the spread of dust by choosing appropriate vegetation like soil binder grasses (Alfalfa)
6. Regulation for oil drilling activities

II. Intensive educational program for awareness and monitoring of disease symptoms, lab test for disease diagnosis and preventive medication

III. Environmental monitoring and building assessment

This includes environmental assessment of building in the area and keeping a record of following data (Vesper et al. 2011):

a. History of moisture intrusion
b. Leakage
c. Seepage

d. Levels of CO_2 if elevated means poor air quality
e. Indoor mold spore levels, if higher than outdoors
f. Sources if any in storage area like stacks of papers, food and others

IV. Clinical evaluation of exposed persons for the presence of

Respiratory symptoms
Spirometry
Sedementation rate (ESR) if elevated
Sneezing, coughing, dizziness, fatigue, headaches, upper respiratory irritation and
rashes for a longer duration

V. Outdoor air monitoring for mold spore concentration

A comparison between outdoor and indoor air sampling data for presence of various mold genera including cocci can help to predict the potential risk to the exposed persons inside or outside the environment.

Airborne fungi are responsible for the majority of fungal infections in humans and animals. Outdoor air markedly influences the prevalence of fungal spore levels in indoor environment. Masoomeh et al. (2014) studied the source of fungal infections in indoor environments of hospitals in Tehran, Iran.

Research near construction area: Continuous survey and air monitoring of outdoor environments of Central valley near Avenal State Prison (Dubey 2010) indicated the presence of a wide range of mold genera. Air samples were collected from a construction zone while digging of soil at different depths of this semi-arid zone. 75–150 liters of air volume was collected at the flow rate of 15 LPM for 5–10 minutes by using a standard air sampler.

Mold spores were identified and counted microscopically by using standard reference material (Smith 1990; Domsch and Gams 1993). These air samples were also examined for any spores showing morphological similarity with those of *Coccidioides* or *Histoplasma*-like species. This research was restricted to spore trap analytical method due to specific lab safety requirements (Biosafety Level III) for culturing of *C. immitis*. About 15% of these samples were observed with structures showing morphological similarities to hyphal and conidial structures of *Coccidioides*. Such structures were identified under a group as "*Coccidioides immitis* or *Histoplasma* type", based on the morphological similarities of the two genera (Table 1) 60% of total air samples had greater fungal diversity (10–20 mold genera/sample). 40% of air samples had a very high concentration of spores (7,000 to 20,000 spores per cubic meter of air volume). High concentration of spores in the air indicated that outdoor air is a potential threat to public health because of harboring a wide array of pathogenic and allergenic airborne fungal spores, which can serve as the main source of contamination of indoor environments of the local residential and commercial buildings.

Interesting part of this research was the presence of higher concentrations of mold spores from deeper digging areas than compared to the lower diversity level and decreased concentrations from surface soil area. A small fraction of soil collected

Table 1. Frequency of distribution of different mold spores in 25 air samples collected from outdoor environments of the central valley region near Avenal state prison.

Mold Genera	% Frequency	Spores/CM3	Mold Genera	% Frequency	Spores/ CM3
Distribution	Out of Total Genera	Range	Distribution	Out of Total Genera	Range
Acremonium sp.	13	20–50	*Epicoccum* sp.	30	70–100
Alternaria sp.	70	20–70	*Nigrospora* sp.	25	70–120
Ascospores sp.	95	120–2000	*Oidium* sp.	15	50–100
Aspergillus/Penicillium type	100	1100–15000	*Rust* sp.	40	70–250
Aureobasidium sp.	20	50–100	*Scopulariopsis* sp.	20	100–740
Basidiospores sp.	90	150–1500	*Smut/Periconia type*	75	50–470
Bipolaris/Dreschlera type	40	20–70	*Spegazzinia* sp.	13	20–Aug
Botrytis sp.	33	20–120	*Stachybotrys* sp.	10	20–100
Chaetomium sp.	50	50–2000	*Stemphyllium* sp.	33	20–150
Cladosporium sp.	95	1100–12000	*Trichocladium* sp.	13	20–50
Coccidioides/ Histoplasma type	15	20–70	*Torula* sp.	33	20–70
Curvularia sp.	40	70–100	*Ulocladium* sp.	50	20–120
Total Genera	28	Temperature range:	100–120 F	Moisture:	50–70%

from the samples area was also analyzed by PCR techniques (other reference lab) to detect the presence of this fungus. Soil analysis indicated the absence of structures of *C. immitis*.

Regular outdoor air monitoring of sensitive area can be a stronger tool to create a dependable database, provided the presence of pathogen can be confirmed by safe and quick molecular techniques like Real Time PCR. Drinking water inside a building can be a major source of opportunistic mold with clinically significant implications.

Regular checking of water distribution system of a building for the presence of arthrospores of *C. immitis* can be an effective control measure

Elias et al. (2003) pointed out that drinking water supplied to hospital building could cause severe health problems due to presence of toxic molds because the recovery of molds in water systems has been reported worldwide in the community and hospital settings as early as in 1982. They suspected that it is likely that our recommendations may be applicable to various communities and hospitals. They had suggested an effective and inexpensive approach to prevent human exposure to waterborne molds inside a building is to provide high-risk patients with sterile (boiled) water for drinking and sterile sponges for bathing (to avoid the aerosolization associated with showering). Such precautions could also be extended to the community.

Cleaning of shower area can reduce the mean air concentration of airborne pathogenic molds

Ellias et al. (2003) also strongly recommended that cleaning the floors of shower facilities in a building can reduce the mean air concentration of mold species like *Aspergillus* species, from 22 CFU/m^3 to 4 CFU/m^3; ($P = 0047$) and many other airborne pathogenic molds.

Continuous research is required to forecast the potential outbreak of disease. There are several issues related to lab research on this specific fungal species. Some of them are mentioned below:

Number of research scientists on this subject is very limited: Most of the existing research team leaders are getting older and new team leaders are not encouraged to continue research in laboratories due to potential health hazards or restrictions and high levels of biosafety requirements for this species.

Lack of safe mold handling training for new researchers: New research students and technicians may not be trained enough to handle this species safely and are scared of dealing with this species in culture. New techniques and guidelines must be included to monitor the disease cure and prevention.

Expensive handling of this mold in lab and expensive methods to cure this disease: Cost-effective methods are needed to control the disease. This can be possible by interaction with other countries where coccidiomycosis issue is prevailing or emerging.

Lack of sufficient outdoor environment research data: Additional data on concentration of mold from suspected area may be helpful for eradication of fungus from surrounding outdoor environments. Eradication of sources can be possible by biological or chemical remediation techniques.

Simulation studies: Computer models can be generated to forecast the disease outbreaks among the communities. This should be based on frequent (weekly) air monitoring data for spore concentrations along with moisture and temperature conditions. Dubey (2010) based on her research outcome had also proposed to prepare an annual database from Central Valley Area. Frequent recording of local data can be more helpful for the safety warning available to the local employees and employers. This could be similar to simulation computer models used for agricultural disease forecast system available to the local farmers to reduce the cost of spraying fungicides on diseased crops (Raposo et al. 1993).

Major Research Centers

Research funds have been provided for various research centers in different universities of Arizona and California based on qualifications of researchers and need of the local communities.

Following are major research centers and their contributions to control the disease:

Valley Fever Center for Excellence

The Valley Fever Center for Excellence (VFCE) was established in 1995 and is located at the Southern Arizona VA Healthcare System's facility (Tucson VA Medical Center) and is jointly sponsored by the University of Arizona and the Southern Arizona VA Healthcare System (http://www.valley-fever.org/valley_fever_org_research.html).

The Valley Fever Center for Excellence is the only center of its type and is a regional, national and international resource. It provides a multi-faceted awareness program to the public, medical practitioners and health care workers. The Valley Fever Center for Excellence operates an informational HOTLINE [(520) 629-4777] for information about the disease, its diagnosis and treatment. The Valley Fever Center for Excellence maintains a website (English and Spanish) and answers questions by electronicmail [vfever@email.arizona.edu].

The Valley Fever Center for Excellence has also developed a Valley fever syllabus for medical practitioners (available in English and Spanish), which has been transferred into electronic media as part of an internet, self-study CME course for healthcare professionals.

The Valley Fever Center for Excellence provides information to the public, physician consultations with Valley Fever Center for Excellence physicians and promotes research into all aspects of the disease. Valley fever evaluation and treatment clinics are operated each week at the Southern Arizona VA Health Care System and at St. Luke's Clinic, University Medical Center. The Valley Fever Center for Excellence facilitates, stimulates and fosters research by providing a public profile for Valley Fever and by sponsoring collaboration among clinicians and research investigators. Primary areas of research are in the fields of Valley Fever causing risk factors, immunology, fungal growth, and fungus adaptation within soil or during infection as well as on the development of a vaccine.

University of Arizona, Phoenix has two major researchers—John Galgiani, and Marc Orbach. Under supervision of John Galgiani is an infectious-disease physician, who supervises the Valley Fever Center For Excellence at The University of Arizona in Tucson (https://www.vfce.arizona.edu/). His research is supporting a company for producing a molecular byproduct of bacteria and streptomyces, called NIKKOMYCIN Z for the marketplace. It works by destroying the spherule's ability to make chitin, which forms the protective wall; without it, the disease can't progress (Galgiani et al. 2005). Marc Orbach, is another famous fungal geneticist who works in the Bio Safety3 lab in Tucson, Arizona. He is investigating the genes in cocci that activate when it enters a host, he discovered several genes that were involved in spherulation. Some of these genes can be modified to use as vaccine. Orbach plans to try the vaccine in dogs, which are intensely susceptible and are subject to the more lenient regulations of the USDA.

UC Berkeley: Professor John Taylor and his team use animal studies, environmental studies, and genomic studies to examine the development of fungal hyphae and Arthroconidia of *C. immitis.* He focuses on wild old rodent research and new comparative genomic research in dead animals. His major focus is on Molecular Natural History of the San Joaquin Valley fever fungus *Coccidioides* through DNA variation and natural recombination (Greene et al. 2000). His results support a life cycle closely tied to the population of small native mammals and provide a tool to identify the genes responsible for pathogenicity in fungi.

UC Davis, CA: George Thompson, an assistant professor of infectious diseases that specializes in the care of patients with invasive fungal infections and helps direct the Coccidioidomycosis Serology Laboratory at UC Davis.

Stanford University Research Team: Center for Valley Fever research was established by George John Galgiani as a Post Doc researcher. He studied cure and production of vaccine to control the disease outbreak. In 1978, he moved to University of Arizona in Tucson.

University of California, SanFrancisco: Dr. Anita Sil is a major researcher on basic biology and virulence of thermally dimorphic fungus, *Histoplasma capsulatum* and intends to study the genomic analysis of *C. immitis*. She also specializes in environmental signals and fungal genes that promote transition between soil and host forms of *Histoplasma.*

UC Merced in Merced, CA: Professor David Ojcius is a researcher studying the host response and risk of *C. immitis* as well as organism diversity (genomics and virulence) to identify risk factors for disseminated disease (http://Valleyfever.UCMERCED.EDU/exploreresearch topics/Valleyfeverfacts).

UCSF, FRESNO and Treatment Research : Dr. M. Peterson uses serology, CT biopsy, and Bronchoscopy to distinguish lung nodules caused by this fungus or due to other reasons like lung cancer. He has been helping to establish a network of health care providers and a task force with study team at the NIH and CDC to enhance the clinical trial studies design.

Detection research: At Community Regional Medical Center, Dr. Dominic T. Dizon and Dr. Marlilyn Mitchell have developed a PCR test to identify the presence of fungal species *C. immitis.*

Government support and research funding:

California Representative Mr. McCarthy first time convened a symposium on Valley fever in Bakersfield in September 2013. After sharing the stories of local population, he mustered considerable political power, including the director of the National Institute of Health and the director of the C.D.C. That afternoon, the three men announced the most significant public investment in Valley-fever research in many years. The money will fund a large clinical trial, to be held in Bakersfield that will establish treatment guidelines based on scientific evidence.

Research in Bakersfield area: Ramon Guevara, an epidemiologist who works at the L.A. County Department of Health. Guevara has made it a personal mission to educate people about the emergent issue of cocci in his territory. Dana Goodyear (2014) a reporter from the NewYorker also discussed in her article the high spore concentration of this fungus and called the Golden Dust as "Death Dust" (Figure 1).

New prospects of research: Due to increasing public activities in the sensitive desert area of Central Valley, rapid diagnostic test methods and more effective awareness programs are in high demand. Rapid industrial growth of the area includes oil drilling companies, solar power and energy conservation companies, electricity power plants, and emerging real estate companies.

Consequently, the number of employees and employers has also been increased along with enhanced human activity but on the other hand the incidence of this disease did not decrease.

Conclusion

Valley Fever Disease caused by *Coccidioides immitis* has been a challenging issue for a big geographical section of USA and other parts of the world. Due to the dangerous fungal species and safety issues associated with it, lab studies of this species have been limited to highly proficient labs in the USA. Less expensive efforts to control the spread of *C. immitis* and to reduce the number of exposure are highly needed. This can be possible through molecular and environmental approaches.

References

Ampel, N.M. 2010. The diagnosis of coccidioidomycosis. F1000 Med Rep.

Antje Lauerl, Joe Darryl Hugo Baal, Jed Cyril Hugo Baal, Mona Verma and Jeffrey M. Chen. 2012. Detection of Coccidioides immitis in Kern County, California by multiplex PCR. Mycologia. 104(1): 2012, pp. 62–69. DOI: 10.3852/11-127.

Beneke, E. and A. Rogers. 1996. Medical Mycology and Human Mycoses. Star Publishing Co., Belmont, CA.

CDCP 2012. "Coccidiomycosis" Center of Disease Control and Prevention". http://www.cdc.gov/nczved/divisions/dfbmd/diseases/Coccidiomycosis/. Accessed March 20, 2012.

Dana Goodyear. 2014. "Death Dust" The valley fever menace, A Reporter at Large. The New Yorker: January 20.

de Perio, M.A. and G.A. Burr. 2014. Evaluation of coccidioides exposure and coccidiomycosis infection among prison employees. Report H. 2013-01B-3198. January 2014.

Domsch, K.H. and W. Gams. 1993. Compendium of Soil Fungi. Vol. 1. IHW-Verlag.

Dubey, T. 2010. Mold spores from outdoor air of valley fever region of California. AIHC 2010 Annual Conference presentation.

Eileen, S., K.H. Dangman, P. Schenk, R. DeBernardo, C.S. Yang, A. Bracker and M.J. Hodgson. 2004. Guidance for clinicians on the recognition and management of health effects related to mold exposure and moisture indoors. UCHC publication. 09302004.

Elias, J. Anaissie, Shawna L. Stratton, Dignani M. Cecilia, Choon-kee Lee, Richard C. Summerbell, John H. Rex, Thomas P. Monson and Thomas J. Walsh. 2003. Pathogenic molds (including *Aspergillus* species) in hospital water distribution systems: a 3-year prospective study and clinical implications for patients with hematologic malignancies. Blood 101(7): 2542–2546.

Galgiani, J.N., N.M. Ampel and J.E. Blair. 2005. "Coccidioidomycosis". Clin Infect Dis. 41: 1217–23.

Germain, G. St. and R. Summerbell. 1996. Identifying Filamentous Fungi: A Clinical Laboratory Handbook. Star Publishing Co., Belmont, USA pp. 314.

Grant, E. Smith. 1990. Sampling and identifying allergenic pollen and molds. An illustrated identification manual for air samples. Bluestone Press pp. 92.

Greene, D.R., G. Koenig, M.C. Fisher and J.W. Taylor. 2000. Soil isolation and molecular identification of *C. immitis*. Mycologia. 92: 406–410.

Hoog, G.S., J. Guarro, J. Gene and M.J. Figueras. 1995. Atlas of Clinical Fungi. 2nd edition de, Centraalbureau voor Schimmelcultures. The Netherlands/Universitat Rovira I Virgili, 2000, 593–595 pp.
http://www.valley_fever_org_statistics.html-Valley Fever Connections.
https://www.vfce.arizona.edu/
http://Valleyfever.UCMERCED.EDU/exploreresearchtopics/Valleyfeverfacts
Lauer, A., J.D.H. Baal, J.C.H. Baal, M. Verma and J.M. Chen. 2012. Detection of *Coccidioides immitis* in Kern County, California, by multiplex PCR. Mycologia 104(1): 62–69.
Lauer A., J. Talamantes, L.R.C. Olivares, L.J. Medina, Joe D.H. Baal, K. Casimiro, N. Shroff and K.W. Emery. 2014. Combining Forces—The use of landsat TM satellite imagery, soil parameter information, and multiplex PCR to detect *Coccidioides immitis* growth sites in Kern County, California. PLoS One e111921. doi:10.1371/ journal.pone.0111921.
Maddy, K. 1957. Ecological factors possibly relating to the geographic distribution of *Coccidioides immitis*. Public Health Serv Publ. 575: 144–157.
McPhee, S.J. 2012. Current Medical Diagnosis and Treatment, 51st ed. The McGrew Hill Company, New York.
Nguyen, C., B.M. Barker, S. Hoover, D.E. Nix, N.M. Ampel, J.A. Frelinger, M.J. Orbach and J.N. Galgiani. 2013. Recent advances in our understanding of the environmental, epidemiological, immunological, and clinical dimensions of coccidioidomycosis. Clin Microbiol Rev. 26(3): 505–525.
Raposo, R., D.S. Wilks and W.E. Fry. 1993. Evaluation of potato late blight forecasts modified to include weather forecasts: a simulation analysis. Phytopathology 83: 103–108.
Rodricks, J., C.W. Heseltine and M.A. Mehlmn. 1977. Mycotoxin in human and animal health. Park Forest South Illinois Pathotox Publishers pp. 807.
Scott, J.A. 2001. Studies on indoor fungi, a thesis submitted in conformity with the requirements for the degree of Doctor of Philosophy in Mycology, Graduate Department of Botany in the University of Toronto, Canada.
Shams-Ghahfarokhi, M., S. Aghaei-Gharehbolagh, N. Aslani and M. Razzaghi-Abyaneh. 2014. Investigation on distribution of airborne fungi in outdoor environment in Tehran, Iran. J Environ Health Sci Engin. 12: 54.
Steckelberg, J.M. 2012. Expert opinion, Mayo Clinic, Rochester Minn. April 10th issue.
Stevenson, W.R. 1981. Analysis of potato late blight epidemiology by simulation modeling. www.ipm.ucdavis.edu/DISEASE/DATABASE/potatolateblight.html
Synergist. 2014. Valley Fever Disease, AIHA News Letter April issue. http://www.valley-fever.org/valley_fever_org_research.htm
Vesper, S., J. Wakefield, P. Ashley, D. Cox, G. Dewalt and W. Friedman. 2011. Geographic Distribution of Environmental Relative Moldiness Index Molds in USA Homes. J Environ Public Health 242457.
Volk, T.J. 2002. http://botit.botany.wisc.edu/toms_fungi/jan2002.html.

SECTION IV

Fungal Pathogenesis in Biofilm and Allergy

CHAPTER 12

Fungal Biofilms: Formation, Resistance and Pathogenicity

Janaina de Cássia Orlandi Sardi, Nayla de Souza Pitangui, Fernanda Patrícia Gullo, Ana Marisa Fusco-Almeida and *Maria Jose Soares Mendes-Giannini**

Introduction

Biofilm formation has been detected quite early in the fossil record (~3.25 billion years ago) and is common throughout a diverse range of organisms in both the Archaea and bacteria lineages, including the 'living fossils' in the most deeply divided branches of the phylogenetic tree. The first report of microbial biofilms occurred in the 16th century, when Antonie van Leeuwenhoek noted the presence of "animalcules" on the plaque of his own teeth, but little was known about these microorganisms or how they came to be on his teeth. Bacterial aggregates had also been described by Pasteur in 1864 and by Henrici in 1933 in two independent observations, including one from rocks from an alpine stream (Geesy et al. 1977), which initiated the modern era of biofilms. The concept of biofilms appears with the observations of Jendresen's dental biofilm and with a description of 77 masses of cells of *Pseudomonas aeruginosa* in the sputum and lung tissue of patients with chronic cystic fibrosis Høiby (Høiby et al. 1977). The term biofilm was introduced into medicine in 1985 by Nickel (Nickel et al. 1985).

In the 70s, biofilms gained attention in healthcare and industry. In 1978, the general theory of microbial biofilm predominance was put forth, and quantitative methods were described for the analysis of microorganisms in a biofilm. These methods were developed by Costerton and collaborators in 1978 to quantify the metabolic activity of

Laboratório de Micologia Clínica e Núcleo de Proteomica, Faculdade de Ciências Farmacêuticas, UNESP – Universidade Estadual Paulista Júlio Mesquita Filho, Araraquara, São Paulo, Brasil.

* Corresponding author: giannini@fcfar.unesp.br

biofilms present in a stream of an uninhabited mountain. Before 1978, the concept of biofilms was limited to aquatic systems and was defined as the population of adherent microorganisms, but the knowledge of biofilm formation and cellular behaviour within the biofilm was still scarce. Bacterial numbers and metabolic activity between bacterial cells in the free form (planktonic) and in a biofilm were studied and bacteria present in the biofilm showed superior metabolic activity per cell (Costerton et al. 1978). The quantitative methods from this study were applied to other fields, for example, to industry and medicine. These methods allowed for the understanding of the biofilm formation of different microorganisms in different situations. Overall, when microbial cells are metabolically active, they show high avidity for the adherence to surfaces that are associated with the amount of available nutrients important to the process of replication cells and production of exopolysaccharides, which are important processes in biofilm formation. It should be noted that in 1993, the urgency of differentiating the planktonic mode from the biofilm mode was still being discussed (Costerton et al. 1995). Despite this fact, this discussion advanced to bacteria studies. An important model was introduced by McCoy and colleagues in 1981 that reported, for the first time, biofilm formation in test tubes (McCoy et al. 1981).

The recognition of fungal biofilms brought attention to nosocomial infections, and now it is believed that several fungal genera are capable of organising into a biofilm (Fridkin and Jarvis 1996; Sardi et al. 2014).

In the early 90s, interest in studying biofilms increased in the area of healthcare because problems arose associated with chronic infections and recurrent infections due to microorganisms colonising medical devices. From the 20th century, evidence was shown supporting the fact that biofilms can form on any surface that is in contact with an aqueous medium, which became the exact definition of a biofilm (Azevedo and Cerca 2012).

Biofilms in human infections had been studied in some areas such as dentistry, which stands out as a pioneer in the inhibition of biofilms on teeth and dental appliances. It is also the area with great prominence in fungal biofilms studies, mainly performed on *Candida albicans* (Sardi et al. 2012; Sardi et al. 2013; Skupien et al. 2013; Soll 2014).

Invasive medical devices are in constant contact with body fluids, which are highly rich in nutrients and are susceptible to biofilm formation. Thus, there is great interest in developing procedures for the prevention of fungal biofilms in prostheses, implants, catheters, contact lenses, heart valves and other medical devices. Often, the decontamination procedure is not sufficiently able to eliminate microorganisms that have a greater capacity to survive in the form of a biofilm, which results in contamination by commensal microorganisms that are faced with a different habitat (Azevedo and Cerca 2012; Skupien et al. 2013).

Biofilms may be formed by one or several species of microorganisms, bacterial and fungal isolates, or both. Multi-species biofilms are generally composed of microorganisms present in the human microbiota in association with pathogenic or nosocomial microorganisms. These are difficult to diagnose, complex to treat, and act as a reservoir of microorganisms (Ellias and Banin 2012; Rendueles and Ghigo 2012; Sardi et al. 2013; Xu et al. 2014).

Currently, biofilm formation has been described for several fungi (yeast and filamentous), and there is a growing understanding of the structural composition of biofilms, metabolic behaviour, resistance mechanisms and new ways of eradicating biofilms through the use of "high throughput" techniques.

General Concepts

There is evidence of biofilm formation in nature, early in the fossil record, and more recently in hydrothermal environments such as hot springs and deep-sea vents (Hall-Stoodley et al. 2004). Fluvial biofilms are the dominant mode of microbial life in streams and rivers, which fulfil critical ecosystem functions (Besemer et al. 2013). It is estimated that more than 90% of the microorganisms found in nature are in the biofilm form, and it is believed that this is the natural presentation and preferred form for fungal growth. In addition, biofilms have been the major cause of persistent infections in human. The characteristics of biofilms growing in diverse environments are extremely similar, demonstrating that important strategies are assembled in part by structural specialisation (Sardi et al. 2014). The nature of the substrate, climatic variables, and geography influence the microbial community structure. The substrate type is the most significant factor in structuring bacterial communities, whereas geographic location is the most influential factor for microbial/eukaryotic communities (Ragon et al. 2012). The cell–cell interactions require close proximity and concentrated nutrients (Besemer et al. 2013).

The definition of a biofilm is an organised community of microorganisms attached to a surface or interface surrounded by a self-secreted polymer. This community of microorganisms is controlled by signalling molecules called "*quorum sensing*" (QS) molecules (Sardi et al. 2013; Sardi et al. 2014; Ramage et al. 2012). A biofilm is considered a distinct form of microorganisms' growth, which differs from planktonic growth and promotes the survivability of microorganisms in poor environments, conferring resistance to microbial cells (Costerton et al. 1995; Ramage et al. 2012). The fact that biofilm formation is an adaptation of microorganisms to different environments leads us to believe that this capacity is essential for the establishment of an infection in the human organism (Azevedo and Cerca 2012).

The process of biofilm formation occurs in five stages (Pitangui et al. 2012). The first stage, characterised by initial adherence in which microorganisms attach to a surface, is influenced by several parameters such as motility through the aqueous medium, cell transport of microorganisms and the characteristics of the surface on which the biofilm is established. During the initial adherence, there is no differentiation of cellular processes but the microorganisms already show alterations in their phenotypes based on the approximation of the response surface. The second stage is the colonization process, which consists of the irreversible attachment of microbial cells to the substrate. Microbial cells overlap each other, whether cells of the same species or different species, and begin to secrete extracellular polysaccharide substances (EPS) that envelop the microbial cells, a fundamental part of establishing a biofilm. In the third stage, the microcolony formation is established as a result of the accumulation and growth of microorganisms associated with the extracellular matrix. At this stage, microbial cells remaining in the biofilm microenvironment are protected by the extracellular matrix from any stress from the external environment. In this microenvironment, the cells show alterations in their cellular growth and there is physiological cooperation between cells. These systems are induced in the biofilm due to cell-cell communication by means of QS molecules. The biofilm microenvironment is formed by polymers, and the self-secretion of EPS is responsible for 90% of the composition of the extracellular matrix (Blankenship and Mitchell 2006; Bonhomme et al. 2013; Cuéllar-Cruz et al. 2012; López-Ribot 2005).

The fourth stage is related to biofilm maturation and the development of genetic alterations in genes that encode proteins involved in translation, metabolism, transport, secretion and the regulation of the cellular membrane. This process is characteristic of a microbial community attached to a surface. The fifth stage corresponds to the dispersion process of biofilm cells, allowing the cells to revert to planktonic cells. This is an active process that allows the colonization of new sites, especially in environments with higher concentrations of nutrients (Srey et al. 2014). The cells dispersed from biofilms are more virulent and cause higher mortality than planktonic cells, which can be explained by the genetic alterations these cells undergo within the biofilm (Chua et al. 2014; Dongari-Bagtzoglou et al. 2009; Nett 2014; Sardi et al. 2014; Uppuluri et al. 2010). Figure 1 illustrates the stages of development of the microbial biofilm. A large part of the metabolism of a biofilm is associated with extracellular proteins. The physical and chemical attractions between polymers forming the extracellular matrix of a biofilm are electrostatic forces, Van der Waals and polar interactions as well as hydrogen bonding. The regulation of genes involved in the synthesis of extracellular polymeric substances (EPS) influences biofilm formation and maturation (Karunakaran et al. 2014).

Microbial cells have a communication mechanism using "*quorum sensing*" molecules (QSM), which functions by releasing these hormone-resembling self-inducing substances. Such substances are released into and accumulate in the external medium of the biofilm, where their interaction with specific receptors coordinate several cellular behaviours such as the secretion of virulence factors, capacity for biofilm formation and the evasion of the host immune response. In pathogenic fungi, lipid molecules such as farnesol, sphingolipids and oxylipins are considered QSMs (Sabra et al. 2013; Sardi et al. 2014; The et al. 2012). Farnesol was the first QS

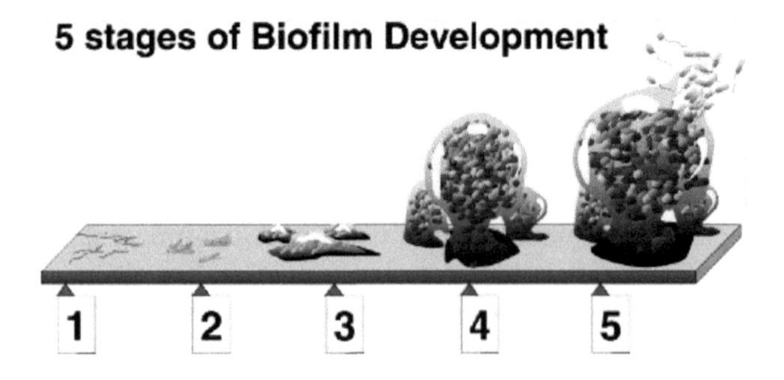

Figure 1. Stages of biofilm development of biofilm: 1. Adherence initial; 2. Colonization process; 3. Micro-colony formation; 4. Biofilm maturation, 5. Dispersion process of biofilm cells. Image adapted from Monroe et al. (2007).

molecule described in fungal biofilms through studies with *C. albicans* (Albuquerque and Casadevall 2012; Ramage et al. 2012). Studies showed that farnesol is important for biofilm development because it is associated with mycelia growth. An increase in farnesol concentrations in the medium causes morphological and cellular density alterations in the biofilm cells. Farnesol is a well-characterised inhibitor of hypha formation (Ganguly et al. 2011; Ramage et al. 2012).

Fungi growing in biofilms have the advantage of cellular protection as well as the metabolic cooperation between the cells and the expression of genes that distinguish fungal cells from biofilms cells (Ramage et al. 2012).

Biofilms can form on tissues or on abiotic surfaces such as medical devices. However, biofilm formation on tissues is a more complex process because compounds such as antibodies present on the endothelial surface can prevent microorganism attachment, which in addition to the phagocytic action of cellular immunity, can inhibit planktonic cells from forming biofilms (Costerton et al. 1995). However, biofilm formation can occur in extremely adverse conditions and depends on many factors including the ability of the virulent fungus to form biofilms; the properties of the material surface (roughness and hydrophobicity) and environmental parameters such as the amount of nutrients, pH, temperature, oxygen, osmolality, the flow of the circulating liquid and immunological factors (Ramage et al. 2002; Sardi et al. 2013; Srey et al. 2014).

The presence of biofilms in industry can be viewed as a positive or negative characteristic. For example, the dense presence of microorganisms in piping increases the efficiency of water and sewage treatment. Another positive example is in stations that produce alcohol, vinegar and citric acid, in which the aggregation of yeast during the fermentation process increases the final yield (Van Houdt and Michiels 2010). However, the formation of biofilms in industry is more closely associated with a production problem. The deposit of undesirable microorganisms in water pipes and in the stages of production generates problems such as thermal resistance of microorganisms (contamination), biocorrosion and significant production damage, creating serious economic problems (Azevedo and Cerca 2012).

The resistance imparted by biofilms is a characteristic of fungi related to better survival. This resistance can be reversible, such as when it is induced as a response of cells to an antifungal agent, or irreversible, when it originates from genetic alterations resulting from prolonged exposure to antifungal agents. This resistance can be explained by different mechanisms: the presence of an extracellular matrix, which limits the diffusion of antifungal agents; the activation of efflux pumps; alterations in or the over expression of targets; persistent cell formation; and changes in cellular density and physiology of persistent cells to biofilms. This resistance can be related to protection via a physical barrier or via the regulation of cellular processes (Ramage et al. 2012). Table 1 shows the resistance mechanisms associated with biofilm formation as well as the effects triggered by these mechanisms in cells growing in a biofilm.

The extracellular matrix, a typical characteristic of a biofilm, promotes the protection of cells from external agents such as host immunity and antifungal agents. Its composition varies among microorganisms, but it is typically comprised of proteins, carbohydrates, hexosamine, phosphorus, uronic acid, DNA and membrane vesicles. Some species of *Candida* exhibit β-1,3 glucan in their extracellular matrix, which is

Table 1. Resistance mechanisms associated with biofilm formation.

Resistance Mechanisms	Effect	References
Cellular density	Quorum sensing	Perumal et al. 2007; Seneviratne et al. 2008
Differential regulation drug target	Changes in target levels, often associated with changes in target structure rendering the drug incapable of binding the target	Borecká-Melkusová et al. 2009; Nailis et al. 2010; White 1998
Upregulation drug efflux pumps	Antifungal is pumped out of cell and can thereby not perform its intracellular function	Nett et al. 2009; Ramage et al. 2002a; Sanglard et al. 1995, 1997
Persister cell	Because of the dormant state of persisters, antifungal targets are inactive	LaFleur et al. 2006
Presence of a matrix	Specific binding of antifungals by β-1,3-glucans, a major matrix component, which prevents antifungals from reaching their targets	Al-Fattani and Douglas 2006; Nett et al. 2007b
Diverse stress responses	Possibly only indirect effects via regulation of other resistance mechanisms	Diez-Orejas et al. 1997; Kumamoto 2005

Adapted from Mathé and Dijick (2013)

present in higher concentrations on the cell wall of yeast in the biofilm form. The β-1,3 glucan acts like a "drug sponge" because these substances are kidnapped, preventing them from acting within biofilms (Mathé and Van Dijck 2013; Ramage et al. 2012). The upregulation of drug efflux pumps has been described as a causative factor in biofilm drug resistance for several biofilm-forming microorganisms and are divided into two groups (Soto 2013). The first group is linked to the ATP-binding cassette (ABC) transporters encoded by CDR-genes, and the second group is composed of the major facilitator (MF) superfamily encoded by the MDR-genes (Mathé and Van Dijck 2013). The activation of drug efflux pumps is associated with the "expulsion" of antifungal agents after biofilm contact. The increased expression of the *CDR1* and *MDR1* genes results in resistance of yeast to azole drugs. Interesting data from biofilm-associated resistance studies have shown that the increased expression of efflux pump genes is observed during the first hour of biofilm formation but not in the mature biofilm (Mathé and Van Dijck 2013). Moreover, the overexpression of genes involved in the ergosterol biosynthesis pathway can change the target structure rendering antifungal agents inactive. Others genes can be altered for reducing the action of these drugs, including genes encoding proteins in the cellular membrane and cell wall (Mathé and Van Dijck 2013). Through the construction of mutants, studies have demonstrated that biofilms confer increased resistance to the commonly used antifungal fluconazole. Biofilms display high levels of resistance to fluconazole, and this antifungal exerts minor effects on dispersion levels. Amphotericin B proved effective in reducing the viability of cells within the biofilms and dispersion but only at high concentrations. *C. albicans* biofilms are generally considered to be resistant to azole antifungal agents but susceptible to echinocandins. Recently, some studies demonstrated that in a sequential therapy regimen, treatment with fluconazole followed by caspofungin leads to a significant decrease in the efficacy of this echinocandin (Sarkar et al. 2014).

Biofilms can be reservoirs for persistent cells derived from previous infections. These cells show phenotypic variations that render the cells resistant to high concentrations of antifungal agents (Mathé and Van Dijck 2013; Ramage et al. 2012). The inability of drugs to eradicate persistent cells is a consequence of a dormant state, in which cells persist for a long period. The first report of fungal persistent cells was described in *C. albicans*, in which tolerance was observed in a subpopulation of cells after treatment with drugs. In this study, the authors observed that 1% of cells belonging to biofilms were resistant and died after treatment with a high concentration of amphotericin B. It was also found that the formation of persistent cells is not due to the ability to form biofilms but due to the capacity to bind to a surface (LaFleur et al. 2006; Mathé and Van Dijck 2013). The resistance of fungal biofilm is explained, first, by the alteration of inoculum as cells have a higher cell density when growing in a biofilm than when growing in the planktonic form (Mathé and Van Dijck 2013). The architecture of the biofilm is established to allow the influx of nutrients and for expulsion of undesirable waste products. Mature biofilms exhibit spatial heterogeneity within microcolonies that allows the cell density to increase and hinder the penetration of antimicrobial agents into the biofilm (Ramage et al. 2012). Therefore, cell density appears to effect the resistance of yeast; however, this mechanism is not limited to biofilm formation because it is possible to find persistent cells in the planktonic form (Mathé and Van Dijck 2013).

In fungal biofilms, other resistance mechanisms related to the activation of signalling pathways occur due to the stress that the fungi suffer when they transition into the adverse conditions of the host organism, including temperature variation, osmolarity, ionic and oxidative stress. When fungi find themselves under the adverse conditions of the host, the stress is detected through receptors that activate signalling pathways and the main pathway, mitogen-activated protein kinase (MAPK), is involved in biofilm formation. The calcineurin pathway is also considered a mechanism of resistance to certain antifungals such as fluconazole. The depletion of heat shock protein Hsp90 reduces biofilm growth and maturation and abolishes the resistance of *C. albicans* biofilms to the most broad-spectrum class of antifungal drugs, the azoles. Elevated expression of heat shock proteins (HSP) promotes *Candida* yeast-hyphae switch, which is a crucial step in biofilm formation. The levels of β-1,3-glucan in the extracellular matrix are reduced, which causes the depletion of Hsp90, reducing the targets of many antifungal (Mathé and Van Dijck 2013; Ramage et al. 2012). Depletion of Hsp90 leads to the reduction of calcineurin and Mkc1 under planktonic but not biofilm conditions, suggesting that Hsp90 regulates drug resistance through different mechanisms in these distinct cellular states (Robbins et al. 2011).

Alternatives for the eradication of biofilms by increasing the efficiency of antifungal agents include fungal growth inhibition, disruption of the biofilm, biofilm eradication, sodium salts, nanoparticles, antibiotics and chitosan derivatives (Sardi et al. 2014). Many studies in biotechnology and public health have attempted to control biofilms. For this to be successful, it is essential to know the mechanisms involved in biofilm growth, but this information is still scarce due to the complexity of biofilms and their unknown metabolic interactions. Thankfully, advances in new methodologies have allowed new insights into these interactions and have led to a new vision of the complex process of the establishment of a biofilm (Sardi et al. 2014).

Fungal Biofilms

Fungal pathogenicity involves complex mechanisms that are still unclear. What is well established is the need for a pathogen to present some virulence factors for infection to be established in the host. One way of defending against microorganisms capable of causing infection on hospital implants is to target their ability to adhere to biotic and abiotic surfaces and thus inhibit biofilm formation (Douglas 2003; Miceli et al. 2012). In this regard, understanding the mechanism of biofilm formation is the first step in determining biofilm function and, consequently, its economic impact on health, industry and the environment (Huq et al. 2008; Siqueira et al. 2011).

Infections associated with biofilm formation have been recognised by the scientific and medical community to be a significant and growing clinical problem, and for this reason, research on the biofilm phenotype has become increasingly important in medical mycology (Jabra-Rizk et al. 2004; Ramage et al. 2012). Indeed, biofilm formation by bacteria and fungi is so prevalent that it must serve as an important and successful survival strategy in different surface environments. Currently, it is known that microorganisms rarely exist as solitary cells, preferring instead to live in the biofilm form (Harding et al. 2009). Recent advances in molecular techniques and confocal microscopy have shown that the formation of biofilms is the preferred form of mould growth in nature and is a major cause of persistent human infection. These formations allow microorganisms to grow in multicellular communities and produce an extracellular matrix that is able to protect them from host defence mechanisms and antifungal drugs (Costerton et al. 1999; Müller et al. 2011).

Because fungi are eukaryotes and therefore more complex than bacteria, fungal infections associated with biofilms are difficult to diagnose and treat properly (Douglas 2003; Harding et al. 2009; Ramage et al. 2012). Historically, *C. albicans*, described in the literature as the main fungal pathogen with the ability to form biofilms, has been extensively studied and is considered the third leading cause of intravascular catheter-related infections (Cuéllar-Cruz et al. 2012; Ramage et al. 2012). Indeed, a wide variety of fungi have demonstrated the ability to colonise surfaces and form biofilms. Several authors have reported the involvement of yeasts and filamentous fungi infections associated with biofilm formation, including *Cryptococcus neoformans*, *Rhodotorula* species, *Aspergillus fumigatus*, *Malassezia pachydermatis*, *Histoplasma capsulatum*, *Pneumocystis* species, *Coccidioides immitis, Fusarium* species, *Saccharomyces cerevisiae*, *Trichosporon asahii*, *Zygomycetes, Blastoschizomyces* and more recently *Paracoccidioides brasiliensis*, *Trichophyton rubrum* and *Trichophyton mentagrophytes* (Cannizzo et al. 2007; Costa-Orlandi et al. 2014; Cushion et al. 2009; Davis et al. 2002; Di Bonaventura et al. 2006; Mowat et al. 2009; Muszkieta et al. 2013; Pitangui et al. 2012; Ramage et al. 2009; Reynolds and Fink 2001; Singh et al. 2011; Walsh et al. 1986). In light of this, the chapter focuses on the latest insights into the fungal biofilms of some of these genera and species.

Much of our knowledge of fungal biofilms was acquired through studies of biofilms *in vitro* and *in vivo* by *Candida* species (Lebeaux et al. 2012). Through investigations focused on *C. albicans* biofilms, it was possible to more clearly elucidate the molecular characteristics of fungal biofilm development (Finkel and Mitchell 2011; López-Ribot 2005; Ramage et al. 2005). Biofilms of *Candida* on medical devices are associated with high mortality rates (approaching 40%) (Fox and Nobile 2012). Recently, the transcriptional network that governs the development of *C. albicans* biofilms was identified. This network consists of six master transcriptional regulators (*Efg1*, *TEC1*, *BCR1*, *Ndt80*, *BRG1* and *Rob1*) and approximately 1000 target genes whose expression is controlled by such regulators. The six master regulators were identified by screening a library of approximately 165 mutant transcripts during the formation of *in vitro* biofilms and observing which mutants exhibited altered biofilm formation. Six deletion mutants that cause defects in biofilm formation were identified: three are new (*Rob1, BRG1* and *Ndt80*) and three were previously known to play a role in biofilm development (*BCR1*, *TEC1* and *Efg1*). All six genes identified were associated with defects in biofilm formation both *in vitro* and *in vivo*. Banerjee et al. (2013) studied the role of UME6 and found it to be a regulator of *C. albicans* hyphae biofilms. Taff et al. (2012) also demonstrated that three enzymes were related to the production of extracellular polysaccharides and that these enzymes are encoded by Bgl2, PHR1 and XOG1. It has been shown that these enzymes are essential for the delivery of β-1,3-glucan for the matrix of the biofilm, mature biomass and accumulation of extracellular matrix. These researchers have therefore proposed that the discovery of inhibitors of these enzymes would provide a promising anti-biofilm therapeutic.

The use of molecular biology tools has helped unravel the mystery of microbial biofilms. Much has been discovered; however, despite all the recent advances and technology, it is still unclear as to how can we completely eliminate this form of life.

The Latest Discoveries in Pathogenic Fungi Biofilms

Recent findings have reported the involvement of new species of pathogenic fungi in biofilm formation and the role they play in infections. In this context, there is a current interest in describing the particular characteristics of biofilms containing *Rhodotorula* sp., *Trichosporon* sp., *Aspergillus fumigatus*, *Malassezia pachydermatis*, *Histoplasma capsulatum*, *Paracoccidioides brasiliensis* and *Trichophyton* species (Costa-Orlandi et al. 2014; Muszkieta et al. 2013; Ramage et al. 2011, 2012; Pitangui et al. 2012).

Rhodotorula sp.

Recently, it was demonstrated that *Rhodotorula* species are able to form biofilms. These yeasts were previously considered avirulent saprophytes and are widely distributed in nature. Their isolation occurs from a variety of environmental sources, including soil, air, aquatic ecosystems, plants and fruits. In humans, these yeasts have been isolated from nails, skin, sputum, urine, faeces and hands (Cordeiro et al. 2010; Galán-Sánchez et al. 1999; Nunes et al. 2013; Obłąk et al. 2013). However, in the last two years, these yeasts have been recognised as opportunistic pathogens (Miceli et al. 2012).

The increase in invasive infections caused by emerging pathogens such as *Rhodotorula* species is related to several factors, including an increase in the occurrence of degenerative diseases and malignancies in different populations, an increasing number of patients undergoing organ transplantation immunosuppressive therapies, the use of broad spectrum antibiotics and invasive medical procedures (Tuon and Costa 2008), and the use of central venous catheters (CVC) and other implantable medical devices that facilitate the formation of biofilms by these pathogens, which can lead to fungemia followed by eye infections, peritonitis and meningitis (De Almeida et al. 2008; Savini et al. 2008; Tuon and Costa 2008; Unal et al. 2009). Nunes et al. (2013) studied several isolates of *Rhodotorula* and noted that this genre is capable of forming biofilms and that this condition could play a role in the pathogenesis of infections caused by these species (Nunes et al. 2013). *Rhodotorula* sp. has been isolated in association with *C. albicans* in subgingival biofilms in patients with severe chronic periodontitis (Canabarro et al. 2013).

Trichosporon sp.

Invasive infections caused by *Trichosporon* spp. have increased considerably in recent years, especially in neutropenic and critically ill patients using catheters and antibiotics. These species present limited susceptibility to some antifungal agents; nevertheless, triazoles are the first choice for treatment. Mature *T. asahii* biofilms (72 h) displayed a complex, heterogeneous three-dimensional structure, consisting of a dense network of metabolically active yeast cells and hyphal elements completely embedded within exopolymeric material (Di Bonaventura et al. 2006). In another study, *T. inkin*, *T. asteroides* and *T. faecale* were capable of forming biofilms and exhibited high adhesion. Once more, the microscopy images showed that *T. asahii* presented mainly hyphae and arthroconidia, whereas *T. asteroides* exhibited mainly short arthroconidia and few filaments (Iturreita-González et al. 2014).

Malassezia pachydermatis

Another pathogen that has been highlighted is *M. pachydermatis* as it is able to form biofilms *in vitro* on devices commonly used in medical practice, including polystyrene microplates and polyurethane catheters (Canizzo et al. 2007). *M. pachydermatis* is a commensal yeast that lives on the skin and mucous membranes of healthy dogs and cats (Cafarchia et al. 2008). However, it has become an important pathogen causing fungemia in humans in intensive care units (Ashbee et al. 2002) and has been isolated from preterm neonates, children and adults. These infections are directly associated with the formation of biofilms on catheters for permanent use in patients receiving parenteral nutrition by lipid formulations (Chryssanthou et al. 2001; Curvale-Fauchet et al. 2004). Studies have been intensified to find new drugs for *M. pachydermatis* biofilms (Figueredo et al. 2013).

Aspergillus fumigatus

The ability of filamentous fungi to form biofilms has increased, and recent reports describe the growth of *Aspergillus fumigatus* as a biofilm (Chotirmall et al. 2014). *A. fumigatus* is naturally present in the soil but has become one of the main pathogens in recent decades, especially in immunocompromised patients (Brakhage and Langfelder 2002; Latgé 1999). This species is responsible for approximately 90% of cases of invasive aspergillosis, a severe infectious disease characterised by high mortality rates (Hogan et al. 1996; Ramage et al. 2011). *Aspergillus* colonization and biofilm formation occurs predominantly in patients with chronic pulmonary functional or genetic abnormalities, such as cystic fibrosis or chronic obstructive pulmonary disease (Moss 2010). *Aspergillus* biofilms can affect many sites including biomaterials such as catheters, prostheses, cardiac pacemakers, heart valves and breast implants (Chotirmall et al. 2014; Escande et al. 2011; Jeloka et al. 2011). Furthermore, a spherical mass of hyphae, called an aspergilloma, can form in the respiratory tract and, less frequently, in the urinary tract (Kaur and Singh 2014; Lee 2010; Martinez-Pajares et al. 2010; Ramage et al. 2012). Under these conditions, all clinical antifungal drugs are significantly less effective, suggesting that there is a need for high doses or combined antifungal therapy for better dissemination of drugs to fungal cells (Manavathu et al. 2014). Bugli et al. (2014) has associated *A. fumigatus* biofilm formation with increased production of gliotoxin, a toxin that has immunosuppressive properties. *A. fumigatus* can produce an extracellular hydrophobic matrix *in vitro* with typical biofilm characteristics. The extracellular matrix (ECM) is composed of galactomannan, α-1,3-glucans, monosaccharides and polyols, melanin and proteins including major antigens and hydrophobins (Beauvais et al. 2014).

Trichophyton rubrum and *T. mentagrophytes*

Recent studies by Costa-Orlandi et al. (2014) showed that the genus *Trichophyton* was able to form biofilms *in vitro*. This genus is responsible for causing dermatophytosis. Dermatophytes are fungi that have the ability to invade keratinised structures of human and animals, producing a condition called dermatophytosis (Weitzman and Summerbell 1995). The soil is a natural reservoir of dermatophytes, and keratin present in the soil is used as a nutrient, so these fungi are adapted to various environments. *T. rubrum* and *T. mentagrophytes* were able to form mature biofilms, and the *T. rubrum* biofilm produced more biomass and extracellular polymeric substances (EPS) and was denser than the *T. mentagrophytes* biofilm (Costa-Orlandi et al. 2014). Burkhart et al. (2002) introduced the concept of biofilms for dermatophytes to explain dermatophytomas, constrained dense white fungal masses living within and under the nail plate, but this has not been characterised. Dermatophytomas are more resistant to traditional therapies, host more than one microorganism species and have living fungal elements strongly adherent to the nail plate that survive in histological findings.

Histoplasma capsulatum

Recently, an *in vitro* study demonstrated the effectiveness of *Histoplasma capsulatum* var. *capsulatum* to form biofilms on abiotic surfaces (Pitangui et al. 2012). *H. capsulatum* is the causative agent of histoplasmosis, a systemic fungal disease that has become a major health problem in Latin America and has also spread worldwide (Nosanchuk et al. 2012). High concentrations of this fungus are isolated from areas where droppings of birds, chickens and bats can be found; caves and urban buildings constitute the main source for the natural spread of this mycosis (Smith and Wang 2013). Through the study by Pitangui et al. (2012), the pattern of *H. capsulatum* infection of epithelial cells was described as characterised by a compact mass of yeast cells, which can possibly lead to the formation of complex three-dimensional architecture of biofilms and consequently promote the internalisation of the yeast by host cells. A previous study by Suárez-Alvarez et al. (2010) represents the first report of the adherence of *H. capsulatum* yeasts in cryosections of different organs of bats. Through this study, it is also possible to observe that the yeasts in the lung parenchyma, spleen, liver and intestines exhibit the same cluster-forming characteristics.

Paracoccidioides spp.

Paracoccidioidomycosis (PCM) is a systemic mycosis of great importance in Latin America, mainly in Brazil, which has the highest concentration of endemic areas and claims 80% of the reported cases. The etiologic agents of PCM are the dimorphic fungi *Paracoccidioides brasiliensis* and *P. lutzii* (Teixeira et al. 2014). Working with soil and crops in rural areas are factors increasing the risk of occupational acquisition of PCM. These fungi have several virulence factors that may cause harm to the host. The adherence and colonization of the fungus are characteristics that enable it to withstand hostile environments and correlate with the development of disease by the host (Mendes-Giannini et al. 2005, 2008; Puccia et al. 2011). Adhesion is a phenomenon that is widely distributed and shared by many microorganisms that enable them to colonise their habitats. Many fungi are able to adhere to host tissues, which is the first step in the biofilm formation process. Recently, it was demonstrated that *P. brasiliensis* is capable of biofilm formation. The tests were performed *in vitro*, and this fungus was able to form a biofilm at low oxygen tensions (data not published).

Interface among Fungal Biofilms and Human Infections

There has been a growing interest in recent years to uncover how fungal biofilms participate in human disease. These formations have acquired an important role in the development of infection because microorganisms that grow in these structures exhibit unique phenotypic characteristics compared to their planktonic counterparts (Ramage et al. 2009). These features include enhanced resistance to host defences and antimicrobial therapy (Martinez and Fries 2010), and once in a structured biofilm, cells can become up to 1000 times more resistant to antifungal treatment as compared to free cells (Ramage et al. 2005).

The adherent biofilm structure in the host can trigger an acute fungemia and/or disseminated infection. This occurs when cell clusters disperse from the initial biofilm structure and begin to occupy new niches not previously colonised (Ramage et al. 2012). Indeed, cells that are detached from the biofilm have a greater association with mortality than do planktonic microorganisms (Uppuluri et al. 2010). Currently, over 65% of human infections involve the formation of biofilms and are mainly related to the increased use of biomaterials in medical practices and to the increasing number of immunocompromised patients (López-Ribot 2005; Ramage et al. 2005). In addition, more than 500,000 deaths per year are due to biofilms-related infections (Müller et al. 2011).

In light of this, biofilms have important effects and are often deleterious to human life. In this context, the formation of fungal biofilms on implants and catheters contributes to the development of nosocomial infections (Vlamakis 2011). Additionally, the persistence of fungal infections occurs due to the ability of fungi to form biofilms on a wide variety of medical devices (Kojic and Darouiche 2004). Once infected, eradication of a biofilm *in vivo* usually requires the administration of toxic concentrations of antimicrobials and often includes removal of the infected device. However, this is a difficult and expensive procedure that can result in several medical complications (Garsin and Willems 2010).

Impact of Biofilms in Cystic Fibrosis

Vascular devices infected with fungal biofilms are a primary source of candidemia in patients with cystic fibrosis (CF) (Munck et al. 2004), an autosomal recessive disorder characterised by the inability to clear inhaled particles, resulting in their persistence, colonization and potential airway infection, inducing a chronic lung disease (Chotirmall et al. 2010). The main fungi involved in CF include *Candida* species, including *C. albicans*, *C. glabrata*, *C. krusei*, *C. parapsilosis* and *C. dubliniensis*, with *C. albicans* being the most common (Tavanti et al. 2005). Among the filamentous fungi, *A. fumigatus* is the most frequently isolated. Fungal biofilm formation on these implanted devices is a major concern related to the treatment of infections in CF patients because these structures are associated with high levels of antifungal resistance. This condition is even worse because bacteria and fungi can co-exist and form multispecies biofilms in the lung tissue of patients with CF, and such interactions have been reported, for example, between *Pseudomonas aeruginosa* and *A. fumigatus* and *P. aeruginosa* and *C. albicans* (Chotirmall et al. 2010). However, the interdependence between these organisms is not clear in the context of CF, but the fact remains that fungal biofilms have become a major clinical and economic problem.

Impact of Biofilms in Ocular Infections

Fungal keratitis is also a medical complication that has been associated with biofilm formation. Pathogens causing keratitis mainly include filamentous fungi *Fusarium solani* and *Fusarium oxysporum*, which trigger a corneal ulcer inducing visual impairment or blindness (Mukherjie et al. 2012). *Fusarium* has demonstrated the ability to form biofilms on contact lenses, and this fact has been associated with outbreaks of keratitis. Between 2005 and 2007, 300 cases of fungal keratitis were reportedly induced by the use of contact lenses and, in some cases, characterised as a serious disease in which patients underwent corneal transplantation or even complete eye removal. Such severity was assigned to improper treatment resulting from misdiagnosis and to the failure of antifungal agents (Donnio et al. 2007) due to the resistance exhibited by the microorganisms in biofilms. However, the real role of *Fusarium* biofilms in fungal keratitis is not clearly understood. To study this, a murine model of contact lens-associated *Fusarium* keratitis has recently been validated to characterise *Fusarium* biofilms on contact lenses and to establish the true role of these structures in *in vitro* and *in vivo* models (Ghannoum et al. 2010).

Impact of Biofilms in the Oral Cavity

Oral lesions that can trigger systemic infections may also be the result of the formation of fungal biofilms on intraoral surfaces. Several fungi have been identified in the oral cavity and most belong to the *Candida* genus, followed by *Cladosporium*, *Aureobasidium*, *Saccharomyces*, *Aspergillus, Fusarium* and *Cryptococcus* (Ghannoum et al. 2010). Fungal biofilms in the oral cavity may be related to endodontic infections, a disease characterised by infection of the tooth pulp and periapical tissues resulting from carious lesions. *Candida* species are typically identified in this type of infection (Pasich et al. 2013). *Candida* biofilms are also associated with diseases of the oral mucosa, including different forms of candidiasis, denture stomatitis, angular cheilitis and median rhomboid glossitis, which develop as acute or chronic infections characterised by lesions formed on creamy white plaques and focal or scattered erythematous spots (Pereira-Cenci et al. 2008).

Furthermore, it is possible that infections of the oral mucosa are not caused by yeast biofilms adhered directly to the affected mucosa but by the fungal growth in mass on adjacent surfaces such as on teeth and dentures (Ramage et al. 2011). This condition can cause, among others, denture stomatitis (Ramage et al. 2002) as well as the colonization of microorganisms and subsequent formation of biofilms on voice prostheses used for the rehabilitation of speech in laryngectomy patients. Biofilms reduce the life of the prosthesis by increasing the resistance to airflow. *C. albicans* and *C. glabrata* were also identified in biofilms of voice prostheses (Ramage et al. 2014).

Biofilms in the oral cavity may also contribute to a further complication associated with mechanical ventilation, a nosocomial infection closely related to instrumentation of the airway with an endotracheal tube and subsequent micro-aspiration of contaminated secretions (Deem and Treggiari 2010). *Candida* spp. isolated from respiratory secretions of patients with suspected ventilator-associated pneumonia is associated with increased mortality compared to bacterial pneumonia (Delishe et al. 2011).

Health professionals warn that the self-care of teeth and dentures is essential for successful treatment, but that other measures to control infections and prevent biofilm formation can be adopted in intensive care patients, such as oral decontamination with chlorhexidine (Caserta et al. 2012).

Future Directions for Anti-Biofilm Therapeutics: Targeting in Pathogenic Fungi

Fungal biofilm-related infections have high medical relevance because of the increased resistance exhibited in this mode of growth, making adherent communities difficult to eradicate and are thus responsible for treatment failures. In view of this, innovative therapies for the treatment of infections triggered by fungal biofilms are urgently needed. Therapeutic anti-biofilm strategies highlight some pathways that can be targeted in the future to combat these resilient infections. Some of these strategies include the search for an antifungal prototype based on biopanning of natural or synthetic substances with anti-biofilm activity, photodynamic therapy for fungal biofilm, the use of monoclonal antibodies (MAbs) that are potentially useful in the clinical treatment involving formed biofilms and nanoparticles that may exhibit antifungal effects, among others.

In this context, natural substances have been highlighted as a source for the discovery and introduction of new and safe antifungal agents. The use of these substances has proven to be a highly effective therapy against fungi biofilms *in vitro*; however, biofilms of *Candida* species are the most investigated. The species *Acorus calamus* Linn., native to North America, Central Asia and Eastern Europe, popularly known as "sweet flag", displays inhibitory activity against *C. albicans* biofilms. This activity is assigned to one of the main active constituents of the rhizome of this species, the β-asarone (Rajput et al. 2013). Bee products such as honey and propolis also stand out as sources of bioactive compounds. *In vitro*, it is proven that the jujube (*Zizyphuss spina-christi*) honey effectively prevents biofilm formation of *C. albicans* and inhibits the establishment of preformed biofilms of this fungal species (Ansari et al. 2013). Other natural substances, such as usnic acid and cinnamon oil, act as anticandidal agents, being effective for the control of biofilms of *C. parapsilosis* and *C. orthopsilosis* (Pires et al. 2011, 2012).

Another promising strategy is the use of photodynamic therapy (PDT), which is a highly selective modality that may have antifungal applications. PDT involves the use of a photosensitiser (PS), a beam of visible light with appropriate wavelengths and oxygen, which when combined, can display selective anti-infective activity because the photosensitiser substances rapidly and selectively bind to fungal cells and induce cell death in only the regions where the light is applied. This technique is based on the formation of reactive oxygen species (ROS) generated by photo activation, and the effects of PDT have been associated with some cellular events, such as the induction of the release of cytoplasmic calcium, activation of heat shock proteins (Hsp) and protein kinases, the induction of apoptosis and the inhibition of cell-cell adhesion and cell-extracellular matrix adhesion (Margaron et al. 1997). Therefore, two major types of damage can be induced at the cellular level, the destruction of membranes and organelles as well as DNA damage (Sardi et al. 2014). PDT has substantial antifungal effects against several pathogenic fungi, including *Candida* spp., *C. neoformans*, *T. rubrum*, *T. mentagrophytes*, *T. tonsurans*, *Microsporum cookei*, *Microsporum gypseum*, *Microsporum canis*, *Epidermophyton floccosum*, *A. nidulans*, *A. fumigatus* and *Fusarium* sp. (Dai et al. 2012). In addition, *C. albicans* and *C. dubliniensis* biofilms are sensitive to applications of PDT *in vitro*, which also has an effect on multi-species biofilms of *C. albicans*, *C. glabrata*, and *Streptococcus mutans* (Quishida et al. 2013). Thus, optimised PDT using optimal concentrations of a photosensitiser and light can cross the relatively thick barrier of the biofilm extracellular matrix, constituting an attractive alternative for treating fungal infections and can improve the efficiency of other inactivation strategies being used in combination with conventional antifungal agents.

Another recent treatment strategy that stands out is using the adaptive humoral immune response by using monoclonal antibodies (MAbs) that have been shown to be potentially useful in inhibiting fungal biofilm formation. One of the latest therapeutic treatments for *Cryptococcus* biofilms involves the use of specific MAbs to *C. neoformans* capsular polysaccharide because the capsule, which is composed primarily of glucuronoxylomannan (GXM), contributes to the virulence of this pathogen, having a critical role in binding to biotic and abiotic surfaces and consequently in biofilm formation, even on an implantable medical device. The mechanism of action of these specific antibodies is associated with the release of the capsular polysaccharides from the fungal cell. This strategy is based on the prevention of biofilm formation by passive administration of antibodies to capsular polysaccharides and/or immunisation with vaccines that induce the production of antibodies against the capsular antigens. In this regard, specific antibodies that adversely affect the formation and the establishment of biofilms can be administered in a prophylactic dose prior to insertion of a medical device and would be useful in the control of infections. Passive immunotherapy using a MAb directed against the capsular polysaccharide of *C. neoformans* is an approach already in clinical phase evaluation in patients with cryptococcal meningitis (Larsen et al. 2005; Martinez and Casadevall 2005). Thus, there is a consensus that immunotherapy as a new therapeutic strategy in biofilm-related infections can be combined with conventional antifungal therapy to optimise treatment outcomes.

Silver nanoparticles (SN) may also be used in the future for the prevention or control of infections related to biofilms formed by pathogenic fungi. In light of this, the antimicrobial potential of SN is well established for several bacteria (Rai et al. 2009), and although little is understood about the impact of engineered nanoparticles in the face of fungal species, this is an emerging area in the field of medicine. Many efforts have been focused on determining the mechanism of action of SN because the innovation involves the development of new bioactives with silver (Ag) on a nanometric scale.

It is known that the bactericidal activity of nanoparticles is partly due to electrostatic attractions between negatively charged bacterial cells and positively charged nanoparticles. On the other hand, Ag interacts with biological molecules, being able to induce a defect in DNA replication, inactivation of cellular proteins and binding to functional groups of proteins leading to denaturation of these proteins (Ciobanu et al. 2013). *In vitro*, SN significantly inhibit *C. glabrata* and *C. albicans* preformed biofilms, both in relation to the number of cultured cells and total biomass, the latter of which refers to a reduction in extracellular polysaccharide matrix (Silva et al. 2013). The biocidal action of SN in contact with *Candida* biofilms is attributed to the change in the composition of the matrix constituent, including proteins, carbohydrates and DNA, as well as the breakdown of the three-dimensional structure of biofilms by inducing damage to the yeast cell wall (Monteiro et al. 2013). In addition, the application of Ag in fungal biofilms is based on its low host toxicity profile and induction of a well-tolerated tissue response; requirements that in the future may facilitate its clinical use.

Conclusion and Future Perspectives

There are many gaps in knowledge about pathogenic fungal biofilms, and the exact mechanisms involved in the formation of these complex structures and implications in human diseases are not clear. The development of highly effective and nontoxic anti-biofilm therapies is based on evidence from *in vitro* and *in vivo* studies. Therefore, novel research approaches are required to improve the antifungal targets against pathogenic fungal biofilms, and, in this context, the omic approaches and the use of new *in vivo* models fit into the future direction of anti-biofilm research.

The "omics" approaches have attracted an increasing interest from the scientific community and are intended for global analysis of biological systems to provide insight into how microorganisms exist in biofilms and how these resilient structures interact with the host cells. These approaches primarily include transcriptomics, proteomics and metabolomics and are designed to illuminate many specific properties related to the phenotypic or physiological state of the fungi. "Omics" measurements enable the identification of intra- and extracellular molecules involved in pathways and cellular processes.

Proteomic and transcriptomic differential analyses between bacterial species such as *P. aeruginosa*, *Desulfovibrio vulgaris* and *Sulfolobus* species grown in the biofilm and planktonic state have revealed differentially expressed proteins and transcripts between these forms of growth (Clark et al. 2012; Koerdt et al. 2012). In addition, the *C. albicans* proteomes of planktonic and biofilm cultures have been established by several authors, and several proteins were differentially expressed by the fungus in biofilms (Thomas et al. 2006). In this context, our research group has discovered a pattern of more than 40 proteins belonging to different metabolic pathways that have different expression levels in biofilms of *H. capsulatum* compared to yeast in dispersed cultures (data not published).

Some research has also focused on the understanding of the metabolic response of pathogenic fungi in certain conditions that aim to characterise the cellular metabolism represented by a complex set of metabolites linked through biochemical reactions. In this context, it was observed that in multispecies biofilms, which are comprised of various pathogenic bacteria and fungi, some secreted molecules coordinate the microbial interactions that occur, particularly in oral biofilms (Demuyser et al. 2014). In view of this, the identification of a transcript, a protein or a metabolite provides important information related to the survival of microorganisms in a biofilm, and thus targets can be further delineated to develop new interventions for the prevention or treatment of infections caused by fungal biofilms.

Finally, an alternative model to animal testing can be used to determine the virulence of a particular pathogen and to uncover specific characteristics of fungi-host interactions or antifungal efficacy of a drug and can thus offer an important step to managing fungal biofilms. The use of invertebrate models such as *Galleria mellonella, Caenorhabditis elegans, Drosophila melanogaster, Lemna minor, Arabidopsis thaliana* and *Dictyostelium discoideum* offers some advantages over the use of vertebrate models, such as reduction in cost and ethical issues, enabling the use of a large number of models to increase the statistical power of an assay, and the elimination of the need for an animal facility. Moreover, invertebrates have evolved and developed an immune system against pathogens that exhibit large structural and functional similarity to the mammalian immune system; this is further support for the use of these models (Fuchs and Mylonakis 2006). These non-mammalian hosts have been validated for investigating fungal biofilms (Edwards and Kjellerup 2012) and are very promising mainly because they are an inexpensive and less laborious system for studying relevant aspects of fungal biofilm virulence and for screening new compounds with antifungal activity.

Despite advances in the study of microbial biofilms, surveys of fungal biofilms have been moving at a slower pace and thus require greater effort on the part of researchers in search of a better understanding regarding fungal biofilms and their role in human infection.

References

Albuquerque, P. and A. Casadevall. 2012. Quorum sensing in fungi—a review. Med Mycol. 50: 337–345.

Ansari, M.J., A. Al-Ghamdi, S. Usmani, N.S. Al-Waili, D. Sharma, A. Nuru and Y. Al-Attal. 2013. Effect of jujube honey on *Candida albicans* growth and biofilm formation. Arch Med Res. 44: 352–360.

Ashbee, H.R., A.K. Leck, J.W. Puntis, W.J. Parsons and E.G. Evans. 2002. Skin colonization by *Malassezia* in neonates and infants. Infect Control Hosp Epidemiol. 23: 212–216.

Azevedo, N.F. and N. Cerca. 2012. Biofilmes. Na Saúde, no Ambiente, na Indústria. Publindustria, Edições Técnica, Porto, Portugal.

Banerjee, M., P. Uppuluri, X.R. Zhao, P.L. Carlisle, G. Vipulanandan, C.C. Villar, J.L. López-Ribot and D. Kadosh. 2013. Expression of UME6, a key regulator of *Candida albicans* hyphal development, enhances biofilm formation via Hgc1- and Sun41-dependent mechanisms. Eukaryot Cell. 12: 224–232.

Beauvais, A., T. Fontaine, V. Aimanianda and J.P. Latgé. 2014. *Aspergillus* cell wall and biofilm. Mycopathologia. 178: 371–377.

Besemer, K., G. Singer, C. Quince, E. Bertuzzo, W. Sloan and T.J. Battin. 2013. Headwaters are critical reservoirs of microbial diversity for fluvial networks. Proc Biol Sci. 280: 20131760.

Blankenship, J.R. and A.P. Mitchell. 2006. How to build a biofilm: a fungal perspective. Curr Opin Microbiol. 9: 588–594.

Bonhomme, J. and C. d'Enfert. 2013. *Candida albicans* biofilms: building a heterogeneous, drug-tolerant environment. Curr Opin Microbiol. 16: 398–403.

Brakhage, A.A. and K. Langfelder. 2002. Menacing mold: the molecular biology of *Aspergillus fumigatus*. Annu Rev Microbiol. 56: 433–455.

Bugli, F., F.P. Sterbini, M. Cacaci, C. Martini, S. Lancellotti, E. Stigliano, R. Torelli, V. Arena, M. Caira, P. Posteraro, M. Sanguinetti and B. Posteraro. 2014. Increased production of gliotoxin is related to the formation of biofilm by *Aspergillus fumigatus*: an immunological approach. Pathog Dis. 70: 379–389.

Burkhart, C.N., C.G. Burkhart and A.K. Gupta. 2002. Dermatophytoma: recalcitrance to treatment because of existence of fungal biofilm. J Am Acad Dermatol. 47: 629–631.

Cafarchia, C., S. Gallo, P. Danesi, G. Capelli, P. Paradies, D. Traversa, R.B. Gasser and D. Otranto. 2008. Assessing the relationship between *Malassezia* and leishmaniasis in dogs with or without skin lesions. Acta Trop. 107: 25–29.

Canabarro, A., C. Valle, M.R. Farias, F.B. Santos, M. Lazera and B. Wanke. 2013. Association of subgingival colonization of *Candida albicans* and other yeasts with severity of chronic periodontitis. J Periodontal Res. 48: 428–432.

Cannizzo, F.T., E. Eraso, P.A. Ezkurra, M. Villar-Vidal, E. Bollo, G. Castellá, F.J. Cabañes, V. Vidotto and G. Quindós. 2007. Biofilm development by clinical isolates of *Malassezia pachydermatis*. Med Mycol. 45: 357–361.

Caserta, R.A., A.R. Marra, M.S. Durão, C.V. Silva, O.F. Pavao dos Santos, H.S. Neves, M.B. Edmond and K.T. Timenetsky. 2012. A program for sustained improvement in preventing ventilator associated pneumonia in an intensive care setting. BMC Infect Dis. 12: 234.

Chandra, J., E. Pearlman and M.A. Ghannoum. 2014. Animal models to investigate fungal biofilm formation. Methods Mol Biol. 1147: 141–157.

Chotirmall, S.H., C.M. Greene, I.K. Oglesby, W. Thomas, S.J. O'Neill, B.J. Harvey and N.G. McElvaney. 2010. 17Beta-estradiol inhibits IL-8 in cystic fibrosis by up-regulating secretory leucoprotease inhibitor. Am J Respir Crit Care Med. 182: 62–72.

Chotirmall, S.H., B. Mirkovic, G.M. Lavelle and N.G. McElvaney. 2014. Immunoevasive *Aspergillus* virulence factors. Mycopathologia. 178: 363–370.

Chryssanthou, E., U. Broberger and B. Petrini. 2001. *Malassezia pachydermatis* fungaemia in a neonatal intensive care unit. Acta Paediatr. 90: 323–327.

Chua, S.L., Y. Liu, J.K. Yam, Y. Chen, R.M. Vejborg, B.G. Tan, S. Kjelleberg, T. Tolker-Nielsen, M. Givskov and L. Yang. 2014. Dispersed cells represent a distinct stage in the transition from bacterial biofilm to planktonic lifestyles. Nat Commun. 21: 4462.

Ciobanu, C.S., S.L. Iconaru, M.C. Chifiriuc, A. Costescu, P. Le Coustumer and D. Predoi. 2013. Synthesis and antimicrobial activity of silver-doped hydroxyapatite nanoparticles. Biomed Res Int. 2013: 916218.

Clark, M.E., Z. He, A.M. Reddin, M.P. Joachimiak, J.D. Keasling, J.Z. Zhou, A.P. Arkin, A. Mukhopadhyay and M.W. Fields. 2012. Transcriptomic and proteomic analyses of *Desulfovibrio vulgaris* biofilms: carbon and energy flow contribute to the distinct biofilm growth state. BMC Genomics 13: 138.

Cordeiro, R.A., R.S. Brilhante, L.D. Pantoja, R.E. Moreira Filho, P.R. Vieira, M.F. Rocha, A.J. Monteiro and J.J. Sidrim. 2010. Isolation of pathogenic yeasts in the air from hospital environments in the city of Fortaleza, northeast Brazil. Braz J Infect Dis. 14: 30–34.

Costa-Orlandi, C.B., J.C. Sardi, C.T. Santos, A.M. Fusco-Almeida and M.J. Mendes-Giannini. 2014. *In vitro* characterization of *Trichophyton rubrum* and *T. mentagrophytes* biofilms. Biofouling 30(6): 719–727.

Costerton, J.W., G.G. Geesey and K.J. Cheng. 1978. How bacteria stick. Sci Am. 238: 86–95.

Costerton, J.W., Z. Lewandowski, D.E. Caldwell, D.R. Korber and H.M. Lappin-Scott. 1995. Microbial biofilms. Annu Rev Microbiol. 49: 711–745.

Costerton, J.W., P.S. Stewart and E.P. Greenberg. 1999. Bacterial biofilms: a common cause of persistent infections. Science 284: 1318–1322.

Cuéllar-Cruz, M., E. López-Romero, J.C. Villagómez-Castro and E. Ruiz-Baca. 2012. *Candida* species: new insights into biofilm formation. Future Microbiol. 7: 755–771.

Curvale-Fauchet, N., F. Botterel, P. Legrand, J. Guillot and S. Bretagne. 2004. Frequency of intravascular catheter colonization by *Malassezia* spp. in adult patients. Mycoses 47: 491–494.

Cushion, M.T., M.S. Collins and M.J. Linke. 2009. Biofilm formation by *Pneumocystis* spp. Eukaryot Cell. 8: 197–206.

Dai, T., B.B. Fuchs, J.J. Coleman, R.A. Prates, C. Astrakas, T.G. St. Denis, M.S. Ribeiro, E. Mylonakis, M.R. Hamblin and G.P. Tegos. 2012. Concepts and principles of photodynamic therapy as an alternative antifungal discovery platform. Front Microbiol. 3: 120.

Davis, L.E., G. Cook and J.W. Costerton. 2002. Biofilm on ventriculo-peritoneal shunt tubing as a cause of treatment failure in coccidioidal meningitis. Emerg Infect Dis. 8: 376–379.

De Almeida, G.M., S.F. Costa, M. Melhem, A.L. Motta, M.W. Szeszs, F. Miyashita, L.C. Pierrotti, F. Rossi and M.N. Burattini. 2008. *Rhodotorula* spp. isolated from blood cultures: clinical and microbiological aspects. Med Mycol. 46: 547–556.

Deem, S. and M.M. Treggiari. 2010. New endotracheal tubes designed to prevent ventilator-associated pneumonia: do they make a difference? Respir Care. 55: 1046–1055.

Delisle, M.S., D.R. Williamson, M. Albert, M.M. Perreault, X. Jiang, A.G. Day and D.K. Heyland. 2011. Impact of *Candida* species on clinical outcomes in patients with suspected ventilator-associated pneumonia. Can Respir J. 18(3): 131–136.

Demuyser, L., M.A. Jabra-Rizk and P. Van Dijck. 2014. Microbial cell surface proteins and secreted metabolites involved in multispecies biofilms. Pathog Dis. 70: 219–230.

Di Bonaventura, G., A. Pompilio, C. Picciani, M. Lezzi, D. D'Antonio and R. Piccolomini. 2006. Biofilm formation by the emerging fungal pathogen *Trichosporon asahii*: development, architecture, and antifungal resistance. Antimicrob Agents Chemother. 50: 3269–3276.

Dongari-Bagtzoglou, A., H. Kashleva, P. Dwivedi, P. Diaz and J. Vasilakos. 2009. Characterization of mucosal *Candida albicans* biofilms. PLoS One 4: e7967.

Donnio, A., D.N. Van Nuoi, M. Catanese, N. Desbois, L. Ayeboua and H. Merle. 2007. Outbreak of keratomycosis attributable to *Fusarium solani* in the French West Indies. Am J Ophthalmol. 143: 356–358.

Douglas, L.J. 2003. *Candida* biofilms and their role in infection. Trends Microbiol. 11(1): 30–6.

Edwards, S. and B.V. Kjellerup. 2012. Exploring the applications of invertebrate host-pathogen models for *in vivo* biofilm infections. FEMS Immunol Med Microbiol. 65: 205–214.

Elias, S. and E. Banin. 2012. Multi-species biofilms: living with friendly neighbors. FEMS Microbiol Rev. 36: 990–1004.

Escande, W., G. Fayad, T. Modine, E. Verbrugge, M. Koussa, E. Senneville and O. Leroy. 2011. Culture of a prosthetic valve excised for streptococcal endocarditis positive for *Aspergillus fumigatus* 20 years after previous *A. fumigatus* endocarditis. Ann Thorac Surg. 91: 92–93.

Figueredo, L.A., C. Cafarchia and D. Otranto. 2013. Antifungal susceptibility of *Malassezia pachydermatis* biofilm. Med Mycol. 51: 863–867.

Finkel, J.S. and A.P. Mitchell. 2011. Genetic control of *Candida albicans* biofilm development. Nat Rev Microbiol. 9: 109–118.

Fox, E.P. and C.J. Nobile. 2012. A sticky situation: untangling the transcriptional network controlling biofilm development in *Candida albicans*. Transcription 3: 315–322.

Fridkin, S.K. and W.R. Jarvis. 1996. Epidemiology of nosocomial fungal infections. Clin Microbiol Rev. 9: 499–511.

Fuchs, B.B. and E. Mylonakis. 2006. Using non-mammalian hosts to study fungal virulence and host defense. Curr Opin Microbiol. 9: 346–351.

Galán-Sánchez, F., P. García-Martos, C. Rodríguez-Ramos, P. Marín-Casanova and J. Mira-Gutiérrez. 1999. Microbiological characteristics and susceptibility patterns of strains of *Rhodotorula* isolated from clinical samples. Mycopathologia. 145: 109–112.

Ganguly, S., A.C. Bishop, W. Xu, S. Ghosh, K.W. Nickerson, F. Lanni, J. Patton-Vogt and A.P. Mitchell. 2011. Zap1 control of cell-cell signaling in *Candida albicans* biofilms. Eukaryot Cell. 10: 1448–1454.

Garsin, D.A. and R.J. Willems. 2010. Insights into the biofilm lifestyle of enterococci. Virulence 1: 219–221.

Geesey, G.G., W.T. Richardson, H.G. Yeomans, R.T. Irvin and J.W. Costerton. 1977. Microscopic examination of natural sessile bacterial populations from an alpine stream. Can J Microbiol. 23: 1733–1736.

Ghannoum, M.A., R.J. Jurevic, P.K. Mukherjee, F. Cui, M. Sikaroodi, A. Naqvi and P.M. Gillevet. 2010. Characterization of the oral fungal microbiome (mycobiome) in healthy individuals. PLoS Pathog. 6: e1000713.

Hall-Stoodley, L., J.W. Costerton and P. Stoodley. 2004. Bacterial biofilms: from the natural environment to infectious diseases. Nat Rev Microbiol. 2: 95–108.

Harding, M.W., L.L. Marques, R.J. Howard and M.E. Olson. 2009. Can filamentous fungi form biofilms? Trends Microbiol. 17: 475–480.

Hogan, L.H., B.S. Klein and S.M. Levitz. 1996. Virulence factors of medically important fungi. Clin Microbiol Rev. 9: 469–488.

Hoiby, N., E.W. Flensborg, B. Beck, B. Friis, S.V. Jacobsen and L. Jacobsen. 1977. *Pseudomonas aeruginosa* infection in cystic fibrosis. Diagnostic and prognostic significance of *Pseudomonas aeruginosa* precipitins determined by means of crossed immunoelectrophoresis. Scand J Respir Dis. 58: 65–79.

Huq, A., C.A. Whitehouse, C.J. Grim, M. Alam and R.R. Colwell. 2008. Biofilms in water, its role and impact in human disease transmission. Curr Opin Biotechnol. 19: 244–247.

Iturrieta-González, I.A., A.C. Padovan, F.C. Bizerra, R.C. Hahn and A.L. Colombo. 2014. Multiple species of *Trichosporon* produce biofilms highly resistant to triazoles and amphotericin B. PLoS One. 9: e109553.

Jabra-Rizk, M.A., W.A. Falkler and T.F. Meiller. 2004. Fungal biofilms and drug resistance. Emerg Infect Dis. 10(1): 14–9.

Jeloka, T.K., S. Shrividya and G. Wagholikar. 2011. Catheter outflow obstruction due to an aspergilloma. Perit Dial Int. 31: 211–212.

Karunakaran, R., S. Somasundaram, M. Gawthaman, S. Vinodh, S. Manikandan and S. Gokulnathan. 2014. Prevalence of dental caries among school-going children in Namakkal district: a cross-sectional study. J Pharm Bioallied Sci. 6(Suppl 1): S160–161.

Kaur, S. and S. Singh. 2014. Biofilm formation by *Aspergillus fumigatus*. Med Mycol. 52: 2–9.

Koerdt, A., S. Jachlewski, A. Ghosh, J. Wingender, B. Siebers and S.V. Albers. 2012. Complementation of *Sulfolobus solfataricus* PBL2025 with an α-mannosidase: effects on surface attachment and biofilm formation. Extremophiles 16: 115–125.
Kojic, E.M. and R.O. Darouiche. 2004. *Candida* infections of medical devices. Clin Microbiol Rev. 17: 255–267.
LaFleur, M.D., C.A. Kumamoto and K. Lewis. 2006. *Candida albicans* biofilms produce antifungal-tolerant persister cells. Antimicrob Agents Chemother. 50: 3839–3846.
Larsen, R.A., P.G. Pappas, J. Perfect, J.A. Aberg, A. Casadevall, G.A. Cloud, R. James, S. Filler and W.E. Dismukes. 2005. Phase I evaluation of the safety and pharmacokinetics of murine-derived anticryptococcal antibody 18B7 in subjects with treated cryptococcal meningitis. Antimicrob Agents Chemother. 49: 952–958.
Latgé, J.P. 1999. *Aspergillus fumigatus* and aspergillosis. Clin Microbiol Rev. 12: 310–350.
Lebeaux, D., B. Larroque, J. Gellen-Dautremer, V. Leflon-Guibout, C. Dreyer, S. Bialek, A. Froissart, O. Hentic, C. Tessier, R. Ruimy, A.L. Pelletier, B. Crestani, M. Fournier, T. Papo, B. Barry, V. Zarrouk and B. Fantin. 2012. Clinical outcome after a totally implantable venous access port-related infection in cancer patients: a prospective study and review of the literature. Medicine 91: 309–318.
Lee, S.W. 2010. An aspergilloma mistaken for a pelviureteral stone on nonenhanced CT: a fungal bezoar causing ureteral obstruction. Korean J Urol. 51: 216–218.
López-Ribot, J.L. 2005. *Candida albicans* biofilms: more than filamentation. Curr Biol. 15: 453–455.
Manavathu, E.K., D.L. Vager and J.A. Vazquez. 2014. Development and antimicrobial susceptibility studies of *in vitro* monomicrobial and polymicrobial biofilm models with *Aspergillus fumigatus* and *Pseudomonas aeruginosa*. BMC Microbiol. 14: 53.
Margaron, P., R. Sorrenti and J.G. Levy. 1997. Photodynamic therapy inhibits cell adhesion without altering integrin expression. Biochim Biophys Acta. 1359: 200–210.
Martinez, L.R. and A. Casadevall. 2005. Specific antibody can prevent fungal biofilm formation and this effect correlates with protective efficacy. Infect Immun. 73: 6350–6362.
Martinez, L.R. and B.C. Fries. 2010. Fungal biofilms: relevance in the setting of human disease. Curr Fungal Infect Rep. 4: 266–275.
Martinez-Pajares, J.D., M.C. Martinez-Ferriz, D. Moreno-Perez, M. Garcia-Ramirez, S. Martin-Carballido and P. Blanch-Iribarne. 2010. Management of obstructive renal failure caused by bilateral renal aspergilloma in an immunocompetent newborn. J Med Microbiol. 59: 367–369.
Mathé, L. and P. Van Dijck. 2013. Recent insights into *Candida albicans* biofilm resistance mechanisms. Curr Genet. 59(4): 251–264.
McCoy, W.F., J.D. Bryers, J. Robbins and J.W. Costerton. 1981. Observations of fouling biofilm formation. Can J Microbiol. 27: 910–917.
Mendes-Giannini, M.J., C.P. Soares, J.L. da Silva and P.F. Andreotti. 2005. Interaction of pathogenic fungi with host cells: molecular and cellular approaches. FEMS Immunol Med Microbiol. 45: 383–394.
Mendes-Giannini, M.J., J.L. Monteiro da Silva, J. de Fátima da Silva, F.C. Donofrio, E.T. Miranda, P.F. Andreotti and C.P. Soares. 2008. Interactions of *Paracoccidioides brasiliensis* with host cells: recent advances. Mycopathologia. 165: 237–248.
Miceli, M.H., S.M. Bernardo, T.S. Ku, C. Walraven and S.A. Lee. 2012. *In vitro* analyses of the effects of heparin and parabens on *Candida albicans* biofilms and planktonic cells. Antimicrob Agents Chemother. 56: 148–153.
Monroe, D. 2007. Looking for chinks in the armor of bacterial biofilms. PLoS Biol. 5: e307.
Monteiro, D.R., S. Silva, M. Negri, L.F. Gorup, E.R. de Camargo, R. Oliveira, D.B. Barbosa and M. Henriques. 2013. Antifungal activity of silver nanoparticles in combination with nystatin and chlorhexidine digluconate against *Candida albicans* and *Candida glabrata* biofilms. Mycoses 56: 672–680.
Moss, R.B. 2010. Allergic bronchopulmonary aspergillosis and *Aspergillus* infection in cystic fibrosis. Curr Opin Pulm Med. 16: 598–603.
Mowat, E., C. Williams, B. Jones, S. McChlery and G. Ramage. 2009. The characteristics of *Aspergillus fumigatus* mycetoma development: is this a biofilm? Med Mycol. 47: 120–126.
Mukherjee, P.K., J. Chandra, C. Yu, Y. Sun, E. Pearlman and M.A. Ghannoum. 2012. Characterization of *Fusarium* keratitis outbreak isolates: contribution of biofilms to antimicrobial resistance and pathogenesis. Invest Ophthalmol Vis Sci. 53: 4450–4457.
Müller, F.M., M. Seidler and A. Beauvais. 2011. *Aspergillus fumigatus* biofilms in the clinical setting. Med Mycol. 49: Suppl 1: 96–100.

Munck, A., S. Malbezin, J. Bloch, M. Gerardin, M. Lebourgeois, J. Derelle, F. Bremont, I. Sermet, M.R. Munck and J. Navarro. 2004. Follow-up of 452 totally implantable vascular devices in cystic fibrosis patients. Eur Respir J. 23: 430–434.

Muszkieta, L., A. Beauvais, V. Pähtz, J.G. Gibbons, V. Anton Leberre, R. Beau, K. Shibuya, A. Rokas, J.M. Francois, O. Kniemeyer, A.A. Brakhage and J.P. Latgé. 2013. Investigation of *Aspergillus fumigatus* biofilm formation by various "omics" approaches. Front Microbiol. 12; 4: 13.

Nett, J.E. 2014. Future directions for anti-biofilm therapeutics targeting *Candida*. Expert Rev Anti Infect Ther. 12: 375–382.

Nickel, J.C., A.G. Gristina and J.W. Costerton. 1985. Electron microscopic study of an infected Foley catheter. Can J Surg. 28: 50–54.

Nosanchuk, J.D., R.M. Zancopé-Oliveira, A.J. Hamilton and A.J. Guimarães. 2012. Antibody therapy for histoplasmosis. Front Microbiol. 3: 21.

Nunes, J.M., F.C. Bizerra, R.C. Ferreira and A.L. Colombo. 2013. Molecular identification, antifungal susceptibility profile, and biofilm formation of clinical and environmental *Rhodotorula* species isolates. Antimicrob Agents Chemother. 57: 382–389.

Obłąk, E., A. Piecuch, A. Krasowska and J. Luczyński. 2013. Antifungal activity of gemini quaternary ammonium salts. Microbiol Res. 168: 630–638.

Pasich, E., A. Bialecka and J. Marcinkiewicz. 2013. Efficacy of taurine haloamines and chlorhexidine against selected oral microbiome species. Med Dosw Mikrobiol. 65: 187–196.

Pereira-Cenci, T., A.A. Del Bel Cury, W. Crielaard and J.M. Ten Cate. 2008. Development of *Candida*-associated denture stomatitis: new insights. J Appl Oral Sci. 16: 86–94.

Pires, R.H., L.B. Montanari, C.H. Martins, J.E. Zaia, A.M. Almeida, M.T. Matsumoto and M.J. Mendes-Giannini. 2011. Anticandidal efficacy of cinnamon oil against planktonic and biofilm cultures of *Candida parapsilosis* and *Candida orthopsilosis*. Mycopathologia. 176: 453–464.

Pires, R.H., R. Lucarini and M.J. Mendes-Giannini. 2012. Effect of usnic acid on *Candida orthopsilosis* and *C. parapsilosis*. Antimicrob Agents Chemother. 56: 595–597.

Pitangui, N.S., J.C. Sardi, J.F. Silva, T. Benaducci, R.A. Moraes da Silva, G. Rodríguez-Arellanes, M.L. Taylor, M.J. Mendes-Giannini and A.M. Fusco-Almeida. 2012. Adhesion of *Histoplasma capsulatum* to pneumocytes and biofilm formation on an abiotic surface. Biofouling 28: 711–718.

Puccia, R., M.C. Vallejo, A.L. Matsuo and L.V. Longo. 2011. The *Paracoccidioides* cell wall: past and present layers toward understanding interaction with the host. Front Microbiol. 2: 257.

Quishida, C.C., J.C. Carmello, E.G. Mima, V.S. Bagnato, A.L. Machado and A.C. Pavarina. 2013. Susceptibility of multispecies biofilm to photodynamic therapy using Photodithazine®. Lasers Med Sci. 3.

Ragon, M., M.C. Fontaine, D. Moreira and P. López-García. 2012. Different biogeographic patterns of prokaryotes and microbial eukaryotes in epilithic biofilms. Mol Ecol. 21: 3852–3868.

Rai, M., A. Yadav and A. Gade. 2009. Silver nanoparticles as a new generation of antimicrobials. Biotechnol Adv. 27: 76–83.

Rajput, S.B. and S.M. Karuppayil. 2013. β-Asarone, an active principle of Acorus calamus rhizome, inhibits morphogenesis, biofilm formation and ergosterol biosynthesis in *Candida albicans*. Phytomedicine 20: 139–142.

Ramage, G., S.P. Saville, B.L. Wickes and J.L. López-Ribot. 2002. Inhibition of *Candida albicans* biofilm formation by farnesol, a quorum-sensing molecule. Appl Environ Microbiol. 68: 5459–5563.

Ramage, G., S.P. Saville, D.P. Thomas and J.L. López-Ribot. 2005. *Candida* biofilms: an update. Eukaryot Cell. 4: 633–638.

Ramage, G., E. Mowat, B. Jones, C. Williams and J. Lopez-Ribot. 2009. Our current understanding of fungal biofilms. Crit Rev Microbiol. 35(4): 340–355.

Ramage, G., R. Rajendran, M. Gutierrez-Correa, B. Jones and C. Williams. 2011. *Aspergillus* biofilms: clinical and industrial significance. FEMS Microbiol Lett. 324: 89–97.

Ramage, G., R. Rajendran, L. Sherry and C. Williams. 2012. Fungal biofilm resistance. Int J Microbiol. 2012: 528521.

Ramage, G., S.N. Robertson and C. Williams. 2014. Strength in numbers: antifungal strategies against fungal biofilms. Int J Antimicrob Agents 43: 114–120.

Rendueles, O. and J.M. Ghigo. 2012. Multi-species biofilms: how to avoid unfriendly neighbors. FEMS Microbiol Rev. 36: 972–989.

Reynolds, T.B. and G.R. Fink. 2001. Bakers' yeast, a model for fungal biofilm formation. Science 291: 878–881.

Robbins, N., P. Uppuluri, J. Nett, R. Rajendran, G. Ramage, J.L. Lopez-Ribot, D. Andes and L.E. Cowen. 2011. Hsp90 governs dispersion and drug resistance of fungal biofilms. PLoS Pathog. 7: e1002257.

Sabra, W., H. Lünsdorf and A.P. Zeng. 2013. Alterations in the formation of lipopolysaccharide and membrane vesicles on the surface of *Pseudomonas aeruginosa* PAO1 under oxygen stress conditions. Microbiology 149: 2789–2795.

Sardi, J.C., C. Duque, J.F. Höfling and R.B. Gonçalves. 2012. Genetic and phenotypic evaluation of *Candida albicans* strains isolated from subgingival biofilm of diabetic patients with chronic periodontitis. Med Mycol. 50: 467–475.

Sardi, J.C., L. Scorzoni, T. Bernardi, A.M. Fusco-Almeida and M.J. Mendes Giannini. 2013. *Candida* species: current epidemiology, pathogenicity, biofilm formation, natural antifungal products and new therapeutic options. J Med Microbiol. 62: 10–24.

Sardi, J.C., N.S. Pitangui, G. Rodríguez-Arellanes, M.L. Taylor, A.M. Fusco-Almeida and M.J. Mendes-Giannini. 2014. Highlights in pathogenic fungal biofilms. Rev Iberoam Micol. 31: 22–29.

Sarkar, S., P. Uppuluri, C.G. Pierce and J.L. Lopez-Ribot. 2014. *In vitro* study of sequential fluconazole and caspofungin treatment against *Candida albicans* biofilms. Antimicrob Agents Chemother. 58: 1183–1186.

Savini, V., F. Sozio, C. Catavitello, M. Talia, A. Manna, F. Febbo, A. Balbinot, G. Di Bonaventura, R. Piccolomini, G. Parruti and D. D'Antonio. 2008. Femoral prosthesis infection by *Rhodotorula mucilaginosa*. J Clin Microbiol. 46: 3544–3545.

Silva, S., P. Pires, D.R. Monteiro, M. Negri, L.F. Gorup, E.R. Camargo, D.B. Barbosa, R. Oliveira, D.W. Williams, M. Henriques and J. Azeredo. 2013. The effect of silver nanoparticles and nystatin on mixed biofilms of *Candida glabrata* and *Candida albicans* on acrylic. Med Mycol. 51: 178–84.

Singh, R., M.R. Shivaprakash and A. Chakrabarti. 2011. Biofilm formation by zygomycetes: quantification, structure and matrix composition. Microbiology 157: 2611–2618.

Siqueira, V.M., H.M. Oliveira, C. Santos, R.R. Paterson, N.B. Gusmão and N. Lima. 2011. Filamentous fungi in drinking water, particularly in relation to biofilm formation. Int J Environ Res Public Health 8: 456–469.

Skupien, J.A., F. Valentini, N. Boscato and T. Pereira-Cenci. 2013. Prevention and treatment of *Candida* colonization on denture liners: a systematic review. J Prosthet Dent. 110: 356–362.

Smith, I. and L.F. Wang. 2013. Bats and their virome: an important source of emerging viruses capable of infecting humans. Curr Opin Virol. 3: 84–91.

Soll, D.R. 2014. The evolution of alternative biofilms in an opportunistic fungal pathogen: an explanation for how new signal transduction pathways may evolve. Infect Genet Evol. 22: 235–243.

Soto, S.M. 2013. Role of efflux pumps in the antibiotic resistance of bacteria embedded in a biofilm. Virulence 1; 4(3): 223–229.

Srey, S., S.Y. Park, I.K. Jahid, S.R. Oh, N. Han, C.Y. Zhang, S.H. Kim, J.I. Cho and S.D. Ha. 2014. Evaluation of the removal and destruction effect of a chlorine and thiamine dilaurylsulfate combined treatment on *L. monocytogenes* biofilm. Foodborne Pathog Dis. 11: 658–663.

Suárez-Alvarez, R.O., A. Pérez-Torres and M.L. Taylor. 2010. Adherence patterns of *Histoplasma capsulatum* yeasts to bat tissue sections. Mycopathologia. 170: 79–87.

Taff, H.T., J.E. Nett, R. Zarnowski, K.M. Ross, H. Sanchez, M.T. Cain, J. Hamaker, A.P. Mitchell and D.R. Andes. 2012. A *Candida* biofilm-induced pathway for matrix glucan delivery: implications for drug resistance. PLoS Pathog. 8: e1002848.

Tavanti, A., A.D. Davidson, M.J. Fordyce, N.A. Gow, M.C. Maiden and F.C. Odds. 2005. Population structure and properties of *Candida albicans*, as determined by multilocus sequence typing. J Clin Microbiol. 43: 5601–5613.

Teixeira, M.M., R.C. Theodoro, G. Nino-Vega, E. Bagagli and M.S. Felipe. 2014. *Paracoccidioides* species complex: ecology, phylogeny, sexual reproduction, and virulence. PLoS Pathog. 10(10).

The, K.H., S. Flint, J. Palmer, P. Andrewes, P. Bremer and D. Lindsay. 2012. Proteolysis produced within biofilms of bacterial isolates from raw milk tankers. Int J Food Microbiol. 157: 28–34.

Thomas, D.P., S.P. Bachmann and J.L. Lopez-Ribot. 2006. Proteomics for the analysis of the *Candida albicans* biofilm lifestyle. Proteomics 6: 5795–804.

Tuon, F.F. and S.F. Costa. 2008. *Rhodotorula* infection. A systematic review of 128 cases from literature. Rev Iberoam Micol. 25: 135–140.

Unal, A., A.N. Koc, M.H. Sipahioglu, F. Kavuncuoglu, B. Tokgoz, H.M. Buldu, O. Oymak and C. Utas. 2009. CAPD-related peritonitis caused by *Rhodotorula mucilaginosa*. Perit Dial Int. 29: 581–582.

Uppuluri, P., A.K. Chaturvedi, A. Srinivasan, M. Banerjee, A.K. Ramasubramaniam, J.R. Köhler, D. Kadosh and J.L. Lopez Ribot. 2010. Dispersion as an important step in the *Candida albicans* biofilm developmental cycle. PLoS Pathog. 26: e1000828.

Van Houdt, R. and C.W. Michiels. 2010. Biofilm formation and the food industry, a focus on the bacterial outer surface. J Appl Microbiol. 109: 1117–1131.

Vlamakis, H. 2011. The world of biofilms. Virulence 2: 431–434.

Xu, W., X. Zhu, T. Tan, W. Li and A. Shan. 2014. Design of embedded-hybrid antimicrobial peptides with enhanced cell selectivity and anti-biofilm activity. PLoS One 19; 9(6): e98935.

Walsh, T.J., R. Schlegel, M.M. Moody, J.W. Costerton and M. Salcman. 1986. Ventriculoatrial shunt infection due to *Cryptococcus neoformans*: an ultrastructural and quantitative microbiological study. Neurosurgery 18: 373–375.

Weitzman, I. and R.C. Summerbell. 1995. The dermatophytes. Clin Microbiol Rev. 8: 240–259.

CHAPTER 13

Fungal Allergens: Recent Trends and Future Prospects

Marta Gabriel,[a] *Jorge Martínez*[b] and *Idoia Postigo*[c,*]

Introduction

Allergic diseases affect millions of people and have shown a marked increase in recent years, particularly in industrialised nations (D'Amato et al. 2010). Allergens that cause disease include environmental allergens such as pollen, food, animal dander, and various fungi. Fungal spores and mycelial cells are two of the main factors causing several allergic diseases, including asthma, rhinitis, hypersensitivity pneumonitis, certain occupational lung diseases, fungal sinusitis, toxic pneumonia and allergic bronchopulmonary mycosis (ABPM) (Vijay et al. 2005). This broad panel of diseases results from the particular biology of each mould. Fungi are very common in the environment; therefore, the exposure to airborne spores is almost constant throughout the year. The inhalation and ingestion of fungal spores and vegetative cells (hyphae) may result in the colonization of the human body, and these particles may damage airways by producing toxins, proteases, enzymes (Kauffman et al. 2000) and volatile organic compounds (Fischer et al. 1999). Thus, moulds have a far greater impact on the immune systems of patients than other allergenic sources (Simon-Nobbe et al. 2008).

Although the exact prevalence of fungal allergies is not known, it is estimated that approximately 6 to 24% of the general population can be predicted to have allergic symptoms to moulds (Tariq et al. 1996; Pulimood et al. 2007). The prevalence of a respiratory allergy to fungi can increase to approximately 44% in atopic individuals

Department of Immunology, Microbiology and Parasitology, Faculty of Pharmacy, University of the Basque Country, Spain. Paseo de la Universidad, 7. 01006-Vitoria. Spain.
[a] E-mail: mgabriel20023@gmail.com
[b] E-mail: jorge.martinez@ehu.es
[c] E-mail: idoia.postigo@ehu.es
* Corresponding author

(Corey et al. 1997) and up to 80% among asthmatics (Lopez and Salvaggio 1985). Various studies suggest that at least 3–10% of adults and children are affected by a fungal allergy (Kurup et al. 2000; Bush and Protnoy 2001); however, skin reactivities ranging from 3–91% have been reported depending upon the population studied, fungal extracts used and species tested (Lehrer et al. 1986; Sprenger et al. 1988; Horner et al. 1995).

Among the allergic diseases caused by fungi, asthma has been found to be more prevalent, and severe reactions are commonly associated with exposure and sensitisation to moulds (Bush et al. 2006). Recent studies regarding asthma included in the Global Initiative for Asthma (GINA) yielded a global figure of 197 million people suffering from the disease, and among these, 4.8 million adults were associated with allergic bronchopulmonary aspergillosis (ABPA) (Denning et al. 2013). This data, referring to one of several clinical manifestations in which fungi are implicated, indicates the relevance of this kingdom in the development of allergic diseases. We consider this data to be only an approximation. The exact prevalence of fungal allergy/asthma is difficult to assess because correlating fungal exposure with asthmatic symptoms remains problematic. One reason is that the majority of the extracts used in allergy testing are not standardised, which makes it difficult to ascertain the true prevalence of fungal sensitisation (Esch 2004). Furthermore, cutaneous responses do not necessarily indicate the presence of disease, although these types of responses are helpful in defining the frequency of sensitisation and could be considered as markers of the severity of asthma or rhinitis (Friedlander and Bush 2005).

Fungi are ubiquitous organisms, and over 80 genera have been shown to induce type I allergies in susceptible persons, whereas allergenic proteins have been only identified in 25 fungal genera (Simon-Nobbe et al. 2008).

The most important fungi involved in allergic diseases include the genera *Alternaria*, *Aspergillus*, *Aureobasidium*, *Bipolaris*, *Botrytis*, *Candida*, *Curvularia*, *Cladosporium*, *Drechslera*, *Epicoccum*, *Fusarium*, *Mucor*, *Penicillium*, *Phoma*, *Saccharomyces*, *Stemphylium*, *Trichophyton* and *Ustilago*. Among these, *Alternaria, Cladosporium*, *Penicillium* and *Aspergillus*, are classically considered the most relevant fungal genera that cause allergic diseases, including allergic asthma/rhinitis, ABPA or hypersensitivity pneumonitis.

From a traditional point of view, the diagnosis of fungal allergy has been based on a very well defined clinical history of the patient, on *in vivo* tests that include a skin prick test (SPT) and inhalation challenge, and on an *in vitro* test to detect and quantify the presence in serum of specific IgE or IgG antibodies directed against the fungal allergenic source. However, this practice involves certain unresolved issues. Many times, the accuracy and correlation of the results from the *in vivo* and *in vitro* assays are not in concordance due to the variability of the fungal extracts used. There is no standardisation in the production of these extracts (Esch 2004; Mari et al. 2003). The source of proteins could be the mycelia or the spores; therefore, the protein expression patterns will be different, which affects the allergenic composition. Moreover, the processes of growing the fungus and extracting the proteins dramatically affect the composition of each extract. The variability in the conditions results in a great variability of the content and proportion of each allergen in the extracts. All of the above may explain why there was no concordance between the *in vivo* and *in vitro* results in many cases.

Nevertheless, these problems with fungal extracts are being overcome by the use of recombinant allergens. The production of these proteins in the laboratory offers several advantages over the production of fungal extracts. On one hand, the recombinant protein preparations are more homogeneous, reproducible and easier to standardise for use as reagents in biological and immunological tests. On the other hand, the recombinant allergens used for diagnostic purposes may be used as clinical and prognostic markers; these allergens can be used as markers of clinical severity as well as to differentiate among co-sensitisation, co-exposure and cross-reactivity. This differentiation is important because the primary sensitising moulds must be known for successful immunotherapy.

Aerobiology: Fungal Allergenic Sources Associated with Allergy

Allergens are biomolecules, primarily proteins that can be found in very different substrates and often from biological origins. These substrates and/or biological material constitute the allergenic sources. The most important and common allergenic sources are the following: mites and mite by-products; pollens and pollen by-products; fungi and fungal by-products; epithelia and epithelial by-products; foods and food by-products; insects and their by-products; and drugs.

Among the above-mentioned allergenic sources, fungi are, without any doubt, one of the three main causes of respiratory allergy. The kingdom Fungi comprises a highly heterogeneous group of living organisms, including eukaryotic heterotrophic organisms that have either a unicellular (yeast-like) or pluricellular (a branched tubular structure of individual units, hyphae, that form the mycelium) vegetative structure and that reproduce via spores (from sexual and/or asexual origin). The most important roles of these organisms are the decomposition and removal of organic matter, parasitism (phyto- and zooparasitism), mutualism (e.g., lichens or mycorrhizae) and mycotoxin production.

Simon-Nobbe et al. (2008), in their excellent review, reported approximately eighty allergenic genera of a fungal nature. Studies of the allergic diseases caused by most of the species included in these genera mainly contain data concerning sensitisation in atopic patients, and only a limited number of genera and species have been studied in depth. *Alternaria*, *Aspergillus*, *Cladosporium* and *Penicillium* belonging to Ascomycota and *Malassezia* belonging to Basidiomycota are the genera included in this restricted group of allergenic fungi, for which not only the identification as an allergenic source and the prevalence of sensitisation but also the allergenic composition, the role in the clinical development of symptoms and the ability to interact with the immune system have been reported upon exhaustively (Breitenbach et al. 2002; Kurup 2005).

Moulds occur in both outdoor and indoor environments, and moulds grow on substrates with very different natures, including non-organic surfaces. The size of airborne fungal spores ranges from 2–3 μm (*Cladosporium, Aspergillus* and *Penicillium*) up to 160 μm (*Helminthosporium*), while some species such as *Alternaria longissima* contain larger spores (500 μm) (Ingold and Hudson 1993). The spore concentration in the atmosphere is high and ranges from 200 spores/m^3 to 10^6 spores/m^3. This concentration represents between 100 to 1,000 times more than pollen concentrations (Lacey 1981; Burge 1989; D'Amato et al. 1995).

The daily analysis of airborne spores shows a quantifiable relationship between each fungal taxon in the atmosphere and allows the production of an airborne calendar of the different allergenic sources found at each place. Each location is associated with a particular panel of airborne spores, depending on the climate, flora, ecological conditions, etc., and this local calendar is an important tool that helps us in defining the panel of allergen extracts used to evaluate allergic sensitisation. With this in mind and as an example, the data obtained in Barcelona, Spain, during the last ten years shows the presence of fungal spores belonging at least to 26 different genera (*Agaricus, Agrocybe, Alternaria, Arthrinium; Aspergillus, Chaetomium, Cladosporium, Coprinus, Curvularia, Drechslera, Epicoccum, Fusarium, Ganoderma, Helminthosporium, Leptosphaeria, Nigrospora, Penicillium, Pithomyces, Pleospora, Polytricium, Puccinia, Stemphylium, Tilletia, Torula, Ustilago*, Venturiaceae and Xylariaceae). However, only a few genera have maximums of more than 100 spores/m^3 (*Agaricus*: 350 spores/m^3; *Alternaria*: 180 spores/m^3; *Aspergillus/Penicillium*: 450 spores/m^3; *Cladosporium*: 4,500 spores/m^3; *Coprinus*: 500 spores/m^3; and *Ustilago*: 750 spores/m^3). The periodic measurements reveal a high variability among the spore concentrations of each genera, and thus the genus *Alternaria* had a maximum airborne spore concentration of 180 spores/m^3 during the period 1994–2013, but the mean value in the same period was 60 spores/m^3 (http://www.lap.uab.cat/aerobiologia/es/). This data can be extended to other studies with similar objectives, which show the data associated with each location.

The composition of fungi growing inside homes depends on the outdoor fungal panel and the species that grow indoors, and the latter are influenced by humidity, ventilation, the content of biologically degradable material, and the presence of pets, plants and carpets (Dharmage et al. 1999). *Alternaria* spores are found in the atmosphere of many different locations as one of the predominant spore types (Newson et al. 2000; Sanchez and Bush 2001; Rizzi-Longo et al. 2009), and several studies performed worldwide indicate that sensitivity to moulds is a primary risk for asthma and that *Alternaria alternata* is the mould species responsible for the highest percentage of mould sensitivities (Gergen and Turkeltaub 1992; Peat et al. 1993; Halonen et al. 1997; Das and Gupta-Bhattacharya 2012), even though most individuals sensitised to moulds respond to several mould species (Postigo et al. 2011).

Although the number of spores in the environment is high, much more than the pollen grains, the occurrence of fungal sensitisation through the airborne counts is low, and the correlation between the number of spores and mould allergy is weak. Therefore, why would the fungi produce lower rates of sensitisation than pollens? Most likely, the availability of the fungal allergens released by the spores is significantly lower than that of the allergens released by the pollen grains. Alternatively, only a few main allergens have been described from fungi compared with pollens. Among these, Alt a 1 is the main marker of fungal sensitisation and the main risk factor for asthma among the fungal allergens (Lizaso et al. 2008; Postigo et al. 2011; Feo-Brito et al. 2012). Despite the fungal major allergens with no cross-reactivity as Alt a 1 or Asp f 1 are considered the main causes of fungal sensitization other fungal allergens eliciting lower sensitization prevalence and limited cross-reactivity, also can play and important role not only in sensitization but also in the development of the allergy.

The prolonged intense exposure to fungal sources mimics the exposure to other perennial allergens, which likely contributes to both the chronicity and severity of asthma in mould- and especially *Alternaria*-sensitive subjects. The clinical development of an allergy from the components of a mycotic nature could be better understood by taking into account other minor allergens such as proteases that are common to a great variety of fungal species and that also induce inflammation and may thereby act as adjuvants in allergic sensitisation (Snelgrove et al. 2014).

Fungal Allergens

The World Allergy Organisation defines allergens as "antigens that trigger an allergy". Most of these allergens that bind IgE antibodies are proteins that are often linked to carbohydrates and are derived from a biological source. In recent decades, a large number of these allergens have been studied from a molecular point of view; the structural and functional characteristics of these allergens have been examined to answer the question "what makes an allergen an allergen?" (Stadler and Stadler 2003; Breiteneder and Radauer 2004; Ferreira et al. 2004; Jenkins et al. 2005; Radauer and Breiteneder 2006; Jenkins et al. 2007). Radauer et al. (2008) confirmed that allergens are distributed among two hundred protein families and that 16% of these families include 50% of the known allergens (Fig. 1). This data suggests that the narrow spectrum of biological functions of the allergens might explain the allergenic profile of these proteins.

There are more than 70,000 fungal species recorded in the literature, but only one hundred genera have been described as allergenic sources (Simon-Nobbe et al. 2008).

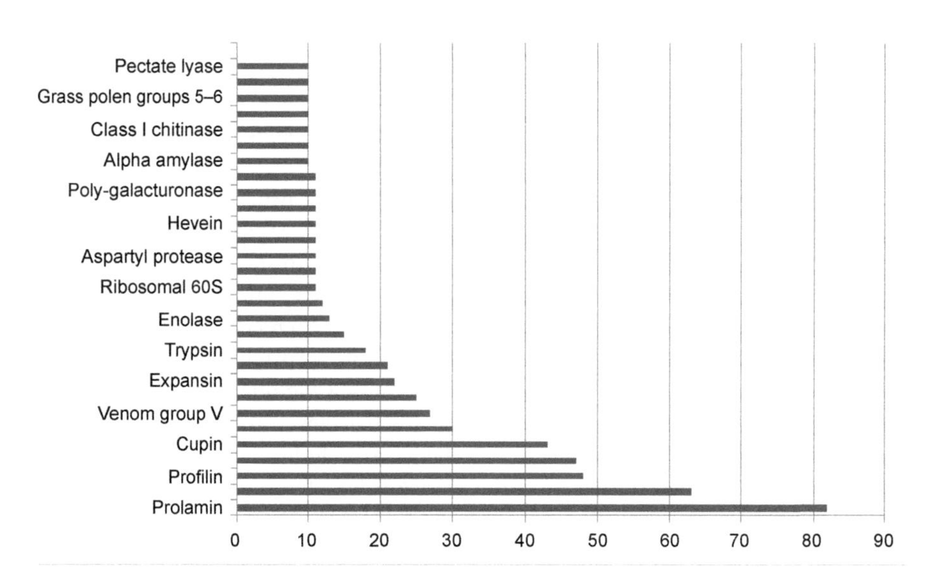

Figure 1. Number of Allergens Belonging to Each Family Protein.

Simon-Nobbe et al. (2008) listed approximately two hundred fungal allergens belonging to 24 fungal genera, and the Nomenclature Subcommittee of the International Union of Immunological Societies (IUIS; www.allergen.org) describes 108 allergens, of which 85 belong to the Ascomycota phylum and 23 belong to the Basidiomycota phylum.

Among the Ascomycota associated with allergic disorders, *Alternaria alternata* is one of the most frequently described species (D'Amato et al. 1997; Zureik et al. 2002; Mari et al. 2003). The major allergen of this species, Alt a 1, is recognised by up to 98% of *A. alternata*-sensitised patients (Asturias et al. 2005), and sensitivity to this allergen was shown to be a risk factor for life-threatening asthma (Lizaso et al. 2008; Postigo et al. 2011; Feo-Brito et al. 2012). The function of Alt a 1 has been recently studied, and this protein seems to act as a competitive inhibitor of the thaumatin-like proteins (Gómez-Casado et al. 2014). Taking into account that several members of the thaumatin protein family display significant *in vitro* inhibition of hyphal growth and sporulation by various fungi, the Alt a 1 could be a relevant fungal protein involved in the fitopathogenicity of *Alternaria* members (Gómez-Casado et al. 2014). Previously, the molecular structure of Alt a 1 was described (Chruszcz et al. 2012), and it was demonstrated that Alt a 1 conforms to a new type of protein family found exclusively in fungi. Sáenz-de-Santamaría et al. (2006) and Martínez et al. (2006) demonstrated that Alt a 1 is expressed by both *Alternaria* and other members of the Pleosporaceae family, including the allergenic species *Stemphylium botryosum* and *Ulocladium botrytis*. Their results showed that Alt a 1 must be considered a species- and family-specific allergen, and Alt a 1 must be used as the main marker for evaluating sensitisation to fungal allergenic species belonging to the Pleosporaceae family.

In addition to Alt a 1, the enolase Alt a 6 from the allergen array described for *Alternaria* also has a high diagnostic value. Several authors demonstrated that this allergen can play an important role in the phenomenon of cross-reactivity among several fungi and among species that belong to different phyla (Simon-Nobbe et al. 2000; Wagner et al. 2000). However, this protein can be considered to be a minor allergen because Alt a 6 is only able to sensitise approximately 15–22% of patients who are allergic to moulds (Simon-Nobbe et al. 2000; Wagner et al. 2000). Moreover, a high proportion of mould-allergic patients seem to be poly-sensitised to several mould species (Horst et al. 1990).

Postigo et al. (2011) evaluated the diagnostic value of Alt a 1 and enolase using the resolved-component diagnosis of allergy in a group of 30 patients who had been defined by a cutaneous and *in vitro* test as only allergic to *Alternaria*. The results revealed that the 23% of the analysed patients were indeed multi-sensitised and that enolase played an important role in the cross-reactivity phenomenon. However, the authors concluded that the panel of allergens used in the study was not sufficient for the accurate diagnosis of a fungal allergy. Among the allergens described for *Alternaria* (Table 1), at least nine have been identified in closely related moulds such as *Cladosporium herbarum*. Other *Alternaria* allergens, including heat shock protein 70, enolase, aldehyde dehydrogenase and glutathione-S-transferase, have been described in both fungal and non-fungal species (Simon-Nobbe et al. 2008). Recently, a new cross-reactive allergen from *Alternaria* has been included in the IUIS allergen list. The new allergen is a manganese-dependent superoxide dismutase (MnSOD) that is highly homologous to the MnSOD of *Aspergillus fumagitus*, Asp f 6, an allergen considered

Table 1. Fungal Allergens of Ascomycota According to their Biological Function.

Biological Function	**Fungal species**							
	Alternaria alternata	***Aspergillus fumigaus***	***A. niger***	***A. oryzae***	***Candida albicans***	***Cladosporium cladosporoides***	***Cl. herbrum***	***Curvularia lunata***
Unknown	Alt a 1 Alt a 2 Alt a 9	Asp f 4 Asp f 7	Asp n 4	Asp o 4 Asp o 7			Cla h 1 Cla h 2	
Heat shock protein 70	Alt a 3							
Disulfide isomerase	Alt a 4							
Ribosomal protein P2	Alt a 5	Asp f 8		Asp o 8			Cla h 5	
Enolase	Alt a 6	Asp f 22					Cla h 6	Cur l 2
YCP4 protein	Alt a 7						Cla h 7	
Mannitol-dehydrogenase	Alt a 8						Cla h 8	
Aldehyde-dehydrogenase	Alt a 10						Cla h 10	
Acid ribosomal protein P1	Alt a 12	Asp f 26	Asp n 26	Asp o 26			Cla h 12	
Glutathione-S-transferase	Alt a 13						Cla h 13	
Manganese superoxide-dismutase	Alt a 14	Asp f 6						Cur l 6
Alkaline serine protease		Asp f 13		Asp o 13				Cur l 1
Alternaria allergen	Alt a 15							
Mitogillin family		Asp f 1						
Peroxysomal protein		Asp f 3	Asp n 3		Can a 3			
Metalloprotease		Asp f 5		Asp o 5				
Aspartate protease		Asp f 10						
Peptidyl-prolyl isomerase		Asp f 11						
Heat shock protein P90		Asp f 12						
Vacuolar serine protease		Asp f 18	Asp n 18	Asp o 18		Cla c 9	Cla h 9	Cur l 4
L3 ribosomal protein		Asp f 23						
Cyclophilin		Asp f 27						
Thioredoxin		Asp f 29						

Table 1. contd....

Table 1. contd.

Biological Function	**Fungal species**							
	Alternaria alternata	***Aspergillus fumigaus***	***A. niger***	***A. oryzae***	***Candida albicans***	***Cladosporium cladosporoides***	***Cl. herbrum***	***Curvularia lunata***
PhiA cell wall protein		Asp f 30						
β-xylosidase			Asp n 14					
3-phytase B			Asp n 25					
TAKA-amylase A				Asp o 21				
Alcohol dehydrogenase					Cand a 1			
Peroxisomal membrane protein A					Cand a 2			
Transaldolase						Cla c 14		
N-acetyl-glucosaminidase								
Calreticulin								
Elongation factor 1 β								
Catalase								
Pectate-lyase								
Extracellular alkaline Mg-dependent exo-desoxy-ribonuclease								
Putative secreted alkaline protease Alp1								
Cerato-platanin		Asp f 15						
Glycosyl-hydrolase		Asp f 16 Asp f 9						
Galacto-mannoprotein		Asp f 17						
Fibrinogen binding protein		Asp f 2						
Cell wall protein		Asp f 34						
Catalase			Asp n 30					
Alpha-amylase				Asp o 21				
Cytochrome C								Cur l 3

Biological Function	***Fungal species***							
	Epicoccum purpurascens	***Fusarium culmorum***	***F. proliferatum***	***Penicillium brevicompactum***	***P. chrysogenum***	***P. citrinum***	***Stachybotrys chartarum***	***Trichophyton rubrum***
Unknown		Fus c 3						
Heat shock protein 70						Pen c 19		
Disulfide isomerase								
Ribosomal protein P2		Fus c 1						
Enolase						Pen c 22		
YCP4 protein								
Mannitol-dehydrogenase								
Aldehyde-dehydrogenase								
Acid ribosomal protein P1				Pen b 26				
Glutathione-S-transferase	Epi p 13							
Manganese superoxide- dismutase						Pen c 6		
Alkaline serine protease	Epi p 1			Pen b 13	Pen ch 13	Pen c 13		Tri r 4
Mitogillin family								
Peroxysomal protein						Pen c 3		
Metalloprotease								
Aspartate protease								
Peptidyl-prolyl isomerase								
Heat shock protein P90								
Vacuolar serine protease					Pen ch 18			
L3 ribosomal protein								
Cyclophilin								
Thioredoxin		Fus c 2						
PhiA cell wall protein								
β-xylosidase								
3-phytase B								

Table 1. contd....

Biological Function	***Fungal species***							
	Epicoccum purpurascens	***Fusarium culmorum***	***F. proliferatum***	***Penicillium brevicompactum***	***P. chrysogenum***	***P. citrinum***	***Stachybotrys chartarum***	***Trichophyton rubrum***
TAKA-amylase A								
Alcohol dehydrogenase								
Peroxisomal membrane protein A								
Transaldolase			Fus p 4		Pen ch 35			
N-acetyl-glucosaminidase					Pen ch 20			
Calreticulin					Pen ch 31			
Elongation factor 1 β						Pen c 24		
Catalase						Pen c 30		
Pectate-lyase						Pen c 32		
Extracellular alkaline Mg-dependent exo-desoxy-ribonuclease							Sta c 3	
Putative secreted alkaline protease Alp1								Tri r 2
Cerato-platanin								
Glycosyl-hydrolase								
Galacto-mannoprotein								
Fibrinogen binding protein								
Cell wall protein								
Catalase								
Alpha-amylase								
Cytochrome C								

Table 2. Fungal Allergens of Basidiomycota According to their Biological Function.

	Fungal species					
Biological Function	***Coprinus comatus***	***Malassezia furfur***	***Malassezia sympodialis***	***Psilocybe cubensis***	***Rhodotorula mucilaginosa***	***Schizophyllum commune***
Unknown	Cop c 3 Cop c 5 Cop c 7		Mala s 1 Mala s 5 Mala s 7 Mala s 8 Mala s 9	Psi c 1		
Heat shock protein 70			Mala s 10			
Enolase				Rho m 1		
Manganese superoxide- dismutase			Mala s 11			
Peroxysomal protein		Mala f 2 Mala f 3				
Vacuolar serine protease					Rho m 2	
Cyclophilin				Psi c 2		
Thioredoxin	Cop c 2		Mala s 13			
Leucine zipper protein	Cop c 1					
Mitochondrial malte dehydrogenase		Mala f 4				
Glucose-methanol-choline oxidoreductase			Mala s 12			
Glucoamylase						Sch c 1

to be a marker of ABPA (Postigo et al. 2011). The clinical relevance of this allergen must be studied, but this MnSOD could have important implications in *Alternaria* sensitisation as a risk factor for the development of ABPA (Jubin et al. 2010).

Aspergillus is an allergenic mould belonging to Ascomycota. This mould is frequently found growing in decaying vegetation or indoors (Kurup et al. 2000; Terr 2004) releasing large quantities of spores, which can reach the terminal airways via inhalation (Kurup and Banerjee 2000; Vijay and Kurup 2004). In some cases, large clusters of spores are deposited in the upper respiratory tract (Kurup 2003), causing disorders ranging from allergic rhinitis, sinusitis or asthma to hypersensitivity pneumonitis and ABPA (Terr 2004).

The IUIS lists 31 allergens belonging to five species of *Aspergillus*, including *A. flavus*, *A. versicolor*, *A. niger*, *A. oryzae* and *A. fumigatus*. Among these species, *A. fumigatus* is implicated in approximately 80% of *Aspergillus*-related infections (Simon-Nobbe et al. 2008). Due to its high prevalence, a large number of allergens from *A. fumigatus* have been cloned, characterised and purified as recombinant allergens (Crameri and Blaser 1996; Crameri et al. 2006).

The main major allergens of *A. fumigatus* are Asp f 1 and Asp f 3 (Table 1). The first, Asp f 1, is a non-glycosylated ribotoxin of 18 kDa that is recognised by 85% of the ABPA patients as well as asthmatic patients with a positive cutaneous test to this mould (Arruda et al. 1990). Arruda et al. (1992) demonstrated that Asp f 1 was abundantly released after germination of the spores and during the early phases of fungal growth; therefore, Asp f 1 was considered to be a virulence factor promoting the colonisation and infection of human tissue. Kurup et al. (1998) analysed the B- and T-cell epitopes of this allergen and demonstrated that the C-terminal region (aa 115–149) was involved in both humoral and cell-mediated immune responses in ABPA, whereas the T-cell epitopes were localised in aa 45–65 and aa 106–125. This allergen is unique and has not been found to share sequence homology with any other known allergen (Bowyer and Denning 2007).

The second main allergen, Asp f 3, is a 19 kDa protein that binds IgE in 72% of the ELISA-determined cases, and it is known that this allergen shares common IgE-binding epitopes with peroxisomal membrane proteins from *Candida boidini* (Hemmann et al. 1997). The clinical relevance of Asp f 3 in *Aspergillus* disorders was demonstrated long ago (Hemmann et al. 1998). The analysis of the B- and T-cell epitopes by Ramachandran et al. (2002) demonstrated the presence of 12 amino acids located at the N-terminus and 8 at the C-terminus that are critical for IgE binding.

Other allergens from *Aspergillus* have been described, including Asp f 2, Asp f 4 and Asp f 6 (Kurup and Banerjee 2000; Vijay and Kurup 2004), and are available as recombinant proteins (Crameri and Blaser 1996). Asp f 4 and Asp f 6 are intracellular proteins; thus, the availability of these proteins as aeroallergens is unlikely. It seems that the process of sensitisation to these allergens occurs when the mould is growing inside the lungs; therefore, some authors have concluded that the use of these two markers is sufficient to allow a precise diagnosis of ABPA (Hemmann et al. 1999).

As can be observed in the annexed tables, a large number of fungal allergens have been described and this number is expected to increase in the coming years. Currently, special attention is being given to several allergens that have protease activity. The biological function of these allergens seems to be closely related to the initiation of the sensitisation process and the pathogenicity of fungal aeroallergens. These proteases act directly on the pulmonary epithelium and disrupt the cells, thereby eliciting robust and rapid inflammation, which allows the massive influx of other allergens and induces the exacerbation of asthma (Snelgrove et al. 2014). These proteases are founded in several fungal genera and it is thought that they are important cross-reactivity allergens.

Diagnosis of Fungal Allergy

Fungi are an important source of airborne allergens associated with asthma and allergic rhinitis. The specific diagnosis of fungal allergy is made by combining aerobiological studies, a clinical history, complementary tests (mainly the determination of specific and total IgE) and a careful analysis of the symptoms in relation to the causative agents, the allergens.

However, to date, the diagnosis of an allergy to fungi has been difficult in many cases and inefficient due to various reasons, such as the high variety of fungal species, the difficulty in species identification, the complexity and variability of the allergen extracts of fungal origin and the lack of quality and standardisation of the most available fungal extracts (Martínez et al. 1994).

Among the patients allergic to moulds, there are commonly a large proportion of individuals sensitised to several fungal species (Zureik et al. 2002; Simon-Nobbe et al. 2008; Postigo et al. 2011). This situation complicates the final diagnosis of these patients. The diagnosis of these cases by implementing individualised allergens has gained increased attention.

The application of genomics and proteomics to the study of allergenic proteins has allowed the development of new diagnostic tools that make the diagnosis of the allergy more effective. The new models are based on the identification of the individual molecules that are involved at the beginning and during the development of the allergenic phenomenon.

At the beginning of the 21st century, Valenta et al. (1999) described the basis of the molecular diagnosis of allergy, which involves using a classical technique to measure the specific IgE levels (ImmunoCAP ®), wherein the complete allergen extracts (allergen sources) coupled to the solid phases are replaced with individualised allergens, either in their native form or as recombinant proteins.

A few years later, the microarray technique was applied to the CRD concept. Small slides spotted with very small quantities of allergens allowed the molecular diagnosis of allergy against hundreds of allergens using a minimal quantity of serum (Jahn-Schmid et al. 2003). This new concept (CDR) has allowed the individualised profiles of sensitisation in the allergic patient to be precisely defined and subsequently associated with the clinical expression of the allergy.

Similarly, concepts such as cross-reactivity, multiple or primary sensitisation, the prediction of the severity of the allergic reactions and the association of the molecular profiles of sensitisation with the development of clinical symptoms are now easier to understand and evaluate, whereas these parameters were previously difficult to assess (Valenta and Dietrich 2002; Casquete et al. 2009; Knol and Knulst 2010).

Currently, there are no doubts regarding the diagnostic effectiveness of replacing allergenic sources with individual allergens. By applying a molecular diagnosis, we can overcome several problems with the standardisation of allergenic extracts, including the variability of the fungal strains, the variability in the batch-to-batch production, the choice of the fungal structure to use as raw material, the type of culture and technology used to prepare the allergen extracts and the autodegradation of the extract once obtained.

A large number of fungal allergens have been identified (Table 1). However, the lack of availability of large quantities of these allergens for widespread use limits their diagnostic application.

The use of CRD and, more specifically, the miniaturised version of CRD, which can simultaneously evaluate a patient's sensitisation against hundreds of allergens belonging to different Phyla, allow us to associate the molecular sensitisation profiles with different stages of the allergic phenomenon and to find molecular markers that predict the severity of the allergic reactions, molecular markers of cross-reactivity and molecules able to differentiate between sensitisation and allergy (Kurup et al. 2005; Ebo et al. 2010a; Ebo et al. 2010b; Nicolaou et al. 2010).

Out of more than 70,000 fungal species described in the literature, Simon-Nobbe et al. (2008) only refer to somewhat less than a hundred genera that have been described as allergenic sources. The same authors report regarding the identification of somewhat more than 200 individualised allergens that belong to 24 genera of fungi (Simon-Nobbe et al. 2008).

In view of this data, we may ask ourselves whether all sources of fungal allergens are equally relevant in the diagnosis of a fungal allergy. Are these 200 fungal allergens enough for the efficient diagnosis of a fungal allergy?

Studies on aerobiology, the cutaneous reaction to fungal allergens and allergenic characterisation, suggest that the minimum number of species that will ensure an acceptable diagnosis of an allergy to fungi would include the following: *A. alternata, Aspergillus fumigatus, Cladosporium herbarum, Epicoccum nigrum, Fusarium roseum, Penicillium crisogenum, Candida* spp. and *Malassezia* spp. (Crameri et al. 2006). Recently Postigo et al. (2011) has stated that *Curvularia lunata* should be another species to be included in the panel of essential fungi used for diagnosis. Additionally, if we consider the results of studies conducted in different geographical areas, we should also include *Trichophyton* spp. (Mari et al. 2003), *Helminthosporium* spp., *Trichoderma* spp. and *Aureobasidium* spp. (Zureik et al. 2002). According to this data, if we look for more relevant species described in the literature or in new studies developed in other geographical areas, it would be very likely that we would have to add new fungal allergenic sources to the above-mentioned panel.

However, taking into account the ratio between the individualised allergens and their allergenic sources, we can assert that the 200 described fungal allergens belong to approximately 20 genera (Simon-Nobbe et al. 2008), and that most of these allergens are related to each other because the allergens belong to the same protein family. Therefore, these allergens could form an allergic unit which would belong to the same group. Thus, from the allergens referred to by Simon-Nobbe et al. (2008), the number of allergens homologous to Alt a 1 described in other species but phylogenetically related to *A. alternata* is approximately fifty.

Alt a 1 is most likely the best allergen to define the diagnosis of a respiratory allergy caused by moulds in our environment (Postigo et al. 2011; Portnoy et al. 2008;

Bartra et al. 2009) and is a very useful tool as a marker of asthma severity (Feo-Brito et al. 2012). An allergen such as fungal enolase is a panallergen and is associated with cross-reactivity. This allergen could explain part of the apparent multiple sensitisations commonly observed in this type of clinic (Postigo et al. 2011). In the same way, it can be observed in Table 1 that there are more allergens belonging to unrelated species, including serine proteases, peroxisomal proteins and the ribosomal proteins P1 and P2, which could be considered to be markers of cross-reactivity between non-phylogenetically related species. These proteins may be responsible for the apparent multi-sensitisations that are frequently observed in fungal allergies (Simon-Nobbe et al. 2008; Horst et al. 1990).

Conclusions

For a long time, the diagnosis of fungal allergy has been difficult due to the complexity of molds and the variability in the allergenic composition of the fungal extracts used. However, nowadays, using biological tools including the genomic and proteomic, we are able to identify new individual allergens belonging to different fungal sources that have been associated with a different diagnostic value in the diagnosis of fungal allergy. Some allergens have been identified as markers of the risk to suffer asthma such as Alt a 1 and the others are implicated in cross-reactivity and/or poly- and co-sensitivity reactions such as the enolase. Currently, it remains necessary to identify which proteins are implicated in the cross-reactivity phenomenon. Using the above-mentioned tools, the genomic and proteomic, we are able to obtain these allergens as individualised molecules. The availability of these allergens would make it possible to study the role of each molecule in the diagnosis and prognosis of a mould allergy, thereby providing a solution for poly-sensitisation and/or cross-reactivity.

The future of the molecular diagnosis of a fungal allergy is moving in two main directions: (1) the identification of new fungal allergens, their structure and bioloogical activities, including markers of both primary sensitisation and cross-reactivity; and (2) the establishment of the relationship between each allergen and the expression of the different clinical profiles that occurs in mold allergy.

References

Arruda, L.K., T.A. Platts-Mills, J.W. Fox and M.D. Chapman. 1990. *Aspergillus fumigatus* allergen I, a major IgE-binding protein, is a member of the mitogillin family of cytotoxins. J Exp Med. 172: 1529–1532.

Arruda, L.K., B.J. Mann and M.D. Chapman. 1992. Selective expression of a major allergen and cytotoxin, Asp f I, in *Aspergillus fumigatus*. Implications for the immunopathogenesis of *Aspergillus*-related diseases. J Immunol. 149: 3354–3359.

Asturias, J.A., I. Ibarrola, A. Ferrer, C. Andreu, E. Lopez-Pascual, J. Quiralte, F. Florido and A. Martinez. 2005. Diagnosis of *Alternaria alternate* sensitization with natural and recombinant Alt a 1 allergens. J Allergy Clin Immunol. 115: 1210–1217.

Bartra, J., J. Belmonte, J.M. Torres-Rodriguez and A. Cistero-Bahima. 2009. Sensitization to *Alternaria* in patients with respiratory allergy. Frontiers in Biosciences 14: 3372–3379.

Bowyer, P. and D.W. Denning. 2007. Genomic analysis of allergen genes in *Aspergillus* spp. The relevance of genome to everyday research. Med Mycol. 45: 17–26.

Breitenbach, M., R. Crameri and S. Lehrer. 2002. Fungal Allergy and Pathogenicity. Krager AG, Basel.

Breiteneder, H. and C. Radauer. 2004. A classification of plant food allergens. J Allergy Clin Immunol. 113: 821–380.

Burge, H.A. 1989. Airborne allergenic fungi classification, nomenclature, and distribution. Immunol Allergy Clin North Am. 9: 307–319.

Bush, R.K. and J.M. Protnoy. 2001. The role and abatement of fungal allergens in allergic diseases. J Allergy Clin Immunol. 107(3 Suppl): S430–440.

Bush, R.K., J.M. Portnoy, A. Saxon, A.I. Terr and R.A. Wood. 2006. The medical effects of mold exposure. J Allergy Clin Immunol. 117: 326–333.

Casquete, E., T. Rosado, I. Postigo, R. Perez, M. Fernandez, H.E. Torres and J. Martinez. 2009. Contribution to molecular diagnosis of allergy to the management of pediatric patients with allergy to pollen. J Invest Allergol Clin Immunol. 19: 439–445.

Chruszcz, M., M.D. Chapman, T. Osinski, R. Solberg, M. Demas, P.J. Porebski, K.A. Majorek, A. Pomés and W. Minor. 2012. *Alternaria alternata* allergen Alt a 1: a unique β-barrel protein dimmer found exclusively in fungi. J Allergy Clin Immunol. 130: 241–247.

Corey, J.P., S. Kaiseruddin and A. Gungor. 1997. Prevalence of mold-specific immunoglobulins in a Midwestern allergy practice. Otolaryngol Head Neck Surg. 117: 516–520.

Crameri, R. 1998. Recombinant *Aspergillus fumigatus* allergens: from the nucleotide sequences to clinical applications. Int Arch Allergy Immunol. 115: 99–114.

Crameri, R. and K. Blaser. 1996. Cloning *Aspergillus fumigatus* allergens by the pJuFo filamentous phage display system. Int Arch Allergy Immunol. 110: 41–45.

Crameri, R., M. Weichel, S. Flückiger, A.G. Glaser and C. Rhyner. 2006. Fungal allergies: a yet unsolved problem. Chem Immunol Allergy 91: 121–133.

D'Amato, G. and F.T. Spieksma. 1995. Aerobiologic and clinical aspects of mould allergy in Europe. Allergy 50: 870–877.

D'Amato, G., G. Chatzigeorgiou, R. Corsico, D. Gioulekas, L. Jäger, S. Jäger, K. Kontou-Fili, S. Kouridakis, G. Liccardi, A. Meriggi, A. Palma-Carlos, M.L. Palma-Carlos, A. Pagan Aleman, S. Parmiani, P. Puccinelli, M. Russo, F.T. Spieksma, R. Torricelli and B. Wüthrich. 1997. Evaluation of the prevalence of skin prick test positivity to *Alternaria* and *Cladosporium* in patients with suspected respiratory allergy. A European multicenter study promoted by de subcommittee on Aerobiology and Environmental Aspects of Inhalant Allergens of the European Academy of Allergology and Clinical Immunology. Allergy 52: 711–716.

D'Amato, G., L. Cecchi, M. D'Amato and G. Liccardi. 2010. Urban air pollution and climate change as environmental risk factors of respiratory allergy: An update. J Investig Allergol Clin Immunol. 20(2): 95–102.

Das, S. and S. Gupta-Bhattacharya. 2012. Monitoring and assessment of airborne fungi in Kolkata, India, by viable and non-viable air sampling methods. Environ Monit Assess. 184: 4671–4684.

Denning, D.W., A. Pleuvry and D.C. Cole. 2013. Global burden of allergic bronchopulmonary aspergillosis with asthma and its complication chronic pulmonary aspergillosis in adults. Med Mycol. 51: 361–370.

Dharmage, S., M. Bailey, J. Raven, T. Mitakakis, F. Thien, A. Forbes, D. Guest, M. Abramson and E.H. Walters. 1999. Revalence and residential determinants of fungi within homes in Melbourne, Australia. Clin Exp Allergy 29: 1481–1489.

Ebo, D.G., C.H. Bridts, M.M. Verweij, K.J. De Knop, M.M. Hagendorens, L.S. De Clerck and W.J. Stevens. 2010a. Sensitization profiles in birch pollen-allergic patients with and without oral allergy syndrome to apple: lessons from multiplexed component-resolved allergy diagnosis. Clin Exp Allergy 40: 339–347.

Ebo, D.G., M.M. Hagendorens, K.J. de Knop, M.M. Verweij, C.H. Bridts, L.S. De Clerck and W.J. Stevens. 2010b. Component Resolved Diagnosis from latex allergy by microarray. Clin Exp Allergy 40: 348–358.

Esch, R.E. 2004. Manufacturing and standardizing fungal allergen products. J Allergy Clin Immunol. 113: 210–215.

Feo-Brito, F., A.M. Alonso, J. Carnés, R. Martín-Martín, E. Fernández-Caldas, P.A. Galindo, T. Alfaya and M. Amo-Salas. 2012. Correlation between Alt a 1 levels and clinical symptoms in *Alternaria alternata*-monosensitized patients. J Invest Allergol Clin Immunol. 22: 154–159.

Ferreira, F., T. Hawranek, P. Gruber, N. Wopfner and A. Mari. 2004. Allergic cross-reactivity: from gene to the clinic. Allergy 59: 243–67.

Fischer, G, R. Schwalbe, M. Moller, R. Ostrowski and W. Dott. 1999. Species-specific production of microbial volatile organic compounds (MVOC) by airborne fungi from a compost. Chemosphere 39(5): 795–810.

Friedlander, S.L. and R.K. Bush. 2005. Fungal allergy. pp. 1–15. *In*: Kurup, V.P. (ed.). Mold Allergy, Biology and Pathogenesis. Research Signpost, Kerala.

Gergen, P.J. and P.C. Turkeltaub. 1992. The association of individual allergen reactivity with respiratory disease in a national sample: data from the Second National Health and Nutrition Examination Survey, 1976–80 (NHANES II). J Allergy Clin Immunol. 90: 579–588.

Gómez-Casado, C., A. Murua-García, M. Garrido-Arandia, P. González-Melendi, R. Sánchez-Monge, D. Barber, L.F. Pacios and A. Díaz-Perales. 2014. Alt a 1 from *Alternaria* interacts with PR5 thaumatin-like proteins. FEBS J. 588: 1501–1508.

Halonen, M., D.A. Stern, A.L. Wright, L.M. Taussing and F.D. Martinez. 1997. *Alternaria* as a major allergen for asthma in children raised in a desert environment. Am J Respir Crit Care Med. 155: 1356–1361.

Hemmann, S., K. Blaser and R. Crameri. 1997. Allergens of *Aspergillus fumigatus* and *Candida boidinii* share IgE-binding epitopes. Am J Respir Crit Care Med. 156: 1956–1962.

Hemmann, S., C. Ismail, K. Blaser, G. Menz and R. Crameri. 1998. Skin-test reactivity and isotypespecific immune responses to recombinant Asp f 3, a major allergen of *Aspergillus fumigatus*. Clin Exp Allergy 28: 860–867.

Hemmann, S., G. Menz, C. Ismail, K. Blaser and R. Crameri. 1999. Skin test reactivity to 2 recombinant *Aspergillus fumigatus* allergens in *A. fumigatus*-sensitized asthmatic subjects allows diagnostic separation of allergic bronchopulmonary aspergillosis from fungal sensitization. J Allergy Clin Immunol. 104: 601–607.

Horner, W.E., A. Helbling, J.E. Salvaggio and S.B. Lehrer. 1995. Fungal allergens. Clin Microbiol Rev. 8(2): 161–179.

Horst, M., A. Hejjaoui, V. Horst, F.B. Michel and J. Bousquet. 1990. Double-blind placebo-controlled rush immunotherapy with a standardized *Alternaria* extract. J Allergy Clin Immunol. 85: 460–472.

Ingold, C.T. and H.J. Hudson. 1993. The Biology of Fungi. Chapman & Hall, London.

Jahn-Schmid, B., C. Harwanegg, B. Bohle, C. Ebner, O. Scheiner and M.W. Mueller. 2003. Allergen microarray: comparison o microarray using recombinant allergens with conventional diagnostic methods to detect allergen specific IgE. Clin Exp Allergy 33: 1443–1449.

Jenkins, J.A., S. Griffiths-Jones, P.R. Shewry, H. Breiteneder and E.N. Mills. 2005. Structural relatedness of plant food allergens with specific reference to cross-reactive allergens: an *in silico* analysis. J Allergy Clin Immunol. 115: 163–170.

Jenkins, J.A., H. Breiteneder and E.N. Mills. 2007. Evolutionary distance from human homologs reflects allergenicity of animal food proteins. J Allergy Clin Immunol. 120: 1399–1405.

Jubin, V., S. Ranque, L.B.N. Stremler, J. Sarles and J.C. Dubus. 2010. Risk factors for *Aspergillus* colonization and allergic bronchopulmonary aspergillosis in children with cystic fibrosis. Pediatr Pulmonol. 45: 764–771.

Kauffman, H.F., J.F. Tomee, M.A. van de Riet, A.J. Timmerman and P. Borger. 2000. Protease-dependent activation of epithelial cells by fungal allergens leads to morphologic changes and cytokine production. J Allergy Clin Immunol. 105: 1185–1193.

Knol, E.F. and A.C. Knulst. 2010. Application of multiplexed immunoglobulin E determination on a chip in CRD in allergy. Clin Exp Allergy 40: 190–192.

Kurup, V.P. 2003. Fungal allergy. pp. 515–525. *In*: Arora, N. (ed.). Handbook of Fungal Biotechnology. Dekker, New York.

Kurup, V.P. 2005. Mold Allergy, Biology and Pathogenesis. Research Sign Post, Kerala.

Kurup, V.P. and B. Banerjee. 2000. Fungal allergens and peptide epitopes. Peptides 21: 589–599.

Kurup, V.P., B. Banerjee, P.S. Murali, P.A. Greenberger, M. Krishnan, V. Hari and J.N. Fink. 1998. Immunodominant peptide epitopes of allergen, Asp f 1 from the fungus *Aspergillus fumigatus*. Peptides 19: 1469–1477.

Kurup, V.P., H.D. Shen and B. Banerjee. 2000. Respiratory fungal allergy. Microbes Infec. 2(9): 1101–1110.

Kurup, V.P., A.P. Knutsen and R.B. Moss. 2005. *Aspergillus* antigens and immunodiagnosis of allergic bronchopulmonary aspergillosis. pp. 137–146. *In*: Kurup, V.P. (ed.). Mold Allergy, Biology and Pathogenesis. Research Signpost, Kerala.

Lacey, L. 1981. The aerobiology of conidial fungi. pp. 123–128. *In*: Cole, G.T. and B. Kendrick (eds.). Biology of Conidial Fungi. Academic Press, New York.

Lehrer, S.B., M. Lopez, B.T. Butcher, J. Olson, M. Reed and J.Z.E. Salvaggio. 1986. Basidiomycete mycelia and spore-allergen extracts: skin test reactivity in adults with symptoms of respiratory allergy. J Allergy Clin Immunol. 78(3): 478–485.

Lizaso, M.T., A.I. Tabar, B.E. García, B. Gómez, J. Algorta, J.A. Asturias and A. Martínez. 2008. Double-blind, placebo-controlled *Alternaria alternata* immunotherapy: *in vivo* and *in vitro* parameters. Pediatr Allergy Immunol. 19: 76–81.

Lopez, M. and J.E. Salvaggio. 1985. Mold-sensitive asthma. Clin Rev Allergy 3: 183–196.

Mari, A., P. Schneider, V. Wally, M. Breitenbach and B. Simon-Nobbe. 2003. Sensitization to fungi: epidemiology, comparative skin tests, and IgE reactivity of fungal extracts. Clin Exp Allergy 33: 1429–1438.

Martínez, J., A. Martínez, G. Gutiérrez, A. Llamazares, R. Palacios and M. Sáenz de Santamaría. 1994. Influencia del proceso de obtención en la actividad alergénica y rendimiento de extractos de *Alternaria alternata*. Rev Iberoam Micol. 11: 60–83.

Martínez, J., A. Gutiérrez-Rodríguez, I. Postigo, G. Cardona and J.A. Guisantes. 2006. Variability of Alt a 1 expression by different strains of *Alternaria alternata*. J Invest Allergol Clin Immunol. 16: 279–282.

Newson, R., D. Strachan, J. Corden and W. Millington. 2000. Fungal and other spore counts as predictors of admissions for asthma in the Trent region. Occup Environ Med. 57: 786–792.

Nicolaou, N., P. Poorafshar, C. Murray, A. Simpson, H. Winell, G. Kerry, A. Härlin, A. Woodcock, S. Ahlstedt and A. Custovic. 2010. Allergy or tolerance in children sensitized to peanut: Prevalence and differentiation using CRD. J Allergy Clin Immunol. 125: 191–197.

Peat, J.K., E. Tovey, C.M. Mells, S. Leeder and A.J. Woolcock. 1993. Importance of house dust mite and *Alternaria* allergens in childhood asthma: an epidemiological study in two climatic regions of Australia. Clin Exper Allergy 23: 812–820.

Portnoy, J.M., C.C.S. Barnes and K. Kennedy. 2008. Importance of mold allergy in asthma. Curr Allergy Asthma Rep. 8: 81–87.

Postigo, I., A. Gutiérrez-Rodríguez, J. Fernández, J.A. Guisantes, E. Suñén and J. Martínez. 2011. Diagnostic value of Alt a 1, fungal enolase and manganese-dependent superoxide dismutase in the component-resolved diagnosis of allergy to pleosporaceae. Clin Exper Allergy 41: 443–451.

Pulimood, T.B., J.M. Corden, C. Bryden, L. Sharples and S.M. Nasser. 2007. Epidemic asthma and the role of the fungal mold *Alternaria alternata*. J Allergy Clin Immunol. 120: 610–617.

Radauer, C. and H. Breiteneder. 2006. Pollen allergens are restricted to few protein families and show distinct patterns of species distribution. J Allergy Clin Immunol. 117: 141–147.

Radauer, C., M. Bublin, S. Wagner, A. Mari and H. Breiteneder. 2008. Allergens are distributed into a few protein families and possess a restricted number of biochemical functions. J Allergy Clin Immunol. 121: 847–852.

Ramachandran, H., V. Jayaraman, B. Banerjee, P.A. Greenberger, K.J. Kelly, J.N. Fink and V.P. Kurup. 2002. IgE binding conformational epitopes of Asp f 3, a major allergen of *Aspergillus fumigatus*. Clin Immunol. 103: 324–333.

Rizzi-Longo, L., M. Pizzulin-Sauli and P. Ganis. 2009. Seasonal occurrence of *Alternaria* (1993–2004) and *Epicoccum* (1994–2004) spores in Trieste (NE Italy). Ann Agric Environ Med. 16: 63–70.

Sáenz de Santamaría, M., I. Postigo, A. Gutiérrez-Rodríguez, G. Cardona, J.A. Guisantes, A. Asturias and J. Martínez. 2006. The major allergen of *Alternaria alternata* (Alt a 1) is expressed in other members of the Pleosporaceae family. Mycoses 49: 91–95.

Sanchez, H. and R.K. Bush. 2001. A review of *Alternaria alternata* sensitivity. Rev Iberoam Micol. 18: 56–59.

Simon-Nobbe, B., G. Probst, A.V. Kajava, H. Oberkofler, M. Susani, R. Crameri, F. Ferreira, C. Ebner and M. Breitenbach. 2000. IgE-binding epitopes of enolases, a class of highly conserved fungal allergens. J Allergy Clin Immunol. 106: 887–895.

Simon-Nobbe, B., U. Denk, V. Pöll, R. Rid and M. Breitenbach. 2008. The spectrum of fungal allergy. Int Arch Allergy Immunol. 145: 58–86.

Snelgrove, R.J., L.G. Gregory, T. Peiró, S. Akthar, G.A. Campbell, S.A. Walker and C.M. Lloyd. 2014. *Alternaria*-derived serine protease activity drives IL-33 mediated asthma exacerbations. J Allergy Clin Immunol. doi: 10.1016/j.jaci.2014.02.002.

Sprenger, J.D., L.C. Altman, C.E. O'Neil, G.H. Ayars, B.T. Butcher and S.B. Lehrer. 1988. Prevalence of basidiospore allergy in the Pacific Northwest. J Allergy Clin Immunol. 82: 1076–1080.

Stadler, M.B. and B.M. Stadler. 2003. Allergenicity prediction by protein sequence. FASEB J. 17: 1141–1143.

Tariq, S.M., S.M. Matthews, M. Stevens and E.A. Hakim. 1996. Sensitization to *Alternaria* and *Cladosporium* by the age of 4 years. Clin Exp Allergy 26: 794–798.

Terr, A.I. 2004. Are indoor molds causing a new disease? J Allergy Clin Immunol. 113: 221–226.

Valenta, R. and K. Dietrich. 2002. From allergen structure to new forms of allergen-specific immunotherapy. Curr Opin Immunol. 14: 718–727.

Valenta, R., J. Lidholm, V. Niederberger, B. Hayek, D. Kraft and H. Gronlund. 1999. The recombinant allergen based concept of component-resolved diagnostics and immunotherapy (CRD and CRIT). Clin Exp Allergy 29: 896–904.

Vijay, H.M. and V.P. Kurup. 2004. Fungal allergens. Clin Allergy Immunol. 18: 223–249.

Vijay, H.M., M. Abebe and V.P. Kurup. 2005. *Alternaria* and *Cladosporium* allergens and allergy. pp. 51–67. *In*: Kurup, V.P. (ed.). Mold Allergy, Biology and Pathogenesis, Research Signpost, Kerala.

Wagner, S., H. Breiteneder, B. Simon-Nobbe, M. Susani, M. Krebitz, B. Niggemann, R. Brehler, O. Scheiner and K. Hoffmann-Sommergruber. 2000. Hev b 9, an enolase and a new cross-reactive allergen from *Hevea* latex and molds. Purification, characterization, cloning and expression. Eur J Biochem. 267: 7006–7014.

Zureik, M., C. Neukirch, B. Leynaert, R. Liard, J. Bousquet and S. Neukirch. 2002. Sensitisation to airborne moulds and severity of asthma: cross sectional study from European Community respiratory health survey. BMJ 325: 411–418.

SECTION V

Novel Diagnostic Methods, Susceptibility Testing and Miscellaneous Mycoses

CHAPTER 14

MALDI-TOF MS: A Rapid and New Approach in Fungal Diagnosis and Susceptibility Testing

Mehmet Ali Saracli

Introduction

Fungal infections can cause high morbidity and mortality, particularly in immunocompromised patients. Identification of the infecting fungal strain and determination of any potential resistance is important for optimal treatment (Nolla-Salas et al. 1997; Denning et al. 2003; Garey et al. 2006). Timely initiation of optimal therapy is required to successfully treat these infections and additionally it also reduces the overall cost of hospital care (Pappas et al. 2009; Iriart et al. 2012).

Fungal identification using conventional methods has several disadvantages, the most important being the extended time period that may stretch from 2 days to several weeks before an identification is obtained. In addition it requires specialized and experienced laboratory personnel. The reliability of identification based on biochemical and morphological (macro and microscopic) features is sometimes low and it is also costly (Ferreira et al. 2013). To cope with the disadvantages of conventional phenotypic methods, many modern genomic approaches including DNA sequencing of target ribosomal RNA genes and various other housekeeping genes have emerged as the 'gold standard' for molecular characterization of fungi over the last 20 years (Weile and Knabbe 2009; Sibley et al. 2012). The genomic approach however is not easily applicable for routine characterization and identification of microorganisms (Posteraro et al. 2013). In contrast to the genomic and transcriptomic studies, proteomics is based

Gulhane Military Medical Academy, Department of Microbiology, 01018 Etlik Ankara, Turkey.
E-mail: saraclima@yahoo.com

on the systematic analysis of the proteome, the protein complement expressed by a genome. Proteins are directly related to function (or phenotype) (Bhadauria et al. 2007).

Many protein separation methods coupled with various mass spectrometry technologies have evolved in recent years as the dominant tools in the field of protein identification and peptide mass fingerprinting (Figeys et al. 2001). Matrix-assisted laser desorption ionization-time of flight mass spectrometry (MALDI-TOF MS) is one of the most common types of "soft ionization" mass spectrometry used to perform peptide mass fingerprinting (Kim et al. 2007). The problems arising from low volatility and thermal instability of proteins are circumvented by the evolution of ionization techniques and thus the most characteristic biomarkers became accessible to the MALDI-TOF MS analysis of intact microorganisms, without any extraction, separation or amplification processes (Fenselau and Demirev 2001). MALDI-TOF MS has made it possible to volatilize and ionize large biomolecules, such as peptides and proteins. A protein-based spectral profile or "fingerprint" is generated that is unique for a given species (Buchan and Ledeboera 2013; Posteraro et al. 2013). Protein fingerprints can be obtained easily with minimal efforts and costs by MALDI-TOF MS, and can be analyzed efficiently by using software (Freiwald and Sauer 2009). The 2002 Nobel Prize for chemistry was awarded to Koichi Tanaka for the use of MALDI-TOF MS with biological macromolecules (Posteraro et al. 2013). Since then, MALDI-TOF MS has become an important analytical method to detect biomarker profiles, and contributes to diagnosis in many medical fields such as cancer, rheumatoid arthritis, Alzheimer's disease, and allergy through the identification of specific biochemical markers (Marvin et al. 2003). It has also emerged as a promising new method for the identification of microorganisms including fungi. It is an efficient diagnostic tool because of its potential for high-throughput, low operating costs and universal, flexible, specific and rapid methodology (Tan et al. 2012; Pinto et al. 2011; Kliem and Sauer 2012; Lagace-Wiens et al. 2012). It was designed as the "molecular-phenotypic-based revolutionary technique" in 2009 (Seng et al. 2009). It has been applied increasingly in many disciplines of microbiology, excluding viruses due to their relatively low protein content (Dubois et al. 2012; Kliem and Sauer 2012).

Principle of MALDI-TOF MS

A typical MALDI-TOF MS device consists of an ion source, a detector and a mass analyzer (Kim et al. 2007; Posteraro et al. 2013). In short, the method enables desorption/ionization of either whole cells or fragmented proteins from microbial cells that have been embedded in an excess of small, acidic molecules, known as the 'matrix'. The matrix is usually an aromatic compound or a weak acid. When the cells in the matrix are excited with a laser beam, the matrix strongly absorbs the laser energy at the wavelength of the laser which leads to ionization of the cellular components such as proteins, nucleic acids and ions (Sauer and Kliem 2010; La Scola 2011). In practice, biological samples are applied with the matrix on a 'target' plate that can be automatically introduced into the MALDI-TOF mass spectrometer. The target plate is either disposable or reusable. While disposable ones can be discarded in the regular laboratory waste, the reusable ones can be cleansed with chemical and mechanical steps. A quick cleaning protocol with 5 min of 70% ethanol and subsequent mechanical cleaning with detergent and cloth is sufficient for regular workup (Wieser et al. 2012). The sample on the target plate is exposed to brief laser pulses resulting in energy transfer from the matrix to the nonvolatile analyte molecules. Short laser pulses excite the crystallized matrix containing analyte molecules and causes ionization and desorption (removal) of embedded analyte from the target plate into the gas phase. The ionized and vaporized biomolecules are accelerated before they enter the flight tube through an electrostatic field, created by a potential of about 20 kV. They cross the instrument in a perfect vacuum flight tube environment in a pulsed manner based on their mass to charge (m/z) ratio so that heavier molecules do not reach the detector before the lighter ones. The degree of ionization as well as the mass of the proteins determines their individual time-of-flight (TOF) meaning the time required for the particles to reach the detector. The TOF of a particle correlates with its m/z ratio. Because the charge value is 1 in most cases, m/z value is equal to mass of the ionized molecule. The data generated in the detector is processed and a final mass spectrum of peaks is produced. The spectrum for species level identification of microorganism usually contains peaks in a range of 2000–20,000 m/z (or 2–20 kDa). The spectra in this range are very stable and accurate, and are only minimally influenced by growth conditions of the microorganism. Each plate should be calibrated and validated with a control strain before analyzing the samples. Comparison of the test spectrum with the spectra of well-characterized microorganisms stored in a reference database is the final step. The reference spectra are species-specific 'fingerprints' and are highly reproducible. The measured spectra may have method-inherent noise and they are never exactly identical for another isolate in the database. A similarity score between spectrum of the test strain and the reference spectra in database is produced by using software. The similarity score is a numerical value based on the similarities between the observed and stored data sets (Kim et al. 2007; La Scola 2011; Wieser et al. 2012; Croxatto et al. 2012; Buchan and Ledeboera 2013; Posteraro et al. 2013).

The level of reliability for the identification is classified based on breakpoints described by the manufacturer; such as "reliable", "low discrimination", "genus level", "species level", or "unidentified". The Saramis software of Vitek MS (bioMerieux, France) uses a standard confidence score ranging between 0 and 100%. Values above 90% are coded in green and considered to be "reliable". Scores between 85% to 90% are coded in yellow, whereas scores below 85% are coded in white. On the other hand, the Bruker Biotyper (Bruker Daltronics, Germany) uses a scoring method ranging from 0 to 3.0. Confidence scores above 2.0 are acceptable enough for species level identification, scores between 1.7 and 2.0 are considered genus level identification, and scores below 1.7 are considered unreliable identification (Buchan and Ledeboera 2013). However, some authors reported that lowering the recommended identification threshold by Bruker Biotyper from 2.0 to 1.8 increased the percentage of correct species level identification from 87% to 92% to above 99%, without changing the identification accuracy of 100% (Stevenson et al. 2010; Dhiman et al. 2011).

One of the commercial mass spectrometer systems available today for clinical microbiology applications is the MALDI BioTyper by Bruker Corporation (Bruker Daltronics, Germany) which includes hardware, software, and database. The Axima mass spectrometer by Shimadzu Corporation (Japan) uses Launchpad software and

Saramis database (Anagnos Tec GmbH), which was acquired by bioMérieux (France) and has been redeveloped and is known as VITEK MS. The third system is the Andromas system (France). The Andromas provides a different type of database and software for routine bacteriology compatible with Bruker and Shimadzu hardware. However, the list may change because this field is evolving rapidly. Both the Biotyper and the Saramis databases contain many mass spectra of bacterial, yeast and mold species. The method by which the spectra are produced by each system is somewhat different. The Bruker system concentrates on a maximum of 100 peaks exceeding a minimum signal-to-noise ratio, while the Shimadzu system focuses on peaks shared by a minimum number of strains to build a fingerprint called the SuperSpectrum. On the other hand, the Andromas SAS (Paris, France) software running on both Bruker and Shimadzu instruments uses species-specific peaks with a higher intensity than a predefined threshold (Emonet et al. 2010; Bille et al. 2012).

The limiting factor of MALDI-TOF MS based microbial identification is coverage of the reference library used. When there is no reference spectrum in the database for a given genus or species, MS may fail in identifications of that genus or species and will produce low scores or unacceptable identifications (Buchan and Ledeboera 2013). Other important reasons for failure to identify or misidentification are insufficient representation of intraspecies variability by multiple reference spectra in database, especially for new and emerging taxa (Putignani et al. 2011; Mancini et al. 2013) and improper sample preparation (Theel et al. 2011; Alanio et al. 2011). Working with fresh cultures can therefore, significantly improve outcome. Inclusion of not only a wide variety of different species but also several strains of each species in databases is important in constructing a spectra database because of strain-to-strain variability within a species and the output is higher-quality identification (Hettick et al. 2008; Pinto et al. 2011; Buchan and Ledeboera 2013). For that reason, MALDI-TOF MS methodology for fungal analyses will definitely involve a continuous updating of current commercial databases provided by vendors as a part of building up new databases for specialized research purpose in future. In a study, a higher rate of correct identifications at the species level was obtained when an in-house-extended database was used instead of a commercial database (100% versus 84.3%) (Mancini et al. 2013). In another study, with the inclusion of 60 successfully identified clinical *Candida* spp. into the original commercial database, the identification rates at species level of 347 clinical isolates of 6 species were increased from 91.6% to 99.1% (Yang et al. 2014). However, the unsystematic and unregulated enrichment of databases could generate misidentifications due to improper entries into the database (Theel et al. 2011). To avoid such errors, the Food and Drug Administration (FDA) requires that *in vitro* diagnostic (IVD) systems need to be closed to the user, and that the addition of strains/species must be cleared by FDA prior to inclusion into the IVD database in the USA (Westblade et al. 2013). However, laboratories may have a research-use-only database in addition to the commercial IVD database.

Speed is an important feature of MALDI-TOF MS based species identification. While conventional definitive identification of a yeast strain can take more than 2 days, a single strain can be identified at the species level in 5–7 min by MALDI-TOF MS, and it takes approximately 20 min per sample including the ethanol-formic acid sample preparation procedure (van Veen et al. 2010; McTaggart et al. 2011;

Wieser et al. 2012). Results of a target plate containing 96 isolates can be obtained in about 1 hr starting from the time-point when the first sample is loaded onto the plate. It was reported in a study of 3480 yeasts including *Candida albicans*, *non-albicans Candida* spp., *Cryprococcus neoformans* and *S. cerevisiae* that the isolates could be identified approximately in an hour by MALDI-TOF MS including sample preparation while it took 24–72 hr by conventional methods (Pinto et al. 2011). When compared to conventional identification methods, the included strains *Candida albicans* (n:52) other *Candida* sp. (n:56), and other yeasts (n:8) were identified by MALDI-TOF MS 0.04, 1.93, and 3.75 days earlier, respectively (Tan et al. 2012).

MALDI-TOF MS is also a cheaper alternative method for microbial identification when the initial cost of a MALDI-TOF MS instrument ($180,000 to $250,000) is excluded. Sample preparation for MALDI-TOF MS analysis requires only minimal, inexpensive reagents that equate to approximately $0.50 per identification (Dhiman et al. 2011; McTaggart et al. 2011; Pinto et al. 2011). For a yearly comparison, the estimated cost of phenotypic conventional identification of 3480 yeasts was reported approximately to be $5200 while it was $3401 by MALDI-TOF MS. It is also anticipated that use of the MALDI-TOF MS for identification can reduce reagent and labor costs by $102, 424 or 56.9% within 12 months for moderate-to high-volume laboratories (Tan et al. 2012). Estimated time until return on investment can be as little as 2 to 3 years for the laboratories performing 20 to 30 yeast identifications daily (Buchan and Ledeboera 2013).

Sample Preparation for MALDI-TOF MS based Fungal Identification

Currently used MALDI-TOF MS techniques are nearly unaffected by culture conditions, and have a high tolerance against sample impurities such as salts and detergents (Chalupová et al. 2014; Wieser et al. 2012). Fungi can be cultured on any nutrientmedia for MALDI-TOF MS; however, it is better not to use inhibitory media such as Mycosel agar. Contrarily, fungal pigments interfere with the ionization of biomolecules (Dong et al. 2009; Sulc et al. 2009). Thus for pigment producing fungi, the use of media causing pigment production should be avoided. Additionally, the quantity of contaminant agar needs to be minimum when punching mycelium and conidia/hyphae because of production of non-specific signal in the fingerprinting profile, which potentially affects the identification (Del Chierico et al. 2012). Isolates have to be analyzed under the same experimental conditions for comparable results because MALDI-TOF MS assay is also a phenotype-based method (Dhiman et al. 2011; La Scola 2011; Posteraro et al. 2013). This means that not only the growth and sample preparation steps but also the instrumental setup for spectra acquisition should be optimized and standardized in order to get good and comparable identification results. The database should include spectra obtained under different culture conditions including media composition, incubation temperatures and duration of incubation (Alanio et al. 2011; Lau et al. 2013).

Sample preparation is probably the most important keystep of an MALDI-TOF MS assay in the whole analytical procedure (Wieser et al. 2012). The diversity of protein's abundance, molecular weight, charge, hydrophobicity, post-translational processing and modifications and complexation with other molecules prevent availability of a single effective protein extraction protocol (Bhadauria et al. 2007). Thus, both the selection of a proper matrix compound and the choice of an optimal sample preparation technique are still rather empiric and various sample preparation techniques including mechanical or chemical pretreatment steps have been introduced to increase the release of cellular proteins (Sulc et al. 2009). Better protein extraction provides a higher amount of characteristic peptides in mass spectrometric profiles and better sample identification, which also requires more time. Today, there is no one standardized protocol applicable to a broad range of microorganisms and sample types. Although transferring a small amount of intact bacterial cells from the agar plate onto the MALDI-TOF MS target plate and overlaying the cells with a small amount of matrix solution will sufficiently lyse the cells, it is not enough for fungal cells due to their stronger, more complex and rigid cell wall which contains glucans, chitin and mannoproteins (Bader 2013; Chalupová et al. 2014).

In the shortest extraction procedure for yeast, one colony is picked from the agar plate and streaked on the target plate. Then 1 μl of 70% formic acid is added. After the sample has been dried for 1 min, the spot is overlaid with matrix (α-cyano-4-hydroxycinnamic acid in 50% acetonitrile and 2.5% trifluoroacetic acid), and is allowed to dry for 1 min before being analysed. This on-plate extraction protocol takes approximately 3-min. By using this protocol, it was reported that 97. 6% of the isolates showed a correct identification in comparison to conventional identification methods and 82.8% of them had a log-score above 1.7 (Van Herendael et al. 2012). If an isolate has a log score below this threshold required for species identification, or in case of a species such as *C. neoformans* that is difficult to identify by the on-plate extraction protocol due to its rigid cell wall/capsule which is difficult to break up, a standard protocol is preferred. Briefly, one or two colonies were suspended in 70% ethanol, resuspended in 70% formic acid and followed by acetonitrile addition for protein extraction, and incubated at room temperature for protein solubilization. The debris are removed by centrifugation, and the supernatant is spotted onto the target plate. After being dried at room temperature, the spot is covered with fresh matrix solution (Bader 2013). In the standard method, the growth should be no older than 48 hr because increased cultivation time causes weaker and less distinguishable peaks in the spectra (Wieser et al. 2012). In a study on a total of 90 yeast species analyzed using the Bruker Microflex LT/SH Biotyper system, the percentage of species identified at the genus level using direct on-plate extraction with formic acid (95.6%) was higher than with tube extraction method (87.8%; $P = 0.020$), while species-level identification rates were equivalent regardless of preparatory protocol (Theel et al. 2012). The length of incubation period for yeasts before sample preparation for MALDI-TOF MS has been evaluated and it has been found that all preincubation periods of 24 hr, 48 hr, or 72 hr yielded comparable results without affecting the accuracy of the identification, which is a considerable advantage in routine practice (Goyer et al. 2012).

For filamentous fungi, extraction can be performed from spores, hyphae, or both. Despite sample preparation from prewashed spores of *Aspergillus*, *Penicillium* and *Fusarium* which seem enough when mixed on MALDI-TOF MS target with a proper matrix solution (Chalupová et al. 2014), various extraction procedures including heating, sonication, bead-beating, or chemical lysis have been used for molds with stronger cell walls (Kim et al. 2007; Hettick et al. 2008; Cassagne et al. 2011). Because the filamentous fungi are more heterogeneous than yeasts in cell structure, there isn't a common extraction protocol that can be applied to all molds.

There are various matrix solutions such as dihydroxybenzoic acid, ferulic acid, sinapinic acid, and α-cyano-4-hydroxy-cinnamic acid (α-CHCA or α-HCCA). The best signal/noise ratio and the narrowest peaks with little signal suppression were obtained in the CHCA matrix (Sulc et al. 2009), and it is the choice of matrix used preferentially for detection of protein markers (Cassagne et al. 2011; Posteraro et al. 2013; Chalupová et al. 2014). The saturated solution of α-CHCA is prepared in 50% acetonitrile and 2.5% trifluoroacetic acid. For cells that still do not lyse efficiently by the standard method, additional measures such as mechanical cell disruption by using a bead-beater can be applied (Kim et al. 2007; Bader 2013; Mancini et al. 2013).

MALDI-TOF based Identification of Fungi

Conventional mycological identification based on biochemical methods is frequently slow, and requires extensive experience especially for molds because it is based on morphological features of the isolate such as macroscopic and microscopic appearance. Since MALDI-TOF MS requires less experience, it has potential advantages over conventional identification. Additionally, the results are obtained earlier, are highly reliable and cheaper. In addition to its rapid run time, MALDI-TOF MS doesn't require that the colonies reach the morphological characteristics used for conventional identification. It can be applied when the colonies are visible on plate (Ferreira et al. 2013).

a) Yeast identification

Delayed diagnosis and therapy of invasive fungal infections contribute significantly to high mortality rates, whereas early intervention with antifungal drugs may result in more effective management of high-risk patients (Morrell et al. 2005; Murali and Langston 2009). *Candida* species are the 4th cause of nosocomial bloodstream infections and are associated with a high mortality rate (Pfaller and Diekema 2010). While *C. albicans* is still the major species associated with candidemia (Falagas et al. 2010), the frequency of isolation of non-*albicans Candida* species differs with geographical location. MALDI-TOF MS shortens the time necessary for identification as compared to use of phenotypic methods for routine identification of yeasts (Tan et al. 2012). Furthermore, closely related yeast species such as *Candida dubliniensis/albicans* cannot be discriminated easily by conventional methods (Hof et al. 2012), other similar yeast species include *Candida glabrata/bracarensis/nivariensis* (Santos et al. 2011; Pinto et al. 2011), *Candida ortho-/meta-/parapsilosis* (Hendrickx et al. 2011; Pinto et al. 2011; Sendid et al. 2012), *Candida haemulonii* complex (Cendejas-Bueno et al. 2012), or the phenotypically similar species *Candida palmioleophila* (Jensen and Arendrup 2011), *Candida famata*, and *Candida guilliermondii* cannot be differentiated by commonly used biochemical methods (Castanheira et al. 2013). A study that included 150 *C. parapsilosis*,

7 *C. metapsilosis*, and 5 *C. orthopsilosis* strains, which were identified by ITS sequencing as a reference method, showed that MALDI-TOF MS can discriminate between them even if the strains were identified with log scores under 1.7 and even if the identification was marked by the software as 'unreliable' (Hendrickx et al. 2011). The authors speculated that the reason for unreliable identification of *C. metapsilosis* and *C. orthopsilosis* strains is probably because more reference profiles of *C. parapsilosis* are available in the database provided. In another study, 264 clinical or reference strains of *Candida* and non-*Candida* yeasts including *Cryptococcus*, *Geotrichum*, *Rhodotorula* and some other genera were identified by using the Bruker Biotyper database library in comparison with DNA sequence analysis. MALDI-TOF MS could identify correctly 80% of 30 reference strains, 96% of 167 clinical isolates, and 94% of 67 prospective clinical strains with scores above 1.7, and also could distinguish between close species such as *Candida parapsilosis*, *C. orthopsilosis* and *C. metapsilosis*, and *C. glabrata*, *C. nivariensis* and *C. bracarensis*. Other important findings in the study were that the direct colony smear method used for bacterial sample preparations is not enough for yeasts for species level identification and the manufacturers described protocols for fungal cell wall disruption that yield more reliable spectra in most cases. The authors explained the mis/no identifications by variability in protein expression within species, suboptimal protein extraction, insufficient database entries for rare species, poor extraction techniques, inadequate drying of the yeast pellet or inadvertent inclusion of agar into the extract. They emphasized that lowering the score threshold for species-level identification even ≥1.70 can still be used to provide reliable identification at species level (Pinto et al. 2011).

There are few studies comparing commercial MALDI-TOF MS systems in respect of correct identification/misidentification percentages (Mancini 2013; Rosenvinge et al. 2013; Hamprecht et al. 2014). Although some authors reported better performance for the Bruker's MALDI-TOF MS system, this is probably related to the extent of the databases provided by the manufacturers and will be improved in advance by inclusion of new spectra. Thus, comparative studies need to mention version information of the databases compared. There are also speed differences reported among commercial systems. In a study, VitekMS (bioMérieux) was found to be slower than Microflex LT Biotyper (Bruker) (19.8 vs. 8.0 min for 10 samples) but had the advantage of a more effective direct transfer protocol with less need for an additional extraction step (Hamprecht et al. 2014). Lohmann et al. (2013) compared the performances of the Biflex III-Biotyper (Bruker Daltonics) and the Axima (Shimadzu)-SARAMIS (AnagnosTec) systems for the identification of 312 yeasts isolated from clinical specimens (including *Candida*, *Saccharomyces*, *Rhodotorula*, *Cryptococcus*, *Trichosporon*, *Pichia*, *Geotrichum* and *Sporopachydermia* genera). They found that both systems for yeast identification were good and comparable under routine clinical conditions, despite their differences in sample preparation and spectrum analysis methods and database contents. However, the identification performance was lower for genera *Cryptococcus* and *Trichosporon* and for *Geotrichum capitatum* probably because of the relatively low representation of these strains in the two databases, or due to their cell wall structure which makes protein extraction more difficult (Lohmann et al. 2013). Similarly, Pinto et al. (2011) also observed technical difficulties with extraction for *Cryptococcus* spp. requiring repeat analysis (Pinto et al. 2011). There are differences in epidemiology, virulence, and antifungal susceptibility characteristics of *C. neoformans* and *C. gattii* isolates. Thus, identification of *Cryptococcus neoformans* and *C. gattii* at species or even at subspecies level is important (Kwon-Chung and Bennet 1984; Trilles et al. 2012). By using tube extraction method, *C. neoformans* and *C. gattii* species, eight major molecular types of their species complex (*C. neoformans* VNI to VNIV and *C. gattii* VGI to VGIV), and AD hybrids could be distinguished by MALDI-TOF (Firacative et al. 2012; McTaggart et al. 2011). Beside *Candida* and *Cryptococcus* species, MALDI-TOF was found successful in identification of arthroconidial yeasts species, including basidiomycetous (*Trichosporon* and *Guehomyces* spp.) and ascomycetous ones (*Galactomyces*, *Geotrichum*, *Saprochaete*, and *Magnusiomyces* spp.) (Kolecka et al. 2013).

b) Mold identification

Routine mold identification relies mostly on phenotypic methods such as macroscopic and microscopic observation of growth by highly skilled mycologists. DNA sequence analysis of rRNA or other protein-coding genes are used for identification of phenotypically indistinguishable strains (Cassagne et al. 2011). MALDI-TOF MS assay of fungal microorganisms began around the year 2000 via first experiments with mould species (Welham et al. 2000; Chalupová et al. 2014). However, there is less data available for the differentiation of molds than yeasts. Sample preparation from molds is more difficult than yeasts because of their more complicated morphology and stronger cell walls which prevent proper extraction. Additionally, factors such as the existence of various growth forms of molds (mycelium and conidia), the presence/absence of sporulation, agar contaminants in the analyte due to vegetative hyphae sampling and degree of maturation make variations in numbers and identities of mass peaks obtained and cause difficulties in spectral analysis (Wieser et al. 2012; Bader 2013). Thus, optimization in growth conditions, protein extraction procedure and analytical steps of the assay are needed to enhance the performance of MALDI-TOF MS-based identification for routine diagnosis of molds. In order to avoid variability of spectra due to culture age, a definitive growth time should be employed for each genus or species. Otherwise, early stage spectra could result in unreliable proteomic phenotyping profiles when compared with mature ones (Del Chierico et al. 2012). Additionally, the cultivation of some fungi such as dermatophytes takes a relatively longer time, thus limiting the potential benefits of MALDI-TOF MS (Wieser et al. 2012). Furthermore, the presence of fungal pigments in molds such as *Fusarium* spp. or *Aspergillus niger* may inhibit ionization of the analyte and cause lack of diagnostic spectra. In all probability, the pigment molecules directly compete with the organic acid matrix for irradiating photons (Buskirk et al. 2011). Pigment production can be inhibited by use of liquid cultures, or by culturing on media supplemented with an agent blocking melanin synthesis. Pigment can also be removed by preanalytical washing steps (Dong et al. 2009). Protein extracts can be prepared by scraping mycelial cells directly from agar plates or from liquid cultures. Although additional attempts for cell disruption such as bead-beating (Hettick et al. 2008; Buskirk et al. 2011), sonication (Sulc et al. 2009) or boiling of the samples can be applied (Cassagne et al. 2011), MALDI-TOF MS analysis can be performed for some filamentous fungi without prior protein extraction that excluded the formic acid lysis (Iriart et al. 2012), or following a tube extraction step (Verwer et al. 2014).

MALDI-TOF MS based Identification of Fungi Directly from Blood Culture

Currently, blood culture is the method of choice for the diagnosis of blood stream infections (Murray et al. 1999). When the blood culture medium yields a growth, it is usually subcultured on plates in order to identify the infecting microorganism (Spanu et al. 2012). Definitive identification of yeasts at species level from a positive blood culture by using conventional phenotypic methods requires a minimum 48 hours, and sometimes it may take several days (Lavergne et al. 2013). Fluorescence *in situ* hybridization (Gherna et al. 2009), oligonucleotide array hybridization (Hsiue et al. 2010), and multiplex PCR (Chang et al. 2001) have been evaluated for fungal identification directly from blood culture bottles. However, these methods require expertise, and are often time-consuming (Spanu et al. 2012). In this respect, MALDI-TOF MS is a promising alternative in rapid identification of fungal pathogens in blood cultures and to initiate specific antifungal therapy earlier (Lavergne et al. 2013).

The most important problem in MALDI-TOF MS assay direct from blood cultures are spectral contaminants that can interfere with microbial peaks. Especially, the highly abundant substances such as charcoal and cations other than H+ available in blood, and blood components such as erythrocytes, leukocytes, and serum proteins, including hemoglobin, that can mask protein peaks from the microorganism. Thus, the concentration of the microorganism in the sample should be increased above identification capacity of MALDI-TOF system, and these spectral contaminants need to be eliminated (La Scola 2011; Bader 2013; Buchan and Ledeboera 2013). Additional to these overlapping interferant peaks, growth of multiple species in the same culture bottle can cause low identification scores or ambiguous results (Bader 2013). To circumvent these problems various methods such as the use of multiple centrifugation steps (Spanu et al. 2012), membrane filtration (Fothergill et al. 2012), saponine forced microbial release (Ferroni et al. 2010), several washing steps with distilled water (Marinach-Patrice et al. 2010), low concentrations of detergents such as SDS (Marinach-Patrice et al. 2010) or Tween 80 (Spanu et al. 2012) have been used. Performance of MALDI-TOF MS based fungal identification has been evaluated in spiked cultures, in patient cultures, or both. Reported correct identification percentages of MALDI-TOF MS assay varied between 96% and 100% for *Candida albicans* strains in studies (Ferroni et al. 2010; Spanu et al. 2012; Lavergne et al. 2013; Pulcrano et al. 2013). However, non-*albicans Candida* species have lower correct identification percentages (Spanu et al. 2012). Lower identification score values were observed mainly with *C. guillermondii*, *C. krusei*, *C. lusitaniae*, and *C. lipolytica* isolates (Spanu et al. 2012; Pulcrano et al. 2013). The main weaknesses of the method are its poor performance in polyfungal infections (Spanu et al. 2012), and in cases where the yeast cells existed lower than 10e6 CFU/ml in positive culture bottles (Pulcrano et al. 2013).

MALDI-TOF MS based fungal identification directly from blood cultures significantly reduces the time of diagnosis at species level because the results are available in 30 minutes and physicians can expect to receive species-level ID data for *Candida* isolates within 24 hr after the blood culture is drawn (Spanu et al. 2012; Lavergne et al. 2013).

MALDI-TOF MS based Identification of Microorganisms Directly from Urine Samples

As described before, microorganisms can easily be identified in pure cultures on agarose plates by using MALDI-TOF MS system and with application of some additional purification steps from positive blood cultures. However, MALDI-TOF MS based identification of microorganisms directly from patient samples is limited to urine. Clinical materials except urine require a culture based enrichment step because they are most often rich in host proteins and sometimes are contaminated by normal flora. They include proteins originating from host or microorganisms in a patient's flora, which can mask spectra of infecting agents. Clinical samples from body regions without a normal flora with limited host proteins are good candidates for direct identification of infecting agents by using MALDI-TOF. In this respect, urine is a good candidate as it does not contain normal flora and there are minimal host proteins in the sample. The best results in urine samples are obtained with Gram negative organisms with a bacterial count above 10e5 CFU/ml (Ferreira et al. 2010). Despite the fact that MALDI-TOF MS based identification of bacterial pathogens directly in urine samples provides a rapid alternative method, continuing necessity for antibitotic susceptibility testing by conventional methods is a reality limiting the use of MALDI TOF based microbial identification in urine samples (Wieser et al. 2012).

Antifungal Susceptibility Testing by Using MALDI-TOF MS

Conventional antifungal susceptibility testing is based on measuring growth of a microorganism in the presence of different drug concentrations to determine the minimum inhibitory concentration. Despite the fact that reference phenotypic methods are robust and reproducible, they require additional time following the initial subculture of the organisms, and are not objective enough due to variations in end-point readings (Revankar et al. 1998). Rapid initiation of appropriate antifungal therapy, however, is crucial to reducing mortality (Morrell et al. 2005; Taur et al. 2010). Given these limitations, there is a clear need for the development of an equally robust methodology, with faster turn-around times and endpoints determination that is objective.

As an important antifungal drug class, major resistance mechanisms to azoles include alteration of antifungal targets on ergosterol biosynthesis pathway and overexpression of efflux pumps (CDR and MDR). Similarly, echinocandin resistance is also mediated by target mutations (glucan synthase) (Pfaller 2012). Since MALDI-TOF MS assay allows comparison of reference spectra deposited in a well-characterized library (Dhiman et al. 2011), variations between proteome composition of resistant and susceptible strains can be detected reproducibly by MALDI-TOF MS when the strains are challenged with antifungal drugs. However, the difference in protein expression level, or the amount of the modified protein should be abundant enough to be discriminated from other "background" proteins expressed in each cell. This technology has rarely been applied as a tool for antifungal drug susceptibility testing (Rogers et al. 2006; Marinach et al. 2009; Vella et al. 2013). Marinach et al. (2009) compared the protein profiles of *C. albicans* strains cultured at various fluconazole concentrations with its absence, and termed the lowest drug concentration at which changes in the protein spectrum becomes statistically significant as "the minimum profile change concentration". They reported that this concentration was in 94% agreement within a one doubling dilution range with broth microdilution method (Marinach et al. 2009). By using the same methodology, more than 94% categorical agreement between the MALDI-TOF MS and reference broth microdilution methods was observed for *C. albicans*, *C. glabrata*, *C. krusei*, and *C. parapsilosis* isolates in response to caspofungin in another study. They also reported a reduction in incubation time required for valid results to as little as 15 hr (de Carolis et al. 2012).

An alternative method was described by Vella et al. (2013). They compared composite correlation index (CCI) values of the spectra at "breakpoint" concentration of caspofungin with concentrations of 0 and 32 μg/ml. The tested isolates were classified as susceptible or resistant to caspofungin if the CCI values of the spectra at breakpoint and 32 μg/ml were respectively higher or lower than the CCI values of the spectra at breakpoint and 0 μg/ml. By using this method, they showed that discrimination of the susceptible and resistant isolates of *Candida albicans* was possible after a 3-hr incubation in the presence of "breakpoint" level drug concentrations of caspofungin (Vella et al. 2013). It should be remembered that timely species identification of the infecting strain might be also an important measure because the susceptibility pattern to antifungal agents varies depending on the genus and the species (Lohmann et al. 2013). For example, correct species identification of *Candida orthopsilosis*, *C. metapsilosis* and *C. parapsilosis* by MALDI-TOF MS may have a therapeutic impact, because these species have significant differences among the antifungal susceptibility profiles, and are phenotypically indistinguishable (Quiles-Melero et al. 2012). Thus, MALDI-TOF MS makes an important contribution in decreasing mortality and morbidity rates through early initiation of appropriate therapy without using it as a tool for susceptibility testing (Tan et al. 2012).

Other Applications of MALDI-TOF MS

Patient or environmental samples isolated from outbreaks can be analysed by MALDI-TOF MS. Because MALDI-TOF MS can show differences between protein spectra of individual strains easily, it is a cheap and quick alternative in differentiation of isolates below the species level when compared to application of molecular methods in phylogenetic and epidemiological analysis (Pulcrano et al. 2012; Padovan et al. 2013). In a study, MALDI-TOF MS was evaluated on a fungemic outbreak of 19 *C. parapsilosis* strains isolated from a neonatal intensive care unit. It was found useful in monitoring the spread of strains, identification of similarities among the isolates and studying microevolutionary changes in the population in comparison with molecular genotypic methods (Pulcrano et al. 2012).

MALDI-TOF MS is also a good alternative in taxonomic studies for revision of species complexes and discrimination of the species within the complexes, as shown for *Candida rugosa* complex (Padovan et al. 2013).

MALDI-TOF MS can show differences between close species and discriminate easily the more virulent species. *Cryptococcus neoformans* and *Cryptococcus gattii* are pathogenic basidiomycetous yeasts causing meningoencephalitis in human hosts. *Cryptococcus neoformans* has two varieties, var. *grubii* and var. *neoformans*. *Cryptococcus neoformans* var. *grubii* includes the molecular types VNI and VNII, var. *neoformans* includes type VNIV, AD hybrids include type VNIII and *C. gattii* includes the types VGI, VGII, VGIII, and VGIV. As molecular types of *C. neoformans* and *C. gattii* have differences in *in vitro* susceptibility to antifungal agents, especially azoles, its genotype is important in deciding on the treatment options for cryptococcosis (Chong et al. 2010). Although molecular methods are the method of choice in genotyping of *C. neoformans* and *C. gattii* strains, MALDI-TOF MS can be applied as a practical tool in differentiation and genotyping these closely related species (Chong et al. 2010; Iqbal et al. 2010). In a study, spectrum analysis of 72 *C. neoformans* and 10 *C. gattii* strains by MALDI-TOF MS yielded same molecular types for 81 (98.8%) of the isolates as in agreement with the DNA-based typing results (Posteraro et al. 2013).

Another study on paracoccidioidomycosis showed that some extracellular proteins were important for virulence and pathogenesis of *Paracoccidioides*, the causative agent of paracoccidioidomycosis, by using MALDI-TOF MS in combination with two-dimensional electrophoresis (Weber et al. 2012).

Conclusion

Automated identification systems are being used increasingly in clinical microbiology laboratories in order to improve workflow, decrease expenses and mistakes made by laboratory workers. Although MALDI-TOF MS has only recently been introduced into the field of microbiological diagnostics, it's rapidly increasing IVD-approved strain database points to a promising future (Wieser et al. 2012). MALDI-TOF MS will certainly improve routine mycological diagnostics with further studies on drawbacks, validation and standardization of sample preparation methods, extending robust databases, and inclusion of new applications (Posteraro et al. 2013).

One of its major advantages for identification of fungal species is its high discriminatory power, accuracy and superiority over morphological and/or biochemical tests. Another major advantage is a shorter turnaround time, which is approximately 10 min, rather than hours or days required for genotypic or phenotypic characterizations (Posteraro et al. 2013). MALDI-TOF MS is a cost-effective alternative for identification of fungal isolates because of its inexpensive reagents and high identification rates as compared to current methods when the initial cost of a MALDI-TOF MS instrument is excluded (Dhiman et al. 2011; McTaggart et al. 2011; Pinto et al. 2011). MALDI-TOF MS is also a useful tool for outbreak investigations, epidemiological studies, research on virulence and taxonomic and phylogenetic evaluations (Pulcrano et al. 2012; Padovan et al. 2013).

However, there are several drawbacks associated with the MALDI-TOF MS. One of the current limitations of the MALDI- TOF system is the inability to detect pathogens directly from patient material with the exception of urine, and from growth-positive blood cultures. Additionally, there are only limited works about its use to determine antifungal drug susceptibility although strain identification at species level may have implications for antifungal therapy (Posteraro et al. 2013). Finally, it is important to note that the capability of MALDI-TOF MS to successfully identify a wide variety of isolates is directly related to the extent of the reference library used for comparison (Buchan and Ledeboera 2013).

References

Alanio, A., J.L. Beretti, B. Dauphin, E. Mellado, G. Quesne, C. Lacroix, A. Amara, P. Berche, X. Nassif and M.E. Bougnoux. 2011. Matrix-assisted laser desorption ionization time-of-flight mass spectrometry for fast and accurate identification of clinically relevant *Aspergillus* species. Clin Microbiol Infect. 17: 750–755.

Bader, O. 2013. MALDI-TOF-MS-based species identification and typing approaches in medical mycology. Proteomics 13: 788–799.

Bhadauria, V., W.S. Zhao, L.X. Wang, Y. Zhang, J.H. Liu, J. Yang, L.A. Kong and Y.L. Peng. 2007. Advances in fungal proteomics. Microbiol Res. 162: 193–200.

Bille, E., B. Dauphin, J. Leto, M.E. Bougnoux, J.L. Beretti, A. Lotz, S. Suarez, J. Meyer, O. Join-Lamber, P. Descamps, N. Grall, F. Mory, L. Dubreuil, P. Berche, X. Nassif and A. Ferroni. 2012. MALDI-TOF MS Andromas strategy for the routine identification of bacteria, mycobacteria, yeasts, *Aspergillus* spp. and positive blood cultures. Clin Microbiol Infect. 18(11): 1117–1125.

Buchan, B.W. and N.A. Ledeboera. 2013. Advances in identification of clinical yeast isolates by use of matrix-assisted laser desorption ionization–time of flight mass spectrometry. J Clin Microbiol. 51(5): 1359–1366.

Buskirk, A.D., J.M. Hettick, I. Chipinda, B.F. Law, P.D. Siegel, J.E. Slaven, B.J. Green and D.H. Beezhold. 2011. Fungal pigments inhibit the matrix-assisted laser desorption/ionization time-of-flight mass spectrometry analysis of darkly pigmented fungi. Anal. Biochem. 411: 122–128.

Cassagne, C., S. Ranque, A. Normand, P. Fourquet, S. Thiebault, C. Planard, M. Hendrickx and R. Piarroux. 2011. Mould routine identification in the clinical laboratory by Matrix-Assisted Laser Desorption Ionization Time-Of-Flight Mass Spectrometry. PLoS ONE 6(12): e28425.

Castanheira, M., L.N. Woosley, D.J. Diekema, R.N. Jones and M.A. Pfaller. 2013. *Candida guilliermondii* and other species of *Candida* misidentified as *Candida famata*: Assessment by the Vitek 2, DNA-sequencing analysis and MALDI-TOF MS in two global antifungal surveillance programs. J Clin Microbiol. 51: 117–124.

Cendejas-Bueno, E., A. Kolecka, A. Alastruey-Izquierdo, B. Theelen, M. Groenewald, M. Kostrzewa, M. Cuenca-Estrella, A. Gómez-López and T. Boekhout. 2012. Reclassification of the *Candida haemulonii* complex; *C. duobushaemulonii* sp. nov. (*C. haemulonii* group II) and *C. haemulonii* var. *vulnera* var. nov.: two multiresistant human pathogenic yeasts. J Clin Microbiol. 50: 3641–3651.

Chalupová, J., M. Raus, M. Sedlářová and M. Šebela. 2014. Identification of fungal microorganisms by MALDI-TOF mass spectrometry. Biotechnol Adv. 32: 230–241.

Chang, H.C., S.N. Leaw, A.H. Huang, T.L. Wu and T.C. Chang. 2001. Rapid identification of yeasts in positive blood cultures by a multiplex PCR method. J Clin Microbiol. 39(10): 3466–3471.

Chong, H.S., R. Dagg, R. Malik, S. Chen and D. Carter. 2010. *In vitro* susceptibility of the yeast pathogen *Cryptococcus* to fluconazole and other azoles varies with molecular genotype. J Clin Microbiol. 48: 4115–4120.

Croxatto, A., G. Prod'hom and G. Greub. 2012. Applications of MALDI-TOF mass spectrometry in clinical diagnostic microbiology. FEMS Microbiol Rev. 36(2): 380–407.

De Carolis, E., A. Vella, A.R. Florio, P. Posteraro, D.S. Perlin, M. Sanguinetti and B. Posteraro. 2012. Use of matrix-assisted laser desorption ionization-time of flight mass spectrometry for caspofungin susceptibility testing of *Candida* and *Aspergillus* species. J Clin Microbiol. 50: 2479–2483.

Del Chierico, F., A. Masott,i, M. Onori, E. Fiscarelli, L. Mancinelli, G. Ricciotti, F. Alghisi, L. Dimiziani, C. Manetti, A. Urbani, M. Muraca and L. Putignani. 2012. MALDI-TOF MS proteomic phenotyping of filamentous and other fungi from clinical origin. J Proteomics. 75: 3314–3330.

Denning, D.W., C.C. Kibbler and R.A. Barnes. 2003. British Society for Medical Mycology proposed standards of care for patients with invasive fungal infections. Lancet Infect Dis. 3: 230–240.

Dhiman, N., L. Hall, S.L. Wohlfiel, S.P. Buckwalter and N.L. Wengenack. 2011. Performance and cost analysis of matrix-assisted laser desorption ionization-time of flight mass spectrometry for routine identification of yeast. J Clin Microbiol. 49: 1614–1616.

Dong, H., J. Kemptner, M. Marchetti-Deschmann, C.P. Kubicek and G. Allmaier. 2009. Development of a MALDI two-layer volume sample preparation technique for analysis of colored conidia spores of *Fusarium* by MALDI linear TOF mass spectrometry. Anal Bioanal Chem. 395: 1373–1383.

Dubois, D., M. Grare, M.F. Prere, C. Segonds, N. Marty and E. Oswald. 2012. Performances of the Vitek MS matrix-assisted laser desorption ionization-time of flight mass spectrometry system for rapid identification of bacteria in routine clinical microbiology. J Clin Microbiol. 50(8): 2568–2576.

Emonet, S., H.N. Shah, A. Cherkaoui and J. Schrenzel. 2010. Application and use of various mass spectrometry methods in clinical microbiology. Clin Microbiol Infect. 16(11): 1604–1613.

Falagas, M.E., N. Roussos and K.Z. Vardakas. 2010. Relative frequency of albicans and the various non-albicans *Candida* spp. among candidemia isolates from inpatients in various parts of the world: a systematic review. Int J Infect Dis. 14: e954–966.

Fenselau, C. and P.A. Demirev. 2001. Characterization of intact microorganisms by MALDI mass spectrometry. Mass Spectrom Rev. 20(4): 157–171.

Ferreira, L., F. Sanchez-Juanes, M. Gonzalez-Avila, D. Cembrero-Fucinos, A. Herrero-Hernandez, J.M. Gonzalez-Buitrago and J.L. Munoz-Bellido. 2010. Direct identification of urinary tract pathogens from urine samples by matrix-assisted laser desorption ionization-time of flight mass spectrometry. J Clin Microbiol. 48(6): 2110–2115.

Ferreira, L., F. Sánchez-Juanes, S. Vega, M. González, M.I. García, S. Rodríguez, J.M. González-Buitrago and J.L. Muñoz-Bellido. 2013. Identification of fungal clinical isolates by matrix-assisted laser desorption ionization-time-of-flight mass spectrometry. Rev Esp Quimioter. 26(3): 193–197.

Ferroni, A., S. Suarez, J.L. Beretti, B. Dauphin, E. Bille, J. Meyer, M.E. Bougnoux, A. Alanio, P. Berche and X. Nassif. 2010. Real-time identification of bacteria and *Candida* species in positive blood culture broths by matrix-assisted laser desorption ionization–time of flight mass spectrometry. J Clin Microbiol. 48(5): 1542–1548.

Figeys, D., D. Linda, L.D. McBroom and M.F. Moran. 2001. Mass spectrometry for the study of protein–protein interactions. Methods 24: 230–239.

Firacative, C., L. Trilles and W. Meyer. 2012. MALDI-TOF MS Enables the rapid identification of the major molecular types within the *Cryptococcus neoformans/C. gattii* species complex. PLoS ONE. 7(5): e37566.

Fothergill, A., V. Kasinathan, J. Hyman, J. Walsh, T. Drake and Y.F.W. Wang. 2012. Rapid identification of bacteria and yeasts from positive BacT/ALERT blood culture bottles by using a lysis-filtration method and MALDI-TOF Mass Spectrum analysis with SARAMIS database. J Clin Microbiol. 51(3): 805–809.

Freiwald, A. and S. Sauer. 2009. Phylogenetic classification and identification of bacteria by mass spectrometry. Nat Protoc. 4(5): 732–742.

Garey, K.W., M. Rege, M.P. Pai, D.E. Mingo, K.J. Suda, R.S. Turpin and D.T. Bearden. 2006. Time to initiation of fluconazole therapy impacts mortality in patients with candidemia: a multi-institutional study. Clin Infect Dis. 43: 25–31.

Gherna, M. and W.G. Merz. 2009. Identification of Candida albicans and Candida glabrata within 1.5 hours directly from positive blood culture bottles with a shortened peptide nucleic acid fluorescence *in situ* hybridization protocol. J Clin Microbiol. 47: 247–248.

Goyer, M., G. Lucchi, P. Ducoroy, O. Vagner, A. Bonnin and F. Dalle. 2012. Optimization of the preanalytical steps of MALDI-TOF MS identification provides a flexible and efficient tool for identification of clinical yeast isolates in medical laboratories. J Clin Microbiol. 50: 3066–3068.

Hamprecht, A., S. Christ, T. Oestreicher, G. Plum, V.A.J. Kempf and S. Göttig. 2014. Performance of two MALDI-TOF MS systems for the identification of yeasts isolated from bloodstream infections and cerebrospinal fluids using a time-saving direct transfer protocol. Med Microbiol Immunol. 203: 93–99.

Hendrickx, M., J.S. Goffinet, D. Swinne and M. Detandt. 2011. Screening of strains of the *Candida parapsilosis* group of the BCCM/IHEM collection by MALDI-TOF MS. Diagn. Microbiol Infect Dis. 70: 544–548.

Hettick, J.M., B.J. Green, A.D. Buskirk, M.L. Kashon, J.E. Slaven, E. Janotka, F.M. Blachere, D. Schmechel and D.H. Beezhold. 2008. Discrimination of *Aspergillus* isolates at the species and strain level by matrix-assisted laser desorption/ionization time-of-flight mass spectrometry fingerprinting. Anal Biochem. 380: 276–281.

Hof, H., U. Eigner, T. Maier and P. Staib. 2012. Differentiation of *Candida dubliniensis* from *Candida albicans* by means of MALDI-TOF mass spectrometry. Clin Lab. 58: 927–931.

Hsiue, H.C., Y.T. Huang, Y.L. Kuo, C.H. Liao, T.C. Chang and P.R. Hsueh. 2010. Rapid identification of fungal pathogens in positive blood cultures using oligonucleotide array hybridization. Clin Microbiol Infect. 16: 493–500.

Iqbal, N., E.E. DeBess, R. Wohrle, B. Sun, R.J. Nett, A.M. Ahlquist, T. Chiller, S.R. Lockhart and for the *Cryptococcus gattii* Public Health Working Group. 2010. Correlation of genotype and *in vitro* susceptibilities of *Cryptococcus* gattii strains from the Pacific Northwest of the United States. J Clin Microbiol. 48: 539–544.

Iriart, X., R.A. Lavergne, J. Fillaux, A. Valentin, J.F. Magnaval, A. Berry and S. Cassaing. 2012. Routine identification of medical fungi by the new Vitek MS matrix-assisted laser desorption Ionization–time of flight system with a new time-effective strategy. J Clin Microbiol. 50(6): 2107–2110.

Jensen, R.H. and M.C. Arendrup. 2011. *Candida palmioleophila*: characterization of a previously overlooked pathogen and its unique susceptibility profile in comparison with five related species. J Clin Microbiol. 49: 549–556.

Kim, Y., M.P. Nandakumar and M.R. Marten. 2007. Proteomics of filamentous fungi. Trends Biotechnol. 25(9): 395–400.

Kliem, M. and S. Sauer. 2012. The essence on mass spectrometry based microbial diagnostics. Curr Opin Microbiol. 15(3): 397–402.

Kolecka, A., K. Khayhan, M. Groenewald, B. Theelen, M. Arabatzis, A. Velegraki, M. Kostrzewa, M. Mares, S.J. Taj-Aldeen and T. Boekhout. 2013. Identification of medically relevant species of arthroconidial yeasts by use of matrix-assisted laser desorption ionization–time of flight mass spectrometry. J Clin Microbiol. 51(8): 2491–2500.

Kwon-Chung, K.J. and J.E. Bennett. 1984. Epidemiologic differences between the two varieties of Cryptococcus neoformans. Am J Epidemiol. 120: 123–130.

La Scola, B. 2011. Intact cell MALDI-TOF mass spectrometry-based approaches for the diagnosis of bloodstream infections. Expert Rev Mol Diagn. 11(3): 287–298.

Lagacé-Wiens, P.R., H.J. Adam, Karlowsky, J.A. Kimberly A. Nichol, P.F. Pang, J. Guenther, A.A. Webb, C. Miller and M.J. Alfa. 2012. Identification of blood culture isolates directly from positive blood cultures by use of matrix-assisted laser desorption ionization-time of flight mass spectrometry and a commercial extraction system: analysis of performance, cost, and turnaround time. J Clin Microbiol. 50(10): 3324–3328.

Lau, A.F., S.K. Drake, L.B. Calhoun, C.M. Henderson and A.M. Zelazny. 2013. Development of a clinically comprehensive database and a simple procedure for identification of mold s from solid media by matrix-assisted laser desorption ionization-time of flight mass spectrometry. J Clin Microbiol. 51: 828–834.

Lavergne, R.A., P. Chauvın, A. Valentin, J. Fillaux, C. Roques-Malecaze, S. Arnaud, S. Menard, J.F. Magnaval, A. Berry, A. Cassaing and X. Iriart. 2013. An extraction method of positive blood cultures for direct identification of *Candida* species by Vitek MS matrix-assisted laser desorption ionization time of light mass spectrometry. Med Mycol. 51: 652–656.

Lohmann, C., M. Sabou, W. Moussaoui, G. Prévost, J.M. Delarbre, E. Candolfi, A. Gravet and V. Letscher-Bru. 2013. Comparison between the Biflex III-Biotyper and the Axima-SARAMIS systems for yeast identification by matrix-assisted laser desorption ionization–time of flight mass spectrometry. J Clin Microbiol. 51(4): 1231–1236.

Mancini, N., E. De Carolis, L. Infurnari, A. Vella, N. Clementi, L. Vaccaro, A. Ruggeri, B. Posteraro, R. Burioni, M. Clementi and M. Sanguinetti. 2013. Comparative evaluation of the Bruker Biotyper and Vitek MS matrix-assisted laser desorption ionization–time of flight (MALDI-TOF) mass spectrometry systems for identification of yeasts of medical importance. J Clin Microbiol. 51(7): 2453–2457.

Marinach, C., A. Alanio, M. Palous, S. Kwasek, A. Fekkar, J.Y. Brossas, S. Brun, G. Snounou, C. Hennequin, D. Sanglard, A. Datry, J.L. Golmard and D. Mazier. 2009. MALDI-TOF MS-based drug susceptibility testing of pathogens: The example of *Candida albicans* and fluconazole. Proteomics. 9: 4627–4631.

Marinach-Patrice, C., A. Fekkar, R. Atanasova, J. Gomes, L. Djamdjian, J.Y. Brossas, I. Meyer, P. Buffet, G. Snounou, A. Datry, C. Hennequin, J.L. Golmard and D. Mazier. 2010. Rapid species diagnosis for invasive candidiasis using mass spectrometry. PLoS ONE 5(1): e8862.

Marvin, L.F., M.A. Roberts and L.B. Faya. 2003. Matrix-assisted laser desorption/ionization time-of-flight mass spectrometry in clinical chemistry. Clin Chim Acta. 337(1-2): 11–21.

McTaggart, L.R., E. Lei, S.E. Richardson, L. Hoang, A. Fothergill and S.X. Zhang. 2011. Rapid identification of *Cryptococcus neoformans* and *Cryptococcus gattii* by matrix-assisted laser desorption ionization–time of flight mass spectrometry. J Clin Microbiol. 49(8): 3050–3053.

Morrell, M., V.J. Fraser and M.H. Kollef. 2005. Delaying the empiric treatment of *Candida* bloodstream infection until positive blood culture results are obtained: a potential risk factor for hospital mortality. Antimicrob Agents Chemother. 49: 3640–3645.

Murali, S. and A. Langston. 2009. Advances in antifungal prophylaxis and empiric therapy in patients with hematologic malignancies. Transpl Infect Dis. 11: 480–490.

Murray, P.R., E.J. Baron, M.A. Pfaller, F.C. Tenover and R.H. Yolken (eds.). 1999. Manual of Clinical Microbiology, 7th ed. ASM Press, Washington, DC.

Nolla-Salas, J., A. Sitges-Serra, C. León-Gil, J. Martínez-González, M.A. León-Regidor, P. Ibáñez-Lucía and J.M. Torres-Rodríguez. 1997. Candidemia in non-neutropenic critically ill patients: analysis of prognostic factors and assessment of systemic antifungal therapy. Intensive Care Med. 23: 23–30.

Padovan, A.C.B., A.S.A. Melo and A.L. Colombo. 2013. Systematic review and new insights into the molecular characterization of the *Candida rugosa* species complex. Fungal Genet Biol. 61: 33–41.

Pappas, P.G., C.A. Kauffman, D. Andes, D.K. Benjamin, T.F. Calandra, J.E. Edwards, S.G. Filler, J.F. Fisher, B.J. Kullberg, L. Ostrosky-Zeichner, A.C. Reboli, J.H. Rex, T.J. Walsh and J.D. Sobel. 2009. Clinical practice guidelines for the management of candidiasis: 2009 update by the Infectious Diseases Society of America. Clin Infect Dis. 48: 503–535.

Pfaller, M.A. 2012. Antifungal drug resistance: mechanisms, epidemiology, and consequences for treatment. Am J Med. 125(1 Suppl): S3–13.

Pfaller, M.A. and D.J. Diekema. 2010. Epidemiology of invasive mycoses in North America. Crit Rev Microbiol. 36: 1–53.

Pinto, A., C. Halliday, M. Zahra, S. van Hal, T. Olma, K. Maszewska, J.R. Iredell, W. Meyer and S.C. Chen. 2011. Matrix-assisted laser desorption ionization-time of flight mass spectrometry identification of yeasts is contingent on robust reference spectra. PLoS One. 6(10): e25712.

Posteraro, B., A. Vella, M. Cogliati, E. De Carolis, A.R. Florio, P. Posteraro, M. Sanguinetti and A.M. Tortorano. 2012. Matrix-assisted laser desorption ionization–time of flight mass spectrometry-based method for discrimination between molecular types of *Cryptococcus neoformans* and *Cryptococcus gattii*. J Clin Microbiol. 50(7): 2472–2476.

Posteraro, B., E. De Carolis, A. Vella and M. Sanguinetti. 2013. MALDI-TOF mass spectrometry in the clinical mycology laboratory: identification of fungi and beyond. Expert Rev Proteomics 10(2): 151–164.

Pulcrano, G., E. Roscetto, V.D. Iula, D. Panellis, F. Rossano and M.R. Catania. 2012. MALDI-TOF mass spectrometry and microsatellite markers to evaluate *Candida parapsilosis* transmission in neonatal intensive care units. Eur J Clin Microbiol Infect Dis. 31: 2919–2928.

Pulcrano, G., D. Vitalula, A. Vollaro, A. Tucci, M. Cerullo, M. Esposito, F. Rossano and R.M. Catania. 2013. Rapid and reliable MALDI-TOF mass spectrometry identification of *Candida* non-albicans isolates from bloodstream infections. J Microbiol Methods 94: 262–266.

Putignani, L., F. Del Chierico, M. Onori, L. Mancinelli, P. Bernaschi, L. Coltella, B. Lucignano, L. Pansani, S. Ranno, C. Russo, A. Urbani, G. Federici and D. Menichella. 2011. MALDI-TOF mass spectrometry proteomic phenotyping of clinically relevant fungi. Mol Biosyst. 7: 620–629.

Quiles-Melero, I., J. García-Rodríguez, Gómez-López and J. Mingorance. 2012. Evaluation of matrix-assisted laser desorption/ionization time-of-flight (MALDI-TOF) mass spectrometry for identification of *Candida parapsilosis*, *C. orthopsilosis* and *C. metapsilosis*. Eur J Clin Microbiol Infect Dis. 31: 67–71.

Revankar, S.G., W.R. Kirkpatrick, R.K. McAtee, A.W. Fothergill, S.W. Redding, M.G. Rinaldi and F. Patterson. 1998. Interpretation of trailing endpoints in antifungal susceptibility testing by the National Committee for Clinical Laboratory Standards method. J Clin Microbiol. 36: 153–156.

Rogers, P.D., J.P. Vermitsky, T.D. Edlind and G.M. Hilliard. 2006. Proteomic analysis of experimentally induced azole resistance in *Candida glabrata*. J Antimic Chemother. 58: 434–438.

Rosenvinge, F.S., E. Dzajic, E. Knudsen, S. Malig, L.B. Andersen, A. Løvig, M.C. Arendrup, T.G. Jensen, B. Gahrn-Hansen and M. Kemp. 2013. Performance of matrix-assisted laser desorption-time of flight mass spectrometry for identification of clinical yeast isolates. Mycoses. 56: 229–235.

Santos, C., N. Lima, P. Sampaio and C. Pais. 2011. Matrix-assisted laser desorption /ionization time-of-flight intact cell mass spectrometry to detect emerging pathogenic *Candida* species. Diagn Microbiol Infect Dis. 71: 304–308.

Sauer, S. and M. Kliem. 2010. Mass spectrometry tools for the classification and identification of bacteria. Nat Rev Microbiol. 8(1): 74–82.

Sendid, B., P. Ducoroy, N. Francois, G. Lucchi, S. Spinali, O. Vagner, S. Damiens, A. Bonnin, D. Poulain and F. Dalle. 2012. Evaluation of MALDI-TOF mass spectrometry for the identification of medically important yeasts in the clinical laboratories of Dijon and Lille hospitals. Med Mycol. 51: 25–32.

Seng, P., M. Drancourt, F. Gouriet, B. La Scola, P.E. Fournier, J.M. Rolain and D. Raoult. 2009. Ongoing revolution in bacteriology: routine identification of bacteria by matrix-assisted laser desorption ionization time-of-flight mass spectrometry. Clin Infect Dis. 49(4): 543–551.

Sibley, C.D., G. Peirano and D.L. Church. 2012. Molecular methods for pathogen and microbial community detection and characterization: current and potential application in diagnostic microbiology. Infect Genet Evol. 12(3): 505–521.

Spanu, T., B. Posteraro, B. Fiori, T. D'Inzeo, S. Campoli, A. Ruggeri, M. Tumbarello, G. Canu, E.M. Trecarichi, G. Parisi, M. Tronci, M. Sanguinetti and G. Fadda. 2012. Direct MALDI-TOF Mass Spectrometry Assay of blood culture broths for rapid identification of *Candida* species causing bloodstream infections: an observational study in two large microbiology laboratories. J Clin Microbiol. 50(1): 176–179.

Stevenson, L.G., S.K. Drake, Y.R. Shea, A.M. Zelazny and P.R. Murray. 2010. Evaluation of matrix-assisted laser desorption ionization-time of flight mass spectrometry for identification of clinically important yeast species. J Clin Microbiol. 48(10): 3482–3486.

Sulc, M., K. Peslova, M. Zabka, M. Hajduch and V. Havlicek. 2009. Biomarkers of *Aspergillus* spores : strain typing and protein identification. Int J Mass Spectrom. 280: 162–168.

Tan, K.E., B.C. Ellis, R. Lee, P.D. Stamper, S.X. Zhang and K.C. Carrolla. 2012. Prospective evaluation of a matrix-assisted laser desorption ionization–time of flight mass spectrometry system in a hospital clinical microbiology laboratory for identification of bacteria and yeasts: a bench-by-bench study for assessing the impact on time to identification and cost-effectiveness. J Clin Microbiol. 50(10): 3301–3308.

Taur, Y., N. Cohen, S. Dubnow, A. Paskovaty and S.K. Seo. 2010. Effect of antifungal therapy timing on mortality in cancer patients with candidemia. Antimicrob Agents Chemother. 54: 184–190.

Theel, E.S., L. Hall, J. Mandrekar and N.L. Wengenack. 2011. Dermatophyte identification using matrix-assisted laser desorption ionization-time of flight mass spectrometry. J Clin Microbiol. 49: 4067–4071.

Theel, E.S., B.H. Schmitt, L. Hall, S.A. Cunningham, R.C. Walchak, R. Patel and N.L. Wengenac. 2012. Formic acid-based direct, on-plate testing of yeast and *Corynebacterium* species by Bruker Biotyper matrix-assisted laser desorption ionization–time of flight mass spectrometry. J Clin Microbiol. 50(9): 3093–3095.

Trilles, L., W. Meyer, B. Wanke, J. Guarro and M. Laze´ra. 2012. Correlation of antifungal susceptibility and molecular type within the *Cryptococcus neoformans/C. gattii* species complex. Med Mycol. 50(3): 328–32.

Van Herendael, B.H., P. Bruynseels, M. Bensaid, T. Boekhout, T. De Baere, I. Surmont and A.H. Mertens. 2012. Validation of a modified algorithm for the identification of yeast isolates using matrix-assisted laser desorption/ionisation time-of-flight mass spectrometry (MALDI-TOF MS) Eur J Clin Microbiol Infect Dis. 31: 841–848.

van Veen, S.Q., E.C. Claas and E.J. Kuijper. 2010. High-throughput identification of bacteria and yeast by matrix-assisted laser desorption ionization-time of flight mass spectrometry in conventional medical microbiology laboratories. J Clin Microbiol. 48(3): 900–907.

Vella, A., E. De Carolis, L. Vaccaro, P. Posteraro, D.S. Perlin, M. Kostrzewa, B. Posteraro and M. Sanguinettia. 2013. Rapid antifungal susceptibility testing by matrix-assisted laser desorption ionization time-of-flight mass spectrometry analysis. J Clin Microbiol. 51(9): 2964–2969.

Verwer, P.E.B., W.B. van Leeuwen, V. Girard, V. Monnin, A. van Belkum, J.F. Staab, H.A. Verbrugh, I.A.J.M. Bakker-Woudenberg and W.W.J. van de Sande. 2014. Discrimination of *Aspergillus lentulus* from *Aspergillus fumigatus* by Raman spectroscopy and MALDI-TOF MS. Eur J Clin Microbiol Infect Dis. 33: 245–251.

Weber, S.S., A.F.A. Parente, C.L. Borges, J.A. Parente, A.M. Bailao and C.M.A. Soares. 2012. Analysis of the secretomes of *Paracoccidioides* mycelia and yeast cells. PLOS ONE 7(12): e52470.

Weile, J. and C. Knabbe. 2009. Current applications and future trends of molecular diagnostics in clinical bacteriology. Anal Bioanal Chem. 394(3): 731–742.

Welham, K.J., M.A. Domin, K. Johnson, L. Jones and D.S. Ashton. 2000. Characterization of fungal spores by laser desorption/ionization time of flight mass spectrometry. Rapid Commun Mass Spectrom. 14: 307–310.

Westblade, L.F., R. Jennemann, J.A. Branda, M. Bythrow, M.J. Ferraro, O.B. Garner, C.C. Ginocchio, M.A. Lewinski, R. Manji, A.B. Mochon, G.W. Procop, S.S. Richter, J.A. Rychert, L. Sercia and C.A.D. Burnhama. 2013. Reply to "Risks of 'Blind' automated identification systems in medical microbiology". J Clin Microbiol. 51(11): 3912.

Wieser, A., L. Schneider, J. Jung and S. Schubert. 2012. MALDI-TOF MS in microbiological diagnostics-identification of microorganisms and beyond (mini review). Appl Microbiol Biotechnol. 93: 965–974.

Yang, S.A., Y. Jin, G. Zhao, J. Liu, X. Zhou, J. Yang, J.A. Wang, Y. Cui, X. Hu, Y. Li and H. Zhao. 2014. Improvement of matrix-assisted laser desorption/ionization time-of-flight mass spectrometry for identification of clinically important *Candida* species. Clin Lab. 60: 37–46.

CHAPTER 15

Medical Mycology in Iran: Past, Present and Future#

Mohammadhassan Gholami-Shabani,[1] *Masoomeh Shams-Ghahfarokhi,*[2] *Mohammadreza Shidfar*[3] and *Mehdi Razzaghi-Abyaneh*[1,*]

Introduction

Modern medical education was started in Iran on December 21, 1851 when Darolfonoun, the first modern academy of Iran, was inaugurated in Tehran through the endeavors of then Prime Minister Mirzataghikahn Amirkabir. The first report of a fungal infection in Iran goes back to around 80 years ago in 1937 when Habibi reported a case of rhinosporidiosis as a subcutaneous fungal involvement of nasal mucosa (Habibi 1939). Until 1962, when modern mycology was introduced in Iran, mycological studies were restricted to Tinea capitis, a dermatophytic infection of scalp and hair. At that time, patients who referred to the Razi hospital in Tehran were treated using only traditional methods according to the clinical signs without any mycological examination of the lesions. Responsibility of the education of medical mycology was denoted to Dr. Mohammadali Faghih when it was restricted to courses about clinical signs of tinea capitis and observation of microscopic slides of scalp hair involvement,

[1] Department of Mycology, Pasteur Institute of Iran, Tehran 13164, Iran.
[2] Department of Mycology, Faculty of Medical Sciences, Tarbiat Modares University, Tehran, 14115-331, Iran.
[3] Department of Mycology and Parasitology, Faculty of Public Health, Tehran University of Medical Sciences, Tehran, Iran.
* Corresponding author: mrab442@yahoo.com; mrab442@pasteur.ac.ir

Dedicated to the memory of Prof. Shamsoddin Mofidi, who established modern Medical Mycology in Iran in 1962.

i.e., Favus (Acorion), Endothrix (Trichophyti) and Ectothrix (Microspori). In order to acquire confirmation of identity, Dr. Faghih referred patients suspected of fungal infections from Razi hospital to the Parasitology and Malariology Institute (the name was later changed to Parasitology Institute) which had a central role in the monitoring of contagious and non-contagious infections in Iran. In 1962, the dean of Parasitology Institute, Prof. Shamsoddin Mofidi, selected Dr. Manoochehr Asgari for passing a course on fungal infections outside Iran. Dr. Asgari passed the necessary courses in the diagnosis and treatment of fungal diseases at CDC and Duke University in the United States for two years and returned to Iran in 1964 when he established the Medical Mycology course at the Institute of Public Health. Education and research on fungal infections was started and conducted at that time by a research team which included Dr. Asgari, Dr. Dorang, Dr. Jaber-Ansari, Dr. Sheiban, Dr. Shidfar, Dr. Falati and Dr. Alilou. During 1965 to 1969, extensive research works on diagnosis and treatment of dermatophytic infections were carried out in Tehran, Mazandaran and Isfahan provinces. After establishing the Faculty of Public Health in 1973, Dr. Masoud Emami who had passed educational courses on medical mycology at CDC and Pasteur Institute of Paris joined the other mycologists and the golden age of medical mycology was started by expanding the research programmes to fungal infections beyond dermatophytosis. At present, a wide array of fungal diseases from superficial to life-threatening invasive ones have been diagnosed and reported by a large number of expert mycologists who are involved in education and research in different universities and research institutes around the country and new reports on such infections are added to the literature every year.

Nowadays, there is no doubt about the increasing importance of fungi as emerging human and animal pathogens worldwide. Despite the development of new therapeutic strategies for control of opportunistic fungal infections, the rate of these infections has dramatically elevated in the last decade due to an increasing rate of predisposing factors such as autoimmune diseases, hematological malignancies, organ transplantation and immunodeficiency syndromes like AIDS. There is an expanding list of opportunistic fungal pathogens which are capable of inducing life-threatening infections in debilitated and immunocompromised patients. Fungal infections from superficial to invasive are real public health hazards in Iran. Among various types of human infections, dermatophytosis, pityriasis versicolor and superficial candidiasis (onychomycosis) are the most important mycoses affecting a large portion of the population. Although there is no doubt that systemic fungal infections such as invasive aspergillosis and disseminated candidiasis are important diseases in hospitalized patients especially those who receive organ transplants, the situation of these infections in Iran is not clear due to the lack of a center for monitoring invasive fungal infections. In this review, we try to highlight the distribution and diversity of fungal infections all over the country with special focus on the etiologic agents, identification methods, prevention and treatment strategies. A general map (Fig. 1) has been designed to show the distribution of important fungal diseases in 32 townships and each disease has been described in details under independent headlines. Special attention has been paid to dermatophytosis and pityriasis versicolor as they are the most prevalent fungal diseases in Iran.

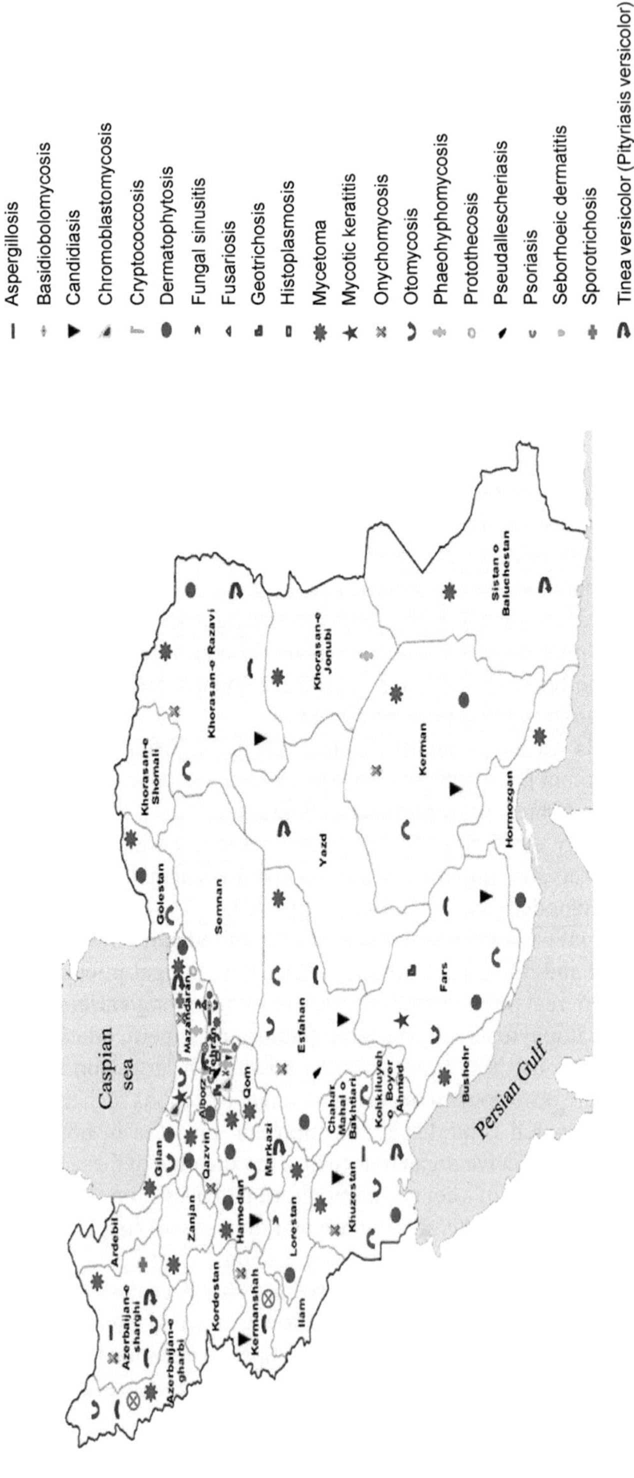

Figure 1. Geographic Distribution of Fungal Infections in Iran.

Superficial Fungal Infections (SFIs)

SFIs are the most prevalent types of mycoses in Iran. Dermatophytosis, candidiasis and pityriasis versicolor are the most important SFIs distributed all over the country. Table 1 represents detailed information about the incidence and prevalence of various types of SFIs. In this section, each SFI is discussed in details with special attention paid to the prevalence and distribution patterns among the affected population.

Dermatophytosis

Dermatophytosis (tinea, ring worm) is a superficial infection of the keratinized tissue of skin, nails and hair with worldwide distribution induced by a group of specialized keratinophilic fungi named "dermatophytes". They contain around 40 species classified into three main genera *Microsporum* (skin and hair), *Trichophyton* (skin, hair and nail) and *Epidermophyton* (skin and nail). Dermatophytes reside in soil (geophilic), human body (anthropophilic) and animals (zoophilic). They are true pathogens capable of infecting various body parts either directly or indirectly. Major species involved in the etiology of dermatophytosis are *T. rubrum, T. interdigitale, T. tonsurans, M. canis,* and *T. violaceum*.

It has been estimated that around 20% to 25% of the world population is affected by dermatophytes and the disease incidence is increasing every year. The severity of the disease depends on the strain or the species of dermatophytes involved and host conditions. Depending on the body site affected, dermatophytosis is named as tinea capitis (scalp ringworm), tinea corporis (ringworm of the smooth skin), tinea cruris (jock itch), tinea pedis (athlete's foot), tinea unguium (nail infections), etc. (Fig. 2). All persons are in contact with dermatophytes during their lifetime but only a small

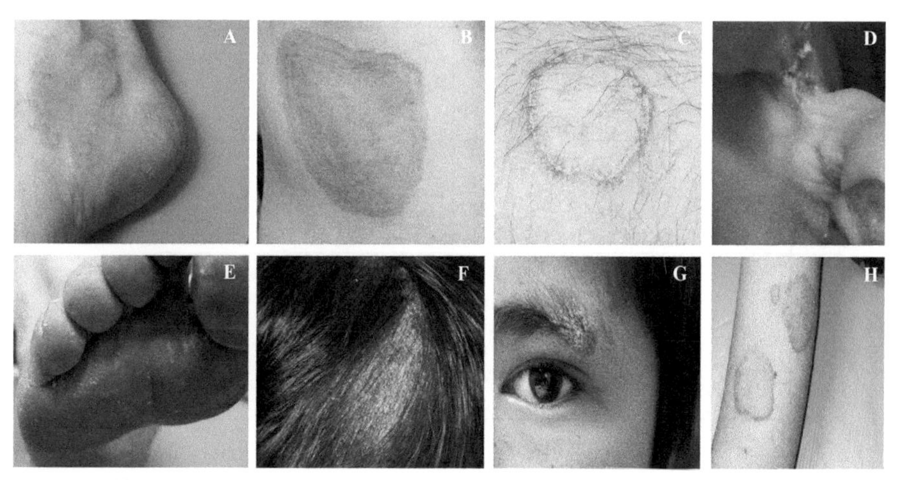

Figure 2. Different Types of Dermatophytosis in Patients Admitted to the Clinical Mycology Laboratory of the Pasteur Institute of Iran (A, D and E, Tinea pedis; B, C and H, Tinea corporis; F, Tinea capitis; G, Tinea faciei).

Table 1. Clinical Features and Geographic Distribution of Superficial Fungal Infections in Iran.

Disease	Place (Area)	Number of patients	Frequency (%)	Sex	Age	Affected body part (%)		Major Etiologic Agents (%)		Reference
Dermatophytosis	Tehran	17573	38.0	Men and Women	ND	ND	ND	*E. floccosum*	32	Bassiri-Jahromi and Khaksari 2009
								T. rubrum	26	
								T. mentagrophytes	19	
	Isfahan	16578	13.3	ND	ND	Trunk	23.8	*T. verrucosum*	32.8	Chadeganipour et al. 1997
						Head	54.1	*E. floccosum*	17.6	
						Feet	8.9	*T. mentagrophytes*	16.2	
						Groin	6.8	*M. canis*	12.3	
						Nails	3.5	*T. verrucosum*	9.5	
						Manus	2.6	*T. schoenleinii*	5.4	
						Beard	0.3	*T. rubrum*	3	
	Different areas of Iran	12150	78	ND	1–50	Head	35.60	*M. canis*	19.4	Khosravi et al. 1994
						Trunk	19.89	*T. rubrum*	16.5	
						Feet	13.90	*E. floccosum*	14.9	
						Groin	9.60	*T. mentagrophytes*	14.6	
								T. verrucosum	11.5	
						Manus	15.09	*T. violaceum*	8.7	
						Nails	5.13	*T. schoenleinii*	5.5	
								T. tonsurans	1.3	
						Beard	0.75	*T. erinacei*	0.8	
								M. gallinae	0.2	

	Hamadan	7495	38.0	ND	0–40	Head	62.9	*T. verrucosum*	54.1	Omidynia et al. 1996
						Trunk	10.4	*T. schoenleinii*	33.3	
						Manus	7.3	*M. canis*	5.5	
						Groin	7.3	*E. floccosum*	3.5	
						Beard	5.4	*T. mentagrophytes*	1.4	
						feet	5	*M. gypseum*	1.4	
						Nails	1.5	*T. tonsurans*	0.7	
	Tehran	4871	65.7	Men (2734)	10–80	Feet	933	*T. rubrum*	28.7	Sadeghi et al. 2011
						Groin	731	*E. floccosum*	25	
						Manus	287	*T. mentagrophytes*	23.8	
				Women (2137)		Trunk	230	*T. tonsurans*	11.5	
						Nails	226	*T. verrucosum*	7.8	
						Facial	133	*M. canis*	1.4	
						Head	95	*T. violaceum*	1.3	
	Tehran	3976	15.2	ND	1–75	Manus	34.7	*T. mentagrophytes*	28.8	Bassiri-Jahromi 2012
						Trunk	28.9	*T. verrucosum*	9.3	
						Facial	13.8			
						Head	8.8	*M. canis*	2.2	
						Feet	8.4			
	Khozestan (Ahvaz)	428	ND	Men (233)	10–50	Groin	38.78	*E. floccosum*	39.25	Rassai et al 2011
						Trunk	21.26	*T. verrucosum*	27.33	
						Manus	12.38	*T. rubrum*	8.41	
						Head	10.98	*T. mentagrophytes*	7.71	
				Women (195)		Feet	7.71	*M. canis*	2.80	
						Facial	6.07	*T. violaceum*	1.63	
						Nails	2.80	*M. gypseum*	0.46	

Table 1. contd....

Table 1. contd.

Disease	Place (Area)	Number of patients	Frequency (%)	Sex	Age	Affected body part (%)		Major Etiologic Agents (%)		Reference
Dermatophytosis	Tehran	1568	13.3	ND	0–50	Head	13.3	*T. violaceum*	37.3	Bassiri-Jahromi and Khaksar 2006a
								T. schoenleinii	21.5	
								M. canis	18.6	
								T. verrucosum	14.8	
								T. tonsurans	5.3	
								T. rubrum	1	
								M. gypseum	1	
								T. mentagrophytes	0.5	
	Tehran	1254	13.5	Men	20–29	Trunk	31.4	*E. floccosum*	31.4	Falahati et al. 2003
						Groin	20.7	*T. rubrum*	18.3	
						Manus	15.4	*T. mentegrophytes*	17.2	
				Women		Head	12.4	*T. violaceum,*	16.6	
						Feet	10.6	*M. canis*	6.5	
						Facial	7.1	*T. verrucosum*	4.7	
						Nails	2.4	*M. gypseum*	4.1	
	Tehran	1254	30.4	Men (226)	2–16	Head	39.6	*T. violaceum*	28.3	Rastegar-Lari et al. 2005
								M. canis	15.1	
						Trunk	30.2	*E. floccosum*	15.1	
						Facial	18.9	*T. rubrum*	13.2	
				Women (156)		Manus	7.5	*T. mentagrophytes*	11.3	
								M. gypseum	7.5	
	Isfahan	12000	10.8	ND	ND	Head	45.6	*T. verrucosum*	43.85	Chadegani et al. 1987
						Trunk	32.9	*T. schoenleinii*	15.78	
						Groin	7.36	*T. mentagrophytes*	11.22	
						Feet	11.92	*T. violaceum*	10.17	
						Nails	2.1	*E. floccosum*	8.77	

	Golestan (Gorgan)	1108	31.6	ND	ND	ND	ND	*E. floccosum*	70.4	Dehghan et al. 2009
								T. rubrum	14.5	
								M. audouinii	7.2	
	Qazvin	1023	34.0	ND	1–69	Groin	31.9	*E. floccosum*	32.8	Aghamirian and Ghiasian 2008
						Trunk	20.7	*T. rubrum*	18.1	
						Feet	19	*T. verrucosum*	17.2	
						Nails	11	*T. mentagrophytes*	22.4	
	Tehran	893	7.1	ND	6–42	Trunk	10.8	*T. tonsurans*	92.6	Bassiri-Jahromi and Khaksar 2003
								T. rubrum	2.8	
								T. mentagrophytes	1.75	
								E. floccosum	1.75	
								T. violaceum	0.43	
								T. verrucosum	0.43	
								M. canis	0.2	
	Alborz (Karaj)	750	21.0	Men (560)	0–50	Head	ND	*T. verrucosum*	11	Hashemi et al. 2005
						Trunk	ND			
						Manus	ND	*M. canis*	10	
						Beard	ND			
				Women (190)		Feet	ND			
						Nails	ND	*T. mentagrophytes*	9	
						Groin	ND			
	Khozestan (Ahvaz)	593	3.1	ND	ND	ND	ND	*T. rubrum*	ND	Rafiei and Amirrajab 2010
								T. verrucosum	ND	
								E. floccosum	ND	

Table 1. contd....

Table 1. contd.

Disease	Place (Area)	Number of patients	Frequency (%)	Sex	Age	Affected body part (%)		Major Etiologic Agents (%)		Reference
Dermatophytosis	Khorasanerazavi (Mashhad)	560	29.6	Men (330)	0–40	Trunk	33.1	*T. violaceum*	27.7	Naseri et al. 2013b
						Head	32.5	*T. mentagrophytes*	21.6	
						Manus	17.5	*E. floccosum*	21.6	
						Groin	10.2	*T. schoenleinii*	12.8	
				Women (230)		Feet	5.4	*M. vanbreuseghemii*	11.4	
						Nails	0.6	*T. rubrum*	1.4	
						Beard	0.6	*T. verrucosum*	1.4	
	Tehran	549	10.5	ND	21–60	Nails	46	*T. mentagrophytes*	ND	Zaini et al. 2009
								T. rubrum	ND	
	Markazi (Arak)	435	31.7	ND	0–29	ND	ND	ND	ND	Hashemi et al. 2014
	Hormozgan (Bandarabbas)	402	74.4	ND	ND	Groin	37.5	*T. mentagrophytes*	35.8	Mahboubi et al. 2006
						Head	22.5	*T. rubrum*	25.1	
						Trunk	20.7	*E. floccosum*	22.4	
	Mazandaran (sari)	324	20.1	Men	9–20	Body	ND	*T. tonsurans*	ND	Hedayati et al. 2007
	Lorestan (Khorramabad)	294	58.5	ND	ND	Trunk	25.6	*E. floccosum*	ND	Sepahvand et al. 2009
						Groin	25	*T. verrucosum*	ND	
	Qazvin	270	ND	ND	5–55	Body	19.2	*T. tonsurans*	82.7	Aghamirian and Ghiasian 2011
								T. rubrum	5.8	
								T. mentagrophytes	3.8	
								E. floccosum	3.8	

	Khozestan (Ahvaz)	279	41.2	Men (172)	1–72	Feet	16.5	*T. mentagrophytes*	25.2	Mahmoudabadi 2005
						Trunk	14.8	*E. floccosum*	21.7	
						Nails	13	*T. verrucosum*	16.5	
						Head	11.3	*M. canis*	13	
				Women (107)		Facial	11.3	*T. rubrum*	9.5	
						Manus	7	*M. gypseum*	9.5	
						Beard	1.7	*T. violaceum*	3	
	Gilan	217	42.9	ND	ND	Groin	47.2	*T. mentagrophytes*	ND	Alizadeh et al. 2004
						Feet	14	*E. floccosum*	ND	
						Head	12.9	*T. tonsurans*	ND	
	Mazandaran (Babol)	200	53.8	Men (104)	13–30	Groin	24.5	*T. mentagrophytes*	ND	Rezvani et al 2010
						Feet	20			
						Trunk	17			
						Nails	15.5			
				Women (96)		Head	10	*E. floccosum*	ND	
						Beard	5			
						Facial	4			
						Manus	4			
	Isfahan (Kashan)	137	34.6	Men (53)	ND	Nails	65.4	*T. violaceum*	ND	Asadi et al. 2009
								T. mentagrophytes	ND	
								T. rubrum	ND	
				Women (84)		Feet	34.6	*T. verrucosum*	ND	
								E. floccosum	ND	
	Fars (Shiraz)	122	49.2	ND		Trunk	ND	*T. rubrum*	ND	Nikpoor et al. 1978
								T. mentagrophytes	ND	
						Head	ND	*T. schoenleinii*	ND	

Table 1. contd....

Table 1. contd.

Disease	Place (Area)	Number of patients	Frequency (%)	Sex	Age	Affected body part (%)		Major Etiologic Agents (%)		Reference
Dermatophytosis	Hamedan	56	ND	Men (26)	0–70	Trunk	32.1	*T. verrucosum*	28.6	Ansar et al. 2011
						Facial	26.8	*T. rubrum*	12.5	
						Groin	14.3	*E. floccosum*	10.7	
				Women (27)		Manus	12.5	*M. canis*	8.9	
						Feet	8.9	*T. violaceum*	3,6	
						Head	5.4	*T. schoenleinii*	1.8	
	Khorasanerazavi (Mashhad)	2	ND	Men	7	Head	ND	*M. vanbreuseghemii*	ND	Naseri et al. 2012a
	Khorasanerazavi (Mashhad)	1	ND	Women	8	Trunk	ND	*M. persicolor*	ND	Naseri et al. 2012b
	Tehran	1	ND	Men	42	ND	ND	*T. violaceum*	ND	Khosravi et al. 2000
	Khozestan (Ahvaz)	1	ND	Men	42	Facial	ND	*M. ferrugineum*	ND	Mahmoudabadi 2006b
	Khozestan (Ahvaz)	1	ND	Men	25	Trunk	ND	*T. simii*	ND	Mahmoudabadi and Yaghoobi 2008
	Alborz (Karaj)	3475	25.5	ND	1–74	Trunk	22.77	*E. floccosum*	ND	Pakshir and Hashemi 2006
						feet	20.7			
						Head	11,3			
						Nails	9.8	*T. mentagrophytes*	ND	
						Manus	7.3			
						Beard	3.3	*T. verrucosum*	ND	
						Facial	0.5			
	Tehran	1	ND	Man	30	Groin	ND	*T. mentagrophytes*	ND	Khosravi et al. 2008
	Yasoj	1	ND	Men	20	Trunk	ND	*M. fulvum*	ND	Nouripour-Sisakht et al. 2013
	Kerman	1	ND	Women	10	Body	ND	*T. concentricum*	ND	Ayatollahi-Mousavi et al. 2009

Candidiasis	Tehran	4871	54.5	Men (2734)	10–80	Groin	12.9	*C. albicans*	29.20	Sadeghi et al. 2011
						Hands	1.2			
				Women (2137)				*Candida* sp.	70.71	
						Glabrous skin	4.9			
						Feet	2.4			
	Hamedan	540	17.2	Women	15–20	Vagina	ND	*C. albicans*	ND	Shobeiri et al. 2006
	Khozestan (ahvaz)	461	55.7	ND	1–51	Groin	21	*C. albicans*	49.4	Mahmoudabadi 2006a
								C. tropicalis	27.4	
								C. parapsilosis	9.6	
								C. krusei	2.5	
								C. guilliermondii	2.1	
								C. pseudotropicalis	1.7	
								C. humicola	1.3	
								C. lipolytica	0.4	
	Fars (Shiraz)	410	86.3	ND	ND	Skin	ND	*C. albicans*	48	Badiee et al. 2011
						Urine	ND	*C. krusei*	16.1	
						Sputum	ND	*C. glabrata*	13.5	
						Esophagous	ND			
						Oropharyngeal area	ND	*C. kefyr*	7.4	
						Vagina	ND	*C. parapsilosis*	4.8	
						Lung	ND	*C. tropicalis*	1.7	

Table 1. contd....

Table 1. contd.

Disease	Place (Area)	Number of patients	Frequency (%)	Sex	Age	Affected body part (%)		Major Etiologic Agents (%)		Reference
Candidiasis	KhorasaneRazavi (Sabzevar)	231	26.8	ND	ND	Vagina	ND	*C. albicans*	38.7	Moallaei et al. 2007
								C. kefyr	17.7	
								C. glabrata	8	
								C. krusei	3.2	
								C. rugosa	6.4	
								C. lipolytica	6.4	
	Isfehan (Kashan)	150	ND	ND	16–36	Vagina	ND	*C. albicans*	ND	Rasti et al. 2014
	Tehran	137	ND	ND	1–89	Skin	61.8	*C. albicans*	72.3	Razzaghi-Abyaneh et al. 2014
						Vagina	16.2	*C. parapsilosis*	11.5	
						Sputum	11.6	*C. glabrata*	4.6	
						Oral cavity	4			
						Lung	3.5	*C. krusei*	4.6	
						Eye	0.6	*C. tropicalis*	2.9	
						Stomach	0.6	*C. guillermondii*	2.9	
						Urine	0.6			
						Tissue biopsy	0.6	*C. sake*	0.6	
						Urinary tract	0.6	*C. intermedia*	0.6	
	Tehran	150	ND	ND	31–50	ND	ND	*C. albicans*	50.2	Katiraee et al. 2010
								C. glabrata	22	
								C. dubliniensis	4.4	
								C. tropicalis	3.4	
								C. kefyr	3.4	
								C. parapsilosis	2.9	
	Kermanshah	100	32	Women	25–75	Vagina	ND	*C. albicans*	62.5	Faraji et al. 2012
								C. glabrata	18.7	
								C. tropicalis	9.4	
								C. parapsilosis	9.4	

	Khozestan (Ahvaz)	20	59.4	Men (14) Women (6)	36–75	Oral cavity	ND	*C. albicans*	ND	Azizi and Rezaei 2009
								C. tropicalis	ND	
								C. krusei	ND	
	Kerman	1	ND	Men	13	Skin of the nose	ND	*C. albicans*	ND	Moghaddami et al.1991
Onychomycosis	Tehran	4871	54.5	Men (2734)	10–80	Finger nails	72	*C. albicans*	29.20	Sadeghi et al. 2011
				Women (2137)		Toenails	6.7	*Candida* sp.	70.71	
	Kermanshah	2402	45.2	ND	30–40	Nails	ND	Yeasts	78.54	Mikaeili and Karimi 2013
								Dermatophytes	18.50	
								Moulds	2.94	
	Tehran	1268	32.33	ND	15–72	Fingernails, Toe nails	ND	Dermatophytes	40.5	Bassiri-Jahromi and Khaksar 2010
								Yeasts	48	
								Moulds	11.5	
	Mazandaran	1100	56.8	ND	1–88	Fingernails, Toe nails	ND	ND	ND	Afshar et al. 2014
	Tehran	549	47.9	ND	1–83	Fingernails, Toe nails	ND	*C. albicans*	16.73	Zaini et al. 2009
								Candida sp.	22.43	
								T. mentagrophytes	10.26	
								T. rubrum	4.94	
								A. flavus	11.78	
								A. fumigatus	6.08	
								A. niger	3.04	
								Aspergillus spp.	1.90	
								Penicillium spp.	2.66	
								Rhizopus spp.	1,52	
								Cladosporium spp.	1.52	
								Scopulariopsis spp.	1.14	
								Acremonium spp.	0.76	

Table 1. contd....

Table 1. contd.

Disease	Place (Area)	Number of patients	Frequency (%)	Sex	Age	Affected body part (%)		Major Etiologic Agents (%)		Reference
Onychomycosis	Tehran	504	42.8	ND	40–60	Fingernails, Toenails	ND	Yeasts	59.7	Hashemi et al. 2010
								Dermatophytes	21.3	
								Moulds	19	
	Isfahan	488	39.8	ND	1–80	Fingernails, Toenails	ND	Yeasts	57.7	Chadeganipour et al. 2010
								Dermatophytes	13.9	
								Moulds	28.4	
	Khozestan (Ahvaz)	461	55.7	ND	1–51	Toenails	14.8	*C. albicans*	49.4	Mahmoudabadi 2006a
								C. tropicalis	27.4	
								C. parapsilosis	9.6	
								C. krusei	2.5	
								C. guilliermondii	2.1	
								C. Pseudotropicalis	1.7	
								C. humicola	1.3	
								C. lipolytica	0.4	
	Gazvin	308	40.2	ND	ND	Fingernails, Toenails	ND	*Candida* spp.	46.8	Aghamirian and Ghiasian 2010
								Dermatophytes	50	
								Moulds	3.2	
	Yazd	262	6.9	ND	10–60	Fingernails, Toenails	ND	ND	ND	Kafaie and Noorbala 2010
	Tehran	252	50.4	ND	0–70	Fingernails, Toenails	ND	Dermatophytes	12.3	Gerami-shoar et al. 2002
								Candida spp.	21	
								Moulds	5.6	
	Tehran	187	ND	ND	10–65	Fingernails, Toenails	ND	Yeasts	43.3	Khosravi and Mansouri 2000
								Dermatophytes	48.4	
								Moulds	8.2	

	Tehran	137	ND	ND	1–89	Skin & nails	61.8	*Candida* spp.	72.3	Razzaghi-Abyaneh et al. 2014
	Isfahan (Kashan)	137	19	ND	ND	Skin and nails	ND	*T. violaceum*	11.6	Asadi et al. 2009
								T. mentagrophytes	7.7	
								T. rubrum	7.7	
								T. verrucosum	3.8	
								E. floccosum	3.8	
								C. albicans	26.9	
								Candida spp.	15.4	
								A. flavus	7.7	
								A. fumigatus	7.7	
								Scopulariopsis brevicalis	3.8	
								Fusarium	3.8	
	Tehran	100	35	ND	20–39	Fingernails	80	*C. albicans*	51.4	Khosravi et al. 2008
								Candida spp.	17.1	
								C. glabrata	11.4	
						Toenails	20	*C. tropicalis*	8.6	
								C. parapsilosis	5.7	
								C. kefyr	2.9	
								C. guillermondii	2.9	
	Khozestan (ahvaz)	1	ND	ND	60	Nails	ND	*A. flavus*	ND	Mahmoudabadi and Zarrin 2005
	Tehran	1	ND	ND	60	Toenails	ND	*A. candidus*	ND	Ahmadi et al. 2012
	Khorasanerazavi (Mashhad)	1	ND	ND	44	Fingernails	ND	*Tritirachium oryzae*	ND	Naseri et al. 2013a
	Kerman	1	ND	ND	51	Fingernails	ND	*Fusarium* spp.	ND	Mousavi et al. 2009

Table 1. contd....

Table 1. contd.

Disease	Place (Area)	Number of patients	Frequency (%)	Sex	Age	Affected body part (%)		Major Etiologic Agents (%)		Reference
Otomycosis	Isfahan (Kashan)	910	5.71	ND	15–60	External ear	ND	*A. niger*	61.5	Mogadam et al. 2009
								C. albicans	13.5	
								A. fumigatus	5.8	
								A. flavus	5.8	
								Aspergillus spp.	5.8	
								Mucor spp.	3.8	
								Rhizopus spp.	1.9	
	Fars (Shiraz)	486	10.28	ND	ND	External ear	ND	*Penicillium* spp.	32	Pakshir et al. 2008
								Cladosporium spp.	20	
								Alternaria spp.	8	
								A. flavus	8	
								Dematiaceous fungi	8	
								Rodotorula spp.	8	
								Exophiala spp.	4	
								C. albicans	4	
								C. tropicalis	4	
	Markazi (Arak)	435	0.45	ND	0–29	External ear	ND	ND	ND	Hashemi et al. 2014
	Mazandaran (Babol)	305	ND	ND	ND	External ear	ND	*A. niger*	ND	Sefidgar et al. 2002
	Khozestan (Ahvaz)	293	ND	ND	20–39	External ear	ND	*A. niger*	67.2	Saki et al. 2013
								A. flavus	13	
								C. albicans	11.6	
								A. fumigatus	6.2	
								Penicillium sp.	2	

	Tehran	200	50.9	ND	30–50	External ear	ND	*A. niger*	89.4	Nowrozi et al. 2014
								A. fumigatus	5.3	
								C. albicans	4.4	
								Penicillium sp.	0.09	
	Isfahan	171	69	ND	21–40	External ear	ND	*A. flavus*	49	Barati et al. 2011
								A. niger	41.6	
								C. albicans	7.6	
								A. fumigatus	5.5	
								A. nidulans	3.7	
								C. parapsilosis	0.9	
	Gilan (Rasht)	100	43	ND	8–81	External ear	ND	*A. niger*	16	Nemati et al. 2014
								A. fumigatus	7	
								C. albicans	12	
	North-Western Area of Iran	89	ND	ND	20–40	External ear	ND	*A. niger*	ND	Kazemi and Ghiasi 2005
								A. terreus	ND	
								A. flavus	ND	
								A. fumigatus	ND	
	North-Western Area of Iran	87	ND	ND	20–40	External ear	ND	*Aspergillus* spp.	79	Ghiyasi, 2001
								C. albicans	15	
	Khozestan (Ahvaz)	57	45.6	ND	4–85	External ear	ND	*A. niger*	30.8	Mahmoudabadi et al. 2010
								A. flavus	23.1	
								A. terreus	7.7	
								A. persicolor	3.8	
								C. albicans	11.5	
								C. parapsilosis	7.7	
								Penicillium spp.	3.8	
								Malassezia spp.	11.5	
	Khozistan	52	ND	ND	1–60	External ear	ND	*A. niger*	ND	Sheikh et al. 1993

Table 1. contd....

Table 1. contd.

Disease	Place (Area)	Number of patients	Frequency (%)	Sex	Age	Affected body part (%)		Major Etiologic Agents (%)		Reference
Tinea versicolor (Pityriasis Versicolor)	Tehran	100	ND	ND	20–30	Skin	ND	*M. globosa*	53.3	Tarazooie et al. 2004
								M. furfur	25.3	
								M. sympodialis	9.3	
								M. obtusa	8.1	
								M. slooffiae	4	
	Azarbaijane sharghi (Tabriz)	1023	ND	Men (352)	1–80	Skin	ND	*Malassezia* spp.	ND	Kazemi et al. 2013
				Women (671)						
	Sistano Baluchestan (Zahedan)	800	42.25	Men (400)	ND	Skin	ND	*Malassezia* spp.	ND	Ebrahimzadeh 2009
				Women (400)						
	Khozestan (Ahvaz)	500	30.6	ND	6–66	Skin	ND	*Malassezia* spp.	ND	Mahmoudabadi et al. 2009
	Markazi (Arak)	435	1.8	ND	0–29	Skin	ND	*Malassezia* spp.	ND	Hashemi et al. 2014
	Khorasane Razavi (Mashhad)	215	ND	ND	ND	Skin	ND	*Malassezia* spp.	ND	Maleki and Fata 2005
	Yazd	200	50	Men (105)	19–70	Skin	ND	*M. globosa*	38.3	Jafari et al. 2013
								M. furfur	29.4	
				Women (95)				*M. sympodialis*	14.9	
								M. pachydermatis	9.6	
								M. slooffiae	5.3	

	Tehran	166	ND	ND	2–66	Skin	ND	*M. globosa*	31.3	Rasi et al. 2009
								M. furfur	20.5	
								M. pachydermatis	7.2	
								M. restricta	7.2	
								M. slooffiae	3.6	
	Mazandaran (Sari)	134	74.6	ND	5–45	Skin	ND	*M. globosa*	ND	Afshar et al. 2013
								M. furfur	ND	
								M. slooffiae	ND	
								M. restricta	ND	
								M. sympodialis	ND	
	Khozestan (Ahvaz)	110	24.5	Men (64)	ND	Skin	ND	*M. globosa*	40.7	Mahmoudabadi et al. 2013
								M. pachydermatis	22.2	
								M. furfur	11.1	
				Women (46)				*M. restricta*	7.4	
								Malassezia species	18.5	
	Tehran	25	ND		13–55	Skin	ND	*M. globosa*	ND	Khosravi et al. 2009
								M. furfur	ND	
								M. sympodialis	ND	
								M. pachydermatis	ND	
								M. obtusa	ND	
	Yazd	1	ND	Men	11	Knees	ND	*M. furfur*	ND	Akaberi et al. 2009
Psoriasis	Tehran	110	ND	ND	20–30	Skin	ND	*M. globosa*	ND	Zomorodian et al. 2008
								M. furfur	ND	
								M. restricta	ND	
	Khorasane Razavi (Mashhad)	50	22	Men (30)	ND	Skin	ND	*Malassezia* spp.	ND	Javidi et al 2007
				Women (20)						

Table 1. contd....

Table 1. contd.

Disease	Place (Area)	Number of patients	Frequency (%)	Sex	Age	Affected body part (%)		Major Etiologic Agents (%)		Reference
Seborhoeic dermatitis	Mazandaran (Sari), Tehran	100	ND	ND	12–65	Skin	ND	*M. globosa*	ND	Hedayati et al. 2010
								M. furfur	ND	
								M. restricta	ND	
								M. sympodialis	ND	
								M. japonica	ND	
	Tehran	81	ND	ND	21–30	Skin	ND	*M. globosa*	ND	Saghazadeh et al. 2010
								M. restricta	ND	
								M. furfur	ND	
								M. sympodialis	ND	
								M. obtusa	ND	
								M. slooffiae	ND	
Mycotic Keratitis	Mazandaran (Sari)	22	ND	ND	15–83	Cornea	ND	*A. fumigatus*	ND	Shokohi et al. 2006
								Fusarium spp.	ND	
	Fars (Shiraz)	ND	ND	ND	ND	Cornea	ND	*Aspergillus* spp.	ND	Berenji and Elahi 2003
								Fusarium solani	ND	

ND: Not determined

percentage of the population proceeds to show clinical symptoms. Given the prevalence of these organisms, researchers have come to the conclusion that dermatophytosis is not a contagious disease. They believe that unknown factors are involved in disease manifestation.

Table 1 represents detailed information about the incidence and prevalence of dermatophytosis over the entire country. Overall, it has been shown that the disease prevalence is around 2.1% to 74% in different parts of Iran (Mahboubi et al. 2006; Rafiei and Amirrajab 2010). Men were affected more than women. In recent years, efforts at improving the standard of health and increased awareness about the disease have led to a dramatic reduction in the disease incidence.

Chadegani et al. (1987) reported 12,000 patients with skin diseases from Isfahan, out of which 10.8% were affected with dermatophytoses. Among the 10.8% group, lesions of tinea capitis were most common (72.1%) and *T. verrucosum* was the most frequent (43.8%) dermatophyte isolated. Omidynia et al. (1996) reported a total of 7495 individuals of which 681 (9%) were suspect of having cutaneous mycoses. Dermatophytoses were the most common infections (259/681 = 38%). Among 259 individuals affected by dermatophytes, tinea capitis were observed in 163 (62.9%); tinea corporis in 27 (10.4%); tinea mannum and tinea cruris in 19 (7.3%) each; tinea barbae and faciei in 14 (5.4%); tinea pedis in 13 (5%) and tinea unguium in 4 (1.5%). A total of 144 patients yielded dermatophyte cultures. The frequency of the isolated species in decreasing order was *T. verrucosum*, 78 (54.1%); *T. schoenleinii*, 48 (33.3%); *M. canis*, 8 (5.5%); *E. floccosum*, 5 (3.5%); *T. mentagrophytes* and *M. gypseum*, 2 (1.4%) each; and *T. tonsurans*, 1 (0.7%). Mahmoudabadi (2005) examined 279 patients suspected of fungal infections in Ahvaz. Skin scrapings, hair samples and nail clippings were collected from patients. Direct and culture examinations were performed for all samples. About 115 cases of the examined subjects had dermatophytosis. Dermatophytosis occurred mainly in adult males (20–29 years). Tinea cruris (24.3%) was the most common type of dermatophytosis followed by tinea pedis (16.5%), tinea corporis (14.8%), tinea unguium (13%), tinea capitis (11.3%), tinea faciei (11.3%), tinea mannum (7%) and tinea barbae (1.7%). *T. mentagrophytes* was the most prevalent species followed by *E. floccosum*. Hashemi et al. (2005) reported 750 suspected samples (including 560 male and 190 female) from Karaj, in which 157 cases (21%) were suffering from dermatophytosis and out of them, 100 cases were culture positive. Out of the isolated dermatophytes, 69% were anthropophilic, 30% zoophilic and 1% were geophilic. Zoophilic agents were *T. verrucosum* (11%), *M. canis* (10%) and *T. mentagrophytes* (9%). The most affected age group was 0–9 years old.

Overall, these data show that dermatophytosis is distributed across various parts of Iran and affects healthy individuals and those suffering from underlying diseases. Bassiri-Jahromi and Khaksar (2006a) reported a total of 1568 patients with suspected tinea capitis in Tehran for causative fungal agents between 1994 and 2001. Laboratory examination confirmed tinea capitis in 209 patients. Males were affected more frequently (67.5%) than females (32.5%) and in both sexes, those who were 3–11 years old, were more infected. *T. violaceum* was the most common etiological agent (37.3%) followed by *T. schoenleinii* (21.5%), *M. canis* (18.6%), *T. verrucosum* (14.8%), *T. tonsurans* (5.3%), *T. rubrum* (1%), *M. gypseum* (1%) and

T. mentagrophytes (0.5%). Hedayati et al. (2007) studied 324 wrestlers (aged 9–20 years) from 7 active clubs in Sari. Skin scrapings were obtained from 135 wrestlers suspected of having tinea gladiatorum. The scraped skin samples were evaluated with potassium hydroxide. Pleated carpet sterile fragments (5 × 5 cm) were used for survey of wrestling mat contamination. Sabouraud dextrose agar with and without chloramphenicole and cycloheximide was used to culture scrapings and wrestling mat samples. The dermatophytes were identified by routine laboratory techniques. This study showed that of the 324 wrestlers, 65 (20.1%) had tinea gladiatorum. Most lesions were on the trunk and head. All the wrestling mat samples were positive for dermatophytes. *T. tonsurans* was isolated from all the scrapings and wrestling mat samples. Dehghan et al. (2009) examined specimens from 1108 patients clinically suspected of fungal infections during 2003 to 2007 in Gorgan. Specimens collected from hair, nails and skin were investigated through direct examination and culture. Fungal colonies were identified by macroscopic and microscopic examination. Around 351 out of 1108 samples were positive for dermatophytes of which 277 had positive cultures. *E. floccosum* was the most frequent species (70.4%) followed by *T. rubrum* (14.5%) and *M. audouinii* (7.2%). Regarding the location of the lesions, groin and nails were the most frequent sites that developed dermatophytosis in the majority of the patients. Sepahvand et al. (2009) reported 294 patients suspected of dermatophytosis in Khorramabad. The age average was 23.5 years and the number of affected men was more than that of women. The most frequent types of dermatophytosis were Tinea corporis (25.6%) and Tinea cruris (25%). *Epidermophyton floccosum* and *T. verrucosum* were the most common dermatophytes isolated. In a comprehensive study by Sadeghi et al. (2011) in Tehran, a total of 12,461 individuals suspected of superficial mycoses were examined. Of these, 4871 cases were positive for fungal infections. Plucked hairs, skin and nail scrapings were examined and identified via direct microscopy and culture. From 4871 patients, 4015 strains were identified including dermatophytes (2635 cases, 65.7%), yeasts (1210 cases, 30.1%) and molds (170 cases, 4.2%). Among the dermatophytes, the most common pathogen isolated was *T. rubrum* (28.7%), followed by *E. floccosum* (25.0%), *T. mentagrophytes* (23.8%) and *T. tonsurans* (11.5%). Among the yeast-like fungi, a predominance of *Candida* spp. (54.5%) was observed. Of these, 29.3% were *C. albicans*, *Aspergillus* spp. was the most prevalent isolated mold (71.8%).

Candidiasis

Candidiasis includes a spectrum of diseases from cutaneous to mucosal, systemic, or multisystem disseminated disease caused by the dimorphic yeasts belonging to the genus *Candida*. This disease is considered to be the most important nosocomial fungal infection with high mortality in disseminated forms in immunocompromised patients. *Candida* species are opportunistic pathogens, causing disease primarily in debilitated or immunocompromised patients. The most common etiologic agent is *Candida albicans*, an endogenous commensal member of the mucosal microbiota of humans which isolates from 30%–50% of healthy human beings (Reiss et al. 2011).

There are five *Candida* species of major medical importance: *C. albicans*, *C. glabrata*, *C. tropicalis*, *C. parapsilosis*, and *C. krusei*.

Superficial candidiasis is a common problem in Iran (Fig. 3). It has been reported from most parts of the country mainly affecting the nails and is defined as onychomycosis (Table 1). As indicated in Table 1, other forms of superficial candidiasis like as vaginal candidiasis in outpatients are frequent as well.

Moghaddami et al. (1991) reported a case of chronic mucocutaneous candidiasis due to *C. albicans* in a 13 year-old boy.

Sadeghi et al. (2011) examined 4871 patients suspected of superficial fungal infections in Tehran. They reported 30.1% candidiasis induced by *C. albicans* (29.29%) and *Candida* spp. (70.71%). The most affected body parts were fingernails (72%) and groin (12.9%).

Badiee et al. (2011) studied 410 patients suspected of candidiasis in Shiraz. They isolated 354 *Candida* species from the samples. The most abundant species was *C. albicans* (48.6%) followed by *C. krusei* (17.5%), *C. glabrata* (11.3%), *C. kefyr* (11.3%), *C. parapsilosis* (5.1%), *C. tropicalis* (1.7%) and *C. dubliniensis* (1.7).

Faraji et al. (2012) studied 100 clinical cases suspected of candidiasis in Kermanshah. About 12% (12 cases) in direct microscopy test and 20% (20 cases) by cultivation on SDA were identified as vaginal candidiasis. The frequencies of the isolated *Candida* species include *C. albicans* with 62.5% (20 cases), *C. glabrata* with 18.7% (6 cases), *C. tropicalis* with 9.4% (3 cases) and *C. parapsilosis* with 9.4% (3 cases). Vulvovaginal candidiasis was more prevalent in women without blood glucose level control than those with blood glucose level control. *C. albicans* was, by far, the most predominant yeast isolated.

In a comprehensive study by Razzaghi-Abyaneh et al. (2014) in Tehran, 173 patients suspected of candidiasis were examined. Clinical samples including skin and

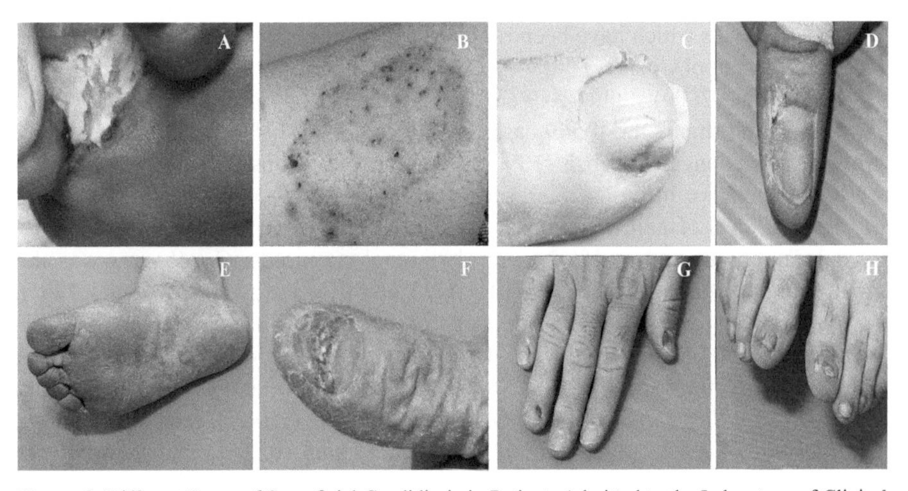

Figure 3. Different Types of Superficial Candidiasis in Patients Admitted to the Laboratory of Clinical Mycology of the Pasteur Institute of Iran (A, B and E show skin involvement; remaining figures show toenail and fingernail candidiasis).

nail scrapings (107; 61.8%), vaginal discharge (28; 16.2%), sputum (20; 11.6%), oral swabs (7; 4.0%), bronchoalveolar lavage (6; 3.5%) and 1 specimen (0.6%) of each eye tumor, gastric juice, urine, biopsy and urinary catheter were taken and confirmed as candidiasis via direct microscopy, culture and histopathology. Susceptibility patterns of the isolated *Candida* species were determined using the disc diffusion and broth microdilution methods. Among 173 *Candida* isolates, *C. albicans* (72.3%) was the most prevalent species followed by *C. parapsilosis* (11.5%). Other identified species were *C. glabrata, C. krusei, C. tropicalis, C. guilliermondii, C. intermedia* and *C. sake.*

Erami et al. (2014) reported a 40-year-old Iranian man from Kashan who had developed a painful swelling on the left knee for a year. Infectious arthritis due to *Candida glabrata* is very rare.

Onychomycosis

Onychomycosis is a frequent fungal disease of nails which is triggered by dermatophytes, yeasts and non-dermatophyte molds. Although non-dermatophyte filamentous fungi are commonly found in soil as well as plant debris and mostly considered as plant pathogens, they are also known as the causative agents of onychomycosis and therefore, are the main focus of considerable studies in this field. It has been estimated that 18%–40% of the nail changes and 30% of all fungal infections are cases of onychomycosis. Dermatophytes are the fungi most commonly responsible for onychomycosis in the temperate western countries; while *Candida* and non-dermatophytic molds are more frequently involved in the tropics and subtropics with a hot and humid climate.

The disease has been reported from various cities of Iran including Tehran, Kermanshah, Mazandaran, Esfehan, Qhazvin, Yazd, Kashan, Mashhad, and Khozestan (Table 1). Here, we shall discuss all types of onychomycosis including those caused by dermatophytes which have been reported under the term "Tinea unguium" in the section "Dermatophytosis".

Zaini et al. (2009) studied 549 cases suspected of nail infections during 2004 to 2005 in Tehran, Iran. Out of 549 cases examined, 263 (47.9%) were mycologically proven cases of onychomycosis (139 fingernails, 124 toenails), of which 33 (6.09%) were only positive in direct microscopic examination. From an etiological point of view, 21.85% of nail infections were caused by yeasts, 10.55% were infected by dermatophytes and 15.5% by non-dermatophyte molds. *C. albicans* was the most common yeast causative agent (16.73%) followed by *A. flavus* (11.78%), *T. mentagrophytes* (10.26%), *C. parapsilosis* (9.12%), *C. tropicalis* (8.74%), *A. fumigatus* (6.08%), *T. rubrum* (4.94%), *A. niger* (3.04%), *Penicillium* spp. (2.66%), *Aspergillus* spp. (1.90%), each of *Rhizopus* spp. and *Cladosporium* spp. (1.52%), *C. guilliermondii* (1.14%), *Scopolariopsis* spp. (1.14%), each of *C. famata, C. glabrata, C. krusei, S. lusitania, Acremonium* spp. (0.76%) and *C. homicola* (0.38%), *T. rubrum* (4.94%). *Candida* species were most common responsible agent for onychomycosis in female hands (74.1%) followed by 17.26% non-dermatophyte molds. Dermatophytes caused tinea unguium of hand (8.63%) and peduum (37.1%) in males. Onychomycosis of fingernails was most prevalent in females while toenail infection was common in male patients. Hashemi et al. (2010) studied nail samples from 504 patients with prediagnosis of onychomycosis during 2005 in Tehran via direct microscopy and culture. Of 504 cases examined, 216 (42.8%) were mycologically proven cases of onychomycosis (144 fingernails, 72 toenails). Among the positive results, dermatophytes were diagnosed in 46 (21.3%), yeasts in 129 (59.7%) and non-dermatophytic molds in 41 (19%). *T. mentagrophytes* was the most common causative agent (n = 22), followed by *T. rubrum* (n = 13), *C. albicans* (n = 42), *Candida* spp. (n = 56) and *Aspergillus* spp. (n = 21).

Bassiri-Jahromi and Khaksar (2010) examined 1268 patients with nail abnormalities in Tehran. In 410 cases (190 involving toenails and 220 involving fingernails), the etiologic agents of onychomycosis were established after repeated cultural examinations. From 410 culture-positive specimens, 47 (11.5%) yielded non-dermatophytic molds, 166 (40.5%) dermatophytes, 197 (48%) yeasts and 9 (2.2%) mixed (two different fungi) growth. All patients with mold onychomycosis had subungual hyperkeratosis. *Aspergillus* spp. was the etiologic agent most responsible for non-dermatophytic onychomycosis, reported from a total of 28 patients (59.6%). Other causative agents were *Acremonium* spp. (17%), *Fusarium* spp. (12.7%), *Geotrichum* spp. (4.2%), *Trichosporun* spp. (4.2%), and *Scopulariopsis* spp. (2.1%). Mold onychomycosis developed mainly in toenails (74.5%). Chadeganipour et al. (2010) evaluated 488 patients suspected of onychomycosis in Isfahan. Direct microscopy of the nail clips was positive in 194 (39.8%) cases. Fingernail onychomycosis was recognized in 141 (72.7%) and toenail onychomycosis in 53 (27.3%) cases. As agents of onychomycosis, yeasts were detected in 112 (57.7%), dermatophytes in 27 (13.9%) and non-dermatophyte fungi in 55 (28.4%) patients. Of the samples cultured, *C. albicans* (84%) was the most prevalent yeast. Among dermatophytes, *T. mentagrophytes* var. *interdigitale* was found to be the most commonetiological agent (8.6%) followed by *E. floccosum* and *T. rubrum*. Among the non-dermatophyte molds, *A. flavus* was the most prevalent species (13%). Moreover, nine samples with positive direct microscopy yielded no growth. Females were affected more frequently with fingernail *Candida* infections than males, and children under 7 years of age were predominantly infected with *Candida paronychia*.

Aghamirian and Ghiasian (2010) studied nail scrapings collected over a 4-year period in Ghazvin. The microscopic and/or cultural detection of fungi was positive in 40.2% of the samples. The most common clinical type noted was distolateral subungual onychomycosis in 48.4% of the cases. Etiological fungal agents were 50% dermatophytes, 46.8% yeasts, and 3.2% saprophytic molds. The most frequently detected dermatophyte species were *T. rubrum* (48.4%) and *T. mentagrophytes* (41.9%). Among yeasts, *C. albicans* (58.6) was the most common species, followed by *C. parapsilosis* (17.2%), *C. glabrata* (10.3%), *C. krusei* and *C. tropicalis* (each 6.9%). *A. niger* and *A. flavus* were the most frequent saprophytic molds. Females were affected more frequently than males, and in both sexes those most infected were between 40–49 years of age. Fingernails were affected more frequently than toenails.

Kafaie and Noorbala (2010) examined 262 (123 men and 139 women) diabetic type 2 patients which were referred to the Yazd diabetic clinic during 2008 to 2009. Onychomycosis was diagnosed clinically in 18 patients (6.9%) and proved by culture in 10 patients (3.8%). 70% of cases were men and 60% were over 60 years old. No significant relation was found between the frequency of onychomycosis and the diabetic duration. In patients who had a weaker control of diabetes, onychomycosis was more frequent.

Ahmadi et al. (2012) reported a case of a toenail infection caused by *A. candidus* in a healthy 60-year-old woman. Based on macroscopic and microscopic characteristics of the culture as well as nucleotide sequencing of 28S region of DNA, the causative agent was identified as *A. candidus*. Mikaeili and Karimi (2013) studied onychomycosis in 2402 patients in Kermanshah Province, western Iran during 1994 to 2010. Direct microscopy of the nail clips was positive in 1086 (45.2%) and fingernail and toenail onychomycosis were recognized in 773 (71.1%) and 313 (28.8%) patients, respectively. Yeasts were detected in 853 (78.5%), dermatophytes in 201 (18.5%) and non-dermatophyte fungi in 32 (2.9%) patients. The results of fungal culture showed *C. albicans* isolated from 384 (45.0%) cases and other *Candida* spp. isolated from 361 (54.0%) cases as the most common agents of onychomycosis while among dermatophytes, *T. rubrum* was found in 63 (37.0%) of the cases as the main dermatophytic agent followed by *T. mentagrophytes* 32 (15.9%) and *E. floccosum* 30 (17.6%). Among the non-dermatophyte molds, *A. flavus* was the most prevalent species (37.5%) followed by *A. niger* (25.0%) and *A. fumigatus* (12.5%). About 139 (12.8%) samples with positive direct microscopy yielded no growth. The highest rate of onychomycosis was found in patients between 30–40 years of age.

Naseri et al. (2013a) reported the first case of *Tritirachium oryzae* as the etiologic agent of onychomycosis in Mashhad. A 44-year-old woman with a lesion in her fingernail was examined directly via the microscope, revealing fungal filaments. Inoculation of portions of the nail clippings on media cultures yielded *T. oryzae* after 8 days. The isolate was identified as *Tritirachium* spp. on the basis of gross morphological characteristics of the fungal colony and microscopic characterization of slide cultures. The diagnosis of *T. oryzae* was confirmed by PCR sequencing of the internal transcribed spacer domain of the rDNA gene. *In vitro* antifungal susceptibility test demonstrated that the fungus was susceptible to itraconazole and posaconazole. The patient was treated with oral itraconazole. Afshar et al. (2014) examined 1100 patients suspected of onychomycosis (398 males and 702 females, aged 1–88 years), during 2003 to 2012. Onychomycosis was diagnosed in 625 (56.8%) cases in Mazandaran. Among cases of onychomycosis, laboratorial confirmation was reached through direct examination with positive cultures in 464 samples (74.3%), while only by positive direct examination in 114 cases (18.2%) or just positive culture in 47 cases (7.5%). The results of fungal culture revealed *Candida* spp. as the most common agent of onychomycosis while among dermatophytes, *T. mentagrophytes* was found in 17.7% followed by *T. rubrum* (1.7%), *E. floccosum* (0.7%), *T. violaceum* (0.2%), *T. verrucosum* (0.2%), *T. tonsurans* (0.2%) and *M. gypseum* (0.2%). Among the non-dermatophyte molds, *Aspergillus* spp. was the most prevalent species by 14.2% frequency.

Otomycosis

Otomycosis is a superficial mycotic infection of the outer ear canal. The infection may either be acute or subacute and is characterized by inflammation, pruritis, scaling, feeling of fullness and severe discomfort. Occasionally a fungus is considered as the primary pathogen in otitis externa, and otomycosis accounts for about 10% of otitis externa cases. Otomycosis occurs worldwide but it is more common in tropical and subtropical areas. The disease prevalence is influenced by a number of predisposing factors such as climate (extremely moist and hot environments), chronic bacterial otitis externa, swimming, dermatomycoses, insertion of foreign subjects and wearing head clothes.

As summarized in Table 1, the disease has been reported from Tehran, Esfehan, Kashan, Khozestan, Babol, Sari, Fars and Tabriz in Iran (Table 1).

Mogadam et al. (2009) studied 910 patients suffering from external otitis in Kashan and reported that 52 patients (5.7%) had otomycosis including 16 males (30.8%) and 36 females (69.2%). *Aspergillus niger* was the most common etiologic agent (61.5%).

Pakshir et al. (2008) studied a total of 486 ear samples of 243 healthy individuals (100 females and 143 males) in Shiraz and found that 10.28% of the samples were positive for fungal growth. The isolated fungal species were distributed in eight genera including *Penicillium* (8 cases, 32%), *Cladosporium* (5 cases, 20%), *Candida* (3 cases, 12%), *Aspergillus* (2 cases, 8%), *Alternaria* (2 cases, 8%), Dematiaceous fungi (2 cases, 8%), *Rodotorula* (2 cases, 8%) and *Exophiala* (one case, 4%). Sixty percent of cases with positive fungal cultures were males and 27.52% had a history of otalgia. Around 34% of the individuals had a history of water remaining in their external auditory canal after bathing or swimming of which 10.97% were positive for fungal agents.

Sefidgar et al. (2002) examined 305 patients (216 females and 89 males) for otomycosis in Babol. The most common fungus isolated was *A. niger* (59.9%).

Saki et al. (2013) studied 293 cases, comprising 162 women and 131 men in Ahvaz. They showed that 20–39 year age group had the highest prevalence of otomycosis. The seasonal distribution of cases was reported as summer, 44.7%; autumn, 28.7%; winter, 14.7%; and spring, 11.9%. The fungal agents isolated were *A. niger* (67.2%), *A. flavus* (13%), *C. albicans* (11.6%), *A. fumigatus* 6.2%) and *Penicillium* species (2%).

Nowrozi et al. (2014) reported a total of 200 patients including 73 male (36.5%), 127 female (63.5%) with clinical diagnosis of otitis externa in Tehran. Around 114 patients were positive for fungi. The most common isolated fungus was *A. niger* with 102 cases (89.4%) followed by *A. fumigatus*, *C. albicans* and *Penicillium.* The seasonal distribution of isolated fungi was reported as 45% in summer, 25% in autumn, 18% in winter and 12% in spring. All patients with otomycosis were immunocompetent which meant that they did not use any antibiotics or steroid drugs. Only 1 patient had bilateral otomycosis.

Barati et al. (2011) reported 171 cases with clinical diagnosis of otomycosis including 86 (50.3%) females and 85 (49.7%) males in Isfahan. The average age of patients was 35.8 (range: 9–78) years old. Patients in their fourth decade of life made up the biggest group (30.4%) followed by 21–30 age group (22.2%). Construction workers and farmers (working in dry, dusty environments) made up the biggest group (61.1%) while among male and female patients, housewives and farmers were the biggest group (73.2%). The seasonal distribution of otomycosis was 36.8% in autumn, 30.4% in summer, 18.1% in winter and 14.6% in spring. Otomycosis was diagnosed in 118 patients (69%) by a positive fungal isolation in culture. The most common fungal isolates belonged to *Aspergillus* accounting for 91.5% of all fungal isolates. Out of 108 *Aspergillus* positive samples, *A. flavus* was the most common species (49%), followed by *A. niger* (41.6%), *A. fumigatus* (5.5%) and *A. nidulans* (3.7%). Species of *Candida* accounted for 8.5% of the fungal isolates.

In a recent report by Nemati et al. (2014) from Rasht, 100 patients with diagnosis of otomycosis/otitis externa were studied. Otomycosis was confirmed in 43% of patients by positive culture. The most prevalent fungal pathogen was *A. niger* which was sensitive to clotrimazole, fluconazole, ketoconazole.

Kazemi and Ghiasi (2005) studied 89 patients; 64 male and 25 female and reported *A. niger* (51 cases) in the North-Western area of Iran. *Aspergillus terreus* (9 cases), *A. flavus* (7 cases), *A. fumigatus* (6 cases), *Eurotium* (prefect stage of *A. glaucus*) (2 cases), *Penicillium* sp. (2 cases), *C. albicans* (8 cases), *Epicoccum* sp. (1 case), *Mucor* sp. (1 case), *Rhizopus* sp. (1 case) and *M. canis* (1 case) were reported as the etiologic agents of otomycosis.

Mahmoudabadi et al. (2010) reported 57 cases of otomycosis in Ahvaz. The most common fungal pathogens were *A. niger* (30.8%) and *A. flavus* (23.1%).

Overall, these results clearly show that otomycosis is a serious health problem in Iran which has been reported from nearly all major cities of the country.

Tinea versicolor

Tinea versicolor (*Pityriasis versicolor*) is a chronic, superficial mycosis affecting the stratum corneum of the smooth skin, usually on the upper chest, back and arms, representing as discolored spots which slowly enlarge and can become confluent patches. The causative agents are lipophilic yeasts of the genus *Malassezia* (Reiss et al. 2011). Besides tinea versicolor, *Malassezia* species are believe to be involved in the etiology offolliculitis, atopic dermatitis, psoriasis and seborrheic dermatitis. The prevalence of this disease is varying in the world with a rate of 5% to 50% and a high rate of 20% to 50% in tropical and subtropical regions. This disease is more prevalent in males than females and exogenous and endogenous predisposing factors initiate the disease by converting the fungus from yeast to a pathogenic mycelial form. These factors, such as age, sex, climate, local environmental factors, malnutrition, and genetic factors influence the course of the disease.

The disease has been reported from various parts of Iran including Tehran, Khozestan, Sari, Mashad, Yazd and Tabriz (Table 1). Iran is located in a subtropical region and several reports show that tinea versicolor is more prevalent in various provinces with a higher rate in north and south, which have warm and humid climate (Fig. 4). Frequency of pityriasis versicolor in Iran varies from 4.4% to 57.7% as indicated in different reports (Mahmoudabadi et al. 2009).

Tarazooie et al. (2004) reported a total of 100 cases of tinea versicolor from Tehran. The most frequently isolated species was *M. globosa* (53.3%), followed by *M. furfur* (25.3%), *M. sympodialis* (9.3%), *M. obtusa* (8.1%) and *M. slooffiae* (4.0%). The most frequently isolated species in the skin of healthy individuals were *M. globosa*, *M. sympodialis*, *M. furfur*, *M. slooffiae* and *M. restricta* which comprised 41.7%, 25.0%, 23.3%, 6.7% and 3.3% of the isolated species.

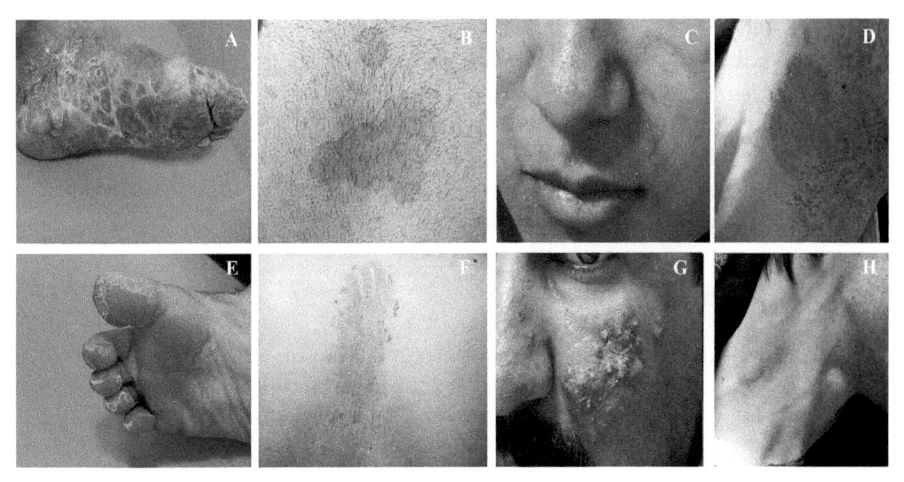

Figure 4. Clinical Features of Tinea Versicolor in Patients Admitted to the Clinical Laboratory of the Pasteur Institute of Iran. Involvement of Feet (A and E), Face (C and G), Neck (H) and Trunk (B, D, and F) are shown.

In a 3 year study between 2009 and 2011, Kazemi et al. (2013) studied 1023 patients, 671 females (66%) and 352 males (34%) who suffered from skin lesions suspected to be tinea versicolor. The disease was more prevalent in the 21–40 year old age group in both genders. The most infected anatomical regions were the posterior surface, the body trunk (shoulder, supra scapula and lumbar region), anterior thorax and abdomen.

Ebrahimzadeh (2009) evaluated 800 student volunteers from universities in Southeast Iran and reported a prevalence of 42.25% for tinea versicolor. The highest prevalence of tinea versicolor (60%) was seen in the students who were 20–21 years old.

Mahmoudabadi et al. (2009) studied 500 patients, 62.1% males, and 37.9% females suspected of tinea versicolor in Ahvaz and found that 30.6% of patients were positive for tinea versicolor. The highest prevalence of tinea versicolor (70.6%) was seen in patients who were 17–28 years old.

Jafari et al. (2013) reported a total of 200 cases of tinea versicolor in Yazd. The most commonly isolated species were *M. globosa* (38.3%), *M. furfur* (29.4%), *M. sympodialis* (14.9%), *M. pachydermatis* (9.6%) and *M. slooffiae* (5.3%).

Rasi et al. (2010) reported a total of 116 patients with positive culture for *Malassezia* species in Tehran. *M. globosa* was found in 52 (31.3%) cases, *M. furfur* in 34 (20.5%) cases, *M. pachydermatis* in 12 (7.2%) cases, *M. restricta* in 12 (7.2%) cases, *M. slooffiae* in 6 (3.6%) cases.

Afshar et al. (2013) studied 134 skin scraping samples in Sari and showed that 116 patients (86.5%) were positive for tinea versicolor. *Malassezia* spp. was isolated from 100/116 (86.2%) patients. *M. globosa* was found in 54 cases, *M. furfur* in 32 cases, *M. sympodialis* in 6 cases, *M. restricta* in 6 cases and *M. slooffiae* in 2 cases.

Mahmoudabadi et al. (2013) reported that 27 of 110 (24.5%) Seborrheic dermatitis patients in Ahvaz had positive cultures for *Malassezia* species of which 17 (63%) were male and 10 (37%) were female. The most commonly identified *Malassezia* species was *M. globosa* (40.7%) followed by *M. pachydermatis* (22.2%), *M. furfur* (11.1%) and *M. restricta* (7.4%).

Saghazadeh et al. (2010) studied 100 patients with seborrheic dermatitis including females (60%) and males (40%) in the age range of 12 to 65 years from Tehran. *Malassezia globosa* was the most commonly isolated species (55.8%). This species was most frequent in scalp and face lesions.

Mycotic Keratitis

It has been indicated that although corneal infections have worldwide distribution, the incidence of these infections is higher in tropical and semitropical areas and is much more frequent in developing countries. It has also been mentioned that these infections follow trauma, especially with plant debris. Mycotic keratitis (Keratomycosis) is a suppurative, ulcerative, and sight-threatening infection of the cornea that can sometimes lead to loss of the eye. Worldwide, the reported incidence of mycotic keratitis is 17% to 36%. The distribution of fungi is not only affected geographically, but also changes with time. Despite advances in diagnosis and medical treatment of the disease, 15% to 27% of the patients require surgical intervention such as keratoplasty, enucleation, or evisceration because of either failed medical treatment or advanced disease at presentation. It is, therefore, very important to recognize the prevalence and etiology of corneal ulcer (Shokohi et al. 2006).

As indicated in Table 1, mycotic keratitis has been reported from Tehran, Shiraz and Sari in Iran.

Over a period of 6 years in Shiraz, Berenji and Elahi (2003) investigated the prevalence of mycotic keratitis among patients suffering from cornea infection. Sampling was done by an ophthalmologist and fresh smears and cultures were prepared for each patient. *Aspergillus* spp. and *Fusarium solani* were isolated from 55% and 7% of the cultures respectively.

Shokohi et al. (2006) studied a total of 22 patients in Sari, 10 (45.5%) females and 12 (54.5%) males. The mean age of patients was 61.5 ± 17.7 (range: 15–83) years. Of 22 patients with corneal ulcer surveyed during the period of May 2004 to March 2005 (12 months), fungal keratitis was identified as the principal etiology of corneal ulceration in 7 (31.8%) patients (5 males and 2 females). The mean age of patients with fungal keratitis was 60.4 ± 12.1 (range: 39–73) years. Three (42.9%) patients with fungal keratitis were farmers; one (14.3%) an animal husbandman, one (14.3%) a laborer, and 2 (28.6%) were housewives. All patients with fungal keratitis lived in rural areas. The mean interval between the onset of symptom and diagnosis was 26.4 (range: 1–93) days.

Subcutaneous Fungal Infections

Subcutaneous fungal infections are infections of the cutaneous and subcutaneous tissues that are caused by direct inoculation of fungal pathogens. The etiologic fungus remains at the site of inoculation locally or slowly developes into adjacent tissues. The fungi that cause these diseases mainly live in nature as saprophytes. In some forms of the disease, the infection may spread through the lymphatic channels (Sporotrichosis) and in some instances; disease may spread the infection through blood and lymphatics (Chromoblastomycosis).

Mycetoma

Mycetoma (Madura foot, Maduramycosis) is a chronic infection that develops very slowly. The disease is frequently seen in tropical and sub-tropical countries and is considered to be an occupational disease. The most affected area are the lower extremities especially the foot which swells and develops multiple draining sinuses from which pus, blood and seed of fungal mycelium called granules (Grains) are discharged. Others sites are hand, knee, arm, leg, head and neck, thigh and the perineum. Granules with white, yellow, brown, red or black colour depending on the causative agents have diagnostic value. There are two types of infections based on the etiologic agent including actinomycotic mycetoma (produced by actinomycetes) and eumycotic mycetoma (induced by fungi). In some areas of the world including Sudan, Mexico and India, mycetoma is a very common disease and in other areas such as Greece, Italy, Romania, Guatemala, Iran and some parts of South America, significant number of patients are reported every year.

As summarized in Table 2, the disease has been reported from most cities of Iran. Reports from Iran indicate that actinomycetoma has a frequency of about 84.5%, while eumycetoma comprises only 15.5% of the cases. Actinomycetoma is caused mainly by *Nocardia asteroides* and *Actinomadura madurae*, while eumycetoma is less common with a maximum number of cases due to *Pseudallescheria boydii*. Various actinomycetes including *Nocardia asteroids*, *Actinomadura Madura*, *Nocardia caviae*, *Actinomyces Israeli*, *Nocardia brasiliensis*, *Nocardia* and *Streptomyces somaliensis* (1.8%) have been reported as the etiologic agents of actinomycetoma in Iran. The disease is more prevalent in North (Gillan, Mazandaran and Golestan provinces) and South (Khuzestan, Bandar Bushehr and Bandar Abbas) of Iran.

In a report by Nematian (1984), 13 cases of mycetoma were reported from the cities Mazandaran, Gilan, Arak, Bushehr, Zanjan and Ghom. Of these 13 patients, 11 cases were actinomycetoma and 2 cases were diagnosed as eumycetoma. *Actinomadura madurae*, *Actinomyces israelii* and *Streptomyces somaliensis* were isolated as the main fungal pathogens.

Moghaddami et al. (1989) reported 13 cases of mycetoma from the Mazandaran, Golestan, Lorestan, Hormozgan, Zanjan, Hamedan, Kerman, Tehran, Gilan and Rasht. Of these 13 patients, 12 cases were actinomycetoma and one was eumycetoma. Fungal agents isolated were *Nocardia asteroids*, *Actinomadura madurae* and *Pseudallescheria boydii.*

Fata and Boloursaz (2000) reported 9 cases of mycetoma from the Khorasane Razavi province in east Iran. Of these 9 patients, 8 cases were actinomycetoma and one was eumycetoma. *Nocardia* spp., *Actinomadura* spp. and *Fusarium* sp. were diagnosed as etiologic agents.

Table 2. Geographic Distribution and Clinical Features of Subcutaneous Fungal Infections in Iran.

Disease	Place (Area)	Number of patients	Age	Sex	Involve-d body part	Major Etiologic Agents	Type of infection	Dissemination to other organs	Emputation	Reference
Mycetoma	Mazandaran, Gilan, Arak, Bushehr, Zanjan, Ghom	13	9–60	Man (7)	Foot (12)	*Actinomadura madurae*	Actinomycetoma (11)	ND	ND	Nematian 1984
						Actinomyces israelii				
						Streptomyces somaliensis	Eumycetoma (2)			
				Woman (6)	Thigh (1)	*Actinomyces israelii*				
						Streptomyces somaliensis				
	Gilan, Mazandaran (Babol), Zanjan, Golestan (Gorgan), Hormozgan (Bandar abas), Mazandaran (Ghaemshahr), Kerman, Hamedan, Gilan (Rasht), Tehran, Lorestan	13	20–65	Man (8)	Tarsus (3)	*Nocardia asteroids*	Actinomycetoma (12)	ND	ND	Moghaddami et al. 1989
					Foot (5)					
				Woman (5)	Shank	*Actinomadura madurae*				
					Elbow		Eumycetoma (1)			
					Finger	*Pseudallescheria boydii*				
					Arm (2)					
	Khorasane Razavi (Mashhad)	9	29–67	Man (6)	Foot	*Nocardia* spp.	Actinomycetoma (8)	Bones	No	Fata et al. 2000
						Actinomadura sp.				
				Woman (3)		*Fusarium* sp.	Eumycetoma (1)			
	Behshahr, Ghazvin, Mazandaran (Amol), Golestan (Gorgan), Hormozgan (Abadan), Mazandaran (Babol), Azarbaijane sharghi (Miyaneh)	8	25–50	Man (3)	Toe (2)	*Actinomadura madurae*	Actinomycetoma (6)	Adjacent tissues	No	Asgari and Alilou 1972
					Knee (2)	*Nocardia caviae*				
				Woman (3)	Metatarsus (2)	*Pseudallescheria boydii*	Eumycetoma (2)			
					Forehead	*Madurella mycetomatis*				
					Leg					

	Southern Iran	7	ND	ND	Foot	*Actinomyces israelii*	Actinomycetoma (5)	No	No	Griffiths et al. 1975
						Nocardia astroides				
						Actinomadura madurae	Eumycetoma (2)			
						Pseudallescheria boydii				
	North of Iran	6	ND	Man (5)	Foot	*Nocardia astroides*	Actinomycetoma	Yes	Yes	Ravaghi and Aflatouni 1978
				Woman (1)						
	Tehran, Mazandaran, Isfahan	4	33, 38, 42	Man (2)	Knee	*Nocardia caviae*	Actinomycetoma	No	No	Forouzesh et al. 1993
					Foot					
				Woman (2)	Elbow					
	Khuzestan	4	31, 42, 52, 56	Man (2)	Right hand (1)	*Nocardia* spp.	Actinomycetoma	Yes (lytic lesions and osteomyelitis)	No	Yaghoobi et al. 1999
				Woman (2)	Left foot (3)	*Streptomyces* spp.				
						Actinomyces spp.				
	Gilan, Bushehr, Mazandaran	3	34, 35, 44	Man (2)	Foot	*Actinomadura madurae*	Actinomycetoma (3)	ND	ND	Mahmoudabadi and Zarrin 2008
				Woman (1)						
	Isfahan(2) Caspian Sea (1)	3	23, 33, 52	Man	Foot	*Nocardia astroides*	Actinomycetoma	ND	ND	Shayban 1989
	Khuzestan	2	42, 56	Man	Left foot	*Nocardia brasiliensis*	Actinomycetoma	No	No	Zarrin and Mahmoudabadi 1997
				Woman						
	Kerman	2	38,41	Man	Foot	ND	Actinomycetoma	Yes (bone involvement)	No	Shamsoddin et al. 2004
	Kerman	2	38,41	Man	Right foot	ND	eumycetoma	Yes (osteomyelitis)	No	Shamsadini et al. 2007
					Right ankle					

Table 2. contd....

Table 2. contd.

Disease	Place (Area)	Number of patients	Age	Sex	Involve-d body part	Major Etiologic Agents	Type of infection	Dissemination to other organs	Emputation	Reference
Mycetoma	Golestan (Gorgan, Minoodasht)	2	39, 44	Man	Hand	ND	Actinomycetoma	No	No	Golsha et al. 2009
					Foot					
	Tehran	1	56	Woman	Arm	*Pseudallescheria boydii*	Eumycetoma	No	No	Moghaddami and Alikhani 1989
	Sistan and Baluchestan (Zahedan)	1	41	Man	Foot	*Actinomadura madurae*	Actinomycetoma	No	No	Ghamgosha et al. 2013
	Hormozgan (Bandar abas)	1	25	Woman	Foot	*Aspergillus flavus*	Eumycetoma	No	No	Hashemi and Gerami 2001
	Khuzestan (Ahvaz)	1	60	Woman	Left leg	*Actinomadura madurae*	Actinomycetoma	No	No	Mahmoudabadi and Yaghoobi 2007
	Tehran	1	60	-	Right foot	*Nocardia asteroids, Actinomadura madurae*	Actinomycetoma	ND	ND	Suodbakhsh et al. 2011
	Isfahan	1	48	ND	Hips	*Paecilomyces* spp.	Eumycetoma	No	No	Emami-Naeini et al. 2002
	Khorasane Jounobi (Birjand)	1	32	Man	Left Leg	*Nocardia* spp.	Actinomycetoma	Yes (Osteomyelitis)	No	Ziaei and Azarkar 2003
	Khuzestan	1	52	Man	Foot	*Nocardia* spp.	Actinomycetoma	No	No	Ebrahimzade 1976
	Tehran	1	24	Man	Arm	*Nocardia astroides*	Actinomycetoma	No	No	Khosravi and Emami 1991
					Foot					
	Isfahan	1	54	Man	Hips	ND	ND	No	No	Nourollohi et al. 1998
	Azarbaijane Gharbi (Urmia)	1	71	Man	Left ankle	*Nocardia brasiliensis*	Actinomycetoma	No	No	Azimi et al. 2000
	Bushehr	1	33	Man	Palm	*Nocardia asteroides*	Actinomycetoma	No	No	Shoar et al. 2003

	Isfahan	1	40	Woman	Right elbow	*Pseudallescheria boydii*	Eumycetoma	ND	No	Enshaieh et al. 2006
	Khuzestan	1	50	Man	Right foot	*Nocardia astroides*	Actinomycetoma	No	No	Mahmoudabadi 1995
Sporotrichosis	Tehran	1	56	Man	Face, Forearm	*Sporothrix schenckii*	Subcutaneous sporotrichosis	ND	ND	Ghodsi et al. 2000
	Azarbaijane Sharghi (Tabriz)	1	23	Man	Left arm	*Sporothrix schenckii*	Subcutaneous sporotrichosis	No	No	Kazemi and Razi 2007
	Tehran	1	36	Woman	Right arm	*Sporothrix schenckii*	Subcutaneous sporotrichosis	No	No	Zain 1984
Chromoblastomycosis	Mazandaran (Babol)	1	23	Woman	Right leg	*Phialophora verrucosa*	ND	No	No	Hasanjani Roushan et al. 200[illegible]
	Tehran	1	27	Man	Palate, Chest	Dematiaceous fungi	Oral chromoblastomycosis	ND	no	Fatemi and Bateni 2012
Rhinosporidiosis	Tehran, Ardebil and Kermanshah	74	20–30	Men (62)	Eye, Nose	*Rhinosporidium seeberi*	Subcutaneous rhinosporidiosis	ND	ND	Firouz-Abadi et al. 1971
				Women (12)						

ND: Not determined

Asgari and Alilou (1972) identified 8 cases of mycetoma in the Behshahr, Ghazvin, Mazandaran, Golestan, Hormozgan, Mazandaran, and Azarbaijane sharghi. Of these 8 patients, 6 cases were actinomycetoma and 2 eumycetoma. Etiologic agents were *Actinomadura madurae, Nocardia caviae, Madurella mycetomatis* and *Pseudallescheria boydii.*

Griffths et al. (1975) reported 7 cases of mycetoma in Southern Iran. Of these 7 patients, 5 cases were actinomycetoma and 2 eumycetoma. Involved pathogens were *Actinomadura madurae, Actinomyces israelii, Nocardia astroides* and *Pseudallescheria boydii.*

Ravaghi and Aflatouni (1978) identified 7 cases of mycetoma in the North of Iran. All cases were actinomycetoma caused by *Nocardia astroides.*

Sporotrichosis

Sporotrichosis is a subacute or chronic granulomatous fungal infection seen in humans, as well as in a wide range of domestic and wild animals. It is caused by the pathogenic soil dimorphic fungus *Sporothrix schenckii.* Infection is the result of traumatic skin inoculation through small cuts or punctures from soil, woody fragments of plants and other organic matter contaminated with *S. schenckii.* In most cases, clinical progression is slow and a skin bump of sporotrichosis can appear within 1–3 weeks after fungal exposure. The fungus spreads via local deep tissue invasion from the inoculation site and the infection usually remains localized, spreading slowly to adjacent tissue and eventually to the lymphatic vessels. Blood vessel dissemination has been rarely reported (Kazemi and Razi 2007). In the Americas, the incidence of this infection, especially in the highlands of Peru, is approximately 1 case per 1000 people. Zoonotic transmission of disease is rare, but small outbreaks can occur from infected cats, from horses with extensive skin lesions and amongst handlers in other cases of animal husbandry. The use of gloves and long sleeves was advised when handling pine seedlings, rose bushes, hay bales or other plants that could cause minor skin breaks.

Despite worldwide distribution of sporotrichosis, it has been rarely diagnosed in Iran. Sporotrichosis was unknown until the 1984's when it was reported by Zaini for the first time (Zaini 1984). The disease has been reported from the cities of Tehran, Mazandaran and Tabriz (Table 2). The low number of sporotrichosis cases in Iran could be due to its rare occurrence or perhaps to the laboratory unfamiliarity with this fungal pathogen. Service professionals who are active in areas endemic for leishmaniasis (such as Northwest Iran) must be made aware about the potential occurrence of sporotrichosis, particularly taking into account the at risk populations, such as farmers and horticulturalists.

Zaini (1984) reported the first case of sporotrichosis in a 36-year-old female housewife in Tehran, Iran. Ghodsi et al. (2000) reported one case of sporotrichosis in a 65-year-old woman with multiple ulcers on the face and the forearm in Tehran, Iran. At the time of admission two lesions were present on the right eyebrow and one on the right proximal forearm. Causative agent was *Sporothrix schenckii.* Kazemi and Razi (2007) reported a 23-year-old male florist gardener diagnosed with subcutaneous sporotrichosis caused by the dimorphic pathogenic soil fungus *Sporothrix schenckii* in Tabriz, Iran. The patient had several small skin lesions over the left upper arm with ascendant chains of enlarged lymph nodes. *Sporothrix schenckii* was detected from clinical samples via direct microscopy, culture and by its ability to switch from mould to yeast form at 37°C. The patient was successfully treated with long-term potassium iodide and advised to wear gloves and long sleeves when handling any kind of plant material.

Chromoblastomycosis

Chromoblastomycosis is a chronic fungal infection caused by a diverse range of soil saprophytic fungi. It is generally accepted that chromoblastomycosis usually affects one leg or foot. In some rare instances the disease begins on the hand or wrist and involves the entire upper extremity. It may also begin on the face. It begins as a small papule or warty growth and slowly spreads via the growth of satellite lesions. The affected area is usually swollen and there is a slow progression of the disease. Plaque-like and cicatricial types of lesions also occur. The lesions that may or may not ulcerate are characterized by round, brown bodies that reproduce by equatorial splitting. The infection is commonly seen among barefooted farm laborers, predominantly in males in the 20–50-year age range. It has been reported from different countries (Fatemi and Bateni 2012).

As indicated in Table 2, chromoblastomycosis has been reported only in the cities Babol and Tehran in Iran as two separate case reports.

Hasanjani et al. (2001) reported a case of chromoblastomycosis caused by *Phialophora verrucosa* in Babol. The patient was a 23-year-old female suffering from a chronic ulcer, like a wart, in her right leg following a leach bite received while working in the paddy field. The patient was treated with 5-fluorocytosine for 3 months. Lymph nodes regressed and the leg ulcer subsided. Fatemi and Bateni (2012) reported a case of a rare primary chromoblastomycosis of the palate and chest in a 27-year-old man in Tehran, who was successfully treated with surgical resection and combined drug therapy, and eventually free tissue transfer reconstructive surgical procedure to cure the palatine defect.

Rhinosporidiosis

Rhinosporidiosis is a rare, chronic, granulomatous infection, caused by an organism once believed to be a sporozoan, but now considered to be a fungus named *Rhinosporidium seeberi*. The vegetable nature of *Rhinosporidium* was recognized by Ashwort, who classified it among the Phycomycetes, suborder Chytridinese. Probably the first case had been observed by Malbran from Argentina in 1892. Seeber (Argentina 1900) reported two cases of nasal polypi. However, he did not name the organism. It is a disease of the mucous membrane, frequently involving the nasal passages and ocular tissue, conjunctiva, lachrymal sac, canali-culus, sclera and lid. However sporadic involvement has been observed in the pharynx, paranasal sinus, maxillary sinus, lip, tongue, palate, epiglottis, tonsil, trachea and bronchus, aural canal, skin, parotid gland, urethra, penis, vagina, bone, wrist and foot. Generalized rhinosporidiosis is an extremely rare occurrence. Although the disease has a worldwide distribution, approximately 90% of recorded cases are from India, Ceylon, U.S.A., South America, and Iran. Sporadic cases have been reported from Singapore, Philippines, Indonesia, Vietnam, Thailand, Malaysia, Pakistan, Israel, Turkey, U.S.S.R., Italy, East Africa and South Africa.

In Iran, the first case of rhinosporidiosis was seen in 1937 by Habibi (Habibi 1939). He had reported 30 cases up to 1947 (Habibi 1947).

Firouz-Abadi et al. (1971) reported 74 cases of rhinosporidiosis in Tehran, Ardebil and Kermanshah, 55 were nasal (including nasopharynx), 16 had eye involvement and 3 had multiple polyps of the nose and eye. The disease was seen in all age groups, but mostly in the age groups from 20 to 30. Sixty-two were male and twelve were female. Attention has been drawn to the geographic distribution of these cases, and very high incidence was observed in the Azerbaijan region.

Systemic Fungal Infections (SFIs)

Due to lack of an integrated center for invasive fungal infections, little has been documented about SFIs in Iran. According to the unpublished reports from different hospitals, SFIs are serious public health threats especially in hospitalized patients who are immunocompromised by various reasons or receive organ transplantation. Here, we discuss some proven cases of SFIs which have been reported sporadically by medical mycologists from different geographic regions.

Zygomycosis

Zygomycosis (syn. Mucormycosis) is a fulminant infection which is being reported with increasing frequency in recent years. This disease is a rare but highly invasive fungal infection. It usually affects patients who are predisposed by diabetes ketoacidosis, severe burning, malignancy, corticosteroid use, immunosuppression therapy for solid organ transplant or bone marrow transplant, neutropenia or neutrophil dysfunction associated with leukemia or lymphoma. Zygomycosis is the name given to several different diseases caused by fungi of the order mucorales. Depending on the site of pathogen entry, various types of zygomycosis involving nose, face, brain, lung, gastrointestinal tract, skin, and may occur also occur in other organs. Etiologic fungi especially *Rhizopus oryzae* have a predilection to invade blood vessels causing embolism and necrosis of surrounding tissue. It has been estimated that around 500 mucormycosis cases/year will occur in the United States.

In Iran, zygomycosis has been reported from Kermanshah, Lorestan, Fars, Khorasane Razavi, Kashan, Urmia, Tabriz and Tehran (Table 3).

Moghaddami et al. (1993) reported 13 cases of mucormycosis in ten diabetics in Tehran, one leukemic, and one burns patient, and one in an apparently normal person. The presenting forms were rhinocerebral in 11 cases and cutaneous in two cases.

Foroutan and Mashayeki (1997) reported a case of gastrointestinal mucormycosis in a young woman in Tehran-Iran. A 33-year-old woman was admitted to the hospital with icterus. There was a six-month history of intermittent jaundice with mild right

Table 3. Geographic distribution and clinical features of systemic fungal infections in Iran.

Diseases	Place (Area)	Number of patients	Frequency (%)	Age	Affected body part (%)		Major Etiologic Agents (%)		Reference
Candidiasis	Fars (Shiraz)	410	86.3	ND	ND	ND	*C. albicans*	48	Badiee et al. 2011
							C. krusei	16.1	
							C. glabrata	13.5	
					ND	ND			
							C. kefyr	7.4	
							C. parapsilosis	4.8	
							C tropicalis	1.7	
	Isfehan (Kashan)	1	ND	40	Left knee	ND	*C. glabrata*	ND	Erami et al. 2014
Zygomycosis	Tehran	7132	0.31	ND	Kidney	ND	Zygomycetes	ND	Ahmadpour et al. 2009
	Kermanshah	ND	ND	ND	ND	ND	*Rhizopus oryzae*	ND	Mikaeili and Karimi 2014
	Tehran	30	ND	20–70	Rhinocerebral	ND	Zygomycetes	ND	Barati et al. 2010
	Tehran	61	44.26	ND	ND	ND	Zygomycetes	ND	Mirzaie et al. 2011
	Tehran	ND	ND	19–71	ND	ND	Zygomycetes	ND	Einollahi et al. 2011
	Tehran	ND	ND	32–67	ND	ND	Zygomycetes	ND	Einollahi et al. 2008
	Tehran	7	ND	ND	Kidney	ND	Zygomycetes	ND	Aslani et al. 2007
	Tehran	13	ND	10–70	Rhinocerebral	ND	*Rhizopus* sp.	ND	Moghaddami et al. 1993
	Fars (Shiraz)	3	ND	2, 2, 18	ND	ND	Zygomycetes	ND	Geramizadeh et al. 2007
	Khorasane Razavi (Mashhad)	2	ND	31,58	Kidney	ND	Zygomycetes	ND	Tayyebi-Meybodi et al. 2005
	Tehran	1	ND	33	Gastrointestinal tract	ND	Zygomycetes	ND	Foroutan and Mashayeki 1997
	Tehran	1	ND	51	Kidney	ND	Zygomycetes	ND	Miladipour et al. 2008
	Tehran	1	ND	47	Rhino-orbitocerebral	ND	*Rhizopus*	ND	Sanavi et al. 2013
	Fars (Shiraz)	1	ND	42	Kidney	ND	Zygomycetes	ND	Azarpira et al. 2012

Table 3. contd....

Table 3. contd.

Diseases	Place (Area)	Number of patients	Frequency (%)	Age	Affected body part (%)		Major Etiologic Agents (%)		Reference
Zygomycosis	Isfahan (Kashan)	1	ND	24	Rhinocerebral	ND	*Rhizopus oryzae*	ND	Erami et al. 2013
	Tehran	1	ND	43	Rhino-orbital-cerebral	ND	*Rhizopus oryzae*	ND	Mohebbi et al. 2011
	West Azerbaijan(Urmia)	1	ND	57	Lung	ND	Zygomycetes	ND	Mohammadi et al. 2012
	Fars (Shiraz)	1	ND	75	Breast	ND	Zygomycetes	ND	Baezzat et al. 2011
	Fars (Shiraz)	1	ND	2	Eye	ND	Zygomycetes	ND	Badiee et al. 2012
	Tehran	1	ND	12	Nose	ND	Zygomycetes	ND	Fahimzad et al. 2008
	Azarbaijane sharghi (Tabriz)	1	ND	24	Neck	ND	Zygomycetes	ND	Hashemzadeh et al. 2008
	Fars (Shiraz)	1	ND	40	Kidney	ND	Zygomycetes	ND	Geramizadeh et al. 2012a
	Fars (Shiraz)	1	ND	63	Stomach	ND	Zygomycetes	ND	Geramizadeh and Azizi 2001
	Tehran	1	ND	1	ND	ND	Zygomycetes	ND	Ebadi et al. 2013
	Khorasane Razavi (Mashhad)	1	ND	29	Skin	ND	Zygomycetes	ND	Bojdy et al. 2013
	Khorasane Razavi (Mashhad)	1	ND	32	ND	ND	Zygomycetes	ND	Heydari et al. 2013
	Tehran	1	ND	19	Cornea	ND	Zygomycetes	ND	Feizi et al. 2012
	Fars (Shiraz)	1	ND	36	Stomach	ND	Zygomycetes	ND	Paydar et al. 2010
Basidiobolomycosis	Fars (Shiraz)	14	ND	1-52	Gastrointestinal	ND	*Basidiobolus ranarum*	ND	Geramizadeh et al. 2012b
	Tehran	1	ND	12	Gastrointestinal	ND	*Basidiobolus ranarum*	ND	Arjmand et al. 2012
Histoplasmosis	Tehran	1	ND	17	Lung	ND	ND	ND	Aslani et al. 2001

Cryptococcosis	Tehran	1	ND	26	Meninges	ND	*Cryptococcus neoformans*	ND	Moghadami et al. 1988
	Tehran	1	ND	21	Lung	ND	*Cryptococcus neoformans*	ND	Shafaghi et al. 2010
	Mazandaran (Sari)	1	ND	43	Meninges	ND	*Cryptococcus neoformans*	ND	Ghasemian et al. 2011
	Tehran	1	ND	35	Meninges	ND	*Cryptococcus neoformans*	ND	Razin et al. 2011
Aspergillosis	Fars (Shiraz)	82	ND	4–57	Bone marrow	ND	*Aspergillus* spp.	ND	Badiee and Alborzi 2010
	Tehran	49	ND	4–45	Lung	ND	*A. fumigatus, A. flavus, A. niger*	ND	Sarrafzadeh et al. 2010
	Tehran	12	ND	18–60	Lung	ND	*Aspergillus* spp.	ND	Marjani et al. 2008
	Tehran	7	ND	2–24	Lung, Chest, Liver, Brain, Lymph nodes	ND	*A. fumigatus, A. flavus*	ND	Mamishi et al. 2007
	Azarbaijane sharghi (Tabriz)	1	ND	60	Sinus tract	ND	*A. fumigatus*	ND	Ghotaslou et al. 2008

ND: Not determined

upper abdominal quadrant pain and progressive symptoms of general malaise, weakness and anorexia. The patient had never taken any medication or alcohol.

Tayyebi-Meybodi et al. (2005) reported 2 cases of mucormycosis in a 31-year-old man and a 58-year-old woman with kidney transplantation in Mashhad. Aslani et al. (2007) reported 7 cases (5 males and 2 females) of zygomycosis in patients from Tehran, who had undergone transplantation. The final diagnosis was pulmonary mucormycosis in 4 cases, rhino-cerebral mucormycosis in 2 cases and disseminated mucormycosis in the latter case. Barati et al. (2010) reported 30 cases (17 males and 13 females) of zygomycosis in Tehran, with a mean age of 49.4 ± 20.3 years. The lag time between onset of symptoms referable to zygomycosis and commencement of amphotericin B was 1 to 90 days. An association between delayed treatment and mortality was found ($p = 0.01$). Visual loss was observed in 53.3% of cases. The ethmoid (86.6%) and maxillary sinuses (66.6%) were the most commonly involved body parts. Eighteen patients had underlying diabetes mellitus (60%). All patients received medical treatment, while 28 (93.3%) underwent surgical intervention. Twenty three patients (76.7%) had orbital involvement with a mortality rate of 43.5%. The overall mortality rate was around 70%. Patients with higher doses of amphotericin B and multiple surgical interventions had lower mortality rate. Factors such as age, gender, orbital involvement, multi-sinus involvement, and white blood cell count had no impact on survival rate. Miladipour et al. (2008) reported a case of mucormycosis primarily affecting the paranasal sinuses in a 51-year-old man with a kidney allograft from Tehran. The patient presented with headache, left facial and orbital pain, nasal discharge, and elevation of serum creatinine 18 months after kidney transplantation. The patient was successfully treated via discontinuation of cyclosporine and mycophenolate mofetil, initiation of systemic amphotericin B, and aggressive surgical debridement. In a retrospective study by Ahmadpour et al. (2009), 7132 Iranian renal transplant recipients were examined for zygomycosis in eight transplant centers from January 1990 to June 2008 in Tehran. They showed that a total of 22 patients who had received kidneys from living donors were involved with zygomycosis. Mirzaie et al. (2011) reported 27 cases of zygomycosis in 61 cases suspected of zygomycosis during 2003 to 2009 in Tehran. Rhino-orbito-cerebral type (100%) was the only presentation of disease which was categorized as nasal-paranasal sinuses (77.8%), orbital (11.1%) and nasal-paranasal sinuses-orbital (11.1%) involvement. Diabetes mellitus (in 55.7% of cases) was the most common underlying condition followed by hematologic malignancy (22.2%).

Baezzat et al. (2011) reported primary zygomycosis of breast in a 75-year-old female. The patient had no underlying disease and was successfully treated with a simple mastectomy and intravenous antifungal therapy. Badiee et al. (2012) reported a 2-year-old healthy boy in Shiraz, who some days after the entry of dust particle in his left eye presented with swelling and redness of the eye. The results of biopsy and tissue culture led to a diagnosis of orbital zygomycosis. Azarpira et al. (2012) reported from Shiraz, a 42-year-old man who had undergone renal transplant showing unilateral bloody nasal discharge. Zygomycosis of the maxillary sinus was detected. Surgical ablation of the infected parts, along with antifungal treatment, restricted extension of the infection. Mohammadi et al. (2012) reported from Urmia, a case of zygomycosis in a 57-year-old man with 8-year history of diabetes mellitus under treatment with oral hypoglycemic medication. This case reinforces the concept that occasionally simple intervention such as dental extraction in immunocompromised patients can causes fatal complications, so awareness of potentially fatal complications such as zygomycosis may help in early diagnosis and prevention of disease dissemination. Sanavi et al. (2013) reported in Tehran, a 47-year-old man with crescentic glomerulonephritis on maintenance prednisolone therapy. He had earlier received steroid and cyclophosphamide pulse therapies. Renal functions improved following immunosuppressive treatment. In the third month of maintenance therapy, he presented to us with left-sided facial swelling and bloody nasal discharge. He had high blood sugar and acidic blood pH (ketoacidosis), probably due to steroid therapy. Magnetic resonance imaging of the head and sinuses showed inflammation and mass in the ethmoid sinus and nose with partial septal destruction, proptosis, global destruction of the left eye, brain infarction and carotid artery obliteration. Endoscopic biopsy of the sinuses revealed severe tissue necrosis. Samples of nasal discharge and biopsy tissue showed aseptate hyphae on light microscopy and culture, compatible with *Rhizopus*. The patient was treated with amphotericin B and multiple wound debridements along with ethmoidectomy and enucleation of the left eye. He was discharged in good general condition but with mild right hemiparesis. On follow-up examination at one year, there were no signs of fungal infection or renal dysfunction. Erami et al. (2013) reported from Kashana case of rhino-orbito-cerebral zygomycosis in a 24-year-old female with diabetes mellitus as the underlying disease. On the basis of morphological characteristics and sequence analysis of fungal DNA extracted from pure cultures, the fungus was tentatively identified as *Rhizopus oryzae*.

Basidiobolomycosis

Entomophthoramycosis is a fungal disease consisting of both conidiobolomycosis and basidiobolomycosis. Historically, they have been known to cause skin and soft-tissue infections in otherwise healthy individuals in tropical areas of Africa, South America, and Asia. Basidiobolomycosis is a rare fungal infection caused by the fungus *Basidiobolus ranarum*. This fungus is an environmental saprophyte found worldwide in soil, decaying organic matter, and gastrointestinal tracts of some animals. Because gastrointestinal basidiobolomycosis is a rare disease with non-specific manifestations, its clinical presentation can be readily confused with commoner gastrointestinal diseases such as infectious, inflammatory and infiltrative diseases of the bowel. Although basidiobolomycosis is known through its skin lesions, its visceral involvement has been rarely reported. Basidiobolomycosis of the gastrointestinal tract is an uncommon event, and less than 20 cases, mostly from Saudi Arabia, Kuwait, Nigeria, USA, and Brazil have been reported. Basidiobolomycosis of the alimentary tract can manifest as abdominal pain, nausea, vomiting, diarrhea, or abdominal mass. Unfortunately, the clinical impressions are often neoplasm or chronic infections (rhabdomyosarcoma of pelvis, gastrointestinal stromal tumor and lymphoma as well as tuberculosis).

As indicated in Table 3, basidiobolomycosis has been reported only from Fars and Tehran in Iran.

Arjmand et al. (2012) reported a 12-year-old boy who presented with abdominal pain, bloody diarrhea, fever and vomiting in Tehran. He was treated for amebiasis, but due to treatment failure and deterioration of his condition, he underwent a laparatomy. Histological examination of the excised bowel in the second look revealed *Basidiobolus ranarum*, a fungus belonging to the order Entomophthorales. The signs, symptoms, treatment and diagnosis of the present case indicated that fungal infections must be considered not only in immunocompromised patients with abdominal pain and mass, but also in apparently immunocompetent ones. Geramizadeh et al. (2012b) identified 14 cases of gastrointestinal basidiobolomycosis in Shiraz; all of them were diagnosed after surgery by characteristic histopathological findings. Diagnosis of this disease requires a high index of suspicion in patients presenting with abdominal symptoms, fever, gastrointestinal mass and eosinophilia accompanied by a high erythrocyte sedimentation rate accounting the most important clinical findings of these patients. The most common presenting symptom was abdominal pain (100%), but other symptoms such as diarrhea (21.4%), constipation (14.3%) and abdominal distension (14.3%) were also notified.

Histoplasmosis

Histoplasmosis (Reticuloendothelial cytomycosis, Cave disease, Darling disease) is a disease that is caused by a dimorphic fungus named *Histoplasma capsulatum*. This disease initiates when the spores of the fungus are inhaled by people in endemic areas. Soil that is moist and rich in the droppings of bats and birds like pigeons, chickens, and blackbirds, provides an ideal environment for the survival of spores over a long period of time. For this reason, the infection is more prevalent in people who live near river valleys. Once the spores enter the human body through the respiratory tract, they change into yeast form. This disease is not contagious and therefore does not spread from one person to another. Initially, histoplasmosis of the lungs does not manifest with any signs or symptoms. The one way to detect the disease in the initial phase is by the histoplasmin skin test. When the patient starts to develop symptoms of the disease, he usually complains of fever with chills, muscle pain, a runny nose, and cough. Clinically, the symptoms are very similar to those of influenza. Hence, getting investigated to confirm the diagnosis is always advisable. Although most cases resolve on their own, some severe infections that occur in people with poor immunity can lead to complications like pneumonia, lung collapse, pleurisy, and the formation of reddish nodules over the skin.

In Iran, only one case of histoplasmosis has been reported from Tehran (Table 3). Aslani and Jeyhounian (2001) reported a 17-year-old diabetic man referred with a history of coughing and fever in Tehran. Open lung biopsy was diagnostic for histoplasmosis. ELISA was positive for *H. capsulatum*. The patient had histoplasmosis pneumonia and received ketoconazole, 400 mg daily, for six weeks. He was then asymptomatic, and the chest radiograph was normal after treatment.

Cryptococcosis

Cryptococcosis is a chronic, subacute, and rarely acute infection and usually involves lungs and the central nervous system. The disease is caused by the yeast-like encapsulated fungus *Cryptococcus neoformans* which is one of the most important pathogens affecting HIV patients. The fungus has two varieties, i.e., *C. neoformans* var. *gattii* and *C. neoformans* var. *neoformans*. It survives in birds droppings and may infect humans especially immunocompromised individuals after inhalation of infective propagules, i.e., yeast cells or sexual basidiospores. The most common sites of involvement are the central nervous system and lungs.

As indicated in Table 3, the disease has been reported in Sari and Tehran in Iran. It seems that the first case of disease has been reported by Moghaddami et al. in 1988 as cryptococcal meningitis.

Moghadami et al. (1988) reported a case of cryptococcal meningitis in a 26-year-old man in Tehran. After direct examination and culture of cerebrospinal fluid, *Cryptococcus neoformans* was isolated as the etiologic agent. Shafaghi et al. (2010) reported cryptococcosis in an asymptomatic lung transplant in a 21-year-old man with cystic fibrosis and end-stage bronchiectasis from Tehran. Ghasemian et al. (2011) reported a case of cryptococcal meningitis in an immunocompetent 43-year-old man in Sari. He experienced a 5-week delay in diagnosis of cryptococcal meningitis which led to blindness of the right eye despite a successful treatment of disease.

Razin et al. (2011) reported cryptococcal meningitis in a 35-year-old housewife from Tehran. The patient was a 35-year-old woman who was admitted to hospital due to fever, headache and changes of mental status. Physical examination revealed neck stiffness and positive Kerning's and Brudsinsky's signs. Cerebrospinal fluid analysis showed lymphocytic pleocytosis and culture examination revealed *Cryptococcus neoformans*.

Aspergillosis

Aspergillosis is an opportunistic invasive mycosis caused by select members of the genus *Aspergillus*. This disease may be caused by food poisoning or inhalation of fungal conidia and presents as aspergilloma, granulomatous invasive disease, necrotizing inflammation of the lungs and other organs. Invasive aspergillosis is associated with the immunocompromised host, especially those who have hematologic malignancies, stem cell or lung transplants. *Aspergillus fumigatus* accounts for the majority of cases, but other species such as *A. flavus*, *A. terreus* and *A. niger* are also involved as the etiologic agents.

As summarized in Table 3, the disease has been reported from Fars, Tabriz and Tehran in Iran. Although superficial forms of the disease mainly as onychomycosis are frequently reported from Iran, confirmed cases of systemic aspergillosis are very less commonly noticed.

Mamishi et al. (2007) reported seven chronic granulomatous disease patients with invasive aspergillosis from Tehran. The sites of infection were lungs, chest wall, brain, liver and lymph nodes. In five cases, *A. fumigatus* was isolated, in one case

A. flavus and in the latter case *Aspergillus* sp. were isolated. Marjani et al. (2008) reported aspergillosis in 8 lung, 3 kidney, and one heart recipient, with overall mean age of 40.6 years from Tehran. Seven cases of tracheobronchitis were diagnosed in lung transplant recipients, all of them in the first 6 months after transplantation. Five cases of invasive pulmonary aspergillosis were also reported. Three patients survived in response to antifungal treatment with voriconazole alone or in combination with caspofungin. Ghotaslou et al. (2008) reported a 60-year-old male diabetic patient with mediastinitis caused by *A. fumigatus* following an open heart surgery that was successfully treated in Tabriz. They concluded that aspergillosis should be considered in the differential diagnosis of mediastinitis after cardiac surgery, especially in a clinical setting of an unexplained sepsis or a non-healing wound infection despite apparently adequate treatment. Sarrafzadeh et al. (2010) reported 49 patients with invasive pulmonary aspergillosis from Tehran. They were diagnosed as 'proven' 16 (32.7%), 'probable' 18 (36.7%) and 'possible' 15 (30.6%) invasive pulmonary aspergillosis. The most common risk factor was solid tumors in 17 (34.7%) cases. Clinical presentations of disease were cough 49 (100%), excessive sputum 46 (93.9%), and fever 43 (87.8%). *Aspergillus fumigatus* was responsible for 44 (89.7%) cases and other less frequent fungal agents were *A. flavus* and *A. niger*, accounting for 3 (6.1%) and 2 (4.1%) cases respectively. Badiee and Alborzi (2010) examined blood specimens (n = 993) from patients (n = 82) scheduled for bone marrow transplantation in Shiraz. The specimens were tested using an *Aspergillus*-specific real-time PCR assay. *Aspergillus* DNA was positive in 94 sequential blood samples from 13 patients with clinical and radiological signs of infection. Samples from three of these patients were PCR-positive for *Aspergillus* in the first week of admission, prior to transplantation. Four patients with aspergillosis were cured with antifungal agents and nine died.

Miscellaneous Fungal Infections

Actinomycosis

Actinomycosis is a rare granulomatous suppurative infection caused by anaerobic or microaerophilic Gram-positive bacteria named actinomycetes, primarily classified in the genus *Actinomyces*. Actinomycetes could be normal flora in oral cavity, gastrointestinal tract and female genitourinary tract. Actinomycosis could be seen more with poor dental hygiene, dental and oral surgery and manipulation, oral mucus damage due to radiation, previous abdominal surgery and pulmonary infection due to aspiration. In up to 15%–20% of the cases the throat is involved and in 10%–20% cases the pelvic area is involved. Cervicofacial involvement could be seen with soft tissues swelling, abscess, mass or ulcerative lesion that can be mistaken with malignancy. Actinomycosis can be seen at any age but it is most prevalent at middle age. As summarized in Table 4, the disease has been reported from most parts of Iran.

Mobedi et al. (2006) studied 150 patients with some kinds of blood cells dyscrasia who underwent the bone marrow aspiration in Tehran. Twenty one samples were positive for actinomycetes infections via involvement of *Actinomyces naeslundii* as the etiologic agent. Ashraf et al. (2011) studied 204 patients from Shiraz, who had

Table 4. Geographic Distribution and Clinical Features of Miscellaneous Fungal Infections in Iran.

Disease	Place (Area)	Number of patients	Frequency (%)	Sex	Age	Affected Body Part	Major Etiologic Agents	Reference
Actinomycosis	Fars (Shiraz)	204	40.6	ND	3–72	Oral cavity	*Actinomyces* spp.	Ashraf et al. 2011
	Tehran	150	14	Men and Women	20–80	Bone marrow	*Actinomyces naeslundii*	Mobedi et al. 2006
	Tehran	91	ND	ND	ND	Oral-servicofacial, Thoracic, Abdominal	*Actinomyces* spp.	Daie-Ghazvini et al. 2013
	Tehran	38	ND	Men and women	0–80	Oral-servicofacial, Thoracic, Abdominal	*A. naeslundii, A. israelii, A. viscosus, A. bovis*	Khodavaisy et al. 2014
	Isfahan	2	ND	Men	22	Midline of chin	*Actinomyces* spp.	Farhad et al. 2012
	Golestan	2	ND	Men	39, 44	Hand, Foot	ND	Golsha et al. 2009
	Fars (Shiraz)	2	ND	ND	42, 47	Tongue	ND	Sodagar and Kohout 1972
	Kerman	2	ND	Men	38, 41	Oral cavity	ND	Shamsadini et al. 2007
	Khorasane Razavi (Mashhad)	1	ND	ND	54	Oral cavity	Actinomycetes	Mellati and Habibi 2008
	Fars (Shiraz)	1	ND	ND	24	Middle ear	*Actinomyces* spp.	Shishegar et al. 2009
	Tehran	1	ND	Women	31	Breast	*A. israelii*	Daie-Ghazvini et al. 2003
	Isfahan	1	ND	Men	23	Cervicofacial area	*Actinomyces* spp.	Avijgan et al. 2010
	Tehran	1	ND	Men	18	Fingernails	*A. bovis*	Mansouri et al. 2011
	Fars (Shiraz)	1	ND	Men	14	Oral cavity	*A. israelii*	Khademi et al. 2011
	Tehran	1	ND	Women	48	Oral cavity	*Actinomyces* spp.	Yadegarynia et al. 2013
	Tehran	1	ND	Men	22	Hand	*A. viscosus*	Daie-Ghazvini et al. 2009
	Tehran	1	ND	Men	56	Kidney	*A. israelii*	Pourmand et al. 2012
	Khuzestan (Ahvaz)	1	ND	Women	48	Breast	*A. israelii*	Salmasi et al. 2010

Table 4. contd....

Table 4. contd.

Disease	Place (Area)	Number of patients	Frequency (%)	Sex	Age	Affected Body Part	Major Etiologic Agents	Reference
Actinomycosis	Fars (Shiraz)	1	ND	Men	12	Spleen	*Actinomyces* spp.	Azarpira and Ghasemzadeh 2005
	Chaharmahal and Bakhtiari (Shahrekord)	1	ND	Women	30	Breast	*A. israelii*	Akhlaghi and Daei-Ghazvini 2009
	Fars (Shiraz)	1	ND	Women	9	Thoracic area	*A. israelii*	Alborzi et al. 2006
	Fars (Shiraz)	1	ND	Men	11	Bone marrow	*A. naeslundii*	Norouzi et al. 2013
	Kerman	1	ND	ND	31	Bone	*A. israelii*	Fariabi and Dabiri 2004
Fungal Sinusitis	Mazandaran	50	ND	Men (26)	10–50	Nose, Sinuses	*A. flavus, A. niger, A. terreus, Trichosporon,* Sterile hyphae, *Penicillium, Alternaria*	Hedayati et al. 2010
				Women (24)				
	Tehran	23	ND	ND	ND	Nose, Sinuses	*Aspergillus* spp.	Dehghan et al. 2008
	Tehran	150	14	Men (101)	10–60	Nose, Sinuses	*Alternaria* spp.	Naghibzadeh et al. 2011
							A. fulvous	
				Women (49)			*Paecilomyces*	
	Tehran	4	ND	ND	28, 32, 69, 15	Nose, Sinuses	*A. fumigatus, C. albicans, Rhizopus* sp. and *Alternaria* sp.	Kordbacheh et al. 2004
	Lorestan (Khoramabad)	83	ND	Men (34)	9–74	Nose, Sinuses	*Mucor* sp., *A. fumigatus, A. flavus, C. albicans*	Sabokbar et al. 2011
				Women (49)				
Phaeohyphomycosis	Tehran	1	ND	Men	18	Nose, Brain, Chest	*Cladosporium bantianum*	BasiriJahromi et al. 2002
	Tehran	1	ND	Men	17	Brain	*Nattrassia mangiferae*	Geramishoar et al. 2004

	Mazandaran (Sari)	1	ND	Men	27	Chest, Neck, Face	*Alternaria malorum*	Mirhendi et al. 2013
	Khorasane Jonobi (Birjand)	1	ND	Women	66	Neck	*Chaetomium* spp.	Najafzadeh et al. 2014
Pseudallescheriasis	Tehran	18	5.5	Men (12)	17–58	Sinus area	*Pseudallescheria boydii*	Bassiri-Jahromi and Khaksar 2006b
				Women (6)				
	Isfahan	1	ND	Women	40	Trunk, Upper and lower extremities, Scalp	*Pseudallescheria boydii*	Enshaieh et al. 2006
geotrichosis	Tehran	52	1.92	Men (29)	1–55	Lung	*Geotrichum candidum*	Zaini and Bassiri-Jahromi 1994
				Women (23)				
	Fars (Shiraz)	1	ND	Women	9	Stomach, Intestine	*Geotrichum candidum*	Vasei and Imanieh 1999
Protothecosis	Mazandaran (Caspian sea)	1	ND	Women	17	Foot	*Prototheca*	Nabai and Mehregan 1974
Fusariosis	Tehran	1	ND	Men	15	Ankle, Buttock area	*Fusarium* sp.	Basiri-Jahromi et al. 1998

ND: Not determined

undergone tonsillectomy for recurrent tonsillitis and for sleep apnea without a history of recurrent tonsillitis. The prevalence of tonsillar colonization with actinomycetes was higher in patients who had undergone tonsillectomy for recurrent tonsillitis (43.9%) than who had undergone tonsillectomy for obstructive sleep apnea (26.3%). Histopathological analysis of the respective tonsils did not show active tissue infection. There was a statistically significant relationship between the presence of actinomycosis and age. Mansouri et al. (2011) reported the first case of cutaneous actinomycosis caused in Tehran by *Actinomyces bovis* in an 18-year-old male patient with common variable immunodeficiency. Yadegarynia et al. (2013) reported sinus infection in a 48-year-old Iranian woman presenting with a spontaneous discharging sinus on the hard palate. It was the first report of actinomycotic infection of sinus tract of the hard palate in Tehran, Iran. Akhlaghi and Daei-Ghazvini (2009) reported one case of breast actinomycosis in a 48-year-old woman caused by *Actinomyces israelii* in Shahrekord. Khodavaisy et al. (2014) reported 92 cases of actinomycosis (57 males, 35 females) during 1972 to 2014 in Tehran. They included 21 cases of oral-servicofacial (23.1%), 7 cases of thoracic (7.7%), 17 cases of abdominal (18.7%), 21 cases of disseminated forms (23.1%) and 25 cases of other forms (27.5%). Findings indicated more common involvement of men (61.5%). *Actinomyces naeslundii* (21 cases) was found as the most common causative agent followed by *A. israelii* (15 cases), *A. viscosus* (3 cases) and *A. bovis* (1 case). Most patients were successfully treated with penicillin and some cases needed surgery along with antibiotic therapy.

Fungal Sinusitis

All saprophytic fungi can cause sinusitis, but in more than 80% of cases with a positive culture result, *Aspergillus* species and dematiaceous fungi are involved. Patients affected by allergic sinusitis, are not immunedeficient and typically have a history of sinusitis. Living in hot and humid climates could be considered as a risk factor for this type of fungal sinusitis.

As indicated in Table 4, the disease has been reported from Mazandaran, Lorestan and Tehran.

Kordbacheh et al. (2004) studied 60 patients who underwent sinus surgery to determine fungal sinusitis. Mycological culture of the biopsy samples yielded pure growth of fungi in 4 cases. Isolated fungi were *Aspergillus fumigatus*, *Candida albicans, Rhizopus* sp. and *Alternaria* sp.

Dehghan et al. (2008) studied fungal sinusitis in Tehran and reported *Aspergillus* isolates belonging to *A. flavus* group as the etiologic agents.

Sabokbar et al. (2011) studied 83 patients in Khoramabad and found an overall frequency of fungal rhinosinusitis around 14.46%. *Aspergillus* and *Candida* species were reported as the most prevalent etiologic agents involved. Naghibzadeh et al. (2011) evaluated 162 cases in Tehran, 52 (32%) women and 110 (68%) men for fungal sinusitis. Twelve samples showed fungal elements of *Aspergillus flavus* (1.2%), *Alternaria* spp. (5.56%) and *Paecilomyces* sp. (0.6%). Results indicated a prevalence of 7.4% for chronic sinusitis in studied population. Hedayati et al. (2010) reported that nasal obstruction in Mazandaran and congestion was the most common (96.0%)

clinical presentation in 50 cases examined for fungal sinusitis. Out of 50 patients, 70.0% and 40.0% were positive for fungi via direct microscopic and culture, respectively. Twelve (24.0%) patients showed allergic fungal rhinosinusitis. Nasal polyps, history of atopy and eosinophilic mucin were seen in all patients. Three (6.0%) patients met criteria for sinus fungal ball. *Aspergillus* was the most prevalent isolated fungi from all categories of fungal rhinosinusitis.

Phaeohyphomycosis

Phaeohyphomycosis is a cosmopolitan infection that is caused by a number of phaeoid fungi. The disease has been reported from most parts of the world. It is characterized by development of dark colored filamentous hyphae in the involved tissues. Phaeohyphomycosis includes a wide spectrum of infections that vary from superficial colonization of the skin or cornea to cutaneous, subcutaneous, cerebral and systemic infections. Infections generally occur in compromised or debilitated hosts. In subcutaneous phaeohyphomycosis, abscesses or verrucous lesions may develop on various body parts. Thelungs and other organs are only rarely involved. As summarized in Table 4, the disease has been reported from Sari, Birjand and Tehran. Bassiri-Jahromi et al. (2002) reported a case with rhino cerebral and chest phaeohyphomycosis caused by *Cladosporium bantianum* in an 18-year-old man with Wegener's granulomatosis in Tehran. Geramishoar et al. (2004) reported a case of cerebral phaeohyphomycosis caused by *Nattrassia mangiferae* in a 17-year-old male with a history of systemic lupus erythematosus in Tehran, which was indicated by renal involvement. Mirhendi et al. (2013) reported from Sari, a 27-year-old patient presented with subcutaneous necrotic lesions with a localized dermatosis affecting the anterior chest, neck and face and diagnosed it as phaeohyphomycosis. *Alternaria malorum* was isolated as the etiologic agent and its identity was confirmed by sequence analysis of the internal transcribed spacer. Najafzadeh et al. (2014) reported from Birjand, a 66-year-old woman with a phaeohyphomycotic cyst with an approximate size of 3 × 2.5 cm on the right lateral side of the neck. It was proven as an implantation phaeohyphomycosis caused by a non-sporulating *Chaetomium* species.

Pseudallescheriasis

Pseudallescheriasis is defined as a broad spectrum of diseases caused by a homothallic fungus called *Pseudallescheria boydii* (*Petriellidium boydii*, *Allescheria boydii*). *Pseudallescheria boydii* is a ubiquitous soil-inhabiting fungus that can be recovered from polluted water and sewage. Human infections can develop after traumatic implantation or aspiration of contaminated water. Disease appears mostly as mycetoma (99%) and occasionally as pulmonary infection, brain abscesses, primary cutaneous involvement and rare systemic infections. Most cases occurred in age range of 20–40 years. As indicated in Table 4, the disease has been reported from Isfahan and Tehran. Bassiri-Jahromi and Khaksar (2006b) reported from Tehran, a fungal sinusitis due to *P. boydii* in a 48-year-old woman presenting with chronic sinusitis. She had a long history of headache and nasal obstruction and discharge. Enshaieh et al. (2006) reported a rare case of cutaneous pseudallscheriasis by *P. boydii* in an atopic woman in Isfahan. The patient was a 40-year-old woman who presented dry, scaly skin from infancy.

Geotrichosis

Geotrichosis appears as pulmonary, oral, cutaneous and rarely gastrointestinal infections. Pulmonary disease is the most common form of the disease. *Geotrichum candidum* is a ubiquitous fungus which is rarely pathogenic in humans as the etiologic agent of geotrichosis. The fungus has been isolated from the dairy products, plant debries and healthy skin. Geotrichosis is reported from Tehran and Fars as shown in Table 4.

Zaini and Bassiri-Djahromi (1994) reported geotrichosis in a-38-year old woman among fifty-two leukemic patients examined for fungal infections in Tehran.

Vasei and Imanieh (1999) examined a nine-year-old girl attending the gastroenterology clinic on account of anorexia and epigastric pain of 3 months' duration in Shiraz. Duodenal colonization by *G. candidum* was established. This observation implies that disseminated geotrichosis in immunocompromised patients may have an intestinal origin. The emergence of this organism as an occasional pathogen in leukemic patients is of interest.

Protothecosis

Davis et al. (1964) reported the first proven case of Protothecosis as a cutaneous granuloma. Human protothecosis is a rare infection caused by members of the genus *Prototheca* which are generally non-pigmented algae found frequently in nature. Protothecosis induces mainly after traumatic inoculation of the organism in subcutaneous tissues.

Nabai and Mehregan (1974) reported the first case of protothecosis in Caspian Sea in a 17-year-old Caucasian girl who had a non-healing ulcerated lesion on her right lower leg. The patient described development of the lesion related to an insect bite acquired while walking barefoot on the shores of the Caspian Sea.

Fusariosis

Fungi of the genus *Fusarium* can cause disease in plants, animals and humans. In humans, there are reports of superficial infections in various sites such as the skin, cornea and nails and of localized organ infections including endophthalmitis arthritis, cystitis, peritonitis, and keratitis. However, systemic infections are rare which occur predominantly in immunocompromised patients. The first description of disseminated fusariosis was reported in 1973 in a child with acute lymphocytic leukemia (Cho et al. 1973).

Bassiri-Jahromi (1998) reported disseminated fusariosis involving the skin and lungs, in a 15-year-old boy with chronic granulomatous disease in Tehran (Table 4). There was a chronic wound in the right ankle and buttock area. Chest x-ray revealed parahilar lesions in both lungs. Tissue biopsy and bronchoalveolar lavage were performed and the specimens were sent to Pasteur Institute. In both specimens, *Fusarium* sp. was recognized as the pathologic agent via direct smear and culture techniques. The patient underwent antifungal therapy receiving amphotericin B and oral ketoconazole. The result of this treatment suggests that aggressive management of fusariosis offers the best chance of survival.

Conclusion

In this chapter, a comprehensive report on fungal infections in Iran with special attention paid to the epidemiological and clinical aspects has been presented. We tried to describe which diseases are common and how they affect the population. On the basis of existing literature, it is evident that superficial fungal infections including dermatophytosis, pityriasis versicolor and candidiasis of nails, groin and vagina are first in the line of investigation and thus, they have been reported from nearly all geographic areas of the country. Unfortunately, despite remarkable increase in the rate of life-threatening systemic fungal infections in recent years following the increased number of immunocompromised diseases and organ transplantation, our report lacks sufficient reliable data about such infections in Iran due to the several reasons. First, many cases of systemic fungal infections are missing every year because of misdiagnosis with infections with similar clinical pictures. Second, majority of medical laboratories especially those that are active in small cities lack experienced personnel and modern equipment needed for accurate diagnosis of fungal diseases. Third, there is not a close relationship between clinicians and medical mycologists, so many cases of suspected systemic fungal infections in hospitalized patients are managed by broad spectrum antifungal agents such as amphotericin B without a confirmatory report of fungal involvement. Taken together, this chapter highlights the importance of fungal infections as emerging microbial diseases affecting a large portion of the population in Iran and indicates that there is an urgent need for a central research laboratory of invasive fungal infections for education, diagnosis and control measures all over the country.

Acknowledgments

This work was supported financially by the Pasteur Institute of Iran. Authors gratefully thank Prof. Manouchehr Asgari for his invaluable information about history of medical mycology in Iran.

References

Afshar, P., M. Ghasemi and S. Kalhori. 2013. Identification of *Malassezia* Species Isolated From Patients With Pityriasis Versicolor in Sari, Iran, 2012. Jundishapur J Microbiol. 6: e8581.

Afshar, P., S. Khodavaisy, S. Kalhori, M. Ghasemi and T. Razavyoon. 2014. Onychomycosis in North-East of Iran. Iran J Microbiol. 6: 98–103.

Aghamirian, M.R. and S.A. Ghiasian. 2008. Dermatophytoses in outpatients attending the dermatology center of Avicenna Hospital in Qazvin, Iran. Mycoses 51: 155–160.

Aghamirian, M.R. and S.A. Ghiasian. 2010. Onychomycosis in Iran: epidemiology, causative agent and clinical features. Nihon Ishinkin Gakkai Zasshi 51: 23–29.

Aghamirian, M.R. and S.A. Ghiasian. 2011. A clinico-epidemiological study on tinea gladiatorum in Iranian wrestlers and mat contamination by dermatophytes. Mycoses 54: 248–253.
Ahmadi, B., S.J. Hashemi, F. Zaini, M.R. Shidfar, M. Moazeni, B. Mousavi and S. Rezaie. 2012. A case of onychomycosis caused by *Aspergillus candidus*. Med Mycol Case rep. 1: 45–48.
Ahmadpour, P., M. Lessan Pezeshki, M. Hassan Ghadiani, F. Pour Reza Gholi, F. Samadian, J. Aslani and N. Nouri Majalan. 2009. Mucormycosis after living fonor kidney transplantation: a multicenter retrospective study. Nephro-Urology Monthly 1: 39–43.
Akaberi, A.A., S.S. Amini and H. Hajihosseini. 2009. An unusual form of tinea versicolor: a case report. Iran J Dermatol. 12: 30–31.
Akhlaghi, M. and R. Daei-Ghazvini. 2009. Clinical presentation of primary actinomycosis of the breast. Breast J. 15: 102–103.
Alborzi, A., N. Pasyar and J. Nasiri. 2006. Actinomycosis as a neglected diagnosis of mediastinal mass. Jpn J Infect Dis. 59: 52–53.
Alizadeh, N., A.S. Sadr, J. Golchai, A. Maboodi and A.A. Falahati. 2004. Descriptive study of dermatophytosis in Guilan. Iran J Dermatol. 7: 255–60.
Ansar, A., M. Farshchian, H. Nazeri and S.A. Ghiasian. 2011. Clinico-epidemiological and mycological aspects of tinea incognito in Iran: a 16-year study. Med Mycol J. 52: 25–32.
Arjmand, R., A. Karimi, A.S. Dashti and M. Kadivar. 2012. A child with intestinal Basidiobolomycosis. Iranian J Med Sci. 37: 134–136.
Asadi, M.A., R. Dehghani and M.R. Sharif. 2009. Epidemiologic study of onychomycosis and tinea pedis in Kashan, Iran. Jundishapur J Microbiol. 2: 61–64.
Asgari, M. and M. Alilou. 1972. Mycetoma in Iran. The first report of eight cases with mycological studies. Ann Soc Beige Med Trop. 52: 287–306.
Ashraf, M.J., N. Azarpira, B. Khademi, B. Hashemi and M. Shishegar. 2011. Relation between Actinomycosis and Histopathological and Clinical Features of the Palatine Tonsils: An Iranian Experience. Iran Red Crescent Med J. 13: 499–502.
Aslani, J. and M. Jeyhounian. 2001. Pulmonary histoplasmosis in Iran: a case report. Kowsar Med J. 6: 21–24.
Aslani, J., M. Eizadi, B. Kardavani, H.R. Khoddami-Vishteh, E. Nemati, S. Hoseini and B. Einollahi. 2007. Mucormycosis after kidney transplantations: report of seven cases. Scandinavian J Infect Dis. 39: 703–706.
Avijgan, M., H. Shakeri and M. Shakeri. 2010. A case report of cervicofacial actinomycosis. Asia. Pacific J Trop Med. 3: 838–840.
Ayatollahi-Mousavi, S.A., S. Salari-Sardoii and S. Shamsadini. 2009. A first case of tinea imbricata from Iran. Jundishapur J Microbiol. 2: 71–74.
Azarpira, N. and B. Ghasem-Zadeh. 2005. Splenic actinomycosis: unusual presentation. Iran J Med Sci. 30: 141–143.
Azarpira, N., M.J. Ashraf, K. Kazemi and B. Khademi. 2012. Rhinomaxillary mucormycosis in a renal transplant recipient: case report. Experiment Clin Transplant 10: 605–608.
Azimi, H., M. Adibpour and M.H. Soroosh. 2000. A case report of Actinomycotic mycetoma. J Med Sci Tabriz 34: 55–58.
Azizi, A. and M. Rezaei. 2009. Prevalence of Candida species in the oral cavity of patients undergoing head and neck radiotherapy. J Dental Res Dental Clin Dental Prospect. 3: 78–81.
Badiee, P. and A. Alborzi. 2010. Detection of *Aspergillus* species in bone marrow transplant patients. J Infect Develop Count. 4: 511–516.
Badiee, P., A. Alborzi, F.S. Shakibae and A. Japoni. 2011. Susceptibility of Candida species isolated from immunocompromised patients to antifungal agents. East Mediterr Health J. 17: 425–30.
Badiee, P., Z. Jafarpour, A. Alborzi, P. Haddadi, M. Rasuli and M. Kalani. 2012. Orbital mucormycosis in an immunocompetent individual. Iran J Microbial. 4: 210–214.
Baezzat, S.R., A. Fazelzadeh, S. Tahmasebi and P.V. Kumar. 2011. Primary breast mucormycosis, a case report. Iran Red Crescent med J. 13: 208.
Barati, B., S.A.R. Okhovvat, A. Goljanian and M.R. Omrani. 2011. Otomycosis in central Iran: a clinical and mycological study. Iran Red Crescent Med J. 13: 873.
Barati, M., M. Talebi-Taher, M. Nojomi and F. Kerami. 2010. Ten-year experience of rhinocerebral zygomycosis in a teaching hospital in Tehran. Archiv Clin Infect Dis. 5: 117–120.
Bassiri-Jahromi, S. 2012. Epidemiological trends in zoophilic and geophilic fungi in Iran. Clin Exp Dermatol. 38: 13–19.

Bassiri-Jahromi, S. and A.A. Khaksar. 2006a. Aetiological agents of tinea capitis in Tehran (Iran). Mycoses 49: 65–67.

Bassiri-Jahromi, S. and A.A. Khaksar. 2006b. Paranasal sinus mycosis in suspected fungal sinusitis. Iran J Clin Infect Dis 1: 25–29.

Bassiri-Jahromi, S. and A.A. Khaksar. 2008. Outbreak of tinea gladiatorum in wrestlers in Tehran (Iran). Ind J Dermatol. 53: 132–136.

Bassiri-Jahromi, S. and A.A. Khaksari. 2009. Epidemiological survey of dermatophytosis in Tehran, Iran, from 2000 to 2005. Ind J Dermatol Venereol Leprol. 75: 142–147.

Bassiri-Jahromi, S. and A.A. Khaksar. 2010. Nondermatophytic moulds as a causative agent of onychomycosis in Tehran. Indian J Dermatol. 55: 140–143.

Bassiri-Jahromi, S., A.A. Khaksar, M. Vaziri-Kashani and S. Arsid. 1998. Disseminated infection due to *Fusarium* sp. in a patient with chronic granulomatous disease. Med J Islam Repub Iran 12: 93–96.

Bassiri-Jahromi, S., A. Khaksar and K. Iravani. 2002. Phaeohyphomycosis of the sinuses and chest by cladosporium bantianum. Med J Islam Repub Iran 16: 55–58.

Berenji, F. and S.R. Elahi. 2003. Mycotic Keratitis among patients referred to mycologic laboratory, Emam Reza Hospital, during 1982–2001. J Mashad Univ Med Sci. 45: 49–54.

Bojdy, A., S.R.H. Shojaie, G.A. Farid and M. Reza. 2013. Cutaneous mucormycosis (zygomycosis) in a kidney transplant recipient: recovery after amphotericin therapy. Ann Biol Res. 4: 275–279.

Chadegani, M., A. Momeni, S.H. Shadzi and M.A. Javaheri. 1987. A study of dermatophytoses in Esfahan (Iran). Mycopathologia 98: 101–104.

Chadeganipour, M., S. Shadzi, P. Dehghan and M. Movahed. 1997. Prevalence and aetiology of dermatophytoses in Isfahan, Iran. Mycoses 40: 321–324.

Chadeganipour, M., S. Nilpour and G. Ahmadi. 2010. Study of onychomycosis in Isfahan, Iran. Mycoses 53: 153–157.

Cho, C.T., T.S. Vats, J.T. Lowman, J.W. Brandsberg and F.E. Tosh. 1973. *Fusarium solani* infection during treatment for acute leukemia. J Pediatr. 83: 1028–1031.

Daie-Ghazvini, R., F. Zaini, E. Zibafar and K. Omidi. 2003. First case report of primary actinomycosis of the breast due to *Actinomyces Israelii* from Iran. Act Med Iranica. 41: 110–112.

Daie-Ghazvini, R., J. Hashemi, M. Abbas-Tabar, S.M. Gerami, E. Zibafar and L. Hosseinpour. 2009. Actinomyces viscosus isolation from skin lesions in Iran (case report). Iran J Publ Health 38: 142–144.

Daie-Ghazvini, R., S.J. Hashemi, E. Zibafar, M. Geramishoar and S. Khodaveisi. 2013. Actinomycosis in Iran. Jundishapur J Microbiol. 206.

Davis, R.R., H. Spencer and P.O. Wakelin. 1964. A case of human protothecosis. Trans Roy Soc Trop Med Hyg. 58: 448–451.

Dehghan, M., S. Hajian, N. Alborzi, A. Borgheyee and A.H. Noohi. 2009. Clinico-mycological profiles of dermatophytosis in Gorgan, north of Iran. Iran J Dermatol. 12: 13–15.

Dehghan, P., F. Zaini, S. Rezaei, A. Jebali, P. Kordbacheh and M. Mahmoudi. 2008. Detection of aflr gene and toxigenicity of *Aspergillus flavus* group isolated from patients with fungal sinusitis. Iran J Publ Health 37: 134–141.

Ebadi, M., S. Alavi, N. Ghojevand, M. Kazemi Aghdam, M.K. Yazdi and A. Zahiri. 2013. Infantile splenorenopancreatic mucormycosis complicating neuroblastoma. Pediatrics International 55: 152–155.

Ebrahimzadeh, A. 1976. Mycetoma pedis. J Jundishapur Med Sch. 3: 5–8.

Ebrahimzadeh, A. 2009. A survey on pityriasis versicolor in the university students in Southeast of Iran. Asian J Dermatol. 1: 1–5.

Einollahi, B., M. Lessan-Pezeshki, V. Pourfarziani, E. Nemati, M. Nafar, F. Pour-Reza-Gholi and J. Aslani. 2008. Invasive fungal infections following renal transplantation: a review of 2410 recipients. Ann Transplant 13: 55–58.

Einollahi, B., M. Lessan-Pezeshki, J. Aslani, E. Nemati, Z. Rostami, M.J. Hosseini and N. Nouri-Majalan. 2011. Two decades of experience in mucormycosis after kidney transplantation. Ann Transplant 16: 44–48.

Emami-Naeini, A.R., P. Dehghan and R. Imani. 2002. Verrucous carcinoma overriding eumycetoma. Arch Iran Med. 5: 59–60.

Enshaieh, S.H., A. Darougheh, A. Asilian, F. Iraji, Z. Shahmoradi and A. Yoosephi. 2006. Disseminated subcutaneous nodules caused by *Pseudallescheria boydii* in an atopic patient. Int J Dermatol. 45: 289–291.

Erami, M., M. Shams-Ghahfarokhi, Z. Jahanshiri, A. Sharif and M. Razzaghi-Abyaneh. 2013. Rhinocerebral mucormycosis due to *Rhizopus oryzae* in a diabetic patient: a case report. J Med Mycol. 23: 123–129.
Erami, M., H. Afzali, M.M. Heravi, A. Rezaei-Matehkolaei, M.J. Najafzadeh, M. Moazeni and L. Hosseinpour. 2014. Recurrent arthritis by *Candida glabrata*, a diagnostic and therapeutic challenge. Mycopathologia 177: 291–298.
Fahimzad, A., Z. Chavoshzadeh, H. Abdollahpour, C. Klein and N. Rezaei. 2008. Necrosis of nasal cartilage due to mucormycosis in a patient with severe congenital neutropenia due to HAX1 deficiency. J Invest Allergol Clin Immunol. 18: 469–472.
Falahati, M., L. Akhlaghi, A.R. Lari and R. Alaghehbandan. 2003. Epidemiology of dermatophytoses in an area south of Tehran, Iran. Mycopathologia 156: 279–287.
Faraji, R., M.A. Rahimi, F. Rezvanmadani and M. Hashemi. 2012. Prevalence of vaginal candidiasis infection in diabetic women. African J Microbiol Res. 6: 2773–2778.
Farhad, A.R., M. Ajami and M.R. Khosravi. 2012. Resolution of a cutaneous lesion associated with a periapical actinomycosis following endodontic surgery: A case report. J Dent Shiraz Univ Med Sci. 13: 40–43.
Fariabi, J. and S. Dabiri. 2004. A case report of periapical Actinomycosis of maxilla. J Kerman Univ Med Sci. 11: 126–130.
Fata, A. and M. Boloursaz. 2000. Study of Mycetoma during past decade in Emam Reza Hospital. M J Mashad Univ Med Sci. 68: 65–71.
Fatemi, M.J. and H. Bateni. 2012. Oral chromoblastomycosis: a case report. Iran J Microbial. 4: 40–43.
Feizi, S., M.R. Jafarinasab and M.R. Kanavi. 2012. A zygomycetes-contaminated corneal graft harvested from a donor with signs of orbital trauma. Cornea 31: 84–86.
Firouz-Abadi, A., M. Moghimi and Y. Azad. 1971. Rhinosporidiosis in Iran (Persia) a study of seventy-four cases. Mycopathol Mycol Applicat. 44: 249–260.
Foroutan, H. and R. Mashayekhi. 1997. Zygomycosis of colon case report and review of the literature. Med J Islam Repub Iran 11: 53–55.
Forozesh, M., K. Motabar, J. Farshy, M. Hassantash and H. Farivar. 1993. Four cases of actinomycetoma in Tehran. Daru Darman 10: 23–27.
Gerami-Shoar, M., K. Zomorodian, M. Emami, B. Tarazouei and F. Saadat. 2002. Study and identification of the etiological agents of Onychomycosis in Tehran, capital of Iran. Iran J Publ Health 31: 100–104.
Geramishoar, M., K. Zomorodian, F. Zaini, F. Saadat, B. Tarazooie, M. Norouzi and S. Rezaie. 2004. First case of cerebral phaeohyphomycosis caused by *Nattrassia mangiferae* in Iran. Jpn J Infect Dis. 57: 285–286.
Geramizadeh, B. and S. Azizi. 2001. Invasive gastric mucormycosis: Report of a case and review of the literature. Med J Islam Repub Iran 14: 397–398.
Geramizadeh, B., M. Modjalal, S. Nabai, A. Banani, H.R. Forootan, F. Hooshdaran and A. Alborzee. 2007. Gastrointestinal zygomycosis: a report of three cases. Mycopathologia 164: 35–38.
Geramizadeh, B., K. Kazemi, A.R. Shamsaifar, A. Bahraini, S. Nikeghbalian and S.A. Malekhosseini. 2012a. Isolated renal mucormycosis after liver transplantation: an unusual case report. Iran Red Crescent Med J. 14: 447.
Geramizadeh, B., R. Foroughi, M. Keshtkar-Jahromi, S.A. Malek-Hosseini and A. Alborzi. 2012b. Gastrointestinal basidiobolomycosis, an emerging infection in the immunocompetent host: a report of 14 patients. J Med Microbial. 61: 1770–1774.
Ghamgosha, M., K. Hassanpour, Z. Bameri, M. Mellat and G. Farnoosh. 2013. A report of *Actinomycetoma* from southeastern Iran. Life Sci J. 10: 374–376.
Ghasemian, R., N. Najafi and T. Shokohi. 2011. Cryptococcal meningitis relapses in an immunocompetent patient. Iran J Clin Infect Dis. 6: 51–55.
Ghiyasi, S. 2001. Survey of otomycosis in north-western area of Iran (1995–1999). Med J Mashhad Univ Med Sci. 43: 85–7.
Ghodsi, S.Z., S. Shams, Z. Naraghi, M. Daneshpazhooh, M. Akhyani, S. Arad and L. Atali. 2000. Case report: an unusual case of cutaneous sporotrichosis and its response to weekly fluconazole Fallbericht. Ein ungewohnlicher Kutaner Sporotrichose-Fall unter Fluconazol-Therapie. Mycoses-Berlin 43: 75–77.
Ghotaslou, R., R. Parvizi, N. Safaei and S. Yousefi. 2008. A case of *Aspergillus fumigatus* mediastinitis after heart surgery in Madani Heart Center, Tabriz, Iran. Prog Cardiovasc Nurs. 23: 133–135.
Golsha, R., L. Njafi, R. Rezaie-Shirazi, M. Vakilinejhad, B. Mortazavi and G. Roshandel. 2009. Actinomycosis may be presented in unusual organs: a report of two cases. J Clin Diag Res. 3: 1938–1941.

Griffiths, W.A., E. Kohout and K. Vessal. 1975. Mycetoma in Iran. Int J Dermatol. 14: 209–213.

Habibi, M. 1939. Etude de TroisCas de *Rhinosporidium seeberi* en Iran. Ann Par Hum Com. 17: 103–107.

Habibi, M. 1947. Infestation avec *Rhinosporidium seeberi;* 16 Cas. Observesen Iran. Ann Parasitot. 22: 81–88.

Hasanjani, R.M., S. Sefidgar, E. Shafigh, M. Shidfar and M. Emami. 2001. The first case report of chromoblastomycosis in Mazandaran. J Mazandaran Univ Med Sci. 11: 74–80.

Hashemi, S.J. and S.M. Gerami. 2001. The case report of the first mycetoma caused by *Aspergillus flavus* in Iran. J Qazvin Univ Med Sci. 19: 64–66.

Hashemi, S.J., A.A. Salami and S.M. Hashemi. 2005. An epidemiological study of human dermatophytosis in Karaj (2001). Arch Razi. 60: 46–54.

Hashemi, S.J., M. Gerami, E. Zibafar, M. Daei, M. Moazeni and A. Nasrollahi. 2010. Onychomycosis in Tehran: mycological study of 504 patients. Mycoses 53: 251–255.

Hashemi, S.J., H.A. Qomi, M. Bayat and I.S. Haghdost. 2014. Mycoepidemiologic study of superficial and cutaneous fungal zoonotic disease in patients who referred to skin clinic of Arak. Eur J Exp Biol. 4: 5–8.

Hashemzadeh, S., R.S. Tubbs, M.B.A. Fakhree and M.M. Shoja. 2008. Mucormycotic pseudoaneurysm of the common carotid artery with tracheal involvement. Mycoses 51: 347–351.

Hedayati, M.T., P. Afshar, T. Shokohi and R. Aghili. 2007. A study on tinea gladiatorum in young wrestlers and dermatophyte contamination of wrestling mats from Sari, Iran. Br J Sports Med. 41: 332–334.

Hedayati, M.T., M. Bahoosh, A. Kasiri, M. Ghasemi, S.J. Motahhari and R. Poormosa. 2008. Prevalence of fungal rhinosinusitis among patients with chronic rhinosinusitis from Iran. J Med Mycol. 20: 298–303.

Hedayati, M.T., Z. Hajheydari, F. Hajjar, A. Ehsani, T. Shokohi and R. Mohammadpour. 2010. Identification of *Malassezia* species isolated from Iranian seborrhoeic dermatitis patients. Eur Rev Med Pharmacol Sci. 14: 63–68.

Heydari, A.A., A. Fata and M. Mojtabavi. 2013. Chronic cutaneous mucormycosis in an immunocompetent female. Iran Red Crescent Med J. 15: 254–255.

Jafari, A.A., H. Zarrinfar, F. Mirzaei and F. Katiraee. 2013. Distribution of *Malassezia* species in patients with pityriasis versicolor compared with healthy individuals in Yazd, Iran. Jundishapur J Microbiol. 6: e6873.

Javidi, Z., M. Maleki, A. Fata, Y. Nahidi, H. Esmaeili and A.R. Hosseini. 2007. Psoriasis and infestation with *Malassezia*. Med J Islam Repub Iran 21: 11–16.

Kafaie, P. and M.T. Noorbala. 2010. Evaluation of onychomycosis among diabetic patients of Yazd diabetic center. J Pak Associ Dermatol. 20: 217–221.

Katiraee, F., A.R. Khosravi, V. Khalaj, M. Hajiabdolbaghi, A. Khaksar, M. Rasoolinejad and M.S. Yekaninejad. 2010. Oropharyngeal candidiasis and oral yeast colonization in Iranian human immunodeficiency virus positive patients. J Med Mycol. 20: 8–14.

Kazemi, A. and S. Ghiasi. 2005. Survey of otomycosis in North-Western area of Iran (1997–2004). J Mazandaran Univ Med Sci. 48: 112–119.

Kazemi, A. and A. Razi. 2007. Esporotricosis en Irán. Rev Iberoam Micologia 24: 38–40.

Kazemi, A., S.A. Ayatollahi-Mousavi, A.A. Jafari and A. Zarei. 2013. Study on pityriasis versicolor in patients referred to clinics in Tabriz. People 18: 22.

Khademi, B., S.H. Dastgheib-Hosseini and M.J. Ashraf. 2011. Vocal cord actinomycosis: a case report. Iran J Otorhinolaryngol. 23: 49–52.

Khodavaisy, S., E. Zibafar, S.J. Hashemi, A.R. Hanar and R. Daie-Ghazvini. 2014. Actinomycosis in Iran: short narrative review article. Iran J Pub Health 43: 556–560.

Khosravi, A.R. and M. Emami. 1991. Study of two cases of cutaneous and subcutaneous nocardiosis. Med J Islam Rep Iran 5: 169–172.

Khosravi, A.R., M.R. Aghamirian and M. Mahmoudi. 1994. Dermatophytoses in Iran. Mycoses 37: 43–48.

Khosravi, A.R., P. Mansouri and M. Moazzeni. 2000. Case report: chronic dermatophyte infection in a patient with vitiligo and discoid lupus erythematosus. Mycoses 43: 317–319.

Khosravi, A.R., H. Shokri, P. Mansouri, F. Katiraee and T. Ziglari. 2008a. Candida species isolated from nails and their *in vitro* susceptibility to antifungal drugs in the department of Dermatology (University of Tehran, Iran). J Med Mycol. 18: 210–215.

Khosravi, A.R., P. Mansouri, Z. Naraghi, H. Shokri and T. Ziglari. 2008b. Unusual presentation of tinea cruris due to *Trichophyton mentagrophytes* var. *mentagrophytes*. J Dermatol. 35: 541–545.

Khosravi, A.R., S. Eidi, F. Katiraee, T. Ziglari, M. Bayat and M. Nissiani. 2009. Identification of different *Malassezia* species isolated from patients with *Malassezia* infections. World J Zool. 4: 85–89.

Kordbacheh, P., F. Zaini, M. Emami, H. Borghei, M. Khaghanian and M. Safara. 2004. Fungal involvement in patients with paranasal sinusitis. Iranian J Publ Health 33: 19–26.

Mahboubi, A., S.H. Baghestani, Y. Hamedi, M. Heydari and M. Vahdani. 2006. Epidemiology of dermatophytosis in Bandar Abbas, Iran (2003–2004). Med J Hormozgan Univ. 9: 227–34.

Mahmoudabadi, A.Z. 1995. Actinomycetoma: A case report. J Kerman Uni Med Sci. 2: 154–157.

Mahmoudabadi, A.Z. 2005. A study of dermatophytosis in South West of Iran (Ahwaz). Mycopathologia 160: 21–24.

Mahmoudabadi, A.Z. 2006a. Clinical characteristics and mycology of cutaneous candidiasis in Ahwaz (Iran). Pakistan J Med Sci. 22: 43–46.

Mahmoudabadi, A.Z. 2006b. First case of *Microsporum ferrugineum* from Iran. Mycopathologia 161: 337–339.

Mahmoudabadi, A.Z. and M. Zarrin. 2005. Onychomycosis with *Aspergillus flavus*: a case report from Iran. Pakistan J Med Sci. 21: 497.

Mahmoudabadi, A.Z. and R. Yaghoobi. 2007. Actinomycetoma in a 60-year-old woman for 20 years. Iran J Med Sci. 32: 245–247.

Mahmoudabadi, A.Z. and M. Zarrin. 2008. Mycetomas in Iran: a review article. Mycopathologia 165: 135–141.

Mahmoudabadi, A.Z. and R. Yaghoobi. 2008. Tinea corporis due to *Trichophyton simii*: a first case from Iran. Med Mycol. 46: 857–859.

Mahmoudabadi, A.Z., Z. Mossavi and M. Zarrin. 2009. Pityriasis versicolor in Ahvaz, Iran. Jundishapur J Microbiol. 2: 92–96.

Mahmoudabadi, A.Z., S.A. Masoomi and H. Mohammadi. 2010. Clinical and mycological studies of otomycosis. Pakistan J Med Sci. 26: 187–190.

Mahmoudabadi, Z.A., M. Zarrin and F. Mehdinezhad. 2013. Seborrheic dermatitis due to *Malassezia* species in Ahvaz, Iran. Iranian J Microbial. 5: 268.

Maleki, M. and A. Fata. 2005. Evaluation of wood s light and direct smear for diagnosis of pityriasis versicolor and erythrasma. Saudi Med J. 26: 1483–1484.

Mamishi, S., N. Parvaneh, A. Salavati, S. Abdollahzadeh and M. Yeganeh. 2007. Invasive aspergillosis in chronic granulomatous disease: report of 7 cases. Eur J Pediat. 166: 83–84.

Mansouri, P., S. Farshi, A. Khosravi and Z.S. Naraghi. 2011. Primary cutaneous actinomycosis caused by *Actinomyces bovis* in a patient with common variable immunodeficiency. J Dermatol. 38: 911–915.

Marjani, M., P. Tabarsi, K. Najafizadeh, F.R. Farokhi, B. Sharifkashani, S. Motahari and D. Mansouri. 2008. Pulmonary aspergillosis in solid organ transplant patients: a report from Iran. Transplant Proceed. 40: 3663–3667.

Mellati, E. and M. Habibi. 2008. Actinomycosis of the tongue. Arch Iran Med. 11: 566–568.

Mikaeili, A. and I. Karimi. 2013. Paper: The incidence of onychomycosis infection among patients referred to hospitals in Kermanshah province, Western Iran. Iran J Publ health 42: 320–325.

Mikaeili, A. and I. Karimi. 2014. The counter immunoelectrophoretic detecting of serum response to *Rhizopus oryzae* and *Candida albicans* in diabetic and non-diabetic subjects. Res J Infect Dis. 2: 1–4.

Miladipour, A., E. Ghanei, A. Nasrollahi and H. Moghaddasi. 2008. Successful treatment of mucormycosis after kidney transplantation. Iran J Kidney Dis. 2: 163–166.

Mirhendi, H., M.J. Fatemi, H. Bateni, M. Hajabdolbaghi, M. Geramishoar, B. Ahmadi and H. Badali. 2013. First case of disseminated phaeohyphomycosis in an immunocompetent individual due to *Alternaria malorum*. Med Mycol. 51: 196–202.

Mirzaie, A.Z., J. Akram, A. Sadeghipour and N. Shayanfar. 2011. Seven years of experience with zygomycosis in Iran: a seasonal disease. Brazilian J Infect Dis. 15: 504–504.

Moallaei, H., H. Ravansalar, M.J. Namazi and A. Akaberi. 2007. Study and identification of various species of *Candida* in candidiasis vaginitis in women admitted to mobini hospital in Sabzevar. J Sabzevar Uni Med Sci. 17: 54–62.

Mobedi, I., F. Daee-Ghazvini, F.A. Nakhjavani, R. Tohidi, M. Ghazanfari, F. Aghakhani and S.R.J. Yusefi. 2006. *Actinomyces naeslundii* in patients with hematological malignancies. Act Med Iranica 44: 345–348.

Mogadam, A.Y., M.A. Asadi, R. Dehghani and H. Hooshyar. 2009. The prevalence of otomycosis in Kashan, Iran, during 2001–2003. Jundishapur J Microbiol. 2: 18–21.

Moghaddami, M. and P. Kordbachae. 1989. Report thirteen cases of mycetoma. Med J Islam Repub Iran 3: 183–186.

Moghadami, M., P. Kordbacheh and M. Emami. 1988. A case report of cryptococcal meningitis. Iranian J Publ Health 17: 61–68.

Moghaddami, M., M. Emami, P. Toosi and L. Nabai. 1991. A case report of chronic mucocutaneous candidiasis in Iran. Iranian J Publ Health 20: 35–42.

Moghaddami, M., M. Mohraz and M.R. Shidfar. 1993. Report of thirteen cases of mucormycosis. Med J Islam Repub Iran 7: 175–178.

Mohammadi, A., A. Mehdizadeh, M. Ghasemi-Rad, H. Habibpour and A. Esmaeli. 2012. Pulmonary mucormycosis in patients with diabetic ketoacidosis: a case report and review of literature. Tuberk Toraks 60: 66–9.

Mohebbi, A., H. Jahandideh and A.A. Harandi. 2011. Rare presentation of rhino-orbital-cerebral zygomycosis: bilateral facial nerve palsy. Case Rep Med. 2011: 1–2.

Mousavi, S.A.A., I. Esfandiarpour, S. Salari and H. Shokri. 2009. Onychomycosis due to *Fusarium* spp. in patient with squamous cell carcinoma: A case report from Kerman, Iran. J Med Mycol. 19: 146–149.

Nabai, H. and A.H. Mehregan. 1974. Cutaneous protothecosis. Report of a case from Iran. J Cutan Pathol. 1: 180–185.

Naghibzadeh, B., E. Razmpa, Sh Alavi, M. Emami, M. Shidfar, Gh Naghibzadeh and A. Morteza. 2011. Prevalence of fungal infection among Iranian patients with chronic sinusitis. Acta Otorhinolaryngol. Italica 31: 35–38.

Najafzadeh, M.J., A. Fata, A. Naseri, M.S. Keisari, S. Farahyar, M. Ganjbakhsh and G.S. Hoog. 2014. Implantation phaeohyphomycosis caused by a non-sporulating Chaetomium species. J Mycol Med. 24: 161–165.

Naseri, A., A. Fata and A.R. Khosravi. 2012a. Tinea capitis due to *Microsporum vanbreuseghemii*: report of two cases. Mycopathologia 174: 77–80.

Naseri, A., A. Fata and A.R. Khosravi. 2012b. The first case of *Microsporum persicolor* infection in Iran. Jundishapur J Microbiol. 5: 362–364.

Naseri, A., A. Fata and M.J. Najafzadeh. 2013a. First case of *Tritirachium oryzae* as agent of onychomycosis and its susceptibility to antifungal drugs. Mycopathologia 176: 119–122.

Naseri, A., A. Fata, M.J. Najafzadeh and H. Shokri. 2013b. Surveillance of dermatophytosis in Northeast of Iran (Mashhad) and review of published studies. Mycopathologia 176: 247–253.

Nemati, S., R. Hassanzadeh, S.K. Jahromi and A.D.N. Abadi. 2014. Otomycosis in the north of Iran: common pathogens and resistance to antifungal agents. Eur Arch Otorhinolaryngol. 271: 953–957.

Nematian, J. 1984. Study on the subcutaneous mycotic infection in Iran. PhD Thesis, Tehran University, School of Public Health, Tehran-Iran.

Nikpoor, N., M.W. Buxton and B.J. Leppard. 1978. Fungal diseases in Shiraz. Pahlavi Med J. 9: 27–49.

Norouzi, F., M. Aminshahidi, B. Heidari and S. Farshad. 2013. Bacteremia due to *Actinomyces naeslundii* in a T cell lymphoma child; a case report. Jundishapur J Microbiol. 6: 306–308.

Nouripour-Sisakht, S., A. Rezaei-Matehkolaei, M. Abastabar, M.J. Najafzadeh, K. Satoh, B. Ahmadi and L. Hosseinpour. 2013. *Microsporum fulvum*, an ignored pathogenic dermatophyte: a new clinical isolation from Iran. Mycopathologia 176: 157–160.

Nourollohi, H., M. Rostami and M. Emranifard. 1999. A case chronic mycetoma. J Isfahan Med School 16: 18–19.

Nowrozi, H., F.D. Arabi, H.G. Mehraban, A. Tavakoli and G. Ghooshchi. 2014. Mycological and clinical study of Otomycosis in Tehran, Iran. Bull Environ Pharmacol Life Sci. 3: 29–31.

Omidynia, E., M. Farshchian, M. Sadjjadi, A. Zamanian and R. Rashidpouraei. 1996. A study of dermatophytoses in Hamadan, the governmentship of West Iran. Mycopathologia 133: 9–13.

Pakshir, K. and J. Hashemi. 2006. Dermatophytosis in Karaj, Iran. Ind J Dermatol. 51: 262–264.

Pakshir, K., B. Sabayan, H. Javan and K. Karamifar. 2008. Mycoflora of human external auditory canal in Shiraz, southern Iran. Iran Red Crescent Med J. 10: 27–29.

Paydar, S., S.R. Baezzat, A. Fazelzadeh and B. Geramizadeh. 2010. A case of gastric zygomycosis in a diabetic patient successfully treated with total gastrectomy. Middle East J Dig Dis. 2: 46–48.

Pourmand, M.R., S. Dehghani, M. Hadjati, F. Kosari and G. Pourmand. 2012. Renal actinomycosis in presence of renal stones in a patient with end stage renal disease. J Med Bacteriol. 1: 62–65.

Rafiei, A. and N. Amirrajab. 2010. Fungal contamination of indoor public swimming pools, Ahwaz, South-west of Iran. Iran J Publ Health 39: 124–128.

Rasi, A., R. Naderi, A.H. Behzadi, M. Falahati, S. Farehyar, Y. Honarbakhsh and A.P. Akasheh. 2010. *Malassezia* yeast species isolated from Iranian patients with pityriasis versicolor in a prospective study. Mycoses 53: 350–355.

Rassai, S., A. Feily, F. Derakhshanmehr and N. Sina. 2011. Some epidemiological aspects of dermatophyte infections in Southwest Iran. Acta Dermatovenerol Croat. 19: 13–15.

Rastegar-Lari, A., L. Akhlaghi, M. Falahati and R. Alaghehbandan. 2005. Characteristics of dermatophytoses among children in an area south of Tehran, Iran. Mycoses 48: 32–37.

Rasti, S., M.A. Asadi, A. Taghriri, M. Behrashi and G. Mousavie. 2014. Vaginal candidiasis complications on pregnant women. Jundishapur J Microbiol. 7: e10078.

Ravaghi, M. and M. Aflatouni. 1978. Mycetoma caused by *Nocardia asteroides*. J Pakistan Med Assoc. 28: 35–36.

Razin, B.N., S.D. Shoaei, A. Family, M. Nabavi and F. Abbasi. 2011. *Mycobacterium tuberculosis* and *Cryptococcus neoformans* co-infection meningitis in a young immuncompetent woman. Arch Clin Infect Dis. 6: 93–94.

Razzaghi-Abyaneh, M., G. Sadeghi, E. Zeinali, M. Alirezaee, M. Shams-Ghahfarokhi, A. Amani and R. Tolouei. 2014. Species distribution and antifungal susceptibility of *Candida* spp. isolated from superficial candidiasis in outpatients in Iran. J Med Mycol. 24: 43–50.

Reiss, E., H.J. Shadomy and G.M. Lyon. 2011. Fundamental Medical Mycology, 1st edition. John Wiley and Sons, New Jersey, 656 p.

Rezvani, S.M., S. Sefidgar and M.R. Hasanjani-Roushan. 2010. Clinical patterns and etiology of dermatophytosis in 200 cases in Babol, North of Iran. Casp J Intern Med. 1: 23–6.

Sabokbar, A., M. Bayat, P. Kordbacheh and B. Baradaran. 2011. Fungal rhinosinusitis in hospitalized patients in Khorramabad, Iran. Middle-East J Sci Res. 7: 387–391.

Sadeghi, G., M. Abouei, M. Alirezaee, R. Tolouei, M. Shams-Ghahfarokhi, E. Mostafavi and M. Razzaghi-Abyaneh. 2011. A 4-year survey of dermatomycoses in Tehran from 2006 to 2009. J Med Mycol. 21: 260–265.

Saghazadeh, M., S. Farshi, J. Hashemi, P. Mansouri and A.R. Khosravi. 2010. Identification of *Malassezia* species isolated from patients with seborrheic dermatitis, atopic dermatitis, and normal subjects. J Med Mycol. 20: 279–282.

Saki, N., A. Rafiei, S. Nikakhlagh, N. Amirrajab and S. Saki 2013. Prevalence of otomycosis in Khouzestan Province, south-west Iran. J Laryngol Otol. 127: 25–27.

Salmasi, A., M. Asgari, N. Khodadadi and A. Rezaee. 2010. Primary actinomycosis of the breast presenting as a breast mass. Breast Care 5: 105–107.

Sanavi, S., R. Afshar and S. Afshin-Majd. 2013. Rhino-orbitocerebral mucormycosis in a patient with idiopathic crescentic glomerulonephritis. Saudi J Kidney Dis Transplant. 24: 768–772.

Sarrafzadeh, S.A., A.H. Rafati, M. Ardalan, D. Mansouri, P. Tabarsi and Z. Pourpak. 2010. The accuracy of serum galactomannan assay in diagnosing invasive pulmonary aspergillosis. Iran. J Allergy Asthma Immunol. 9: 149–155.

Sefidgar, S., K. Kiakojouri, M. Mirzaei and F. Sharifi. 2002. Fungal infections of external ear canal in patients with otomycosis (babol; 1991–2000). J. Babol Univ Med Sci. 13: 25–29.

Sepahvand, A., J. Abdi, Y. Shirkhani, S.H. Fallahi, M. Tarrahi and S. Soleimannejad. 2009. Dermatophytosis in western part of Iran, Khorramabad. Asia. J Biol Sci. 2: 58–65.

Shafaghi, S., M. Pour-Abdollah, P. Tabarsi, F. Ghorbani, S.S.M. Makki, H.R. Khoddami-Vishteh and K. Najafizadeh. 2010. Concomitant cryptococcosis and *Burkholderia* infection in an asymptomatic lung transplant patient with cystic fibrosis. Int J Organ Transplant Med. 1: 183–186.

Shamsoddin, S., S. Shamsi-Meymandi, G.R. Yousefzadeh and V.R. Sepehr. 2004. Report of two cases with mycetoma of foot in Kerman and overview of other reported cases from Iran. J Babol Univ Med Sci. 6: 68–70.

Shamsodini, S., S. Shamsi-Meimandi, S. Sadre-Eshkavari and S. Vahidreza. 2007. Report of two cases of mycetoma in the Islamic Republic of Iran. East Mediterr Health J. 13: 1219–1222.

Sheikh, M.S., B.Y Qazi and B. Rameen. 1993. Otomycosis in Khozistan. Ind J Otolaryngol Head Neck Surg. 45: 73–77.

Shishegar, M., S.H.D. Hosseini, P. Varedi, S. Mahmoodi, M.J. Ashraf and A. Faramarzi. 2009. Actinomycosis of the middle ear. Otol Neurotol. 30: 686–687.

Shoar, M.G., M.R. Shidfar and K. Zomorodian. 2003. A case of actinomycotic mycetoma in hand treated successfully with co-trimoxazole and streptomycin. Pakistan J Med Sci. 19: 310–312.

Shobeiri, F. and M. Nazari. 2006. A prospective study of genital infections in Hamedan, Iran. Literacy 20: 14–19.

Shokohi, T., K. Nowroozpoor-Dailami and T. Moaddel-Haghighi. 2006. Fungal keratitis in patients with corneal ulcer in Sari, Northern Iran. Arch Iran Med. 9: 222–227.

Shyban, Z. 1989. Three cases of actinomycetoma. Daru Darman 6: 13–15.

Sodagar, R. and E. Kohout. 1972. Actinomycosis of tongue as pseudotumor. The Laryngoscope 82: 2149–2152.

Suodbakhsh, A., A. Soleimani and H. Emadi. 2011. Photoclinic actinomycetoma. Arch Iran Med. 14: 299–300.

Tarazooie, B., P. Kordbacheh, F. Zaini, K. Zomorodian, F. Saadat, H. Zeraati and S. Rezaie. 2004. Study of the distribution of *Malassezia* species in patients with pityriasis versicolor and healthy individuals in Tehran, Iran. BMC Dermatol. 4: 5.

Tayyebi-Meybodi, N., S. Amouian and N. Mohammadian-Roashan. 2005. Renal allograft mucormycosis: report of two cases. Urology J. 2: 54–56.

Vasei, M. and M.H. Imanieh. 1999. Duodenal colonization by *Geotrichum candidum* in a child with transient low serum levels of IgA and IgM. Acta Pathol Microbiol Immunol Scandi. 107: 681–684.

Yadegarynia, D., M.A. Merza, S. Sali and A.G. Firuzkuhi. 2013. A rare case presentation of oral actinomycosis. Int J Mycobacteriol. 2: 187–189.

Yaghoobi, R., N. Ranjbari and S. Rasaei. 1999. Actinomycotic mycetoma: Report of 4 cases from Khuzestan province. Iran J Dermatol. 2: 43–48.

Zaini, F. 1984. Sporotrichosis in iran: first report of isolation of *Sporothrix schenckii* from clinical material. Act Med Iran. 26: 33–39.

Zaini, F. and S. Basiri-Djahromi. 1994. Study of fungal infection in patients with leukaemia. Iran. J Publ Health 23: 89–104.

Zaini, F., M. Mahmoudi, A.S.A. Mehbod, P. Kordbacheh and M. Safara. 2009. Fungal nail infections in Tehran, Iran. Iran J Publ Health 38: 46–53.

Zarrin, M. and A.Z. Mahmoudabadi. 1998. Report of two cases of actinomycetoma in Khuzestan. Sci Med J Ahwaz 24: 92–97.

Ziaei, M. and G. Azarkar. 2003. Case report: mycetoma in leg. J Birjand Med Sci. 10: 40–42.

Zomorodian, K., H. Mirhendi, B. Tarazooie, H. Zeraati, Z. Hallaji and K. Balighi. 2008. Distribution of *Malassezia* species in patients with psoriasis and healthy individuals in Tehran, Iran. J Cutan Pathol. 35: 1027–1031.

CHAPTER 16

Culture Collection DPUA: Decades Supporting Diagnostic of Fungal Diseases in Amazonas, Brazil

Maria Francisca Simas Teixeira,[1,*] *Kátia Santana Cruz,*[2]
Iara Maria Bonfim,[3] *Renata de Almeida Lemos,*[3]
Ana Rita Gaia Machado,[3] *Mircella Marialva Alecrim,*[3]
Raimundo Felipe da Cruz Filho,[3] *Nélly Mara Vinhote Marinho*[3]
and *Taciana de Amorim Silva*[3]

Introduction

The Culture Collection DPUA has been established at the Federal University of Amazonas (UFAM), Manaus, Amazonas, Brazil. In 1974, its activities began with the implementation of the Medical Mycology subject. At this time, the fungi were preserved only under refrigeration. From 1987, the collection was restructured, expanded and was opened on December 5, 1988. Currently, the collection has 1566 cultures preserved under mineral oil, Castellani method and silica gel.

The Collection is recorded in the World Data Center for Microorganisms (WDCM) and in the World Federation of Culture Collections (WFCC) under n°715. It's also associated to the Network Culture Collections of Microorganisms of the North and Northeast of Brazil-RENNEBRA.

[1] Curator of Culture Collection DPUA/Federal University of Amazonas-UFAM.
[2] Tropical Medicine Fundation Dr. Heitor Vieira Dourado FMT-HVD/Avenida Pedro Teixeira, 25, Manaus, Amazonas, Brazil, Zip Code: 69040-000.
[3] Federal University of Amazonas-UFAM/Avenida General Rodrigo Octávio, 6200, Coroado I, Manaus, Amazonas, Brazil, Zip Code: 69077-000.
[*] Corresponding author: mteixeira@ufam.edu.br

Since 1988, all the research conducted contributed to the growth of the fungal collection. Some examples are the implementations of projects like "Fungi investigation and systematic" and "Tropical biodeterioration", "Thermotolerant and thermophilic fungi: Isolation, identification and characterization of industrial interest", "Fungi of vegetable waste from the oil region of Urucu, county of the state of Amazonas, Brazil", "Control of contamination by aflatoxins in the production of Brazil nut", "Network Culture Collections of Microorganisms of North and Northeast of Brazil-RENNEBRA, "Recycling of Amazonic agrowaste: Production of edible mushroom for human food and animal feed", "Bioprocessing of *Lentinuscitrinus*: utilization of agroforestry waste from Amazon related to the production of food supplements with medicinal properties" and "Amazonic Actinomycete: Biogenic synthesis of silver nanoparticles and evaluation of antimicrobial activity". All of these projects certainly promoted the development of regional human resources.

The increase of the collection with species for industrial application was favored by exchanges between Brazilian institutions. Many recognized scientists from Universities like UNICAMP and UFPE (Federal University of Pernambuco) contributed to the studies in the area of taxonomy and biotechnology. This interaction between institutions contributed to knowledge dissemination and effectiveness of new technologies increasing the production of scientific research.

Within the scope of Culture Collection DPUA, the supply of diverse cultures for taxonomy, screening of fungi for medical importance, bioprocess and preservation of fungi are performed. This collection constitutes an important heritage of scientific significance Northern Brazil. It gives to the community the opportunity to obtain knowledge about fungi biodiversity through research, technological development, innovation technical and scientific diffusion to the benefit of society.

The aim of this chapter is to demonstrate the contribution of Culture Collection DPUA in Amazon human resource formation in scientific researches about pathogenic fungi infections' diagnosis.

Diagnosis of Fungal Infections

Mycological diagnosis is required to confirm the etiology, identification and effective treatment of pathogenic infections by opportunistic fungi and parasites. There are several techniques that can contribute to the observation of parasites present in the lesion. The clarification of the sample with potassium hydroxide (KOH) and Gram-staining allows the observation of fungal structures but requires culture for confirmation. Another method used is based on the fungi pathogenicity aspects observation and biological response (Brasil et al. 2003; Artal 2004).

The diagnosis of diseases caused by fungi must be done by professionals with training based on practical experiences that complement the theoretical knowledge of Mycology. For the success of these diagnoses, an appropriate patient preparation is essential. It is very important that the patient suspends the use of systemic or topic antifungal medication between 10 to 15 days before the diagnosis. The patient should not use moisturizing lotions or similar products and keep the area as clean as possible. On nail infections, it is not recommended to apply nail polishes, wash, brush or cut in the previous week. If the affected area is the foot then it is important to avoid wearing socks and shoes or using any kind of lotion.

Opportunistic Fungal Pathogens

Fungal species are widely distributed in soil, plant wastes and other organic substrates. They represent around 7% (611,000 species) of the total eukaryotic species on earth and between them, approximately 600 species are human pathogens (Badiee and Hashemizadeh 2014).

Among the fungal pathogens, opportunistic species can proliferate and cause superficial or systemic infections. They predominantly take advantage of the host's weak immune system. Other predisposing factors include systemic diseases, immunosuppressive therapies, prolonged use of antibiotics, transplants of organs or cells, severe burns, malnutrition, etc. (Krishnan 2012; Das and Ranganathan 2012; Tankhiwale 2014).

However, opportunistic fungal infections have also been reported in immunocompetent patients without signs or symptoms of conditions associated with an immunocompromised state, such as *Candida* species that colonize different body sites in healthy hosts, particularly the gastrointestinal tract, genital tract or the skin (Badiee and Hashemizadeh 2014).

Fungi have a versatile life cycle, colonizing different environmental niches freely or as part of the normal flora of humans and animals. This condition favors the contact of man with their mycelial fragment and spores, which can usually be controlled by the host immune system, in particular, alveolar macrophages (d'Enfert 2009; Lewis et al. 2012; Nwako et al. 2014).

There are evidences that inhaled fungi can persist for long periods in macrophages, providing a reservoir for secondary infections when host immunity becomes natural or artificially weakened (Lewis et al. 2012).

The opportunistic fungal infections are rare and can be deadly because of late diagnosis and lack of effective treatment options. That is why these diseases are considered important challenges to modern medicine (Iliev and Underhill 2013; Badiee and Hashemizadeh 2014).

In a review made by Nucci et al. (2010) about the epidemiology from the most clinically relevant opportunistic fungal infections in Latin America the predominance of candidíasis, cryptococcosis, trichosporonosis, aspergillosis and fusariosis was shown. Reports have presented that the incidence of candidemia is 3 to 15 times higher than that reported in studies in North America and Europe. The etiological agents from blood infections are predominately *Candida* species (excepting *Candida albicans*) like *Candida parapsilosis* or *Candida tropicalis* instead of *Candida glabrata.*

Other fungi frequently reported as opportunistic are *Cladophialophora saturnica* [interdigitaltineanigra-like (Badali et al. 2009)]; *Aspergillus fumigatus* [sputum, oral swab, throat swab, vaginal swab, pus, skin scrapings (McCormick et al. 2010)]; *Candida albicans* [osteomyelitis of calcaneus (Fleming et al. 2012)]; *Candida krusei* [neonatal sepsis (Amaral-Lopes and Moura 2012)]; *Pneumocystis jirovecii* [pneumonia (Jose and Brown 2012)]; *Aspergillus niger*, *Aspergillus flavus*, *Fusarium* spp.; *Alternaria* spp.; *Candida* spp. [Skin infection (Pandey et al. 2013)]; *Candida parapsilosis*; *Rhodotorula mucilaginosa*; *Phoma* spp.; *Debaryomyceshansenii*; *Acremonium* spp.; *Aureobasidium pullulanse*; *Aspergillus* spp. [foot skin infection (Chan et al. 2013)]; *Cryptococcus neoformans*, *Paracoccidioides brasiliensis*; *Histoplasma capsulatum*; *Aspergillus fumigatus* [fungal infection obtained in samples collected by fiberoptic bronchoscopy from HIV-positive and HIV-negative patients (Lazzarini-de-Oliveira et al. 1999); *Scedosporiumprolificans*; *Mucor* sp.; *Rhizopus* sp., *Rhizomucor* sp. and *Absidia* sp. [invasive fungal infection (Ramana et al. 2013)].

The occurrence of opportunistic fungal infections are associated with recognized predisposing factors that are cited in this review. The imbalance of environmental conditions has been another cause of onset of these diseases, since the fungi are saprobes that have evolved for decades and have adapted to survive in human tissue and cause infections. Because of this, it becomes necessary to search for efficient techniques to control and monitor fungi.

Superficial Fungal Infections: Comparative Information on 104 Cases

In a research done by Teixeira et al. (2000) it was shown that fungi causing superficial mycoses occurred in 104 patients diagnosed at Culture Collection DPUA from 1986 to 1988 (Group I) and 1997 to 1999 (Group II). Group I had 66 patients, 22 males and 44 females; and Group II had 38 patients, 15 males and 23 females.

Among the patients from Group I, 62.12% expressed clinical symptoms of mycotic infections. The confirmation was made by direct examination and by culture tests showing the presence of fungi. In Group II the diagnosis confirmed, in both tests, that 63.16% of the patients presented superficial mycoses. In the other patients (36.84%) the fungi structures were not observed by direct examination. However, in the culture test, 28.57% of them were considered positive and 71.43% negative.

Among fungal infections diagnosed, the most frequent were yeast infections (Group I and II), dermatophytosis (Group I) and superficial mycoses (keratophytosis) (Group II). In both groups, *Candida* sp. was the genera of yeast infection that predominated in nails and feet.

The dermatophytosis found on patients from Group I (Table 1) was caused by five different species (*Microsporumgypseum*, *Trichophyton mentagrophytes*, *Trichophyton rubrum*, *Trichophyton tonsurans* and *Trichophyton verrucosum*) and among them *T. mentagrophytes* and *T. tonsurans* were responsible for the most of the cases with predominant localization on torso, nails and feet. In Group II (Table 1), *T. rubrum* was the predominant species found especially in the nails.

Piedraiahortae, the etiologic agent of true keratophytosis, was identified in only one patient. *Malassezia* sp. occurred with equal frequency in Group I and Group II. The othermycoses diagnosed were caused by *Aureobasidium pullulans*, *Fusarium* sp. and *Geotrichum* sp., all classified as Hyphomycetes. The first one were classified on Dematiaceae family and the other two on Moniliaceae family. They were detected at low frequency on the feet (*Fusarium* sp.), nails (*Geotrichum* sp.) and scalp (*A. pullulans*) (Table 1).

Table 1. Mycosis isolated from examined patients at the Mycology Laboratory of Federal University of Amazonas-UFAM.

Cutaneous mycosis												
Dermatophytosis												
Areas/ Taxa	**Scalp**		**Trunk**		**Hand**		**Iguinal region**		**Feet**		**Nail**	
	A	**B**	**A**	**B**	**A**	**B**	**A**	**B**	**A**	**B**	**A**	**B**
M. gypseum	0	0	1	0	0	0	0	0	0	0	1	0
T. mentagrophytes	0	0	0	0	0	0	0	0	3	0	1	0
T. rubrum	0	0	0	0	0	0	1	0	0	0	0	4
T. tonsurans	1	0	2	0	0	0	0	0	0	0	1	0
T. verrucosum	0	0	0	0	1	0	0	0	0	0	0	0
Fusariosis, Geotrichosis and Phaeohyphomycosis												
Aureobasidium pullulans	1	0	0	0	0	0	0	0	0	0	0	0
Fusarium sp.	0	0	0	0	0	0	0	0	1	0	0	0
Geotrichum sp.	0	0	0	0	0	0	0	0	0	0	0	1
Yeast Infectons												
Candida spp.	0	1	1	2	3	1	3	2	4	5	4	5
Rhodotorula spp.	1	1	0	0	1	1	0	0	1	0	0	0
Trichosporon spp.	0	0	0	0	1	0	0	0	1	4	2	2
Superfitials mycosis from stratum corneum top layer												
Keratophytosis												
Malassezia spp.	0	0	6	8	0	0	0	0	0	0	0	0
Piedraia hortae	1	0	0	0	0	0	0	0	0	0	0	0
Total	**4**	**2**	**10**	**10**	**6**	**2**	**4**	**2**	**10**	**9**	**9**	**12**

[A: 1986–88 (Group I)] [B: 1997–99 (Group II)]

From 27 female patients (Group I) 28 pathogens were isolated which were represented by *Candida* sp. (42.85%), *Fusarium* sp. (3.58%), *Malassezia* sp. (17.85%), *Microsporum gypseum* (7.14%), *Piedraia hortae* (3.58%), *Rhodotorula* sp. (3.58%), *T. tonsurans* (7.14%) and *Trichosporon* sp. (7.14%). From the 14 male patients, 15 pathogens including *Candida* sp. (20%), *Malassezia* sp. (6.67%), *Rhodotorula* sp. (13.33%), *T. mentagrophytes* (13.33%), *T. rubrum* (6.67%), *T. tonsurans* (13.33%), *T. verrucosum* (6.67%), *Trichosporon* sp. (13.33%) and *Aureobasidium* sp. (6.67%) were isolated.

In Group II, from the 28 patients with mycoses, 15 were females diagnosed with 16 fungi represented by *Candida* spp. (43.75%), *Malassezia* spp. (25.0%), *T. rubrum* (12.5%) and *Trichosporon* spp. (18.75%). Out of 13 male patients 14 fungi were isolated: *Candida* spp. (50.0%), *Geotrichum* sp. (7.14%), *Malassezia* spp. (14.29%), *Rhodotorula* sp. (7.14%), *T. rubrum* (14.29%) and *Trichosporon* spp. (14.29%).

The results showed the presence of different etiologic agents of superficial mycoses in the same patient. The yeasts had higher frequency in the two groups analyzed, especially *Candida* and *Malassezia*, frequently found in females. In male patients, *Candida* spp. was the exclusively found yeast. Comparing the age of the patients and the quantity of isolated yeasts, the highest frequency was observed in 21 year old individuals (48.83%) and 40 year old individuals (56.66%) in both groups, respectively, with no evidence of significant difference. In this age group, *Candida* was more frequent among the identified fungi, thus confirming the predominance of yeasts in the diagnosis.

Prevalence of Mycoses Agents in HIV Virus Patients Treated at a Tertiary Hospital in the City of Manaus, Amazonas, Brazil

Cruz (2005) in a retrospective study of fungal diseases in HIV-postive patients in a tertiary hospital in the city of Manaus-Amazonas Brazil-showed that between 170 survey participants, 73% were male and 27% female. The patients' ages were grouped into 10-year intervals and distributed by gender to facilitate the analysis process.

Based on age (Table 2) the fungal diseases were prevalent in males patients (35.2%), aged 31–40 years and 15.3% in female patients, aged 21–30 years.

The diagnosis of fungal infection involves the detection of biological materials in the organism and the isolation of pathogen in culture. From 170 patients, 214 clinical samples were mycologically examined. Of these 132 cultures were isolated which originated from 74.2% of males and 25.8% females. Table 3 shows the result of identification of the 132 cultures, observing the presence of *Candida* spp. (38.8%); *Trichophyton* spp. (32.6%); *Cryptococcus neoformans* (18.2%); *Histoplasma capsulatum* (6.8%); *Microsporum* spp. (2.2%); *Epidermophyton floccosum* (0.7%) and *Aspergillus* spp. (0.7%). Among these, there was a predominance of yeasts, *Candida* spp. and *Cryptococcus neoformans* (57.0%) and dermatophytes (35.5%). A smaller percentage (7.5%) parasitism was identified by *H. capsulatum* and *Aspergillus* sp.

The mycological laboratory diagnosis allows the definition and identification of opportunistic fungi in patients with HIV. *Candida* spp., *Cryptococcus neoformans* and dermatophytes were the predominant etiological agents and *Histoplasma capsulatum* and *Aspergillus* sp., were lower infrequency. According to this research, Candidosis, Dermatophytosis and Cryptococcosis were the most frequent fungal diseases.

Table 2. Distribution by age group and gender of 170 patients with HIV virus.

Age (years)	Patient/Gender				Total	
	Male		Female			
	No.	%	No.	%	No.	%
0 a 10	3	1.8	3	1.8	6	3.6
11 a 20	1	0.6	0	0.0	1	0.6
21 a 30	32	18.8	26	15.3	58	34.1
31 a 40	60	35.2	10	5.9	70	41.1
41 a 50	20	11.8	5	2.9	25	14.7
≥ 50	9	5.3	1	0.6	10	5.9
Total	125	73.5	45	26.5	170	100.0

Table 3. Prevalence of etiologic agents of superficial and deep mycoses in a tertiary hospital in Manaus, Amazonas, Brazil.

Isolated fungi	**No.**	**%**
Candida spp.	51	38.8
Criptococcus neoformans	24	18.2
Trichophyton spp.	43	32.6
Microsporum spp.	3	2.2
Epidermophyton floccosum	1	0.7
Histoplasma capsulatum	9	6.8
Aspergillus sp.	1	0.7
Total	132	100.0

Occurrence of Yeast Infections in Children Attended to at Health Centers in the City of Manaus

Bonfim et al. (2005) studied to verify the occurrence of *Candida* and *Malassezia furfur* in lesions of atopic dermatitis, seborrheic dermatitis and pityriasis in children. A total of 28 patients, from 0 to 12 years, were examined in localpublic health service (CAIC).

Four samples of each lesion (epidermal scales) were collected for direct examination and culture of the etiological agents. The identification of yeast was performed according to Barnett et al. (1990), Guého et al. (1996), Smith et al. (1996) and Teixeira et al. (1999).

Among the 28 patients, 64.29% were boys and 35.71% girls. A total of 31 clinical samples were analyzed from affected areas like face (32.25%) and trunk (29.03%) (Table 4). In direct examination test the presence of yeasts in 25 clinical samples was detected (80.65%). In the culture test, 34 had been isolated in Sabouraud agar (35.29%) and in Sabouraud agar supplemented with olive oil (64.71%). The diagnostics showed that a direct test with positive or negative result does not always mean the absence or presence of the fungus, respectively (Table 5). These results are probably related to the infeasibility of the parasite or an insufficient number of cells in the clinical sample.

Malassezia globosa was the species more commonly isolated (38.26%), except on the head, but predominantly as the opportunistic pathogen on the neck (14.70%) and trunk (8.82%). Among the other species, *Malassezia furfur* occurred only on face area (5.88%) and trunk (5.88%), while *Candida parapsilosis*, *Malassezia restricta*, *Rhodotorula mucylaginosa* and *Trichosporon mucoides* were detected in 8.82% of the cases. *Cryptococcus laurentii* (5.88%), *Malassezia sympodialis* (2.94%) and *Rhodotorula minuta* (2.94%) were the least representated fungi. This investigation showed that all patients examined had skin diseases with major prevalence in boys from one to four years old.

Table 4. Frequency of examined patients according to age and gender.

Age (years)	Gender					
	Male		Female		Total	
	No.	%	No.	%	No.	%
1–4	12	42.86%	2	7.14%	12	50.00
5–8	5	17.86%	5	17.86%	10	35.72
9–12	1	3.57%	3	10.71%	4	14.28
Sub-Total	18	64.29%	10	35.71%	28	100.00

Table 5. Examined lesions in male and female children.

	Male		Female	
Lesion localization	**No.**	**%**	**No.**	**%**
Face	8	25.81	2	6.45
Scalp	0	0	1	3.23
Trunk	5	16.13	4	12.90
Lower body members	2	6.45	1	3.23
Upper limb member	2	6.45	2	6.45
Neck	2	6.45	1	3.23
Gluteus	0	0	1	3.25
Total	19 (100%)	61.29	12	38.71

Laboratory Diagnosis of Fungal Infections in Patients Treated by Residents from Getúlio Vargas University Hospital HUGV

In a study by Lemos et al. (2008) the diagnosis of 30 clinical samples from patients of Getúlio Vargas University Hospital (HUGV) in Manaus-Amazonas was made. Epidermal scales, skin, hair, nails, and mucous secretion were clarified with a solution of KOH (Potassium Hydroxide) or stained with blue lactophenol. Cerebrospinal fluid diagnosis was made with India ink. Lesions suspected to be pityriasis versicolor were confirmed by direct microscopic examination using the Sellotape method (Sidrim 2004).

Fungal cultures were obtained by inoculating each sample on Sabouraud agar or Micosel agar in petri dishes. The cultures were incubated at 28°C and 37°C for 20 days. The isolated fungi were identified based on morphological characteristics, reproduction and physiological tests through cultures performed on selective media (Raper and Fennell 1977; Barnett et al. 1990).

In this study it was observed that among the 30 clinical samples, in 24 of them the result was positive and an association between these pathologies (1:2) was detected. In the other samples the diagnosis was negative. The following diseases were diagnosed from the samples analyzed: Aspergillosis (7%), systemic cryptococcosis (52%), superficial cryptococcosis (4%), keratophytosis (4%), pneumocystosis (PCP) (26%) and dermatophytosis (7%). More than 50% of samples from patients diagnosed with systemic cryptococcosis and among the clinical samples examined data showed a higher frequency of bronchoalveolar lavage (36.67%), liquor (36.67%) and epidermal scales (16.67%). In all of the clinical samples predominance of female patients was observed (Table 6).

Table 7 shows the fungi identified according to the patients gender. The following species were identified: *Aspergillus flavus*, *Aspergillus tamarii*, *Cryptococcus neoformans*, *Malassezia furfur*, *Pneumocystis jirovecii* and *Trichophyton mentagrophytes*. *Cryptococcus neoformans* was the most frequent (55.56%) microorganism in both genders: 22.22% in males and 33.33% in females. *Pneumocystis jirovecii* was the second most frequent species (25.93%) and occurred in 11.11% of the males and in 14.81% of the females. *Aspergillus flavus*, *A. tamari* and *Malassezia furfur* were least frequent and were identified only in females (3.7%). *Trichophyton mentagrophytes* also showed low frequency (7.41%) occurring in both genders. These data revealed that females were the most affected by fungal infections (64.29%) compared to males (35.71%).

The research data for *C. neoformans* is in agreement with Parmar et al. (2012). Cryptococcosis is a systemic disease caused by *C. neoformans*, the increase of this infection is caused by immunological impairment related to T cells. There are a wide variety of cases registered in Brazil relating this disease with immunosuppression factors such as chronic diseases, use of corticosteroids, antibiotics and transplants of organs (d'Enfert 2009; Parmar et al. 2012).

Peuneumocystis jirovecii is known as an opportunistic etiological agent common in immunocompromised patients, causing pneumonia (Barsotti and Silva 2007). The agent called *Pneumocystis* was first described by Carlos Chagas in 1909 who mistook

Table 6. Frequency of clinical samples from patients (males and females) attended in HUGV.

Biological sample	Gender				Total	
	Male		Female			
	No.	%	No.	%	No.	%
Bronchoalveolar lavage	4	13.33	7	23.33	11	36.67
Liquor	4	13.33	7	23.33	11	36.67
Nail	0	0.00	2	6.67	2	6.67
Epidermal scales	2	6.67	3	10.00	5	16.67
Mucous secretion	1	3.33	0	0.00	1	3.33
Total	**11**	**36.67**	**19**	**63.33**	**30**	**100.00**

Table 7. Distribution of fungi species according to gender of the patients attended to at HUGV.

Fungi	Gender				Total	
	Male		Female			
	No.	%	No.	%	No.	%
Aspergilllus flavus	0	0.00	1	3.70	1	3.70
Aspergillus tamarii	0	0.00	1	3.70	1	3.70
Cryptococcus neoformans	6	22.22	9	33.33	15	55.56
Malassezia furfur	0	0.00	1	3.70	1	3.70
Pneumocystis jirovecii	3	11.11	4	14.81	7	25.93
Trichophyton mentagrophytes	1	3.70	1	3.70	2	7.41
Total	**10**	**35.71**	**17**	**64.29**	**27**	**100.00**

it for a morphological form of *Trypanos somacruzi*. The microorganism was classified as a protozoan and called *Pneumocystis carinii*. Only in 1988 it was shown by DNA analysis that *Pneumocystis* was a fungus and has different genetic sequences in different mammals. For that reason, *Pneumocystis* found in humans is different from the ones that live in association with other mammalian hosts. In 1999 its name was changed to *Pneumocystis jirovecii*, in honor of the Czech parasitologist Otto Jirovec (Tomio and Silva 2005; d'Enfert 2009).

The pneumonia caused by *P. jirovecii* is an opportunistic infection and its transmission is still not fully understood. There is evidence that transmission between individuals is the most likely way to get infected although environmental sources infection may occur (Rivero et al. 2008; Mori and Sugimoto 2012).

This study reports opportunistic mycoses in patients from HUGV in the age range of 10–60 years old. The most frequent diseases diagnosed were the deep mycoses, especially systemic cryptococcosis. The etiologic agents identified were *Aspergillus flavus*, *Aspergillus tamarii*, *Cryptococcus neoformans*, *Malassezia furfur*, *Trichophyton mentagrophytes* and *Pneumocystis jirovecii*, with higher number of *Cryptococcus neoformans* and *Pneumocystis jirovecii*.

Scientific Production Published In Events or Journals

Publications at conferences and journal articles related to the mycological diagnosis are cited in Table 8. These activities demonstrate the participation of Culture Collection DPUA in human resources training in Medical Mycology area under guidance from leading researchers. Moreover, the development of research projects (Table 9) shows the inclusion of undergraduate and graduate students. This is a contribution that

Table 8. Medical Mycology: published papers in journal e-abstracts presented in events.

Studies presented in Conferences		
Authors	**Title**	**Event/Year**
Teixeira, M.F.S.; Vilela, N.A.; Carvalho, S.M.S.	Relato de um caso de onicomicose por *Microsporum gypseum.*	III Congresso Argentino de Micologia, Mar del Plata. de comunicaciones Libres. Mar del Plata, 1987.
Castro, G.B.; Teixeira, M.F.S.; Pecher, S.	Ocorrência de micoses superficiais em escolares do Médio Amazonas.	III Congresso Argentino de Micologia, Mar del Plata, 1987.
Teixeira, M.F.S.; Castro, G.B.; Pecher, S.	Fungos isolados em indígenas mura, tikuna e maku, do médio e alto Amazonas.	Congresso Argentino de Micologia, Mar del Plata. III Congresso Argentino de Micologia Mar del Plata, 1987.
Santos, L.O.; Rodrigues, M.H.; Teixeira, M.F.S.	Micoses superficiais em escolares do Município de Manacapuru-AM	IV Seminário Anual de Iniciação Científica, Manaus. IV Seminário Anual de Iniciação Científica. Manaus, 1991.
Rodrigues, M.H.; Santos, L.O.; Teixeira, M.F.S.	Micoses superficiais em escolares no municipio de Manacapuru-AM.	IX Congresso Brasileiro de Infectologia Pediátrica, Porto Alegre. IX Congresso Brasileiro de Infectologia Pediátrica. Porto Alegre, 1994.
Cristóvão, D.E.; Teixeira, M.F.S.; Matsuura, A.B.J.; Sarquis, M.I.M.	Ocorrência de fungos da família dematiaceae de solo arenoso das praias mais frequentadas de Manaus-AM.	In: XX Congresso Brasileiro de Microbiologia, Salvador. XX Congresso Brasileiro de Microbiologia-resumos. Salvador: SBM, 1999.
Evangelista, N.M.A.; Teixeira, M.F.S.; Matsuura, A.B.J.	Ocorrência de dermatófito e levedura em crianças: dados preliminares	XX Congresso Brasileiro de Microbiologia, Salvador. XX Congresso Brasileiro de Microbiologia - resumos. Salvador: SBM, 1999.
Teixeira, M.F.S.; Oliveira, M.G.; Matsuura, A.B.J.	Fungos queratinolíticos e queratinofílicos isolados das praias mais frequentadas de Manaus, Amazonas	XX Congresso Brasileiro de Microbiologia, Salvador. XX Congresso Brasileiro de Microbiologia - resumos. Salvador: SBM, 1999.
Teixeira, M.F.S.; Costa, P.S.; Lima, C.R.; Santos, K.S.B.	Avaliação morfofisiologica de fungos preservados sob óleo mineral.	XX Congresso Brasileiro de Microbiologia, Salvador. XX Congresso Brasileiro de Microbiologia. 1999.
Teixeira, M.F.S.; Soares, C.S.S.; Souza, V.C.; Santos, K.S.B.	Micoses cutâneas de pacientes atendidos no serviço de micologia da Universidade do Amazonas	III Congresso Brasileiro de Micologia, 2001, Águas de Lindóia,São Paulo
Alves, P.P.; Teixeira, M.F.S.; Santos, S.S.B.; Santos, J.P.; Monteiro, G.B.	Aspergilose Pulmonar Invasiva em Paciente Transplantado Renal-Relato de Caso	VIII Congresso Brasileiro de Clínica Médica e II Congresso Internacional de Medicina de Urgencia, 2006, Gramado – Rio Grande do Sul.
Papers published in journals		
Authors	**Title**	**Journal/Year**
Teixeira, M.F.S.; Vilela, N.A.; Carvalho, S.M.S.	Onicomicose por *Microsporum gypseum.* Relato de um caso.	Revista de Patologia Tropical, Goiana-Goiás 19(1): 115–119, 1990.
Teixeira, M.F.S.; Queiroz, L.A.; Carvalho, S.M.S.	Una tecnica para preparaciones permanentes de leveduras	Boletín Micológico, Valparaiso-Chile 7(1-2): 23–25, 1992.
Teixeira, M.F.; Jackisch-Matsuura, Tavares, M.A.; Santos, K.S.B.; Souza, V.C.	Infecções micóticas superficiais: informe comparativa sobre 104 casos	Revista de Extensão da UA1: 21–29, 2000

Table 9. Culture Collection contribution to health research area in Amazon.

Researcher	Aim
Sandra Regina S.de Menezes Aquino	Onicomicose em pacientes da Fundação Alfredo da Matta [Dissertation (2004)]
Kátia Santana Cruz	Epidemiologia de Micoses em pacientes portadores do vírus HIV [Dissertation (2005)]
Alessandra Alves Drumond	Proteases de levedura para aplicação médica [Dissertation (2004)]
Teresa AlarcónCastillo	Fungos toxigênicos associados à farinha de mandioca [Dissertation (2004)]
Carla Silvana da Silva Soares	Monitoramento de fungos toxigênicos em plantas medicinais comercializadas em Manaus-AM [Dissertation (2004)]
Antônio Batista da Silva	Verificação da atividade fungicida do extrato de *Piper aduncum* L. [Dissertation (2005)]
Raimundo Felipe da Cruz Filho	Potencial biotecnológico de pigmentos produzidos por *Serratia* sp. [Dissertation (2006)]
Hérlon Mota Atayde	Determinação da toxicidade de fungos isolados de ração para peixes [Dissertation (2006)]
Josy Caldas da Silva	Determinação de Atividade Antimicrobiana de Espécies de *Penicillium* [Dissertação (2008)
Carina Toda	Determinação de atividade antimicrobiana de metabólitos fúngicos frente à micro-organismos de interesse odontológico [Dissertation (2010)]
Hérlon Mota Atayde	Aplicação de lipases extraídas de fungo para melhoramento de óleo de pescado Amazônico [Thesis (2014)]

promotes the improvement of Mycology as a promising science and also results in the training of human resources.

Even though the Amazon is known for its dimensions and fungi biodiversity of industrial and medical importance, there are few mycology professionals. The Culture Collections may expand sustainable socioeconomic benefits and support the development of research on taxonomy, biology, physiology, biochemistry and fungi diagnosis.

Conclusion and Future Perspectives

The Culture Collection DPUA aims towards the preservation of biotechnologically and medically important fungal species with the following perspectives: ISO 9001:2008 certification (in progress), species identification by molecular techniques, to become an Amazonic reference center consolidating the development of innovative technological practices.

This work shows the importance of cooperation between doctors and mycology experts as a guarantee of an effective and fast diagnosis which can contribute significantly to the discovery of infectious fungal agents. The researches revealed the predominance of opportunistic pathogenic fungi with higher occurrence of *Candida*, *Cryptococcus neoformans* and *Pneumocystis jirovecii* according to the patient groups analyzed. Species such as *Malassezia* sp., *Rhodotorula* sp., *Trichosporon* sp., *Histoplasma capsulatum*, *Aspergillus flavus* and *Aspergillus tamari* were the other etiological agents associated with fungal diseases. Among the dermatophytes, species of *Microsporum*, *Trichophyton* (*Trichophyton mentagrophytes*, *T. tonsurans* and *T. rubrum*) and *Epidermophytonfloccosum* were also found.

References

Amaral-Lopes, S. and A. Moura. 2012. Neonatal fungal sepsis by *Candida krusei*: a report of three Cases and a literature review. Medical Mycology Case Reports 1: 24–26.

Artal, E.M. 2004. Diagnóstico histopatológico de las micoses. Rev Iberoam Micol. 21: 1–9.

Badali, H., V.O. Carvalho, V. Vicente, D. Attili-Angelis, I.B. Kwiatkowski, A.H.G. Gerrits Van Den Ende and G.S. De Hoog. 2009. *Cladophialophora saturnica* sp. nov., a new opportunistic species of Chaetothyriales revealed using molecular data. Medical Mycology 47: 55–66.

Badiee, P. and Z. Hashemizadeh. 2014. Opportunistic invasive fungal infections: diagnosis and clinical management. Indian J Med Res. 139: 195–204.

Barnett, J.A., R.W. Payne and D. Yarrow. 1990. Yeast: Characteristics and identification, 2 ed. Cambridge University Press, Cambridge 1002 pp.

Barsotti, V. and M.V. Silva. 2007. Pneumocistose em paciente com SIDA. Revista da Faculdade de Ciências Médicas de Sorocaba 9: 19–21.

Bonfim, I.M., MH.C.R. Silva and M.F.S. Teixeira. 2005. Leveduroses em crianças atendidas em centros de saúde do município de Manaus, Amazonas, Brazil.

Brasil, K.W., R.L. Pinheiro and I.C. Pimentel. 2003. Laboratory diagnosis of superficial and cutaneous mycosis: a comparison of the potassium hydroxide and calcofluor white methods. An Bras Dermatol, Rio de Janeiro 5: 547–551.

Castañon-Olivares, L.R., R. Arrguín-Espinosa, G.R.P. Santos and R. López-Martínez. 2000. Frequency of *Cryptococcus* species and varieties in México and their comparison with some Latin American countries. Revista Latinoamericana de Microbiología 42: 35–40.

Chan, G.F., S. Sinniah, T.I.N.T. Idris, M.S.A. Puad and A.Z.A. Rahman. 2013. Multiple rare opportunistic and pathogenic fungi in persistent foot skin infection. Pakistan J. Biol. Sci. 16: 208–218.

Cruz, K.S. 2005. Ocorrência de Micoses em pacientes portadores do virus HIV da Fundação de Medicina Tropical do Amazonas-IMTAM. Mestrado em Doenças Tropicais e Infecciosas. Universidade Estadual do Amazonas-UEA. Manaus, Amazonas, Brasil.

Das, R. and R. Ranganathan. 2012. An overview of changing trends in systemic fungal infections. Webmed Central Microbiology 3: 1–8.

d'Enfert, C. 2009. Hidden killers: persistence of opportunistic fungal pathogens in the human host. Current Opinion in Microbiology 12: 358–364.

Fleming, L., A. Ng, M. Paden, P. Stone and D. Kruse. 2012. Fungal osteomyelitis of calcaneus due to *Candida albicans*: A Case Report. The Journal of Foot and Ankle Surgery 51: 212–214.

Guého, E., G. Midgley and J. Guillot. 1996. The genus *Malassezia* with description of four new species. Antonie van Leeuwenhoek 69: 337–355.

Iliyan, D.I. and M.U. David. 2013. Striking a balance: fungal commensalism versus pathogenesis. Current Opinion in Microbiology 16: 366–373.

Jose, R.J. and J.S. Brown. 2012. Opportunistic and fungal infections of the lung. Medicine 40: 335–339.

Krishnan, P.A. 2012. Fungal infections of the oral mucosa. Indian J Dent Res. 23: 650–659.

Lazzarini-de-Oliveira, L.C., A.A. Arantes and M.J.M. Caiuby. 1999. Utilidade da investigação rotineira de infecção fúngica pela broncoscopia em pacientes infectados ou não pelo HIV em um hospital geral, referência para SIDA. Revista da Sociedade Brasileira de Medicina Tropical 32: 255–261.

Lemos, R.A., M.O. Mafra and M.F.S. Teixeira. 2008. Diagnóstico laboratorial de micses em pacientes atendidos no Hospital Universitário Getúlio Vargas.

Lewis, L.E., J.M. Bain, C. Lowes, N.A.R. Gowb and L.-P. Erwig. 2012. *Candida albicans* infection inhibits macrophage cell division and proliferation. Fungal Genetics and Biology 49: 679–680.

McCormick, A., J. Loeffler and F. Ebel. 2010. *Aspergillus fumigatus*: contours of an opportunistic human pathogen. Cellular Microbiology 12: 1535–1543.

Mori, S. and M. Sugimoto. 2012. *Pneumociystis jirovecii* infection: an emerging threat to patients with rheumatoid arthritis. Rheumatology 51: 2120–2130.

Nucci, M., F. Queiroz-Telles, A.M. Tobo, A. Restrepo and A.L. Colombo. 2010. Epidemiology of opportunistic fungal infections in Latin America. CID 51: 561–570.

Nwako, O.F., G.C. Mbata, E.U. Ofondu, A.B. Nwako, I.N.S. Dozie and C.A. Nwako. 2014. Mycological studies on the sputa of HIV positive clients attending HIV clinic in federal medical centre Owerri, Nigeria. Clin Res. 5: 3–5.

Pandey, A., M. Pandey and M.R.S. Tomar. 2013. Isolation of opportunistic fungi from skin samples at Gwalior, Madhya Pradesh, India. Int J Curr Sci. 7: 139–141.

Ramana, K.V., S. Kandi, V.P. Bharatkumar, C.H.V. Sharada, R. Rao and R. Mani. 2013. Invasive fungal infections: a comprehensive review. American J. Infec. Dis. and Microbiol. 1: 64–69.

Raper, K.B. and D.I. Fennell. 1977. The Genus *Aspergillus*. Robert E. Krieger Publishing Company, New York 686 p.

Rivero, L., C. Horra, M.A. Montes-Cano, A. Rodriguez-Herrera, N. Respaldiz, V. Friaza, R. Morila, S. Gutierrez, J.M. Varela, F.J. Medrano and E.J. Calderó. 2008. *Pneumocistys jirovecii* transmission from immunocompetent carriers to infart. Emerging Infection Diseases 14: 1116–1118.

Sidrim, J.J.C. 2004. Micologia Médica à luz de autores contemporâneos. 1ed. Rio de Janeiro. Guanabara Koogan S.A. pp. 252–273.

Tankhiwale, S.R. 2014. Systemic fungal infection in immunocompromised patients. Vidarbha J. Int. Med. 16: 29–34.

Teixeira, M.F.S., A.B. Jackisch-Matsuura and C.S. da S. Santos-Soares. 1999. Micologia Médica—Manual de Laboratório. Manaus: Editora da UA 111 p.

Teixeira, M.F.S., A.B. Jackisch-Matsuura, M.A. Tavares, K.S.B. dos Santos and V.C. de Souza. 2000. Infecções micóticas superficiais: informe comparativo sobre 104 casos. Revista de extensão da UA 1(1): 21–29.

Tomio, D. and R.M. Silva. 2005. Pneumocistose. Arquivos Catarinenses de Medicina 34(4): 81–96.

Index

For Product Safety Concerns and Information please contact our EU representative GPSR@taylorandfrancis.com Taylor & Francis Verlag GmbH, Kaufingerstraße 24, 80331 München, Germany

T - #0193 - 160425 - C31 - 234/156/20 - PB - 9780367738013 - Gloss Lamination